T0175904

FRESHWATER BIVALVE ECOTOXICOLOGY

EDITED BY
JERRY L. FARRIS AND JOHN H. VAN HASSEL

Other Titles from the Society of Environmental Toxicology and Chemistry (SETAC):

Use of Sediment Quality Guidelines and Related Tools for the Assessment of Contaminated Sediments
Wenning, Batley, Ingersoll, Moore, editors

2005

Ecological Assessment of Aquatic Resources: Linking Science to Decision-Making
Barbour, Norton, Preston, Thornton, editors
2004

Amphibian Decline: An Integrated Analysis of Multiple Stressor Effects
Linder, Krest, Sparling, editors
2003

Metals in Aquatic Systems:
A Review of Exposure, Bioaccumulation, and Toxicity Models
Paquin, Farley, Santore, Kavvadas, Mooney, Winfield, Wu, Di Toro
2003

Silver: Environmental Transport, Fate, Effects, and Models:
Papers from Environmental Toxicology and Chemistry, 1983 to 2002
Gorusch, Kramer, La Point, editors
2003

Contaminated Soils: From Soil–Chemical Interactions to Ecosystem Management
Lanno, editor
2003

Environmental Impacts of Pulp and Paper Waste Streams
Stuthridge, van den Heuvel, Marvin, Slade, Gifford, editors
2003

Porewater Toxicity Testing: Biological, Chemical, and Ecological Considerations
Carr and Nipper, editors
2003

Reevaluation of the State of the Science for Water-Quality Criteria Development
Reiley, Stubblefield, Adams, Di Toro, Erickson, Hodson, Keating Jr, editors
2003

For information about SETAC publications, including SETAC's international journals, *Environmental Toxicology and Chemistry* and
Integrated Environmental Assessment and Management, contact the SETAC Administrative Office nearest you:

SETAC Office
1010 North 12th Avenue
Pensacola, FL 32501-3367 USA
T 850 469 1500   F 850 469 9778
E setac@setac.org

SETAC Office
Avenue de la Toison d'Or 67
B-1060 Brussels, Belgium
T 32 2 772 72 81   F 32 2 770 53 86
E setac@setaceu.org

www.setac.org

Environmental Quality Through Science®

FRESHWATER BIVALVE ECOTOXICOLOGY

EDITED BY
JERRY L. FARRIS AND JOHN H. VAN HASSEL

Coordinating Editor of SETAC Books
Joseph W. Gorsuch
Gorsuch Environmental Management Services, Inc.
Webster, New York, USA

CRC Press
Taylor & Francis Group
Boca Raton London New York

CRC Press is an imprint of the
Taylor & Francis Group, an **informa** business

CRC Press
Taylor & Francis Group
6000 Broken Sound Parkway NW, Suite 300
Boca Raton, FL 33487-2742

First issued in paperback 2019

© 2007 by Taylor & Francis Group, LLC
CRC Press is an imprint of Taylor & Francis Group, an Informa business

No claim to original U.S. Government works

ISBN-13: 978-1-4200-4284-9 (hbk)
ISBN-13: 978-0-367-38989-5 (pbk)

This book contains information obtained from authentic and highly regarded sources. Reasonable efforts have been made to publish reliable data and information, but the author and publisher cannot assume responsibility for the validity of all materials or the consequences of their use. The authors and publishers have attempted to trace the copyright holders of all material reproduced in this publication and apologize to copyright holders if permission to publish in this form has not been obtained. If any copyright material has not been acknowledged please write and let us know so we may rectify in any future reprint.

Except as permitted under U.S. Copyright Law, no part of this book may be reprinted, reproduced, transmitted, or utilized in any form by any electronic, mechanical, or other means, now known or hereafter invented, including photocopying, microfilming, and recording, or in any information storage or retrieval system, without written permission from the publishers.

For permission to photocopy or use material electronically from this work, please access www.copyright.com (http://www.copyright.com/) or contact the Copyright Clearance Center, Inc. (CCC), 222 Rosewood Drive, Danvers, MA 01923, 978-750-8400. CCC is a not-for-profit organization that provides licenses and registration for a variety of users. For organizations that have been granted a photocopy license by the CCC, a separate system of payment has been arranged.

Trademark Notice: Product or corporate names may be trademarks or registered trademarks, and are used only for identification and explanation without intent to infringe.

Library of Congress Cataloging-in-Publication Data

Freshwater bivalve ecotoxicology / editors, Jerry L. Farris and John H. Van Hassel.
    p. cm.
  Includes bibliographical references and index.
  ISBN-13: 978-1-4200-4284-9 (alk. paper)
  ISBN-10: 1-4200-4284-X (alk. paper)
  1. Freshwater mussels--Effect of water pollution on. 2. Indicators (Biology) 3. Water quality bioassay. I. Farris, Jerry L. II. Van Hassel, John H. III. SETAC (Society)

  QL430.6F74 2006
  594'.4195--dc22
                                    2006019525

Visit the Taylor & Francis Web site at
http://www.taylorandfrancis.com

and the CRC Press Web site at
http://www.crcpress.com

SETAC Publications

Books published by the Society of Environmental Toxicology and Chemistry (SETAC) provide in-depth reviews and critical appraisals on scientific subjects relevant to understanding the impacts of chemicals and technology on the environment. The books explore topics reviewed and recommended by the Publications Advisory Council and approved by the SETAC North America, Latin America, or Asia/Pacific Board of Directors; the SETAC Europe Council; or the SETAC World Council for their importance, timeliness, and contribution to multidisciplinary approaches to solving environmental problems. The diversity and breadth of subjects covered in the series reflect the wide range of disciplines encompassed by environmental toxicology, environmental chemistry, and hazard and risk assessment, and life-cycle assessment. SETAC books attempt to present the reader with authoritative coverage of the literature, as well as paradigms, methodologies, and controversies; research needs; and new developments specific to the featured topics. The books are generally peer reviewed for SETAC by acknowledged experts.

SETAC publications, which include Technical Issue Papers (TIPs), workshop summaries, newsletter (*SETAC Globe*), and journals (*Environmental Toxicology and Chemistry* and *Integrated Environmental Assessment and Management*), are useful to environmental scientists in research, research management, chemical manufacturing and regulation, risk assessment, and education, as well as to students considering or preparing for careers in these areas. The publications provide information for keeping abreast of recent developments in familiar subject areas and for rapid introduction to principles and approaches in new subject areas.

SETAC recognizes and thanks the past coordinating editors of SETAC books:

Andrew Green, International Zinc Association
    Durham, North Carolina, USA

C.G. Ingersoll, Columbia Environmental Research Center
    US Geological Survey, Columbia, Missouri, USA

T.W. La Point, Institute of Applied Sciences
    University of North Texas, Denton, Texas, USA

B.T. Walton, US Environmental Protection Agency
    Research Triangle Park, North Carolina, USA

C.H. Ward, Department of Environmental Sciences and Engineering
    Rice University, Houston, Texas, USA

Table of Contents

Chapter 9 Linking Bioaccumulation and Biological Effects to Chemicals
in Water and Sediment: A Conceptual Framework for
Freshwater Bivalve Ecotoxicology 215

Michael H. Salazar and Sandra M. Salazar

Chapter 14   Case Study: Sensitivity of Mussel Glochidia and Regulatory
               Test Organisms to Mercury and a Reference Toxicant 351
Theodore W. Valenti, Donald S. Cherry, Richard J. Neves,
Brandon A. Locke, and John J. Schmerfeld

Preface

Bivalves constitute one of the largest groups, in terms of biomass, of filter-feeding organisms in many freshwater ecosystems. Freshwater bivalves have been used in an increasingly diverse array of ecotoxicological applications over the past 30 years. Among the four families of freshwater bivalves, there are sharply differing reasons for their use in ecotoxicological research. The Corbiculidae, Dreissenidae and, to a lesser extent, the Sphaeriidae, have fulfilled the traditional role of ecotoxicological research organisms. These three bivalve families have supported a large percentage of basic research on contaminant uptake, toxicokinetics, and toxicity testing. The fourth freshwater bivalve family, the Unionidae, has emerged as a critical group for consideration in the field of ecotoxicology over the past 20 years, receiving heightened scientific and regulatory focus as a result of recently documented declines in North America's unionid mussel fauna, with greater than 70 percent of the 297 native species currently considered endangered, threatened, or of special concern.

In the field of freshwater bivalve ecotoxicology, as in any field of scientific endeavor, there is the need for an occasional pause, assessment, and refocus of research accomplishment and communication in terms of our research paradigms. We hope to at least partially address this need for reflection and consideration through this book. The range of test techniques and monitoring methods has been applied to freshwater bivalves is extensive, suggesting a wide-scale search for relevant endpoints and specific indicators representative of this fauna. Despite the oft-repeated promotion of unionid mussels as ideal biomonitors and sensitive indicators of environmental stress, existing knowledge concerning these organisms has not been synthesized, and a certain amount of misinformation regarding the usefulness of mussels as biomonitors and environmental indicators needs review and clarification. Three major sessions on molluscan ecotoxicology have been held at Society of Environmental Toxicology and Chemistry (SETAC) annual meetings (1989, 1994, and 1997), and a fourth session in 2003 previewed many of the chapters found in this book. Seven national meetings on the conservation and management of freshwater mussels have been held since 1992. The need to take the next logical step—a text that synthesizes the burgeoning knowledge in this field—is readily apparent.

This book offers a review, synthesis, and critical assessment of the state-of-the-art of freshwater bivalve ecotoxicology by some of the leading scientists in the field. The intent from the outset, among the authors collaborating on this book about freshwater bivalve ecotoxicology, was to provide a collective review of the techniques and approaches being used to assess suspected contaminant impact upon freshwater bivalves. The recognized need for input from scientists in the field also challenged us to relate current achievements in general monitoring of population responses to stressors, fundamental concepts of toxicology specific to burrowing bivalves, and useful insights that might offer direction and priority for resolving specific problems challenging protection and conservation efforts. Authors representing a wide-ranging field of interests were invited to lend their viewpoints and expertise toward a variety of research topics and management issues that have been shared at various meetings throughout the past several years. Authors were then encouraged to critically analyze respective chapters beyond simple reviews of existing research in their area of expertise. We hope the resulting array of viewpoints provides a valuable tool for those interested in freshwater bivalve ecotoxicology. Such varying perspectives among researcher experiences and concerns have been gathered in one volume to provide a sense of both appropriate applications and developmental needs for assessment techniques. Field and laboratory, physiological and ecological, and impact assessment and toxicological studies are covered, and research needs within each area of study are identified. Specific case histories demonstrating the use and value of various ecotoxicological approaches are included. Most importantly, the chapters provide a critical assessment of gaps and weaknesses in the current state of our knowledge on these subjects, providing direction for future research and management tools involving freshwater bivalves.

Acknowledgments

The editors wish to acknowledge the valuable contributions to this book by:

- Those researchers and managers who have supported the ecotoxicology conferences throughout the years that contributed to the body of work on bivalves reviewed in this book and for their ongoing participation in the developing field
- The authors, without whose expertise, time, and scientific acumen this collaboration would not have been possible
- SETAC and its staff in guidance for proposal and sponsorship of this book, identification of qualified peer reviewers, and subsequent cooperative agreements to ensure publication

We are also especially grateful for the financial support of:

- Arkansas Department of Environmental Quality
- Arkansas State University
- EA Engineering, Science, and Technology
- North Carolina State University
- Presbyterian College
- USGS Upper Midwest Environmental Sciences Center
- USGS Columbia Environmental Research Center

Editors

John H. Van Hassel has more than 25 years of experience in various aspects of environmental research, permitting and regulations. He has published more than two dozen technical papers and given more than 25 technical presentations at scientific meetings on the subjects of water quality-based effluent limits, site-specific permit limits, aquatic toxicology, and biomonitoring. He has formulated and chaired sessions at international scientific meetings on the subjects of whole-effluent toxicity testing and freshwater mussel toxicology.

As a biologist at American Electric Power and, prior to that, a researcher in a university laboratory, Van Hassel has extensive experience in the development of water quality-based effluent limits and site-specific water quality standards. Often, this work involved development of innovative procedures under the guidance of state and federal regulatory agencies where existing protocols had been lacking or inadequate to address site-specific situations. He also has had long-term involvement in the critical analysis and development of comments on state and federal water quality standards and regulations, research on freshwater mussel toxicology and ecology, and the design, implementation, and data analysis of aquatic biomonitoring studies.

Van Hassel holds a bachelor's degree in physical science from Michigan State University, and a master's degree in fisheries and wildlife sciences from Virginia Polytechnic Institute and State University. He is a member of the Freshwater Mollusk Conservation Society (FMCS) and the Society of Environmental Toxicology and Chemistry (SETAC). He is currently chair of the Guidelines and Techniques Committee of FMCS, and has served three years on the editorial board of the scientific journal *Environmental Toxicology and Chemistry*, published by SETAC.

Jerry L. Farris is professor of environmental biology and associate dean of sciences and mathematics at Arkansas State University. He studied aquatic ecology and ecotoxicology and received a PhD in zoology from Virginia Polytechnic Institute and State University. Having worked for years at Virginia Polytechnic Institute and State University's Center for Environmental and Hazardous Materials Studies and Arkansas State's Ecotoxicology Research Facility, he has focused much of his research efforts upon development of biomonitoring techniques in both field and laboratory settings. He has authored or co-authored more than 80 open literature publications on diverse topics, with many of those focusing upon bivalves. He has worked as a consultant, educator, and principal investigator on projects for the past 25 years involving community interactions with disturbed or contaminated habitats to determine relative risks posed to ecosystems and organisms. Most often these investigations have involved aquatic organisms with more recent inclusion of assessment frameworks that include interfacing avian and amphibian populations. His research experience has included propagating endangered mussel species, monitoring mussels using field surveys, caged deployments of mussels and Asian clams, and laboratory toxicity assays of glochidia and juveniles.

Contributors

Mindy Yeager Armstead
Potesta & Associates, Inc.
Charleston, West Virginia

Cristi D. Bishop
EA Engineering, Science, and Technology
Sparks, Maryland

Donald S. Cherry
Biology Department
Virginia Polytechnic Institute
  and State University
Blacksburg, Virginia

W. Gregory Cope
Department of Environmental
  and Molecular Toxicology
North Carolina State University
Raleigh, North Carolina

Heidi L. Dunn
Ecological Specialists, Inc.
St. Peters, Missouri

Jerry L. Farris
Arkansas State University
Jonesboro, Arkanas

Timothy S. Gross
University of Florida
Gainesville, Florida

Robert Hudson
Presbyterian College
Clinton, South Carolina

Christopher G. Ingersoll
Columbia Environmental
  Research Center
U.S. Geological Survey
Columbia, Missouri

Anne Keller
USEPA
Sam Nunn Federal Center
Atlanta, Georgia

Nicola J. Kernaghan
Center for Transnational and
  Global Studies & Center for European Studies
University of Florida
Gainesville, Florida

Brandon A. Locke
Biology Department
Virginia Polytechnic Institute
  and State University
Blacksburg, Virginia

Mike Lydy
Department of Zoology
Fisheries & Illinois Agriculture Center
Southern Illinois University
Carbondale, Illinois

Richard J. Neves
Department of Fisheries & Wildlife Sciences
Virginia Polytechnic Institute
  and State University
Blacksburg, Virginia

Teresa J. Newton
U.S. Geological Survey
Biological Resources Division
Upper Midwest Environmental Sciences Center
LaCrosse, Wisconsin

Andy Roberts
U.S. Fish and Wildlife Service
Ecological Services
Columbia, Missouri

D. Shane Ruessler
U.S. Geological Survey
Gainesville, Florida

Michael H. Salazar
Applied Biomonitoring
Kirkland, Washington

Sandra M. Salazar
Applied Biomonitoring
Kirkland, Washington

John J. Schmerfeld
U.S. Fish & Wildlife Service
Virginia Field Office
Gloucester, Virginia

Damian Shea
Department of Environmental
and Molecular Toxicology
North Carolina State University
Raleigh, North Carolina

David J. Soucek
Illinois Natural History Survey
Center for Economic Entomology
Champaign, Illinois

Waverly A. Thorsen
Department of Environmental
and Molecular Toxicology
North Carolina State University
Raleigh, North Carolina

Theodore W. Valenti
Biology Department
Virginia Polytechnic Institute
and State University
Blacksburg, Virginia

John H. Van Hassel
Environmental Services Division
American Electric Power
Columbus, Ohio

Ning Wang
Columbia Environmental
Research Center
U.S. Geological Survey
Columbia, Missouri

G. Thomas Watters
Department of Evolution,
Ecology, and Organismal Biology
The Ohio State University
Columbus, Ohio

Jessica L. Yeager
Potesta & Associates, Inc.
Charleston, West Virginia

1 Freshwater Bivalve Ecotoxicology

John H. Van Hassel and Jerry L. Farris

SCOPE OF FRESHWATER BIVALVE ECOTOXICOLOGY

All too often, freshwater mussel populations are referred to as objects of study for their noticeable decline (Williams et al. 1993; Biggins, Neves, and Dohner 1995; Neves 1997). A description of this rationale has most often been followed by an assertion of cause attributed to a generalized list of human activities that have rarely been sufficiently assessed for their respective impacts. Any verification of those relative conditions relies largely on survey information from studies that have often lacked clearly defined objectives, coherence with concerns, and consistency of methods directed toward sampling effort or spatial or temporal coverage. Indeed, the lack of coherence and consistency of assessment methods could have at times been considered for its relative contribution to those declines. Many resource managers have found that historical data, while critical to perspective, must be interpreted with caution since most is of low resolution and can contribute only to qualitative generalizations about past ecosystem conditions (Steedman et al. 1996; Metcalfe-Smith et al. 1998). Recent collaboration and communication among biologists, ecologists, resource managers, and ecotoxicologists have established an urgent forum of exchange, given the growing necessity for an aggressive yet conservative approach to evaluate mussel populations for their relative status, contribution to ecosystem processes, or response to natural and anthropogenic stressors. This current dilemma of dwindling resources and ever increasing enthusiasm to share our collective expertise and interests for the assessment of mussels seems well suited for a focused, conceptual framework that can be used to establish priorities for the sake of the resource.

Evaluation of stressor impacts upon freshwater mussel communities has progressed to distinct phases of investigation that now place greater emphasis upon the review, synthesis, and refinement of methodologies that include standardization of applicable techniques among investigators and different geographic locations, whenever possible. The rationale for standardization seems to stem from the need to guide sampling or monitoring programs toward well-suited objectives that offer greater perspective and resolution of mussel community or population status. The need for balanced guidance ranges from field to laboratory consideration of evaluative tests (ASTM 2005), as well as surveying, monitoring, and sampling acceptability (ASTM 2001; Strayer and Smith 2003). These refined methods can then support different monitoring approaches in consideration of water quality parameters, discharge limits, or even impacts of large-scale disturbance. The fact that data from these approaches may then be used in much larger and more focused assessments of biological integrity suggests that these sometimes critical observations have lasting consequences beyond consideration of any single species preservation. Although scientists are noted for their tendency to place greater emphasis upon their preferred organism, community, response of study, or even single indicators of stress (Cairns 2003), the recent attention being given to the possible role of contaminants in freshwater mussel population declines has garnered consideration within the collaborative and multidisciplinary field of ecotoxicology. This field of diverse goals, involving a wide array of investigative tools, is well suited to provide meaningful explanation to malacologists, conservationists, regulators, and resource managers, when risk probability estimates are influenced by the way in which questions are framed by those with a shared understanding of the temporal and spatial scales involved. Else, the precautionary principle becomes rule.

Newman (1998) has provided important perspective as to how ecotoxicologists contribute to current knowledge with overlapping yet distinct goals. New or improved technologies are enabling the detection of causal relationships or reduced conditions, which in turn requires an effective balance between integration of knowledge and practical application of conceptual approaches and emerging technology to characterize risk. Ecotoxicology has been recognized for content and contribution of scientific, technological, and practical goals (Slobodkin and Dykhuizen 1991; Newman 1998) from diverse disciplines. At times, portions of that knowledge base are purposefully excluded from consideration to progress effectively toward respective and more focused questions or prioritized goals. The goal of solving or documenting impacts on freshwater bivalves may require tools with specific steps for either biomonitoring or emphasizing a risk assessment-based approach. Those differing instances may not necessarily require greater emphasis upon understanding the processes of contaminant uptake than on the extent of population loss (valued listed species). Progress seems most probable when those distinct goals and approaches are understood and respected by professionals in any applied science (Newman and Unger 2002). Life history information can be as critical as elaborating the science of nutrient processing or employing a weight-of-evidence approach to assess the need for site remediation. Such seemingly different goals may often require a program shift or more flexible decision matrix to progress a remedy or assess a specific problem such as declining freshwater mussel diversity.

The intent from the outset, among the authors collaborating on this book about freshwater bivalve ecotoxicology, was to provide a collective review of the techniques and approaches being used to assess suspected contaminant impact upon freshwater bivalves. The recognized need for input from a cross-section of active scientists in the field also challenged us to relate current achievements in general monitoring of population responses to stressors, fundamental concepts of toxicology specific to burrowing bivalves, and useful insights that might offer direction and priority for resolving specific problems challenging protection and conservation efforts. Authors representing a diverse field of interests were invited to lend their viewpoints and expertise toward a variety of research topics and management issues that have been shared at various meetings throughout the years. Authors were then encouraged to critically analyze respective chapters beyond simple reviews of existing research in their areas of expertise. We hope the resulting array of viewpoints provides a valuable tool drawn from both internal and external peer reviews. Such varying perspectives among researcher experiences and concerns have been gathered in one volume to provide a sense of both appropriate applications and developmental needs for assessment techniques. Since issues constantly challenge our concepts of ecotoxicology and ecological processes, it is expected that pertinent considerations specifically targeting adverse impacts will likewise continue to demand novel approaches to relate historic information on ecological effects from field surveys to exposure and effects data.

Participants in this book are representative of the varied history of situations and investigations involving the recognized diversity of freshwater bivalves found in the United States. Any bias offered by US perspective is not intended to isolate topics or review material that encompasses the global aspects of freshwater bivalve ecotoxicology, here defined as "the branch of toxicology that studies the toxic effects of natural or artificial substances on freshwater bivalves" (modified from Truhaut 1975). Stressors would be expected to manifest their effects with ecoregion, geochemical, and waterbody modification, and therefore regional aspects of contaminant fate and effect would be expected to act on a broader range and scale than that best documented in North American mussel distributions. Overlooked systems and collection records, once inspected for trends of species losses and changing community composition, have provided compelling evidence of such steady declines in Canada (Metcalfe-Smith et al. 1998) and globally (Bogan 1993). Fundamental concepts regarding cross media and global transport of contaminants would be expected to apply to mechanisms by which toxicants exert their effect broadly on many aquatic organisms. Indeed, the very nature of defining the science of contaminant bioavailability and effects is typically not exclusive to any taxonomic group or hierarchical level of biological organization. Rather, conceptual coherency exists among levels of hierarchical topics ranging from molecular to biospherical as illustrated by Caswell's (1996) consideration of higher-level consequences derived from mechanisms operating at lower levels. The diversity and hierarchical organization of topics currently addressed by ecotoxicology places depend upon a synthetic science drawn from many disciplines (Newman and Unger 2002).

Ecologists and ecotoxicologists have recognized inherent biological traits in organisms that allow species to respond differently to a compound at a given concentration (i.e., different species have different sensitivities). The acknowledgment that species' sensitivities to toxic compounds differ (without explanation of cause) and the description of that variation, usually with a parametric distribution function, yield the species sensitivity distributions (Posthuma, Suter, and Trass 2001). These distributions are now used for both derivation of environmental quality criteria and ecological risk assessment for contaminated ecosystems first in North America and Europe but now elsewhere. They are most often used at the interface of science and regulation so their efficacy is therefore dependent on data quality and quantity and careful interpretation in decision making. The variance in sensitivity among the test species and the means are used to calculate a safe concentration for most species of interest. Since a percentile of that distribution is chosen as a concentration that is protective for most species in a community, there have always been challenges concerning claims of significant taxon decline (as with mussels and amphibians), where unique biological attributes are believed to contribute to their sensitivity. This presents some degree of urgency with groups in decline, since they could be early indicators of irreversible damage if disturbance is severe enough to change ecosystem function (Lubchenco et al. 1991).

Landis and Yu (2004) have provided a framework of three functions representing the basic aspects of environmental toxicology that can also be applied to instances of causality of bivalve impact linked to contaminant exposure:

1. The interaction of the introduced chemical or xenobiotic with the environment controlling the amount of toxicant or available dose to biota
2. The xenobiotic interaction with a particular protein or other biological molecule
3. The above function resulting in the produced effects at a higher level of biological organization such as a reduced mussel community or declining population

Any further recognition or accurate prediction of the effects of pollutants on bivalves requires the identification of these functions as they relate from community level effects back to the specific receptor molecule. A realization of this requirement makes apparent the need for research that examines how such interactions with xenobiotics extend from geochemical processes and cycles through physiological and behavioral responses amidst bivalves functioning in dynamic ecosystem processes.

Many of the topics covered within this volume cover familiar ground for ecotoxicologists: impact assessment, toxicokinetics, biomarkers, pollution tolerance, etc. Once sufficiently introduced to the available knowledge on the life history of the fauna, chapters quickly move towards fundamental concepts surrounding responses that have been measured in freshwater bivalves as a consequence of chemical exposures or accumulated contaminants in target organs or tissues. That such basic processes as contaminant uptake, accumulation, biotransformation, detoxification, and elimination are mediated by an individual's physiological condition, size, sex, or genetics, is reason enough to warrant consideration of the range of bivalve adaptations that are most closely related to their ecological roles, demographic interactions, and trophic positions in freshwater environments.

Far from being adequately described and understood, the activities of unionacean mussels that contribute to particle processing, nutrient release, and sediment mixing are more recently being considered for their contribution to ecosystem function and maintenance of ecological integrity (Vaughn and Hakenkamp 2001; Strayer et al. 2004). Elaboration of feeding processes across environments, species, and life stages (Tankersley 1996; Ward 1996; Raikow and Hamilton 2001; Christian et al. 2004) challenges many earlier concepts linking simple siphoning activity by the imperiled "ideal biomonitor" to chemical uptake. Continuing resolution of trophic position and potential food sources of unionid bivalves based on stable isotope ratios and biochemical markers (Nichols and Garling 2000) will be critical to a better understanding of possible routes of contaminant uptake and bioavailability as they relate to ingestion and assimilation. Depending on their physical and chemical properties, contaminants may be absorbed, biotransformed, or excreted. Not only are these potential stressors acting on bivalves and, in turn, their role in ecosystem processes, but some are modified by the interaction and therefore exert a different potential divergence from homeostatic conditions (i.e., effects of excreted products like ammonia, stability to sediments by density of individuals, and shells as increased substrate for other aquatic organisms) (Figure 1.1). Luoma et al. (2001) have suggested that success in identifying the occurrence and cause of a stressor is often dependent on characterizing the most important driving processes in an ecosystem and by capturing the dynamic stabilities and events driven by those processes. Even with a chemical stressor, the misconception of only a localized exposure to a single medium can quickly elaborate to include distribution among solution, suspended particles, sediments, pore waters, and specific food sources (Table 1.1).

The range of water column and sediment processes affected by freshwater bivalves includes particle processing that may shift from suspension or filter feeding to sediment surface or interstitial feeding upon organic detritus and bacteria (Vaughn and Hakenkamp 2001). Collectively, these processes may in turn affect water clarity and nutrient cycling and retention to a degree dependent on contributing biomass of benthic residents, which more recently in North American systems include significant additional amounts from zebra mussels (*Dreissena polymorpha*) and Asian clams (*Corbicula fluminea*). With such widespread occurrence, Asian clams have been cited for their significant contribution to total benthic respiration and organic matter dynamics (Hakenkamp and Palmer 1999) as well. Zebra mussel impacts upon the status of freshwater mussel populations

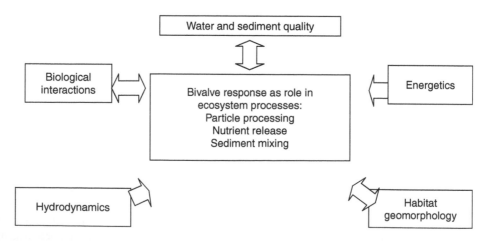

FIGURE 1.1 Array of stressor interactions with bivalves as valued ecosystem components.

TABLE 1.1
Potential Stressors of Freshwater Burrowing Bivalves in an Aquatic Ecosystem

Category	Stressors
Biological interactions	Community succession or displacement
	Colonization, predation, and competition
	Feeding, deposit, and suspension
	Disease and parasitism
	Microbial metabolism
	Effects on benthos
	Reproductive strategies, timing, and type of hosts
	Vegetation as cover, trap, and carbon source
Hydrodynamics	Disturbance as pulse or sustained
	Divergence of stability during base or high flow
	Shear stress or velocity
	Interstitial flow
	System recharge from surface or subsurface
	Runoff with precipitation
	Volume
	Retention time
Energetics	Influencing primary productivity
	Resource availability
	Seasonal patterns
	Availability of algae, bacteria, detritus, and small animals
	Dissolved organic matter
	Biomass
	Behavior allocated to reproduction, avoidance, and stability
	Nutrient and particle retention
Habitat geomorphology	Catchment lithology
	Hyporheic zone characteristics
	Channel morphology
	Cover
	Bank stability and overflow capability
	Temporal heterogeneity
	Gradient
	Magnitude and frequency of meander
Water and sediment quality	Dissolved oxygen
	Macro and micronutrients
	Metals and metalloids
	Organics
	Hardness
	Alkalinity
	Temperature
	pH
	Turbidity
	Compaction or pore spaces
	Alluvial hydraulic conductivity
	Cation exchange capacity
	Salinity

Source: Modified from Foran, J. A. and Forenc, S. A., *SETAC Pellston Workshop on Multiple Stressors in Ecological Risk and Impact Assessment*, 13–18 September 1997, Pellston, Michigan, SETAC, Pensacola, FL, p. 100, 1999. With permission of the Society of Environmental Toxicology and Chemistry [SETAC].

and their habitats throughout the United States continue to be assessed as relative indicators of ecosystem health (Shear et al. 2005). All three freshwater bivalves (clams, zebra mussels, and native unionids) have at times been naively designated as indicator organisms of relative environmental deterioration for their appearance and dominance, and less so for their ability to bioaccumulate or store energy. Temperature, water quality, and current velocity would represent only a few of the interacting factors influencing rates of such related processes as sediment resuspension and chemical mobilization such that excreted nutrients and biodeposited faeces and pseudofaeces could contribute to particle size selection and filtration rates by resident bivalves. This represents a significant gap in the understanding of both basic ecological processes and how prevailing conditions interact with water columns and sediment-associated contaminant exposures. Basic conditions involving the interplay between food availability (Strayer et al. 1999) and energy storage (Naimo et al. 1998) in the face of invasion by an exotic species have been cited as reason enough for the observed variability in unionid response to any disturbance (Vaughn and Hakenkamp 2001), let alone multiple stressors. Uncertainty over the extent and importance of sediment-related ecological processes has potential consequence for the resolution of chemical interaction with the microhabitat scales experienced by mussels. These processes must also be considered in the design of monitoring programs and guidance for toxicity tests with contaminated sediments using various life stages of freshwater unionid bivalves or suitable surrogate organisms. Current progress toward resolving stressor impact has concentrated more on water-borne exposures than on sediment or integrated exposures among different media, in large part for the relative certainty of effects measured in water (Figure 1.2).

Other chapter topics address issues that are particularly relevant to freshwater bivalve research, including difficulties encountered with the laboratory culture of these organisms for the purpose of toxicity testing or other controlled experiments, and the use of surrogate test organisms to relate sensitivities of response and reduce pressure on an often overlooked and already impacted fauna. Innovative field research that has used in situ bivalve toxicity testing and effects-oriented tissue contaminant assessment is reviewed, as well. The book concludes with three specific laboratory or combined field/laboratory ecotoxicological studies of freshwater bivalves.

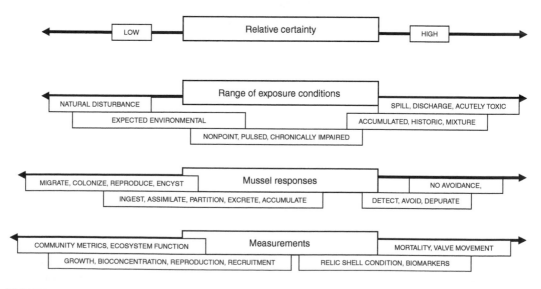

FIGURE 1.2 Scale of relative certainty integrated among exposure conditions and measured mussel responses.

NEED FOR A FRESHWATER BIVALVE ECOTOXICOLOGY BOOK

Many of the current researchers conducting ecotoxicology studies upon freshwater bivalves were first drawn to these organisms because of their touted sensitivity to a variety of environmental perturbants, ease of collection and handling, and/or the lack of reliable information to support conservation and management of imperiled Unionidae. Situations currently prompting assessment of future and past impacts to mussels are diverse and therefore require both site- and stressor-specific approaches that are properly integrated to address the challenging issues (Table 1.2). Declines in unionid mussel diversity and abundance in the United States have been noted in many aquatic systems for at least the past century (Neves 1987; Layzer, Gordon, and Anderson 1993). These declines have become alarmingly frequent and widespread over the past twenty years, with a number of short-term die-offs going largely unexplained but documented to some degree (Neves 1987; Sparks et al. 1990). Better documentation seems to be available for longer-term declines (Neves 1987; Keller and Zam 1990; Busby and Horak 1993; Layzer, Gordon, and Anderson 1993; Williams et al. 1993), although it tends to assume such a generalized notion surrounding:

TABLE 1.2
Situations Prompting Biologically and Stressor-Directed Approaches for the Assessment of Mussel Impact

Situations	Categories
Infrastructure impacts from construction and hydrologic disturbances	Bridges and roadways
	Dams
	Drainage, irrigation, diversion
	Channelization and levees
	Gravel and sand removal
Waterbody use directly	Sport and recreation
	Navigation
	Cooling and hydro energy production
Resource use directly	Commercial shelling
	Historic removal for pearls or button manufacture
	Cultured pearl industry
	Surveys to assess water body condition
Responsibility and oversight	Management of refuges, federal, state, and private
	Resource damage assessments
	Agency obligation for species of concern
	Permitted discharges
	Site remediation
	Treatment of pests
	Registration of products
	Spill, dump, or exceedance response
	Control of nuisance and introduced species
Large scale disturbance	Climate warming
	Flood events
	Fire events
	Sedimentation from above
	Criminal intent or eco/bioterrorism

- Construction of dams on streams and rivers, thereby converting needed riffle and shoal habitat to lentic habitat
- Siltation resulting from both instream (e.g., dredging) and riparian (e.g., bank erosion) activities, thereby altering mussel habitat and interfering with life history functions
- Pollution from a variety of sources that degrades water quality beyond the assimilative capacity of ecosystems and environmental limits for mussel populations
- Commercial harvest for the Japanese cultured pearl industry, in cases where mussel beds have been decimated by over-harvest
- Habitat alteration (e.g., channelization)

Understandably, linking causality to observed losses during localized mussel die-offs has been more difficult to establish. It seems likely that chemical stressors are at least contributing to, if not the cause of, many impacts on freshwater unionid bivalves. However, many populations additionally deal with a host of competitive interactions with invading exotic species (e.g., Asian clams, zebra mussels) and predation, particularly by muskrats (Neves 1987, 1992; Never and Odom 1989; Sparks et al. 1990; Shannon, Biggins, and Hylton 1993). Less likely potential causes for decline include disease or parasites (Neves 1987; Sparks et al. 1990). Many of the scientific and environmental management challenges then depend on determining the significance of potential chemical effects in the presence of other environmental stressors and interacting factors (Dorward-King et al. 2001). Overemphasis on potential chemical effects without such consideration can and most likely has resulted in pollution control investments with relatively few rigorous attempts to document significant improvements in environmental quality (Hart 1994). A growing consensus suggests that the physical alterations of landscapes have a greater influence on ecological resources than do chemicals whose concentrations are near or slightly above toxicity thresholds; this concensus is supported by documented instances of channelization, fragmentation of plant communities, timber harvest, fire control, and increase of impervious surfaces (Burton and Pitt 2001).

The cumulative result of the seemingly accelerating decline of unionid mussel populations is staggering: greater than 70 percent of the 297 native unionid species are currently considered either endangered, threatened, or of special concern (Williams et al. 1993). As many as 18–21 species may already be extinct (Shannon, Biggins, and Hylton 1993; Williams et al. 1993). A number of measures to slow the decline have been adopted, mostly on a state or waterbody-specific basis, including commercial mussel harvest regulations, water quality standards considering mussel responses, refuge establishment and improved management, and riparian conservation measures (Busby and Horak 1993; Cherry et al. 2002). These well-intentioned actions are most often applied to situations without adequate life history information on the affected populations or reliable spatial scales related to the distance and timing of impacts. The urgent sense of "just doing something" through conservation actions without adequate information on causality or status of processes ignores recent discoveries about the nonlinear dynamics, time delays, and multiple interactions that confer upon ecological systems the property of complexity (Matthews, Landis, and Matthews 1996). This is referred to as the foundation of the community conditioning hypothesis, which may well contain historical information critical to the survival of freshwater mussels. Management considerations at this scale would certainly change our current concepts beyond that of the functional "bed" dimensionality.

Impacts upon mussel communities may prompt assessment of situations ranging from acute stressor exposures associated with spills or unusual discharges to chronic conditions of historic or degrading water or habitat quality. If consideration of the temporal scale can be neatly fitted into ecological risk and impact assessment based on before and after measures, respectively, this concept is made all the more challenging by changing conditions that expand beyond the analysis of effects of single chemicals on individuals, populations, or communities. More often, the impact may be attributable to an array of chemical, physical, or biological processes that may be active, either concurrently or sequentially. The formal, qualitative and quantitative assessment of multiple stressors, although difficult, has more recently been provided guidance through ecological risk assessment (Suter 1993; USEPA 1996; Foran and Ferenc 1999; Ferenc and Foran 2000). The above-mentioned risks and impacts can be identified using at least two perspectives that would examine the structure and function of ecological systems (biologically directed approach) and/or through a stressor-directed (risk-based) approach. Given the current attempts to organize the life-history strategies and functional roles of freshwater mussels into conceptual frameworks, both strategies would seem well suited to address impacts encompassing large-scale and multi-stressed systems. More recent efforts to expand ecological risk assessment to more accurately reflect the reality of the structure, function, and scale of ecological structures have developed a regional risk assessment at the landscape scale (O'Neil et al. 1997). This idea of assessment for multiple stressors, historical events, spatial structure, and multiple endpoints (as defined for regional risk assessment by Landis (2005)) more closely resembles the probable range of processes experienced by freshwater mussels.

Recovery of impacted mussel populations, whenever and if it occurs, will likely be a slow process. Due to a life span that may vary from decades to centuries, the pace of temporal change in freshwater mussel populations may be dependent on any number of generational factors related to this life span (Strayer et al. 2004). Interactions of chemical and physical stressors with populations would be expected to alter measures of population size (number of individuals or their summed biomass, equilibrium abundance, stability, and recovery), population structure (tolerance or sensitivity distributions, genetic diversity, and size and age structure), and measures of persistence or viability (time to and probability of extinction, minimum viable population, and quasi-extinction) (Maltby 1999). Recruitment rates even in robust unionid mussel populations are typically low or sporadic, and their complex reproductive cycle or life history strategy is easily disrupted by environmental challenges resulting from both manmade and natural events. As a result, many mussel populations may be dominated by older individuals, with low representation of recent age classes indicating unsuccessful recruitment during recent years (Neves 1992). Mussels may therefore require several years of successful spawns to recolonize areas in which they have been decimated or eliminated (Neves 1987). Monitoring macroinvertebrate and fish response to such change has come to include a much better understanding of how communities can be reset to alternative stable states during recovery phases (Clements et al. 2001). Here again, in order to include a holistic and integrative overview of disturbance and contaminant impact on mussel organization, the appropriate links to biological organization (linking individual-level responses and population or community-level consequences) within the same organism become critical. Such links should include the degree of specificity for fish hosts and, in turn, the consideration for habitat or hydrologic constraints to support such propagation requirements. Disturbance and impact could elicit a range of sensitivities that do not necessarily pair mussels with their hosts. This concept for hierarchic-level considerations in monitoring traits with measured ecological relevance has been outlined for freshwater bivalves by Smolders et al. (2003).

Even with so many unknown factors about the stability, relative condition, and formation of mussel beds (see discussion of negative and positive mechanisms attributed to bed production in Strayer et al. 2004), the recolonization of mussel populations is being reported for certain areas. For example, areas now support resident communities where incremental reductions in point source inputs and refinements in discharge treatment have occurred. Follow-up investigations are still needed to answer whether such recovered beds have comparable community composition and sustainable recruitment or merely a temporarily sustained relic community subject to forces of scale not covered by localized water quality improvement. Comparative colonization and migration abilities may leave the mussels at some disadvantage. Assumptions for the degree of protection offered to mussels by the use of invertebrate or fish models may often overlook the complexity that not only exists within mussel life history strategies, but also the degree of selective pressures at work upon an organism with very specific spatial and temporal dependency on the higher-level system of which it is an integral part.

Overall, unified protocol and concept regarding assessment of freshwater bivalves lags behind that being applied to marine bivalves (DeFur et al. 1999) or other taxon-directed ecotoxicological investigations such as amphibians and reptiles (Sparling, Linder, and Bishop 2000; Linder, Krest, and Sparling 2003), birds (Hart et al. 2001), and earthworms (Sheppard et al. 1998). There is recent consideration for order-specific approaches that are necessary to provide adequate protection of ecological resources. Many of these considerations recognize order-specific trophic interactions, differences among cutaneous coverings or routes of uptake, habitat selection, life history strategies, or other niche-segregating characteristics. Whatever the taxon-specific differences, regulators, risk assessors, managers, and researchers are now encouraged to integrate these comparative differences into the risk assessment framework and eventual decision-making process. For this reason and many others, unionids challenge a traditional consideration of their specific, occupied niche as related to concerns for their protection from impact and potential for conservation management.

Research sessions specifically dedicated to freshwater bivalve ecotoxicology have been held at SETAC annual meetings in Toronto in 1989, Denver in 1994, and Austin in 2003. As a result of these sessions, the need for documentation of where freshwater bivalve ecotoxicology research has been, and where it is headed, became increasingly apparent.

IMPORTANCE OF RESEARCH ON FRESHWATER BIVALVES

Bivalves often constitute one of the largest groups, in terms of biomass, of filter-feeding organisms in many freshwater ecosystems. If ever their ecological position can be categorized as typical, they are most often described as sessile, primary consumers with what would appear to be rather direct and relatively easily measured routes of contaminant exposure. This, however, seems contradictory given the range of life history strategies and feeding regimes being currently described. Bivalves also are typically larger than most filter-feeding organisms, such that they are often easier to sample and handle and provide a larger tissue mass for measurement of exposure and effects endpoints. Among the four families of freshwater bivalves, however, there are sharply differing reasons for their use in ecotoxicological research. The Corbiculidae, Dreissenidae, and to a lesser extent, the Sphaeriidae, have fulfilled the traditional role of ecotoxicological research organisms. These three bivalve families have supported a large percentage of basic research on contaminant uptake, toxicokinetics, and toxicity testing due in part to their broad distribution, abundance, high reproductive potential, and ease of collection and laboratory culture. The fourth freshwater bivalve family, the Unionidae, has emerged as a critical group for consideration in the field of ecotoxicology over the past twenty years for a different reason. The mussel fauna of North America exhibits the greatest variety of species in the world and is mainly concentrated in the southeastern United States (Parmalee and Bogan 1998). The ongoing confirmation of population declines is not a dissimilar situation to the recognized global reduction in mollusks listed as threatened in 2004 as the percentage of species evaluated by the International Union for the Conservation of Nature (IUCN 2004).

An interesting aspect of research on *C. fluminea* and *Dreissena* spp. addresses the widespread proliferation of these bivalves in nonnative environments. Because these organisms were already in common use at the time of their introduction to nonnative waters, there was sufficient knowledge of their life history and culturing requirements, so that new research could proceed directly to critical investigation of the means to control their spread.

ROOTS OF BIVALVE ECOTOXICOLOGY

During the late 1960s, elevated public consciousness of water pollution problems gave rise to government legislation and public and private research funding of pollution research in North America and elsewhere. Subsequently, aquatic toxicological methods development and research on the effects of natural and manmade substances were directed at a broad array of species, including bivalves (e.g., Bedford, Roelofs, and Zabik 1968; Imlay 1971, 1973). Early work on freshwater bivalves was summarized by Wurtz (1956) and later, in more thorough fashion, by Fuller (1974). However, as research on easily obtained and cultured species such as *Lepomis macrochirus* (bluegill), *Pimephales promelas* (fathead minnow), *Daphnia pulex* (daphnid), and *Hyallela azteca* (amphipod) proliferated, freshwater bivalve research failed to be included in attempts to standardize testing protocol, in large part due to the lack of information and success involving propagation and maintenance of laboratory cultures. Marine bivalve research has continued to develop in advance of freshwater research partly because of the ready availability and research utility of the genus *Mytilus*, in addition to members of the bivalve class having been broadly used to determine the existence and quantity of chemical exposure (Jorgensen et al. 2005). Goldberg (1975) is credited for introducing the concept of "mussel watch," which utilizes mollusks broadly in the detection of polluting substances. Similar programs currently exist worldwide beyond the conceptualization in North America. Marine mollusks have been extensively used in experimental studies, and a vast body of data exists on fundamental physiological, biochemical, genetic, and toxicological investigations (Salazar and Salazar 1991; Uhler et al. 1993). Standard methods also exist for testing several marine bivalves and include endpoints ranging from survival to larval settlement (USEPA 1997; ASTM 1999).

Beginning about 1980, freshwater bivalve research activity increased significantly. Much of this increase can be attributed to the burgeoning use of Asian clams in both field and laboratory studies (Doherty 1990), followed by the use of two freshwater *Dreissena* species. At about the same time, unionid species such as *Anodonta* spp. and *Elliptio complanata* became frequently used research organisms (e.g., Balogh and Salanki 1984; Hudson and Isom 1984; Tessier et al. 1984; Hemelraad, Holwerda, and Zandee 1986a; Hemelraad et al. 1986b; Holwerda and Herwig 1986; Campbell and Evans 1987) primarily due to the development of in vitro methods to transform glochidia to juveniles. Methods development and resulting research activity on freshwater bivalves has continued to progress, including the development of a standard ASTM test method for in situ growth studies (ASTM 2001), such that there is currently heightened interest in the testing and monitoring of various bivalve species for impact assessment and regulatory development.

A NEED FOR FRESH PARADIGMS

In the field of freshwater bivalve ecotoxicology, there is the need for an occasional pause, assessment, and evaluation of research accomplishment and communication in terms of our research paradigms. We hope to at least partially address this need for reflection and consideration through this book. Over time, certain generalizations arise as various research findings are published and assimilated by the scientific community. These generalizations often influence the direction of future research, so that it is important to routinely examine the underpinnings of these generalizations to ensure that they do not become "red herrings" that divert our attention from important areas of research. A few examples are examined below.

FRESHWATER BIVALVES AS IDEAL BIOMONITORS

Although the notion of an "ideal" monitor is not as frequently referenced today as it was some years back, the concept warrants the position as a suitable null hypothesis. A primary source of confusion might lie in the variety of definitions that have been assigned to the term "biomonitor." There is a long history of both freshwater and marine bivalve use to provide in situ monitoring of chemical body burdens, as these organisms possess several attributes (e.g., sedentary organisms, large tissue mass, variable exposure routes, and relative tolerance) that make them useful biomonitors (see Chapter 2). Conversely, there are many freshwater habitats that contain no or only limited bivalve populations.

And, there are certainly exposure conditions under which many bivalves fail to provide reliable or accurate measures of contaminant risk because of their ability to "clam up" or because they lack sufficient biotransformation potential, which may limit the use of certain biomarkers as rapid and sensitive endpoints of exposure. In scenarios of sufficient exposure to allow chemical equilibrium being reached in organisms facilitating respiratory function, potentially toxic effects may manifest themselves. Use of freshwater bivalves in other types of biomonitoring, such as in situ measurement of growth, tissue accumulation, or biomarker responses to chemical exposure, continues with careful development among a broad range of bivalve exposures and assessment conditions (Figure 1.3).

FRESHWATER BIVALVES AS THE MOST SENSITIVE SPECIES

There is a persistent, conventional wisdom among many of us working in the field of ecotoxicology that freshwater bivalves, and unionid mussels in particular, are more sensitive to chemical exposures, and a variety of other environmental stressors, than other organismal groups. For the most part, this thinking is based on a combination of the relatively recent and global decline in unionid populations and some preliminary laboratory toxicity data indicating extreme sensitivity of a few unionid species to select contaminants. Therefore, as noted by Cairns and Niederlehner (1987), any notion of extreme sensitivity attributed directly to the organism should carefully weigh the lack of certainty associated with any species for any given combination of chemicals and conditions. An examination of available data bears out this statement. Firstly, there is compelling data that demonstrates sensitivity of some tested unionid species to certain contaminants, particularly copper and un-ionized ammonia. Cherry et al. (2002) tested eight mussel species for acute copper toxicity and found that these eight species (seven genera) ranked among the seven most copper-sensitive aquatic genera contained in the comprehensive USEPA acute toxicity database. Similarly, Newton et al. (2003) showed that some unionid species were more acutely sensitive to un-ionized ammonia than other tested aquatic species. However, testing of some other contaminants, including chlorine (Ingersoll et al. 2003, unpublished data) and malathion (Keller and Ruessler 1997), demonstrated more moderate sensitivity, or even relative tolerance, of unionids compared to other tested species (see also Chapter 7).

Field biotic assessments directed at determining, as quantitatively as possible, the relative tolerance to the array of environmental stressors common to natural systems, have similarly demonstrated a range of unionid sensitivities. The North Carolina Biotic Index (NCDEHNR 1997) included three mussels species: two (*Alasmidonta undulata*, *Elliptio lanceolata*) were rated very sensitive to environmental contaminants compared to other freshwater macroinvertebrates in the state, whereas the third (*E. complanata*) was rated in the mid-range of tolerance.

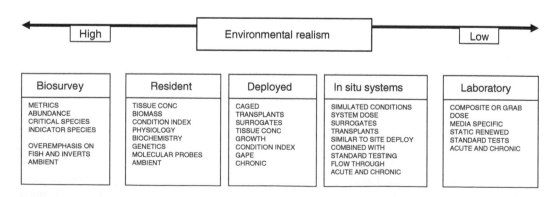

FIGURE 1.3 Relative degree of environmental realism associated with range of bivalve exposure and assessment conditions.

It is important to note from these tolerance ratings that two unionids in the same genus, *Elliptio*, differed widely in relative tolerance in the field.

While it may very well be necessary and desirable to continue research that describes species sensitivity to various contaminants so that appropriate protection can be applied to sensitive ecosystems, it is counterproductive to make ecosystem management decisions based on an automatic, and potentially incorrect, assumption of mussel sensitivity to any environmental stressor. According to Cairns and Niederlehner (1987), "(s)ound judgments in management and regulatory decisions will only be possible if one can extrapolate or predict, based on the results from the tested species, the response of a much larger array of species and behaviors at different levels of biological organization."

BIOMARKERS AS SURROGATES FOR TRADITIONAL RESPONSE MEASURES

Great strides have been made in contaminant-specific laboratory studies of freshwater bivalves, but development of in situ impact monitoring strategies is lagging behind. Recent work has produced a strong basis for in situ studies of bivalve contaminant body burdens (e.g., ASTM 2001); however, there is an additional need for contaminant-specific biomarkers for use in field assessments where traditional measures of survival, growth, or reproduction do not provide answers to cause-effect questions of impact. Biomarkers have the advantage of being relatively quick and easy to measure compared to growth or reproduction and within the appropriate context can be used to investigate causality. As for other groups of organisms, biomarkers have great potential for use with bivalves following an impact event. Bivalves are particularly amenable to this type of assessment because they typically remain in place during impact events, which provides an excellent assessment tool for making before–after or upstream-downstream comparisons. Useful assessments are possible for both short-term events and long-term, chronic impacts.

Biomarkers have already been successfully used in in situ assessments of impact on bivalves (see Chapter 10). Potentially useful biomarkers include metallothionein as an indicator of elevated metal exposure (Couillard, Campbell, and Tessier 1993) and cholinesterase as an indicator of exposure to common classes of pesticides (Fleming, Augspurger, and Alderman 1995). Southwick and Loftus (2003) provide guidance for the use of biomarkers to investigate mollusk kills associated with chemical spills, effluent releases, and runoff events. These measures offer a promising tool for reducing the uncertainty involved in ecologically important assessments of cause and effect.

UNIONIDS AS DRIVERS OF BIVALVE RESEARCH

Some characteristic weaknesses of unionid species as subjects of research (e.g., slow growth, complex reproduction, limited availability or no clear reference populations, and culture difficulties) may be contrasted against similar characteristics of some nonunionid bivalve species that help strengthen their selection for study. We must be careful not to dismiss or downplay particular areas of research based on unionid characteristics but must instead look for ways to integrate and build upon the strengths of various bivalve species. Unionid ecotoxicological research, particularly chemical- or compound-specific toxicity testing, will continue to be important to the protection of ecosystems from water quality impacts. From a practical standpoint, toxics-related bivalve research is currently very dependent upon nonunionids (*Corbicula, Dreissena,* sphaeriids) because of their utility in laboratory and field testing. In order to develop useful models of contaminant uptake, distribution, and elimination by bivalves, as well as a better understanding of the exposure/dose/response triad that incorporates in situ testing (Salazar and Salazar 1998), it is imperative that basic bivalve ecotoxicological research continue on multiple fronts, rather than risk limitation of important discoveries as a result of unidirectional policy or management concerns.

REFERENCES

[ASTM] American Society for Testing and Materials, Guide for conducting static acute toxicity tests starting with embryos of four species of saltwater bivalve mollusks, *Annual Book of Standards,* Vol. 11.05, ASTM, Philadelphia, PA, E724-94, 1999.

[ASTM] American Society for Testing and Materials, Standard guide for conducting in situ field bioassays with caged bivalves, *2003, Annual Book of ASTM Standards,* Vol. 11.05, ASTM, Philadelphia, PA, E2122-02, 2001.

[ASTM] American Society for Testing and Materials, Standard guide for conducting laboratory toxicity tests with freshwater mussels, *Annual Book of Standards,* Vol. 11.05, ASTM, Philadelphia, PA, E2455-05, 2005.

Balogh, K. V. and Salanki, J., The dynamics of mercury and cadmium uptake into different organs of *Anodonta cygnea* L., *Water Res.,* 18, 1381–1387, 1984.

Bedford, J. W., Roelofs, E. W., and Zabik, M. J., The freshwater mussel as a biological monitor of pesticide concentration in a lotic environment, *Limnol. Oceanogr.,* 13, 118–126, 1968.

Biggins, R. G., Neves, R. J., and Dohner, C. K., *Draft National Strategy for the Conservation of Native Freshwater Mussels,* U.S. Fish and Wildlife Service, Washington, DC, 1995.

Bogan, A. E., Freshwater bivalve extinctions (Mollusca: Unionidae): A search for causes, *Am. Zool.,* 33, 599–609, 1993.

Burton, G. A. Jr. and Pitt, R., *Handbook for Assessing Stormwater Effects: A Toolbox of Approaches and Methods for Watershed Managers, Scientists and Engineers,* CRC/Lewis Press, Boca Raton, FL, 2001.

Busby, W. H. and Horak, G., Unionid mussels in Kansas: Overview of conservation efforts and harvest regulations, In *Conservation and management of freshwater mussels. Proceedings of a UMRCC Symposium* 1992, Cummings, K. S., Buchanan, A. C., and Koch, L. M., Eds., St. Louis MO, pp. 50–55, October 12–14, 1993.

Cairns, J., Jr. Biotic community response to stress, In *Biological Response Signatures,* Simon, T. P., Ed., CRC Press, Boca Raton, FL, pp. 13–21, 2003.

Cairns, J., Jr. and Niederlehner, B. R., Problems associated with selecting the most sensitive species, *Hydrobiologia,* 153, 87–94, 1987.

Campbell, J. H. and Evans, R. D., Inorganic and organic ligand binding of lead and cadmium and resultant implication for bioavailability, *Sci. Total Environ.,* 62, 219–227, 1987.

Caswell, H., Demography meets ecotoxicology: Untangling the population level effects of toxic substances, In *Ecotoxicology: A Hierarchical Treatment,* Newman, M. C. and Jagoe, C. H., Eds., CRC/Lewis Publishers, Boca Raton, FL, 1996.

Cherry, D. S., Van Hassel, J. H., Farris, J. L., Soucek, D. J., and Neves, R. J., Site-specific derivation of the acute copper criteria for the Clinch River, Virginia, *Hum. Ecol. Risk Assess.,* 8, 591–601, 2002.

Christian, A. D., Smith, B. N., Berg, D. J., Smoot, J. C., and Findlay, R. H., Trophic position and potential food sources of 2 species of unionid bivalves (Mollusca: Unionidae) in 2 small Ohio streams, *J. N. Am. Benthol. Soc.,* 23(1), 101–113, 2004.

Clements, W. H., Luoma, S. N., Gerritsen, J., Hatch, A., Jepson, P., Reynoldson, T., and Thom, R. M., Stressor interactions in ecological systems, In *Ecological Variability: Separating Natural from Anthropogenic Causes of Ecosystem Impairment,* Baird, D. J. and Burton, G. A., Eds., SETAC, Pensacola, FL, 2001.

Couillard, Y., Campbell, P. G. C., and Tessier, A., Response of metallothionein concentrations in a freshwater bivalve (*Anodonta grandis*) along an environmental cadmium gradient, *Limnol. Oceanogr.,* 38, 299–313, 1993.

DeFur, P. L., Crane, M., Ingersoll, C., and Tattersfield, L., Eds., *Endocrine Disruption in Invertebrates: Endocrinology, Testing, and Assessment,* SETAC, Pensacola, FL, p. 303, 1999.

Doherty, F. G., The Asiatic clam *Corbicula* spp., as a biological monitor in freshwater environments, *Environ. Monit. Assess.,* 15, 143–181, 1990.

Doward-King, E. J., Suter II, G. W., Kapustka, L. A., Mount, D. R., Reed-Judkins, D. K., Cormier, S. M., Dyer, S. D., Luxon, M. G., Parrish, R., and Burton, Jr., G. A., Distinguishing among factors that influence ecosystems, In *Ecological Variability: Separating Natural from Anthropogenic Causes of Ecosystem Impairment,* Baird, D. J. and Burton, G. A., Eds., SETAC, Pensacola, FL, 2001.

Ferenc, S. A. and Foran, J. A., Eds., *Multiple Stressors in Ecological Risk and Impact Assessment: Approaches to Risk Estimation,* SETAC, Pensacola, FL, p. 264, 2000.

Fleming, W. J., Augspurger, T. P., and Alderman, J. A., Freshwater mussel die-off attributed to anticholines-terase poisoning, *Environ. Toxicol. Chem.*, 14, 877–879, 1995.

Foran, J. A. and Ferenc, S. A., Multiple stressors in ecological risk and impact assessment, *SETAC Pellston Workshop on Multiple Stressors in Ecological Risk and Impact Assessment*, 13–18 September 1997, Pellston, Michigan, SETAC, Pensacola, FL, p. 100, 1999.

Fuller, S. L. H., Clams and mussels (Molluska: Bivalvia), In *Pollution Ecology of Freshwater Invertebrates*, Hart, Jr., C. W. and Fuller, S. L. H., Eds., Academic Press, New York, pp. 215–273, 1974.

Goldberg, E. D., The mussel watch—a first step in global marine monitoring, *Mar. Pollut. Bull*, 5, 111, 1975.

Hakenkamp, C. C. and Palmer, M. A., Introduced bivalves in freshwater ecosystems: The impact of *Corbicula* on organic matter dynamics in a sandy stream, *Oecologia*, 119, 445–451, 1999.

Hart, D. D., Building a stronger partnership between ecological research and biological monitoring, *J. N. Am. Benthol. Soc.*, 13, 110–116, 1994.

Hart, B., Balluff, D., Barfknecht, R., Chapman, P. F., Hawkes, T., Joermann, G., Leopold, A., and Luttik, R., Eds., *Avian Effects Assessment: A Framework for Contaminants Studies*, SETAC, Pensacola, FL, p. 216, 2001.

Hemelraad, J., Holwerda, D. A., and Zandee, D. I., Cadmium kinetics in freshwater clams. III. The pattern of cadmium accumulation in *Anodonta cygnea*, *Arch. Environ. Contam. Toxicol.*, 15, 1–7, 1986a.

Hemelraad, J., Holwerda, D. A., Teerds, K. J., Herwig, H. J., and Zandee, D. I., Cadmium kinetics in freshwater clams. II. A comparative study of cadmium uptake and cellular distribution in the Unionidae *Anodonta cygnea*, *Anodonta anatina*, and *Unio pictorum*, *Arch. Environ. Contam. Toxicol.*, 15, 9–21, 1986b.

Holwerda, D. A. and Herwig, H. J., Accumulation and metabolic effects of di-n-butyltin dichloride in the freshwater clam, *Anodonta anatina*, *Bull. Environ. Contam. Toxicol.*, 36, 756–762, 1986.

Hudson, R. G. and Isom, B. G., Rearing juveniles of the freshwater mussels (Unionidae) in a laboratory setting, *Nautilus*, 98(4), 129–137, 1984.

Imlay, M., Bioassay tests with naiads, In *Rare and Endangered Mollusks (Naiads) of the U.S.* Jorgensen, S. E. and Sharp, R. W., Eds., U.S. Department of The Interior, Fish and Wildlife, Service, Bureau of Sport Fisheries and Wildlife, Region 3. Symposium Proceedings, Twin Cities MN: p. 38–41, 1971.

Imlay, M. J., Effects of potassium on survival and distribution of freshwater mussels, *Malacologia*, 12, 97–111, 1973.

Ingersoll, C., Wang, N., Greer, E., Whites, D., Dwyer, J., Roberts, A., Augspurger, T., Kane, C., and Tibbott, C., Acute toxicity of copper, ammonia, and chlorine to glochidia and juveniles of freshwater mussels [presentation], *Society of Environmental Toxicology and Chemistry 24th Annual Meeting*, Austin TX, November 13, 2003.

[IUCN] International Union for the Conservation of Nature, 2004.

Jorgensen, S. E., Xu, F. L., Salas, F., and Marques, J. C., Application of indicators for the assessment of ecosystem health, In *Ecological Indicators for Assessment of Ecosystem Health*, Jorgensen, S. E., Costanza, R., and Xu, F. L., Eds., CRC Press, Boca Raton, FL, pp. 5–104, 2005.

Keller, A. E. and Ruessler, D. S., The toxicity of malathion to unionid mussels: Relationship to expected environmental concentrations, *Environ. Toxicol. Chem.*, 16, 1028–1033, 1997.

Keller, A. E. and Zam, S. G., Simplification of in vitro culture techniques for freshwater mussels, *Environ. Toxicol. Chem.*, 9, 1291–1296, 1990.

Landis, W. G., *Regional Scale Ecological Risk Assessment: Using the Relative Risk Model*, CRC Press, Boca Raton, FL, 2005.

Landis, W. G. and Yu, M. -H, *Introduction to Environmental Toxicology: Impacts of Chemicals upon Ecological Systems*, CRC Press, Boca Raton, FL, 2004.

Layzer, J. B., Gordon, M. E., and Anderson, R. M., Mussels: The forgotten fauna of regulated rivers. A case study of the Caney Fork River, *Reg. Rivers: Res. Manage.*, 8, 63–71, 1993.

Linder, G. L., Krest, S. K., and Sparling, D. W., Eds., *Amphibian Decline: An Integrated Analysis of Multiple Stressor Effects*, SETAC, Pensacola, FL, p. 368, 2003.

Lubchenco, J. et al., The sustainable biosphere initiative: An ecological research initiative, *Ecology*, 72, 371–412, 1991.

Luoma, S. N., Clements, W. H., DeWitt, T., Gerritsen, J., Hatch, A., Jepson, P., Reynoldson, T., and Thom, R. M., Role of environmental variability in evaluating stressor effects, In *Ecological Variability: Separating Natural from Anthropogenic Causes of Ecosystem Impairment*, Baird, D. J. and Burton, G. A., Eds., SETAC, Pensacola, FL, 2001.

Maltby, L., Studying stress: The importance of organism-level response, *Ecol. Appl.*, 9, 431–440, 1999.

Matthews, R. A., Landis, W. G., and Matthews, G. B., Community conditioning: An ecological approach to environmental toxicology, *Environ. Toxicol. Chem.*, 15, 597–603, 1996.

Metcalfe-Smith, J. L., Staton, S. K., Mackie, G. L., and Lane, N. M., Changes in the biodiversity of freshwater mussels in the Canadian waters of the Lower Great Lakes drainage basin over the past 140 years, *J. Great Lakes Res.*, 24(4), 845–858, 1998.

[NCDEHNR] North Carolina Department of Environment, Health and Natural Resources, Standard operating procedures, biological monitoring, Report prep. by Environ. Sci. Branch, Ecosyst. Anal. Unit, Biol. Assess. Grp, p. 52, 1997.

Naimo, T. J., Damschen, E. D., Rada, R. G., and Monroe, E. M., Nonlethal evaluation of the physiological health of unionid mussels: Methods for biopsy and glycogen analysis, *J. N. Am. Benthol. Soc.*, 17, 121–128, 1998.

Neves, R. J., Recent die-offs of freshwater mussels in the United States: An overview, In *Proceedings of the Workshop on Die-offs of Freshwater Mussels in the United States*, Neves, R. J., Ed., 1986 June 23–25; Davenport, IA, 1987.

Neves, R. J., Mollusk research in the Clinch and Powell Rivers. Clinch-Powell Rivers bi-state conference: Program and abstracts, May 28–30, Cumberland Gap TN, pp. 5–6, 1992.

Neves, R. J., Keynote address: A national strategy for the conservation of native freshwater mussels, In *Conservation and Management of Freshwater Mussels II: Initiatives for the Future, Proceedings of an Upper Mississippi River Conservation Committee Symposium*, Cummings, K. S., Buchanan, A. C., Mayer, C. A., and Naimo, T. J., Eds., Illinois Natural History Survey, Champaign, IL, pp. 1–10, 1997.

Neves, R. J. and Odom, M. C., Muskrat predation on endangered freshwater mussels in Virginia *J. Wildl. Manage*, 53, 934–941, 1989.

Newman, M. C., *Fundamentals of Ecotoxicology*, Ann Arbor Press, Chelsea, MI, 1998.

Newman, M. C. and Unger, M. A., Muskrat predation on endangered freshwater mussels in Virginia, J. Wildl. Manage., CRC Press, Boca Raton, FL, 1989.

Newton, T. J., Allran, J. W., O'Donnell, J. A., Bartsch, M. R., and Richardson, W. B., Effects of ammonia on juvenile unionid mussels (*Lampsilis cardium*) in laboratory sediment toxicity tests, *Environ. Toxicol. Chem.*, 22, 2554–2560, 2003.

Nichols, S. J. and Garling, D., Food-web dynamics and trophic-level interactions in a multi-species community of freshwater unionids, *Can. J. Zool.*, 78, 871–882, 2000.

O'Neill, R. V., Hunsaker, C. T., Jones, K. B., Ritters, K. H., Wickham, J. D., Schwartz, P. M., Gooman, I. A., Jackson, B. L., and Baillargeon, W. S., Monitoring environmental quality at the landscape scale, *Bioscience*, 47, 513–519, 1997.

Parmalee, P. W. and Bogan, A. E., *The Freshwater Mussels of Tennessee*, The University of Tennessee Press, Knoxville, TN, 1998.

Posthuma, L., Suter, G. W., and Traas, T. P., *Species Sensitivity Distributions in Ecotoxicology*, Lewis Publishers, Boca Raton, FL, 2001.

Raikow, D. F. and Hamilton, S. K., Bivalve diets in a midwestern U.S. stream: A stable isotope enrichment study, *Limnol. Oceanogr.*, 46, 514–522, 2001.

Salazar, M. H. and Salazar, S. M., Assessing site-specific effects of TBT contamination with mussel growth rates, *Mar. Environ. Res.*, 32, 131–150, 1991.

Salazar, M. H. and Salazar, S. M., Using caged bivalves as part of an exposure-dose- response triad to support an integrated risk assessment strategy, In *Proceedings, Ecological Risk Assessment: A Meeting of Policy and Science*, de Peyster, A. and Day, K., Eds., SETAC Special Publication, SETAC Press, Pensacola, FL, pp. 167-192, 1998.

Shannon, L., Biggins, R. G., and Hylton, R. E., Freshwater mussels in peril: perspective of the U.S. Fish and Wildlife Service, In *Conservation and Management of Freshwater mussels. Proceedings of a UMRCC Symposium*, Cummings, K. S., Buchanan, A. C., and Koch, L. M., Eds., 1992 Oct 12–14, St. Louis MO, pp. 66–68, 1993.

Shear, H., Bertram, P., Forst, C., and Horvatin, P., Development and application of ecosystem health indicators in the North American Great Lakes Basin, In *Ecological Indicators for Assessment of Ecosystem Health*, Jorgensen, S. E., Costanza, R., and Xu, F. L., Eds., CRC Press, Boca Raton, FL, pp. 105–126, 2005.

Sheppard, S. C., Bembridge, J. D., Holmstrup, M., and Posthuma, L., Advances in earthworm exotoxicology, *Proceedings from the Second International Workshop on Earthworm Ecotoxicology*, SETAC, Amsterdam, The Netherlands, Pensacola, FL, 2–5 April 1997, p. 472, 1998.

Slobodkin, L. B. and Dykhuizen, D. E., Applied ecology, its practice and philosophy, In *Integrated Environmental Management*, Cairns, J. Jr. and Crawford, T. V., Eds., Lewis Publishers, Chelsea, MI, pp. 63–70, 1991.

Smolders, R., Bervoits, L., Wepener, V., and Blust, R., A conceptual framework for using mussels as biomonitors in whole effluent toxicity, *Hum. Ecol. Risk Asses.*, 9(3), 741–760, 2003.

Southwick, R. I. and Loftus, A. J., Eds., *Investigation and Monetary Values of Fish and Freshwater Mussel Kills*, American Fisheries Society, Spec. Publ. 30, Bethesda, MD, p. 177, 2003.

Sparks, R. E., Blodgett, K. D., Durham, L., and Horner, R., Determination whether the causal agent for mussel die-offs in the Mississippi River is of chemical or biological origin. Final report. ILENR/RE-WR 90/09, Illinois Department of Energy and Natural Resources, Springfield, IL, p. 27, 1990.

Sparling, D. W., Linder, G., and Bishop, C. A., Eds., *Ecotoxicology of Amphibians and Reptiles*, SETAC, Pensacola, FL, p. 904, 2000.

Steedman, R. J., Whillans, T. H., Behm, A. P., Bray, K. E., Cullis, K. I., Holland, M. M., Stoddart, S. J., and White, R. J., Use of historical information for conservation and restoration of Great Lakes aquatic habitat, *Can. J. Fish. Aquat. Sci.*, 53(Suppl. 1), 415–423, 1996.

Strayer, D. L. and Smith, D. R., *A Guide to Sampling Freshwater Mussel Populations, Monograph 8*, American Fisheries Society, Bethesda, MD, p. 103, 2003.

Strayer, D. L., Caraco, N. F., Cole, J. F., Findlay, S., and Pace, M. L., Transformation of freshwater ecosystems by bivalves, *Bioscience*, 49, 19–27, 1999.

Strayer, D. L., Downing, J. A., Haag, W. R., King, T. L., Layzer, J. B., Newton, T. J., and Nichols, S. J., Changing perspectives on pearly mussels, North America's most imperiled animals, *Bioscience*, 54(5), 429–439, 2004.

Suter II, G. W., *Ecological Risk Assessment*, Lewis, Boca Raton, FL, p. 538, 1993.

Tankersley, R. A., Multipurpose gills: Effect of larval brooding on the feeding physiology of freshwater unionid mussels, *Invertebr. Biol.*, 115, 243–255, 1996.

Tessier, A., Campbell, P. G. C., Auclair, J. C., and Bisson, M., Relationships between the partitioning of trace metals in sediments and their accumulation in the tissues of the freshwater mollusc *Elliptio complanata* in a mining area, *Can. J. Fish. Aquat. Sci.*, 41, 1463–1472, 1984.

Truhaut, R., Ecotoxicology—a new branch of toxicology: A general survey of it aims, methods, and prospects, In *Ecological Toxicology Research*, McIntyre, A. D. and Mills, C. F., Eds., Plenum, New York, pp. 3–24, 1975.

Uhler, A. D., Durell, G. S., Steinhauer, W. G., and Spellacy, A. M., Tributyltin levels in bivalve mollusks from the east and west coasts of the United States: Results from the 1988–1990 national status and trends mussel watch project, *Environ. Toxicol. Chem.*, 12, 139–153, 1993.

[USEPA] U.S. Environmental Protection Agency, Proposed guidelines for ecological risk assessment; notice, *Federal Register*, 61(175), 47552–47631, 1996.

[USEPA] U.S. Environmental Protection Agency, Office of pollution prevention and toxic substances (OPPTS) test guidelines, Series 850 on Ecological effects, Washington, DC, USEPA, 1997.

Vaughn, C. C. and Hakenkamp, C. C., The functional role of burrowing bivalves in freshwater ecosystems, *Freshwat. Biol.*, 46, 1431–1446, 2001.

Ward, J. E., Biodynamics of suspension-feeding in adult bivalve mollusks: Particle capture, processing, and fate, *Invertebr. Biol.*, 115, 218–231, 1996.

Williams, J. D., Warren, M. L. Jr., Cummings, K. S., Harris, J. L., and Neves, R. J., Conservation status of freshwater mussels of the United States and Canada, *Fisheries*, 18(9), 6–22, 1993.

Wurtz, C. B., Fresh-water mollusks and stream pollution, *Nautilus*, 69, 96–100, 1956.

2 A Review of the Use of Unionid Mussels as Biological Indicators of Ecosystem Health

John H. Van Hassel and Jerry L. Farris

INTRODUCTION

The currently heightened concern for freshwater mussel resources has brought about a critical need for accurately monitoring the status of these organisms. In addition, because of their sensitivity to many environmental perturbations, unionid mussels have been used as indicators of both exposure to and effects of these perturbations. This review provides the historical background of unionid mussel use in various biomonitoring applications and critically evaluates the effectiveness of these applications. Recommendations for future directions in mussel biomonitoring and for system-level management practices for the conservation of mussel resources are offered.

An extensive search of published literature and unpublished reports was made, with a cut-off date of May 2003. This literature search was comprehensive, but not exhaustive, and is certainly representative of the work that has been conducted in this field. In all, over 700 published articles, reports, and abstracts were examined for relevance to the scope of this review.

In this review, the term "biomonitoring" is expanded beyond the narrow definition of "a continuing collection of data to establish whether explicitly stated quality conditions are being met" (Cairns and Smith 1994) to address the entire scope of applications of mussel measures that have been, or may be, used to evaluate and protect mussel populations. Previous reviews have focused on specific aspects of mussel biomonitoring, including bioaccumulation and toxicity of contaminants (Havlik and Marking 1987; Elder and Collins 1991; Naimo 1995), metal bioaccumulation and tissue residue effects (Salazar 1997), and biomonitoring using caged mussels (Smolders et al. 2003). In the present review, we attempt to link the considerable literature dealing with the use of unionid mussels as biological indicators of ecosystem health. According to Cairns, McCormick, and Niederlehner (1993), the basic types of biological indicators are: early warning, diagnostic, and compliance indicators. The effectiveness of mussel toxicity testing as a regulatory compliance tool is dealt with in Chapter 5, whereas this chapter deals with the other applications: (1) the effectiveness of mussels as sentinels of environmental perturbations (i.e., the effectiveness of measurements applied to mussels for the purpose of determining that exposure to an environmental insult has occurred), and (2) the effectiveness of mussels as indicators of ecological integrity (i.e., the effectiveness of measurements applied to mussels for the purpose of discriminating effects on spatial and temporal scales), including their use in impact assessment, waterbody status monitoring, and as recorders of environmental history.

REVIEW OF UNIONID MUSSEL BIOMONITORING LITERATURE

MONITORING OF MUSSEL POPULATIONS

Collection Techniques

During the past twenty years, many studies have been conducted with field-collected mussels to measure their response to anthropogenic impacts. In undertaking these studies, a variety of collection techniques have evolved to address the variety of habitat conditions encountered, as well as the particular needs of the study. Table 2.1 provides a summary of the most commonly used methods, the most suitable applications for each, and the individual biases inherent to each method.

TABLE 2.1
Applications, Advantages, and Disadvantages of Mussel Collection Methods

Method	Wadable Stream	Large River	Lake	Quantitative	Qualitative	Advantages	Disadvantages	References
Brail		X	X		X	1,2,3	1,2,3	1,2,4–7,11,12, 14,16,17,20
Hand pick	X				X	2,4	1,2,5	1–5,7,11,17,18
Dip net	X			X	X	2,4	8	5
Rake w/basket	X			X	X	2,4	8	4,5
Mechanical grab		X	X	X	X		3,5,6,9	4,7,11,14,17
Skimmer dredge		X	X	X	X	2,4	5,7,8	2,7,9,11,17,20
Diver transects	X	X	X		X	4	2,4,5,8	1,2,4,7–12,14,17
Quadrats	X	X	X	X	X	5	4,5,6	1–8,10–12, 15,17,18
Shoreline shell collection	X	X	X		X	1,2,3,6	9	2,7,11,13,17,19

Key to advantages		Key to disadvantages	
	1 Useful as an exploratory device		1 Does not provide quantitative data
	2 Inexpensive		2 Size selective
	3 Not time intensive		3 Low catch rate
	4 Provides good qualitative information		4 Expensive
	5 Provides good quantitative information		5 Time intensive
	6 Provides supplemental information		6 Statistical confidence requires large sample size
			7 Habitat intrusive
			8 Difficult to obtain quantitative samples
			9 Data unreliable, qualitatively and quantitatively

References cited: 1. Cawley (1993); 2. Dunn (2000); 3. Hornbach and Deneka (1996); 4. Isom and Gooch (1986); 5. Klemm et al. (1990); 6. Kovalak, Dennis, and Bates (1986); 7. Miller and Nelson (1983); 8. Miller and Payne (1993); 9. Miller, Whiting, and Wilcox (1989); 10. Miller et al. (1993); 11. Nelson (1982); 12. Payne and Miller (1987a); 13. Rothwell (1979); 14. Sickel, Chandler, and Pharris (1983); 15. Smith et al. (2000); 16. Sparks et al. (1990); 17. Strayer and Smith 2003; 18. Vaughn, Taylor, and Eberhard (1997); 19. Watters (1993–1994); 20. Wilcox, Anderson, and Miller (1993).

Discussions of each of these methods and their individual strengths and weaknesses can be found in Nelson (1982), Miller and Nelson (1983), and Isom and Gooch (1986).

The consensus of several evaluations of the techniques listed in Table 2.1 was that crowfoot brails are useful, deep-water, exploratory devices, but have limited usefulness in quantitative studies because of their low catch rate and selectivity against juvenile mussels and small species (Kovalak, Dennis, and Bates 1986; Payne and Miller 1987a; Cawley 1993; Wilcox, Anderson, and Miller 1993). Crowfoot brails consist of wood or iron bars that range from two to twenty feet in length, depending upon the application. Suspended from the bar are several, approximately one-foot, lengths of chain or cord with "crowfoot" hooks attached and with the tips of the hooks beaded to improve retention of captured mussels. The brail is towed very slowly along the bottom and captures mussels when the hook slides between an animal's valves that are slightly agape for filtering, eliciting a reflexive muscle contraction that causes the valves to clamp down on the hook. Klemm et al. (1990) suggest at least six 100-meter brail hauls per site for the collection of qualitative data. Hand picking (without the use of quadrats or transects) of mussels is restricted to shallow waters but can be an inexpensive method of obtaining qualitative information (Nelson 1982). Restricted visibility during hand picking can bias a sample against juvenile or small species of mussels (Cawley 1993). Dip nets and rakes with baskets are also restricted to wadable waters and are another inexpensive way to collect qualitative data. Sampling for quantifiable measures with such devices is difficult, and somewhat habitat limited, as most dip nets and rakes work best in fine-particle substrates with low-debris loads (Miller and Nelson 1983). Transect sampling in deep water (diving) or shallow water (snorkeling) has been used to collect quantitative data in certain applications; however, because of the collector bias inherent to this method (e.g., Payne and Miller 1987a), we recommend that it be used quantitatively only under carefully defined circumstances.

Quantitative samples have primarily been collected using mechanical grabs, skimmer dredges, and hand excavation of defined quadrats. Grab samplers, such as Ponar dredges, are generally not recommended for mussel studies because of the very high number of grabs required to obtain a representative sample and the correspondingly high level of required effort (Sickel, Chandler, and Pharris 1983; Isom and Gooch 1986). Skimmer dredges can offer a cost-effective alternative to diving for deep-water sampling (Miller, Whiting, and Wilcox 1989; Wilcox, Anderson, and Miller 1993), but sampling is difficult to quantify, requires considerable equipment outlays, is destructive to both habitat and mussels, and is effective primarily only on sand-silt substrates (Nelson 1982). For most applications, the method of choice for quantitative sampling is quadrats. The quadrat size has varied considerably among studies, with the most common areas sampled being 0.25, 0.5, and 1.0 m^2 (Salmon and Green 1983; Kovalak, Dennis, and Bates 1986; Miller and Payne 1988; Holland-Bartels 1990; Klemm et al. 1990; Amyot and Downing 1991; Goudreau et al. 1993; Harris et al. 1993; Miller 1993). A quadrat size of 1.0 m^2 or less is recommended because of the necessity of replicate samples for quantitative data. The depth to which quadrats are excavated also varies but is typically in the range of 5–15 cm (Miller and Payne 1988; Holland-Bartels 1990). The number of replicates needed to obtain statistically useful data has been addressed by Cawley (1993), Miller et al. (1993), and Strayer and Smith (2003) for mussels in particular, and by Green (1979) for biological sampling in general. Depending upon the needs of specific studies, quadrat sampling programs can require up to 100 quadrats per site (Klemm et al. 1990; Harris et al. 1993; Miller 1993).

In deciding upon the qualitative versus quantitative needs of a proposed sampling program, insight from past studies can be drawn upon to offer some basic guidelines. These guidelines, provided in Table 2.2, list measures that have been applied in various mussel investigations, the type of sampling required for these measures, and the strengths and weaknesses of each measure. In general, good qualitative samples provide sufficient data for determining species richness, species diversity, evenness, community composition, relative abundance, and presence/absence of rare species. On the other hand, quantitative sampling is recommended for the measurement of density, production, biomass, size demography, and recruitment, as well as for hypothesis

TABLE 2.2
Applications, Advantages, and Disadvantages of Measures Used on Field-Collected Mussels

Measure	Applications	Data Needs	Advantages	Disadvantages	References
Species richness/ diversity/evenness	2,3,4	Qualitative	1,2,3,4	4	3,4,9,30,37–40,44,53
Relative abundance	2,3,4	Qualitative	1,2,3,4	4	9,13,22,25,29,37–39,44
Density/biomass	2,3,4	Quantitative	1,2,3,4	4	3,9,20,27,29,38–40, 43,44,49
Size demography	1,2,3,4,5	Quantitative	1,2,3,4	4	2,9,22,27,37–39,43,44,51
Mortality	1,2,3,4	Quantitative	1,2,3	3	21,32,43,51
Growth	2,3,4	Nonspecific	2 (at times), 3,4	5	5,8,12,13,16,23,24,27,28, 31,33,35,36,40,43,48, 50,51,54
Condition indices	1,2,3,4	Nonspecific	1,2,3 (at times)	2 (at times), 3	26,32,33,40,45–47,51
Behavioral responses	1,2,3,4	Nonspecific	1,2,3	1,3	1,52
Bacteriological parameters	1,2,3,4	Nonspecific		1,2,3	18,52
Physiological parameters (e.g., enzyme activity, blood chemistry)	1,2,3,4	Nonspecific	4	2	6–8,10–12,14,15,17–19, 21,32–34,36,41,42

Key to applications	
	1 Mussel population status/characterization
	2 Impact assessment
	3 Monitoring of spatial/temporal changes
	4 Hypothesis testing
	5 Recruitment success
Key to advantages	1 Ease of use
	2 Data obtainable in the field
	3 Nondestructive
	4 High discriminatory power when quantitative data used
Key to disadvantages	1 Subjective or poorly defined
	2 Destructive
	3 Low discriminatory power
	4 Requires large, quantitative samples to provide adequate statistical power
	5 Requires long-term monitoring or labor-intensive thin-sectioning techniques

References cited: 1. Amyot and Downing (1991); 2. Anderson, Romano, and Pederson (1993); 3. Arbuckle and Downing (2002); 4. Bailey (1988); 5. Beckvar et al. (2000); 6. Berg et al. (1995); 7. Blaise et al. (2002); 8. Blaise et al. (2003); 9. Cawley (1993); 10. Cherry, Farris, and Belanger (1988); 11. Couillard et al. (1995a); 12. Couillard et al. (1995a); 13. Davies (1963); 14. Day, Metcalfe, and Batchelor (1990); 15. Doyotte et al. (1997); 16. Fischer et al. (1993); 17. Fleming, Augspurger, and Alderman (1995); 18. Gagne et al. (2002); 19. Gardner, Miller, and Imlay (1981); 20. Goudreau, Neves, and Sheehan (1993); 21. Haag et al. (1993); 22. Harris et al. (1993); 23. Haukioja and Hakala (1978); 24. Hinch and Green (1989); 25. Holland--Bartels (1990); 26. Hornbach et al. (1996); 27. Houslet and Layzer (1997); 28. Imlay (1982); 29. Isom and Gooch (1986); 30. Klemm et al. (1990); 31. Layzer and Madison (1997); 32. Makela (1995); 33. Makela et al. (1992); 34. Malley, Huebner, and Donkersloot (1988); 35. McCuaig and Green (1983); 36. Milam and Farris (1998); 37. Miller (1993); 38. Miller and Payne (1988); 39. Miller and Payne (1993); 40. Miller et al. (1993); 41. Naimo and Monroe (1999); 42. Naimo et al. (1998); 43. Negus (1966); 44. Payne and Miller (1987a); 45. Payne, Miller, and Lei (1995); 46. Pekkarinen (1993); 47. Roper and Hickey (1994); 48. Rothwell (1979); 49. Russell-Hunter and Buckley (1983); 50. Scott (1994); 51. Sparks and Blodgett (1987); 52. Sparks et al. (1990); 53. Watters (1992); 54. Yokley (1976).

testing, impact assessment, and monitoring of spatial/temporal changes (Payne and Miller 1987a; Cawley 1993; Miller and Payne 1993; Miller et al. 1993; Vaughn, Taylor, and Eberhard 1997; Dunn 2000; Strayer and Smith 2003). For some applications, the success of the sampling program is specifically tied to the type of sampling used, whereas for other measures, success can be achieved by a variety of sampling approaches. For example, in determining the presence/absence of rare species, studies have determined that quantitative methods, such as quadrats, are limited because of the excessive number of samples needed to find rare species. Qualitative methods such as timed searches, or a combined qualitative/quantitative approach, are recommended (Kovalak, Dennis, and Bates 1986; Payne and Miller 1987a; Vaughn, Taylor, and Eberhard 1997; Metcalfe-Smith et al. 2000; Smith et al. 2000, Smith, Villella, and Lemarie 2003; Strayer and Smith 2003). Conversely, most functional measures, such as those listed in the bottom half of Table 2.2, can be applied to a variety of qualitatively or quantitatively collected data, depending upon the needs of the study.

Structural/Functional Indices

Unionid mussels have long been known as important indicators of environmental perturbations. For example, Ortmann (1909) reported that unionids were the first organisms to disappear from streams subjected to pollution stress. However, the use of freshwater mussels for biomonitoring purposes is a mostly recent phenomenon. Although Wurtz (1956) and Ingram (1957) promoted their use as a pollution indicator organism in the 1950s, it was not until the 1970s that mussels were widely used for biomonitoring. Prior to that time, published studies of mussel biomonitoring were relatively few. Early environmental studies involving the use of mussel measurements addressed the effects of temperature and heated effluents (Grier 1920; Davies 1963; Negus 1966), sewage (Baker 1920; Shimek 1935), siltation (Ellis 1936), and impoundment (Bates 1962). Effects measures applied to unionid mussels fulfill most of the criteria provided by Widdows and Donkin (1992):

1. They should be sensitive to environmental levels of pollutants and have a large scope for response throughout the range from optimal to lethal conditions.
2. They should reflect a quantitative and predictable relationship with toxic contaminants.
3. They should have a relatively short response time so that pollution impact can be detected in its incipient stages.
4. The technique should be applicable to both laboratory and field studies to relate laboratory-based, concentration-response relationships to field measurements of spatial and temporal changes in environmental quality.
5. They should provide both an integrated response to the total pollutant load and insight into the underlying cause and mechanism of toxicity.
6. The biological response should have ecological relevance (i.e., is related to deleterious effects on growth, reproduction, or survival of the individual, population, and community).

Several measures have been applied to determine the status of a particular mussel population, or to initially characterize a population. Of these, the most useful are size demography (i.e., analysis of age-size classes of a population) and physiological parameters. Most structural measures (e.g., density, mortality, and condition indices) tend to be insensitive to small or moderate changes within a population and are biased by habitat influences. Size demography analyses provide species-specific information indicative of factors affecting specific segments of the population, such as removal of older individuals by commercial harvest, or lack of recruitment caused by environmental stress (Sparks and Blodgett 1987; Anderson, Romano, and Pederson 1993; Miller 1993; Houslet and Layzer 1997). Studies of mussel recruitment success require a large, quantitative size demography database (Miller and Payne 1988). Anderson, Romano, and Pederson (1993) demonstrated the use of annuli-length regression data to characterize mussel reproductive activity.

Physiological measures provide a means of documentation of mussel condition at the time of sample collection. Although such measures are not well developed for use on unionids, a few investigators have successfully used them and established their attractiveness for future applications requiring increased sensitivity and rapidity over more traditional measures (growth). Haag et al. (1993) applied a variety of physiological measures to a study of zebra mussel (*Dreissena polymorpha*) fouling effects on unionids and determined that cellulolytic enzyme activity and glycogen content were very sensitive measures of mussel stress. Malley, Huebner, and Donkersloot (1988) detected changes in mussel blood-ion composition caused by alum addition to a lake. Gardner, Miller, and Imlay (1981) found that tissue amino acid levels were elevated in mussels from acid mine streams, whereas Day, Metcalfe, and Batchelor (1990) measured a similar response to agricultural and urban runoff. A response in gill antioxidant enzyme activity to a coking effluent was documented by Doyotte et al. (1997). Fleming, Augspurger, and Alderman (1995) detected reduced cholinesterase activity in mussels at and downstream of a suspected pesticide-related mussel kill. A number of related studies of municipal effluents demonstrated measurable effects on several enzyme systems and other cellular and subcellular measures (Blaise et al. 2002, 2003; Gagne et al. 2002). Production of metallothionein in mussels has been related to gradients in cadmium concentration but not to concentrations of copper or zinc, possibly because of homeostatic control of the latter two elements (Couillard, Campbell, and Tessier 1993; Couillard et al. 1999). The same group of researchers proposed that metallothionein alone is insufficient as an effects biomarker but has value in combination with information on metal partitioning to other cytosolic ligands (termed "spillover") (Giguere et al. 1999). Further research on these physiological measures is needed to address the issues of reversibility of effects and links to long-term implications for the health of individual mussels.

All of the measures listed in Table 2.2 have been applied to impact assessment studies, monitoring of spatial/temporal changes, and field hypothesis testing. Of these measures, density and growth have been the most sensitive and frequently used measures. Davies (1963) found that mussels within a thermal discharge attained greater maximum size than mussels outside of the discharge influence. Both Negus (1966) and Rothwell (1979) measured more rapid growth in mussels residing in a heated effluent than in mussels outside of the effluent influence. Yokley (1976) found that mussel growth rates were much greater upstream of dredging compared to immediately downstream. A study of an industrial discharge by AEPSC/APCO (1987) found that mussels upstream of the discharge achieved greater length gains than did mussels within the discharge influence. Imlay (1982) also found decreased growth rates in field-collected mussels under stress. Hinch and Green (1989) found significant differences in growth between reciprocally transplanted mussels among environmentally variable sites. Fischer et al. (1993) discriminated longitudinal differences in growth rates within species (see also Scott 1994). Goudreau, Neves, and Sheehan (1993) used density measurements to show depletion of mussels downstream of sewage treatment plants compared to upstream populations. Cherry, Farris, and Belanger (1988) found significantly depressed cellulolytic enzyme activity in mussels caged for 30 days below an industrial effluent compared to those in upstream cages. Growth was not significantly affected over the 30-day period; however, studies with *Corbicula fluminea* have demonstrated that reductions of cellulolytic enzyme activity are associated with eventual reductions in growth rate (Farris et al. 1989). Similarly, Couillard et al. (1995b) measured reduced growth along with elevated metallothionein and cellular toxicity in mussels transplanted to a cadmium-contaminated lake. Houslet and Layzer (1997) measured reduced growth, recruitment, and density in mussels downstream of a strip-mined area compared to upstream specimens.

Although growth and density measurements have been successfully applied to quantitative field studies, both measures have drawbacks (Roper and Hickey 1994; Layzer and Madison 1997). Because of the patchy distribution of mussels, the ability to discriminate differences in density requires very large, quantitative sample sizes. Because of the slow growth of unionid mussels, direct measurement of growth in situ requires long-term monitoring and, usually, caging or tagging of the mussels. Both caging and tagging have been used successfully in growth studies, and a standardized protocol for in situ growth studies has been developed (ASTM 2001). AEPSC/APCO (1987) used wire cages to hold mussels in situ over a five-month period. Yokley (1976) used a tagging method in a one-year study. Several other studies have employed marking of the shells for repeat-measurement studies (Sparks and Blodgett 1987; Hinch and Green 1989; Miller et al. 1993), and Miller and Nelson (1983) include a discussion of how to mark shells. Waller et al. (1993a) describe a grid device that can be anchored to the substrate to aid in the recovery of mussels under study. Growth rates can also be determined using measurements of annuli or length-frequency histograms. Both techniques require large sample sizes, and the use of length-frequency data can be very problematic for any but the most abundant species. For studies examining annuli measurements, thin sectioning to allow the use of internal bands is highly recommended (Neves and Moyer 1988; Metcalfe-Smith and Green 1992). McCuaig and Green (1983) discuss the use of growth rate parameters in quantitative mussel studies. Further development of physiological measures for mussels, such as various enzyme assays (e.g., cellulase, metallothionein, and cholinesterase) and tissue glycogen, is needed to provide rapid and less labor-intensive methods of quantitative field assessment, to supplement traditional measures and provide a wider array of tools for ecotoxicological investigations.

Contaminant Body Burdens

Much of the monitoring that has been undertaken with freshwater unionid mussels has involved the measurement of contaminant body burdens for various applications. Mussels have been found to be a useful organism for this purpose, and they possess several attributes that are characteristic of useful biomonitors (Phillips 1980; Green, Singh, and Bailey 1985):

1. Most species inhabit shallow, nearshore areas in lentic waters or riffle areas in lotic waters, which are the most productive and the most susceptible to most types of pollution.
2. They are sedentary.
3. They are long-lived.
4. They are often quite abundant.
5. They are large enough to easily provide adequate tissue for contaminant analyses.
6. They are easy to sample.
7. Many species are very sensitive to a variety of contaminants, however, there are also tolerant species that can accumulate contaminants under fairly severe perturbations.
8. They have been demonstrated to accumulate a variety of contaminants to concentrations that correlate well with exposure concentrations and durations.
9. They provide a record of exposure history in their shells.

Most of the applications that have been developed for monitoring contaminant body burdens in freshwater mussels have been based on the large body of work established using marine bivalves. Mix (1984) provided an extensive review of the marine bivalve biomonitoring literature. The marine "mussel watch" program employs an extensive monitoring system that has provided useful biomonitoring data for many years to assess the level of contamination in marine coastal areas (Goldberg 1975). Metcalfe-Smith (1994) stated that this program has been successful because of sound sampling and analytical protocols. Similar protocols are now being applied to freshwater mussel investigations.

Table 2.3 summarizes biomonitoring applications that have measured freshwater mussel contaminant body burdens. A variety of sampling and exposure protocols have been used, depending upon the purpose of the study. Analytical techniques have varied as well, and a

TABLE 2.3
Applications Using Mussel Contaminant Body Burdens, and Advantages of Using Mussels for Those Applications

Applications	Component Analyzed	Advantages	Disadvantages	References
Impact assessment	1,2,3	2,3,4,6	1,2	1,5,6,11,14,16, 18–22,24–26,28,29,35, 39,40,47,54,60,61
Pollution source identification	1,2,3	1,2,3,4,5,6,7	2	4,5,12,23,24,27,28,30–34,43,46–52, 54,56,57,60,63
Sentinel organism	1,2	1,3,4,5,6,7	2,3	2,3,6,13,14,23,24,36,38,42,44, 46–50,52–54,60,62,66,70,71
Bioavailability measure	1,2,3	1,2,6	2	5,7–9,12–15,17,22,25,26,29,33,34, 39–42,44–46,48,53,55,58,59,61, 65–69,71
Environmental history	3	2,4,5,6,7	2,4	10,23,24,37,64

Key to component analyzed   1 Whole body; 2 Specific tissues/organs; 3 Shell/shell layers

Key to advantages
1 Cost-effective, quantitative measure for this application
2 Good exploratory tool for this application
3 Mussels are often sensitive indicator organisms
4 Exposures are integrated over time
5 Mussels are typically long-lived
6 Mussels are sedentary and amenable to caging
7 Mussel shell layers offer potential to date results

Key to disadvantages
1 Does not provide a quantitative measure of impact
2 Body burdens and uptake rate influenced by mussel size, age, growth rate, type of contaminant, long tissue retention times for many contaminants, source water (for transplants), and environmental variations
3 Selected species needs to be pollution-tolerant in some situations
4 Analytical restraints and other confounding influences on results

References cited: 1. Adams, Atchison, and Vetter (1981); 2. Anderson (1977); 3. Balogh (1988); 4. Becker et al. (1992); 5. Beckvar et al. (2000); 6. Bedford, Roelofs, and Zabik (1968); 7. Brungs (1967); 8. Campbell and Evans (1987); 9. Campbell and Evans (1991); 10. Carell et al. (1987); 11. Cherry, Farris, and Belanger (1988); 12. Couillard, Campbell, and Tessier (1993); 13. Couillard et al. (1995a); 14. Couillard et al. (1995b); 15. Couillard et al. (1999); 16. Czarnezki (1987); 17. Dobrowolski and Skowronska (2002); 18. Foster and Bates (1978); 19. Gaglione and Ravera (1964); 20. Gagne et al. (2002); 21. Gardner and Skulberg (1965); 22. Giguere et al. (1999); 23. Green et al. (1989); 24. Green, Singh, and Bailey (1985); 25. Hayer and Pihan (1996); 26. Hayer, Wagner, and Pihan (1996); 27. Herve (1991); 28. Herve et al. (1988); 29. Herve et al. (1991); 30. Hinch and Green (1989); 31. Hinch and Stephenson (1987); 32. Hinch, Bailey, and Green (1986); 33. Kauss and Hamdy (1985); 34. Kauss and Hamdy (1991); 35. Klose and Potera (1984); 36. Leard, Grantham, and Pessoney (1980); 37. Lingard, Evans, and Bourgoin (1992); 38. Lopes et al. (1993); 39. Makela (1995); 40. Makela, Lindstrom-Seppa, and Oikari (1992); 41. Malley, Chang, and Hesslein (1989); 42. Malley, Stewart, and Hall (1996); 43. Manly and George (1977); 44. Marquenie (1985); 45. Mathis and Cummings (1973); 46. Metcalfe and Charlton (1990); 47. Metcalfe and Hayton (1989); 48. Metcalfe-Smith (1994); 49. Muncaster et al. (1989); 50. Muncaster, Hebert, and Lazar (1990); 51. Naimo (1993); 52. Pellinen et al. (1993); 53. Perceval et al. (2002); 54. Phillips (1980); 55. Price and Knight (1978); 56. Pugsley, Hebert, and McQuarrie (1988); 57. Pugsley et al. (1985); 58. Renaud et al. (1995); 59. Renzoni and Bacci (1976); 60. Salanki (1989); 61. Schmitt et al. (1987); 62. Servos et al. (1987); 63. Smith, Green, and Lutz (1975); 64. Sterrett and Saville (1974); 65. Stewart (1999); 66. Storey and Edward (1989); 67. Tessier et al. (1984); 68. Tessier et al. (1993); 69. Tevesz et al. (1989); 70. Turick, Sexstone, and Bissonnette (1988); 71. Wang et al. (1999).

review of these will not be undertaken here; however, several good, standardized protocols exist (e.g., ASTM 1999). For most applications, the analysis of the whole tissue mass of the mussel is sufficient. Analysis of specific tissues is appropriate for certain types of uptake studies, but is not recommended for routine applications because of the additional time necessary to separate the desired tissues and the potential for analytical limitations related to the analysis of small amounts of tissue. Similarly, analysis of shell material is not recommended for most applications because of difficulties in digestion and analytical interferences.

Several studies have used freshwater mussels as a means of pollution source identification and impact assessment. Mussels have been found to be particularly useful in the former application. Studies of body burdens in mussels downstream of contaminant discharge sources compared to uninfluenced areas have clearly shown elevated concentrations related to the discharge of both metals (Foster and Bates 1978; Czarnezki 1987; Schmitt et al. 1987; Cherry, Farris, and Belanger 1988; Salanki 1989; Couillard et al. 1995b; Makela 1995; Giguere et al. 1999; Beckvar et al. 2000; Gagne et al. 2002) and organics (Kauss and Hamdy 1985; Herve et al. 1988; Metcalfe and Hayton 1989; Metcalfe and Charlton 1990; Herve et al. 1991; Makela 1995; Hayer and Pihan 1996; Hayer, Wagner, and Pihan 1996). Mussels have also been a valuable tool for contaminant mapping of a waterbody for both metals (Adams, Atchison, and Vetter 1981; Pugsley, Hebert, and McQuarrie 1988; Becker et al. 1992; Naimo 1993; Beckvar et al. 2000) and organics (Pugsley et al. 1985; Herve et al. 1988; Herve 1991; Kauss and Hamdy 1991). Although contaminant body burdens in mussels effectively reflect exposure to a variety of pollutants, they have not been shown to quantitatively measure impact. In order to demonstrate impact, body burdens of a particular contaminant would have to be unambiguously linked to significant impairment of some aspect of the organism's life cycle (Stewart and Malley 1997). This has not been done for freshwater mussels beyond simple association of tissue contaminant levels with effects on growth, etc. (Couillard et al. 1995a; Makela 1995; Beckvar et al. 2000; Gagne et al. 2002; Perceval et al. 2002).

The same qualities that make mussels useful for pollution source identification also make them good sentinel organisms (Table 2.3). Mussels have been successfully used to monitor waterbodies for a variety of contaminants, including metals, organics, and fecal coliform bacteria (Bedford, Roelofs, and Zabik 1968; Leard, Grantham, and Pessoney 1980; Turick, Sexstone, and Bissonnette 1988; Muncaster et al. 1989; Salanki 1989; Lopes et al. 1992). However, a level of discrimination is necessary to ensure that the most applicable monitor is well suited for the suspected contaminant of impact. Such route-specific uptake or effect is not always apparent with mussels and may, therefore, require the use of other monitored organisms to demonstrate system impact. In comparative studies, Metcalfe and Hayton (1989) found leeches to be better than mussels for monitoring chlorophenols, and Balogh (1988) found that zooplankton were better biomonitors than mussels for metals from a sewage treatment plant. Renaud et al. (1995) attributed differential organochlorine uptake by mussels compared to lampreys to differences in lipid composition between the species. A review by Elder and Collins (1991) stated that metal accumulation in mollusks is typically higher than in fish.

Studies of the bioavailability of various contaminants to freshwater mussels have shown that uptake of metals is likely related to metal concentrations that are easily extractable from sediments or that exist in ionic form, and are not necessarily correlated to total water or sediment concentrations (Tessier et al. 1984, 1993; Marquenie 1985; Czarnezki 1987; Schmitt et al. 1987; Pugsley, Hebert, and McQuarrie 1988; Koenig and Metcalfe 1990; Campbell and Evans 1991; Elder and Collins 1991; Kauss and Hamdy 1991; Couillard, Campbell, and Tessier 1993; Metcalfe-Smith 1994; Perceval et al. 2002). These studies have also helped to define those factors that tend to confound the interpretation of mussel body burden data. Metals have been shown to increase, decrease, or not to vary with the size of the animal, depending upon the metal (Manly and George 1977; Price and Knight 1978; Hinch and Stephenson 1987; Green et al. 1989; Naimo 1993; Balogh and Mastala 1994). Jeffree et al. (1993) eliminated the size-related variability in individual mussel metal concentrations by regression against tissue calcium concentrations.

Organochlorine body burdens have also been shown to be related to mussel size (Muncaster, Hebert, and Lazar 1990). Other biological factors that have been shown to influence contaminant-body burdens include age, reproductive status, growth rate, tissue type, species, lipid content, and the valve closure response (Brungs 1967; Smith, Green, and Lutz 1975; Renzoni and Bacci 1976; Foster and Bates 1978; Phillips 1980; Hinch and Stephenson 1987; Green et al. 1989; Tevesz et al. 1989; Metcalfe-Smith and Green 1992; Balogh and Mastala 1994; Renaud et al. 1995; Stewart and Malley 1997). Environmental factors that have been shown to influence contaminant-body burdens include pH, temperature, humic matter, calcium, organic and inorganic carbon, season, and interactions with other contaminants (Phillips 1980; Tessier et al. 1984, 1993; Campbell and Evans 1987, 1991; Hanna 1992; Metcalfe-Smith 1994; Naimo 1995; Stewart 1999; Perceval et al. 2002). The sex of a mussel has been shown to have relatively little influence on body burdens of both metals (Metcalfe-Smith 1994) and organochlorines (Muncaster, Hebert, and Lazar 1990).

Many applied studies of mussel contaminant body burdens have successfully used transplanted and/or caged mussels (see Chapter 9). Transplanting is useful for the purpose of introducing mussels for biomonitoring of areas where resident mussels do not exist or exist in numbers that are too low for effective sampling, and for the purpose of reducing experimental error by using mussels from a known source or sources. Several biomonitoring studies have successfully used transplanted mussels (Schmitt et al. 1987; Cherry, Farris, and Belanger 1988; Hinch and Green 1989; Salanki 1989; Couillard et al. 1995a; 1995b; Makela 1995; Beckvar et al. 2000). The use of transplanted mussels can be advantageous because of the unknown exposure history of resident mussels and the long retention time for some contaminants (Marquenie 1985); however, a concern involving the use of transplants has been raised by Hinch and Green (1989) and Salanki (1989), who found that metal uptake by transplanted mussels is influenced by their source. A reciprocal transplant study by Englund and Heino (1996) found cage and source-related effects on mussel valve movement measures to be minimal. Investigators should be aware of the potential for source-related effects when planning a transplant study. In particular, the selection of source locations, types and frequency of measurements, number of replicate measurements, and statistical analysis of resulting data all need to account for potential bias due to source.

Many investigators also choose to cage mussels during biomonitoring, in order to (1) increase their chances of retrieval, (2) ensure original mussels are collected for analysis, and (3) protect the mussels from predation and other outside influences. Disadvantages of cages include the potential for vandalism and the potential for fouling of the cages by debris. Many of the studies referenced in Table 2.3 used caged mussels, with exposure periods ranging from two days to one year. Muncaster, Hebert, and Lazar (1990) reported that cages, corrals, and leashes were all equally effective for biomonitoring studies. Salazar et al. (2002) developed a cage for use in long-term in situ studies, and an ASTM (2001) method is available providing the appropriate methodology. A standardized exposure period may not be advisable because of the demonstrated tendency of contaminants to fluctuate over time (Muncaster et al. 1989). Rather, exposure time should be determined by the study objectives, characteristics of the contaminant being measured, and the environmental conditions at the time of the study. Salazar (1997) recommended an exposure period of 60–90 days based on the time necessary to reach chemical equilibrium for most metals and hydrophobic chemicals.

Freshwater mussels have received limited use to date as recorders of environmental history. Lingard, Evans, and Bourgoin (1992) detailed techniques for measuring metal concentrations in the shell nacre. It is recognized that mussels represent a potentially valuable tool in this regard, especially in fluvial systems and other areas where sediments do not often provide a chronological record of contaminant concentrations. Promising results based on the separation and elemental analysis of mussel shell layers as a historic record have been obtained (Sterrett and Saville 1974; Carell et al. 1987, 1995), but further work is needed to explore the relationship between contaminant levels in the environment and levels in shells (Stewart and Malley 1997).

LABORATORY TESTING OF FRESHWATER MUSSELS

Several investigators have undertaken controlled testing of freshwater mussels, with varying degrees of success. As with field monitoring, laboratory studies involving freshwater mussels have lagged behind marine studies in both quantity and quality (Jacobson 1990; Johnson, Zam, and Keller 1990) but have gained significant attention over the past 10–15 years. An in-depth discussion of this topic is provided in Chapter 5 and will only be briefly touched upon here.

A summary of the types of laboratory and artificial stream tests that have been conducted with freshwater mussels is provided in Table 2.4. To date, there is no consensus on test methodologies, and many investigators have developed their own, in-house testing techniques that are customized toward

TABLE 2.4
Laboratory and Artificial Stream Test Methods Used for Freshwater Mussels

Life Stage	Test System	Test Length	References
Glochidia			
In Gravid adults	Artificial stream	30 days	30,36
Isolated	Beaker, culture dish	25 min–6 days	12,20,21,24,34,36,37,40,43,80,82
Encysted	Static-renewal tanks	12–20 days	30,36
Juvenile	Static or static-renewal	1–10 days	5,13,33,36,40–43,51,52,77
	Flow-thru or artificial stream	10–37 days	2,36,55,56
	In situ glass vial	7 days	44,81
Adult	Static or static-renewal	18 h–7 months	1,3,5,7–10,15,18,19,22,23,29,30,43,45–47, 50,53,57–59,61–64,68–76,78,79,83
	Flow-thru tanks	7–300 days	4,6,14,25–28,31,32,35,38,48,54,57,60,65–67
	Artificial streams	96 h–14 months	11,12,16,17,39

References cited: 1. Aldridge, Payne, and Miller (1987); 2. Allran et al. (2002); 3. Baker and Hornbach (1997); 4. Balogh and Salanki (1984); 5. Barfield, Clem, and Farris (1997); 6. Bartsch et al. (2000); 7. Birdsall, Kukor, and Cheney (2001); 8. Black et al. (1996); 9. Cassini et al. (1986); 10. Cheney and Criddle (1996); 11. Cherry, Farris, and Belanger (1988); 12. Cherry, Farris, and Neves (1991); 13. Dimock and Wright (1993); 14. Doran et al. (2001); 15. Doyotte et al. (1997); 16. Ellis (1936); 17. Farris, Cherry, and Neves (1991); 18. Foster and Bates (1978); 19. Gagne et al. (2001a, 2001b, 2001c); 20. Goudreau, Neves, and Sheehan (1993); 21. Granmo and Varanka (1979); 22. Hameed and Raj (1989); 23. Hammer, Merkowsky, and Huang (1988); 24. Hansten, Heino, and Pynnonen (1996); 25. Hemelraad, Holwerda, and Zandee (1986a); 26. Hemelraad et al. (1986b); 27. Hemelraad and Herwig (1988); 28. Hemelraad et al. (1990a); Hemelraad et al. (1990b); Hemelraad et al. (1990c); 29. Henry and Saintsing (1983); 30. Holwerda and Herwig (1986); 31. Holwerda et al. (1988); 32. Holwerda (1991); 33. Hudson and Isom (1984); 34. Huebner and Pynnonen (1992); 35. Imlay (1973); 36. Jacobson (1990); 37. Jacobson et al. (1993, 1997); 38. Jeffree and Brown (1992); 39. Jenner et al. (1991); 40. Johnson, Zam, and Keller (1990); 41. Keller (1993); 42. Keller and Zam (1991); 43. Keller and Ruessler (1997); 44. Klaine and Williams (1992); 45. Koppar, Kulkarni, and Venkatachari (1993); 46. Lagerspetz, Anneli Korhonen, and Tuska (1995); 47. Makela and Oikari (1990); 48. Makela and Oikari (1992); 49. Makela and Oikari (1995); 50. Makela et al. (1991); 51. Masnado, Geis, and Sonzogni (1995); 52. McKinney and Wade (1996); 53. Moulton, Fleming, and Purnell (1996); 54. Naimo, Waller, and Holland-Bartels (1992a); Naimo, Waller, and Holland-Bartels (1992b); 55. Newton et al. (2002); 56. Payne and Miller (1987b); 57. Pellinen et al. (1993); 58. Pynnonen (1990); 59. Pynnonen (1994); 60. Pynnonen, Holwerda, and Zandee (1987); 61. Pynnonen (1995a), (1995b); 62. Rajalekshmi and Mohandas (1993); 63. Rajalekshmi and Mohandas, (1995); 64. Ravinder Reddy (1991); 65. Russell and Gobas (1989); 66. Salanki and Balogh (1989); 67. Salanki and Hiripi (1990); 68. Sauve et al. (2002); 69. Scarpato, Migliore, and Barale (1990); 70. Silverman, McNeil, and Dietz (1987); 71. Sivaramakrishna, Radhakrishnaiah, and Suresh (1991); 72. Streit and Winter (1993); 73. Sumathi (1991); 74. Tessier, Vaillancourt, and Pazdernik (1994a); 75. Tessier, Vaillancourt, and Pazdernik (1994b); 76. Tessier, Vaillancourt, and Pazdernik (1996); 77. Wade, Hudson, and McKinney (1993); 78. Waller et al. (1993b); 79. Waller, Rach, and Luoma (1998); 80. Wang et al. (2003); 81. Warren, Klaine, and Finley (1995); 82. Weinstein (2001); 83. Winter (1996).

the focus of their research. This trend is readily apparent when examining the range of test lengths listed in Table 2.4. There is little doubt that freshwater mussels are useful organisms for applied laboratory studies; however, there are still a number of problems to be overcome. Most testing that is performed depends on field-collected organisms, since long-term culture of unionid mussels is not practical for most laboratories because of time and space requirements (see Chapter 4). Use of field-collected organisms, although an acceptable practice, at times can be problematic because of the often unpredictable availability of appropriate test specimens and the often unknown environmental history of the specimens. Pond culture of commonly used test species is a potential solution that should be pursued. Selection of species for testing should also be carefully evaluated because the easiest cultured mussels may not be the most appropriate (e.g., representative of other mussels or sensitive to contaminant exposure) (Johnson, Zam, and Keller 1990). As with other classes of organisms, mussel sensitivity to contaminants varies among species and among contaminants (Jacobson 1990; Farris, Cherry, and Neves 1991; Keller and Zam 1991; Waller et al. 1993b; Hansten, Heino, and Pynnonen 1996; Sauve et al. 2002).

Laboratory testing of freshwater mussels has encompassed a large variety of applications (Table 2.5). In general, early life stages provide the most sensitive contaminant response within the shortest period of time. Glochidia isolated from gravid female adults and recently transformed juveniles have been shown to be sensitive in short-term tests to pesticides (Varanka 1977, 1978, 1979; Johnson, Zam, and Keller 1990; Keller 1993), metals (Jacobson 1990; Cherry, Farris, and Neves 1991; Keller and Zam 1991; Huebner and Pynnonen 1992; Jacobson et al. 1993, 1997; Masnado, Geis, and Sonzogni 1995; Hansten, Heino, and Pynnonen 1996; Barfield, Clem, and Farris 1997), industrial effluents (Cherry, Farris, and Neves 1991; Keller and Zam 1991; Keller 1993; Wade, Hudson, and McKinney 1993; McKinney and Wade 1996), acidic pH (Huebner and Pynnonen 1992; Dimock and Wright 1993), and ammonia and monochloramine (Goudreau, Neves, and Sheehan 1993; Newton et al. 2002; Wang et al. 2003). Test protocols involving early mussel life stages need further development, as the determination of even the simple endpoint of mortality has not been satisfactorily resolved (Goudreau, Neves, and Sheehan 1993).

Adult mussels are much easier to use in testing but can be problematic in many applications because of their valve closure response to many types of contaminants (Miller and Nelson 1983; Elder and Collins 1991), although valve closure itself represents a useful response measure that could be applied to some types of monitoring, as has been demonstrated with *Corbicula* spp. (Doherty 1990). Conversely, Johnson, Zam, and Keller (1990) have determined that young juvenile mussels cannot close their valves for long periods of time, as adults can. For studies of contaminant uptake and depuration, adult mussels are typically preferred because of their larger size, and, because exposure levels in these studies are typically below acute-effect concentrations, the valve closure response is not a problem. An expanding database of studies exists on the uptake, accumulation, and elimination of contaminants, particularly metals and chlorinated organics. Although several of these studies have demonstrated preferential uptake of contaminants by certain tissues, such as cadmium, lead, tin, and chlorinated phenolics in the kidney (Balogh and Salanki 1984; Holwerda and Herwig 1986; Holwerda et al. 1988; Salanki and Balogh 1989; Makela and Oikari 1990; Jenner et al. 1991; Streit and Winter 1993; Winter 1996) and copper in the mantle (Salanki and Balogh 1989), whole body accumulation provides sufficient resolution for most studies, unless the study objective deals with contaminant dynamics at the tissue, cellular, or subcellular level. Definitive studies of contaminant uptake have been undertaken, resulting in information regarding bioaccumulation models (Tessier, Vaillancourt, and Pazdernik 1994b), the species specificity of accumulation (Hemelraad et al. 1986a,b,c), the effect on uptake and accumulation of confounding environmental factors (Holwerda et al. 1988; Holwerda 1991; Makela et al. 1991; Jeffree and Brown 1992), the lack of metal sequestration by calcium concretions in mussel gills (Pynnonen, Holwerda, and Zandee 1987; Silverman, McNeil, and Dietz 1987), and the role of metallothionein induction in metal accumulation dynamics (Hemelraad et al. 1986a; Streit and Winter 1993; Couillard et al. 1995a; Giguere et al. 1999; Wang et al. 1999).

TABLE 2.5
Measures Applied to Laboratory and Artificial Stream Tests of Freshwater Mussels

Measure	Life Stage	Advantages	Disadvantages	References
Survival/viability	Glochidia	1,2,3	5	11,19,20,23,32,34,35,38,43,81,83
	Juvenile	1,3	5	4,12,34,38,40,42–44,51,52,56,78,82
	Adult	2,3	1,2	5,10,11,16,21,33,43,45,48,53,54,72,79,80
Transformation success	Glochidia	2,3	5	34,35,38,41
Encystment success	Glochidia	3	5	34,35
Snap rate	Glochidia	1,3	2,5	11,20,34,35
Bioaccumulation	Glochidia			83
	Adult	2	4	3,6,8,17,22,24–26,29–31,36,37,47,49,50, 58,62,66,67,71–73,75–77,80,84
Siphoning activity	Adult	1,2,3,4	5	2,3,15,54,67
Growth	Juvenile			56
	Adult	1,4	3,4	10
Enzyme activity	Adult	1,3	2,4	10,13,14,16,27,28,54,65
Blood ion concentration	Adult	1	4,5	18,27–29,48,59–61,69
Metabolic rates	Adult	4	2,4,5	1,2,9,27,55,57,64,72
Shell/condition index	Adult	2,4 (shell)	1,3	2,48,57
Ciliary activity	Adult	1,3,4	5	39,46,74
Other physiological	Adult		2,4,5	1,7,18,21,27,29,46,55,63,65,68,70

Key to advantages		Key to disadvantages	
1 Sensitive		1 Insensitive	
2 Ease of use		2 High variability	
3 Short-term response (1–2 days)		3 Long-term response	
4 Nondestructive		4 Time/labor intensive	
		5 Method development needed	

References cited: 1. Aldridge, Payne, and Miller (1987); 2. Baker and Hornbach (1997); 3. Balogh and Salanki (1984); 4. Barfield, Clem, and Farris (1997); 5. Bartsch et al. (2000); 6. Birdsall, Kukor, and Cheney (2001); 7. Black et al. (1996); 8. Cassini et al. (1986); 9. Cheney and Criddle (1996); 10. Cherry, Farris, and Belanger, (1988); 11. Cherry, Farris, and Neves (1991); 12. Dimock and Wright (1993); 13. Doran et al. (2001); 14. Doyotte et al. (1997); 15. Ellis (1936); 16. Farris, Cherry, and Neves (1991); 17. Foster and Bates (1978); 18. Gagne et al. (2001a, 2001b, 2001c); 19. Goudreau, Neves, and Sheehan (1993); 20. Granmo and Varanka (1979); 21. Hameed and Raj (1989); 22. Hammer, Merkowsky, and Huang (1988); 23. Hansten, Heino, and Pynnonen (1996); 24. Hemelraad, Holwerda, and Zandee (1986a); 25. Hemelraad et al. (1986b); 26. Hemelraad and Herwig (1988); 27. Hemelraad et al. (1990a, 1990b, 1990c); 28. Henry and Saintsing (1983); 29. Holwerda and Herwig (1986); 30. Holwerda et al. (1988); 31. Holwerda (1991); 32. Huebner and Pynnonen (1992); 33. Imlay (1973); 34. Jacobson (1990), (1993); 35. Jacobson et al. (1997); 36. Jeffree and Brown (1992); 37. Jenner et al. (1991); 38. Johnson, Zam, and Keller (1990); 39. Kabeer Ahmad et al. (1979); 40. Keller (1993); 41. Keller and Zam (1990); 42. Keller and Zam (1991); 43. Keller and Ruessler (1997); 44. Klaine and Williams (1992); 45. Koppar, Kulkarni, and Venkatachari (1993); 46. Lagerspetz, Anneli Korhonen, and Tuska (1995); 47. Makela and Oikari (1990); 48. Makela and Oikari (1992); 49. Makela and Oikari (1995); 50. Makela et al. (1991); 51. Masnado, Geis, and Sonzogni (1995); 52. McKinney and Wade (1996); 53. McLeese et al. (1980); 54. Moulton, Fleming, and Purnell (1996); 55. Naimo, Waller, and Holland-Bartels (1992a); Naimo, Waller, and Holland-Bartels (1992b); 56. Newton et al. (2002); 57. Payne and Miller (1987b); 58. Pellinen et al. (1993); 59. Pynnonen (1990); 60. Pynnonen (1994); 61. Pynnonen (1995a, 1995b); 62. Pynnonen et al. (1987); 63. Rajalekshmi and Mohandas (1993); 64. Rajalekshmi and Mohandas (1995); 65. Ravinder Reddy (1991); 66. Russell and Gobas (1989); 67. Salanki and Balogh (1989); 68. Salanki and Hiripi (1990); 69. Sauve et al. (2002); 70. Scarpato, Migliore, and Barale (1990); 71. Silverman, McNeil, and Dietz (1987); 72. Sivaramakrishna, Radhakrishnaiah, and Suresh (1991); 73. Streit and Winter (1993); 74. Sumathi (1991); 75. Tessier, Vaillancourt, and Pazdernik (1994a); 76. Tessier, Vaillancourt, and Pazdernik (1994b); 77. Tessier, Vaillancourt, and Pazdernik (1996); 78. Wade, Hudson, and McKinney (1993); 79. Waller et al. (1993b); 80. Waller, Rach, and Luoma (1998); 81. Wang et al. (2003); 82. Warren, Klaine, and Finley (1995); 83. Weinstein (2001); 84 Winter (1996).

Adult unionid mussels were highly resistant in short-term exposures to metals (Cherry, Farris, and Neves 1991), acidic pH (Makela and Oikari 1992), pesticides (McLeese et al. 1980; Keller and Ruessler 1997), and a variety of potential molluscicides (Waller et al. 1993b). Siphoning activity provides a measurable response to contaminant exposure (Ellis 1936; Balogh and Salanki 1984; Salanki and Balogh 1989), but the response has not been defined in terms of adverse impact on the organism. Growth provides a good, quantifiable measure of impact, but most unionid mussels exhibit such slow growth that sufficient exposure duration for laboratory testing limits the usefulness of this measure. Several physiological measures have been applied in a variety of studies of contaminant effects of mussels (Table 2.5). Many of these measures are highly variable (Naimo, Atchison, and Holland-Bartels 1992a; Naimo, Waller, and Holland-Bartels 1992b) and insufficiently defined in terms of measuring significant impairment of mussel life history functions. Further development of many of the physiological measures is necessary to increase their utility in the assessment of contaminant impacts on mussels. A good review of what is known regarding the role of metal ions in both marine and freshwater mollusk metabolism, including accumulation dynamics, toxic effects, and cellular detoxification systems, has been provided by Simkiss and Mason (1983), and an updated treatment of this topic is provided in Chapter 8.

Recent work involving the response of various mussel enzyme systems to contaminants also show promise for defining impacts on mussels, with the additional advantage of providing a relatively rapid response compared to most other measures (Henry and Saintsing 1983; Cherry, Farris, and Belanger 1988; Hemelraad et al. 1990b; Farris, Cherry, and Neves 1991; Ravinder Reddy 1991; Moulton, Fleming, and Purnell 1996; Doyotte et al. 1997; Doran et al. 2001). However, as with most of the measures listed in Table 2.5, further development of the methodologies is needed to provide definitive relationships to mussel life history functions. For example, Cherry, Farris, and Belanger (1988) and Farris, Cherry, and Neves (1991) have shown that mussel cellulolytic enzyme activity is rapidly and significantly depressed in response to contaminant stress, and have further shown that reductions in the activity of these enzymes can be associated with eventual reductions in the growth rate of *C. fluminea* (Farris et al. 1989). Couillard et al. (1995b) related increased metallothionein induction to cellular toxicity and reduced mussel growth. See Chapter 10 for a detailed discussion of response measures in freshwater mussel research.

EFFECTIVENESS OF MUSSEL BIOMONITORING

MUSSELS AS SENTINELS OF ENVIRONMENTAL PERTURBATIONS

The most common biomonitoring application of unionid mussels has been as sentinels of environmental perturbations. As indicated in Table 2.6, mussels are well suited to this application, although there are some limitations. On the positive side, mussels are large, meaning that they are typically easy to collect and handle, and they provide sufficient tissue for individual analyses (Metcalfe-Smith and Green 1992). Mussels are also sedentary, which is an important attribute for sentinel organisms (Phillips 1980; Green, Singh, and Bailey 1985) because it can generally be assumed that most species of collected mussels have been residing in the same general area for their entire adult life, and transplanted mussels will remain fairly close to where they are placed. The long life span (typically 10–40 years or more) of unionid mussels means that they will accumulate contaminants over any time period needed for a particular study (Metcalfe-Smith and Green 1992). Their long life span also is useful for nondestructive, repeat-measure monitoring of the same individuals.

For monitoring both contaminant body burdens and response measures, mussels respond to a variety of contaminants in direct proportion with exposure concentrations and durations, which provides integration of long-term exposure and evidence of contaminant pulses (Phillips 1980; Mix 1986). One difficulty that arises, however, is that exposure duration may be critical for monitoring of certain types of contaminants under certain conditions. If the monitoring period is too short,

TABLE 2.6
Advantages and Disadvantages of Unionid Mussels as Sentinels of Environmental Perturbations (SEP) or as Indicators of Ecological Integrity (IEI)

Advantages	Disadvantages
Large, easy to use (SEP, IEI)	Some methods inadequately developed (SEP, IEI)
Sedentary (SEP, IEI)	Most measures are destructive (SEP)
Long-lived (SEP, IEI)	Destructive use depletes a declining resource (SEP, IEI)
Widely distributed (IEI)	Confounding factors compromise results (SEP)
Integrate exposure over time (SEP)	Low discriminatory power for some measures (SEP, IEI)
Easily transplanted and caged (SEP)	Uncommon or absent in many waters (SEP)
High discriminatory power when quantified (SEP, IEI)	Habitat specificity confounds data interpretation (IEI)
Nondestructive methods available (IEI)	Large sample size required for some methods (IEI)
Both sensitive and tolerant species available (SEP)	Exposure duration critical for many applications (SEP)
Pollution-sensitive species widespread (IEI)	Long study duration required for some methods (IEI)
Important component of benthos (IEI)	
Measure bioavailable contaminants (SEP)	
Shells provide historical record (IEI)	
Cost-effective for many applications (SEP, IEI)	

variability in both contaminant exposures and uptake may prevent useful interpretation of the data (Jones and Walker 1979). If the monitoring period is too long, a contaminant pulse that resulted in elevated body burdens may not be detected if the body burden returns to background levels before it is measured (Phillips 1980). The contaminant-specific and species-specific uptake and elimination dynamics under various exposure conditions are an important area of study that is addressed in detail in Chapter 8 of this volume.

Monitoring of resident mussel populations is not always possible because mussels are often rare or absent from areas where monitoring is desired (Mix 1986; Elder and Collins 1991); however, mussels are easily transplanted and are amenable to caging. Waller et al. (1995) advised caution when conducting transplant studies because of preliminary evidence that mussels can experience significant latent effects due to handling and removal from water. However, later studies (Bartsch et al. 2000; Chen, Heath, and Neves 2001; Greseth et al. 2003) have shown negligible short-term effects of handling and emersion on the survival or biochemical condition of unionids.

Mussel monitoring has displayed reliable and sensitive discriminatory power in a variety of applications (Havlik and Marking 1987). An overwhelming majority of these applications have used body burdens as the method of contaminant detection. Contaminant body burdens have been successfully used to detect pesticides (Bedford, Roelofs, and Zabik 1968; Fikes and Tubb 1972; Leard, Grantham, and Pessoney 1980), other organic contaminants (Curry 1977; Kauss and Hamdy 1985; Metcalfe and Charlton 1990), metals (Anderson 1977; Foster and Bates 1978; Adams, Atchison, and Vetter 1981; Klose and Potera 1984; Czarnezki 1987; Couillard et al. 1995a; Malley, Stewart, and Hall 1996; Gagne et al. 2002), radionuclides (Gaglione and Ravera 1964; Gardner and Skulberg 1965), and fecal coliforms (Turick, Sexstone, and Bissonnette 1988).

Use of unionid mussel body burdens as sentinels of contamination, despite its fairly common application, still needs further development. First, confounding biological and environmental factors that influence contaminant uptake, accumulation, elimination, and measurement need to be further defined and quantified (Mix 1986; Hinch and Stephenson 1987; Green et al. 1989; Koenig and Metcalfe 1990; Metcalfe and Charlton 1990; Kauss and Hamdy 1991; Balogh and Mastala 1994; Metcalfe-Smith 1994). Second, monitoring frequency needs to be carefully tailored to the type of contaminant being monitored, in terms of the expected duration of exposure and tissue elimination rate, the number of animals that must be sacrificed, and the specific objectives of the monitoring program (Jones and Walker 1979; Phillips 1980; Doran et al. 2001). Finally, the purpose of the monitoring should be carefully considered with respect to the information being provided. Body burden measurement provides evidence of contaminant uptake, but levels of contaminants in tissue have not been defined in terms of adverse effects on mussels. If the goal of a monitoring program is to detect elevated levels of contaminants being introduced to a waterbody, then body burden measurement seems to be a reliable, cost-effective tool. In fact, mussels are considered to be excellent indicator organisms for reflecting bioavailable concentrations of environmental contaminants (Tessier et al. 1984, 1993; Kauss and Hamdy 1985; Marquenie 1985; Mix 1986; Campbell and Evans 1991; Couillard, Campbell, and Tessier 1993). However, if the goal is to demonstrate impact, then a response measure, possibly in addition to body burden measurement, should be used. For the purpose of rapid response monitoring, population measures are probably not practical because of the time and expense necessary to obtain quantitative data. Physiological response measures are called for in this situation; however, to date, such measures are not well developed for use on freshwater mussels. A few investigators have had success using measures of mussel enzyme activity, glycogen content, blood ion chemistry, or tissue amino acid levels to detect environmental stress on mussels (Gardner, Miller, and Imlay 1981; Malley, Huebner, and Donkersloot 1988; Day, Metcalfe, and Batchelor 1990; Scarpato, Migliore, and Barale 1990; Haag et al. 1993; Fleming, Augspurger, and Alderman 1995; Doyotte et al. 1997; Wang et al. 1999; Gagne et al. 2001a, 2001b, 2001c; Blaise et al. 2003). A more thorough development of these methods or others is needed in order to provide reliable, sensitive, and definitive measures of contaminant impact on mussel life history functions.

Another advantage in the use of freshwater unionid mussels as sentinel organisms is the availability of both pollution-sensitive and pollution-tolerant species, depending on the needs of the monitoring application. Although most applications will require sensitive monitors of chronic or episodic levels of contamination, there may be situations where a more tolerant organism is more suitable to provide a record of severe cases of contamination that eliminate sensitive organisms and that could otherwise escape identification (Salanki 1989). Although many species of mussels have been shown to be very sensitive to a variety of contaminants, there are some more tolerant species, particularly among the thick-shelled forms (Miller and Nelson 1983), and the ability of adult mussels to endure acute toxicant exposures by tightly sealing their valves (Elder and Collins 1991) makes them ideal for this type of application.

A major concern with the use of freshwater mussels for biomonitoring is the fact that almost all monitoring methods require that the mussels be sacrificed, potentially adversely impacting an already declining resource. For tissue measurements, promising work has been published advancing the use of a nonlethal biopsy technique (Berg et al. 1995; Naimo et al. 1998; Naimo and Monroe 1999). Other solutions to this problem include the use of nondestructive monitoring techniques (e.g., external condition indices and behavioral responses) and development of mass culture facilities for commonly used species. The use of surrogate bivalve species for rare unionid species should be implemented whenever possible. The Asian clam, *C. fluminea*, has been recommended as one potential surrogate (Foster and Bates 1978; McMahon 1983). Both Cherry, Farris, and Neves (1991) and Milam and Farris (1998) found *Corbicula* to be a useful unionid surrogate in laboratory, artificial stream, and instream studies of metal and effluent impact. However, it is not advisable, and in some jurisdictions it is unlawful, to introduce the Asian clam or other exotic species where populations of these organisms are not already established. As with any surrogate, specific applications using *Corbicula* should be carefully examined for appropriateness. For example, the toxicological sensitivity of *Corbicula* varies depending upon the contaminant of concern and the species being represented. *Corbicula* has been found to have sensitivity on the level of the most sensitive unionids tested for the metals copper and zinc (Cherry,

Farris, and Neves 1991; Milam and Farris 1998) but was substantially less sensitive than unionids in comparative studies of the toxicity of ammonia (Cherry et al. 2005) and rotenone (Chandler 1982).

Comparisons among potential sentinel organisms have generally shown that mussels are as good or better than other types of organisms for most situations, including monitoring of both metals (Salanki 1989; Keller and Pepin 1999) and chlorinated organics (Metcalfe and Charlton 1990). Monitoring of mussels is far superior to monitoring of water or sediments because of their ability to integrate contaminant exposure over time and because mussel uptake of, and response to, contaminants reflects only the bioavailable fraction of those contaminants, which cannot be measured directly from water or sediments.

In summary, mussels are an excellent choice as sentinels of environmental perturbations for the reasons discussed above, but further work is needed to increase their utility in this application. The following areas of research are recommended:

1. Continued research on defining and quantifying biological and environmental factors that confound interpretation of mussel measurements.
2. Continued method development, particularly for short-term, physiological response measures and nondestructive monitoring techniques.
3. Investigation of the feasibility and practicality of developing mass culture facilities (e.g., ponds and artificial production of juveniles in the laboratory) for commonly used species.
4. Further investigation of the use and representativeness of readily available surrogate species, such as *Corbicula*.
5. Further investigation of methods for assessing impact based on body burden data, or a combined approach of body burden and effects monitoring.

MUSSELS AS INDICATORS OF ECOLOGICAL INTEGRITY

Many of the same attributes that make unionid mussels good sentinel organisms also make them well suited to use as indicators of ecological integrity in assessments of environmental impact, waterbody status monitoring, and assessments of environmental history (Table 2.6). Successful applications of mussels in these types of assessments include impacts from river dredging (Yokley 1976), commercial mussel harvest (Harris et al. 1993; Miller 1993), barge traffic (Sparks and Blodgett 1987; Miller and Payne 1993), acid mine drainage (Simmons and Reed 1973; Gardner, Miller, and Imlay 1981), eutrophication (Arter 1989), thermal effluents (Davies 1963; Negus 1966; Rothwell 1979), impoundment (Bates 1962), zebra mussel infestation (Haag et al. 1993), lake liming (Malley, Huebner, and Donkersloot 1988), industrial effluents (AEPSC/APCO 1987; Cherry, Farris, and Belanger 1988; Herve et al. 1988, 1991; Makela 1995; Hayer and Pihan 1996; Hayer, Wagner, and Pihan 1996), municipal effluents (Goudreau, Neves, and Sheehan 1993; Gagne et al. 2002), assessments of contaminant exposure history (Dermott and Lum 1986; Carell et al. 1987, 1995), and population status monitoring (Payne and Miller 1987a; Anderson, Romano, and Pederson 1993; Miller et al. 1993; Pekkarinen 1993). In these applications, the use of mussels was again advantageous for the ease of collection and handling, in addition to their being sedentary and long-lived (Allen 1923; Metcalfe-Smith and Green 1992). Diamond and Serveiss (2001) performed a risk assessment of a riverine mussel population relating species diversity to urban and agricultural land use. To be considered as indicator species, it is also essential that mussels often comprise an important structural and functional component of the benthos (Amyot and Downing 1991). For routine use in assessments of ecological integrity, mussels should be widely distributed (Amyot and Downing 1991); however, because of the narrow habitat niche occupied by many mussel species, their absence in ecological assessments does not necessarily imply impact, and variations in habitat can confound data interpretation in discerning differences between sampling locations. Additionally, the key role of the fish host in the mussel life cycle can complicate ecological assessments because of its influence on mussel distribution.

For assessments of ecological integrity in the form of environmental impact studies or water-body status monitoring, the most useful measures applied to unionid mussel populations have been density, size demography, growth, and physiological measures. Other commonly used structural measures (e.g., species richness and mortality) and body burden data have been less useful because of the frequent lack of sensitivity in detecting changes or differences in mussel populations. Contaminant body burden data must be definitively linked to significant population or individual response measures to provide meaningful impact assessments. To obtain useful, quantitative data, measurements of density and size demography require large, quantitative sample sizes (Miller and Payne 1988; Smith et al. 2000; Strayer and Smith 2003). Such samples can be time and labor intensive but have the advantage of being nondestructive if the mussels are carefully returned to the substrate following counting and measurement. Measurement of mussel growth rates has also been shown to be a sensitive measure but requires long monitoring periods (i.e., two or more months, depending upon species and life stage) to show significant differences in growth rates for most unionid species, and particularly older specimens and those from northern latitudes.

Physiological measures, such as enzyme activity or blood and tissue chemistry, offer excellent potential as a means of short-term response biomonitoring, but applications to date have been limited (e.g., Gardner, Miller, and Imlay 1981; Cherry, Farris, and Belanger 1988; Malley, Huebner, and Donkersloot 1988; Haag et al. 1993; Couillard et al. 1995a, 1995b; Gagne et al. 2001a, 2001b, 2001c, 2002; Blaise et al. 2002). Further work in this area will help provide more rapid, quantitative, nonlabor intensive, and preferably nondestructive methods of measuring impacts on, and the status of, mussels.

Mussels comprise one of the most pollution-sensitive components of any waterbody in which they reside (Ortmann 1909; Simmons and Reed 1973; Miller and Nelson 1983; Green et al. 1989; Amyot and Downing 1991). Several species of unionid mussels have exhibited sensitive responses in comparisons with fish, amphibians, aquatic insects, crayfish, amphipods, isopods, cladocerans, snails, sphaeriids, and Asian clams for contaminants that included metals, organic wastes, ammonia, chlorine, acid mine drainage, siltation, and industrial effluents (Simmons and Reed 1973; Klemm et al. 1990; Cherry, Farris, and Neves 1991; Farris, Cherry, and Neves 1991; Goudreau, Neves, and Sheehan 1993; Masnado, Geis, and Sonzogni 1995; McKinney and Wade 1996; Keller and Pepin 1999). In examining short-term, acute contaminant-exposure episodes, however, adult mussels are not expected to provide a sensitive response because of their ability to limit short-term exposure by tightly closing their valves. The lack of adult mussel sensitivity in short-term studies has been demonstrated in several studies (e.g., Cherry, Farris, and Neves 1991; Makela and Oikari 1992; Waller et al. 1993b; Milam and Farris 1998).

A unique form of monitoring with unionid mussels is the investigation of environmental history using the chronological record of contaminant accumulation available in mussel shell annuli (Sterrett and Saville 1974; Carell et al. 1987; Green et al. 1989; Amyot and Downing 1991; Lingard, Evans, and Bourgoin 1992). Carell et al. (1987, 1995) have provided promising results showing that mussel shell layers can be used to date environmental changes in elemental concentrations and α-radioactivity. However, further development of this technique is needed (Green et al. 1989) to address, in particular, the concern that levels of contaminants in shells may not provide a direct reflection of environmental exposure, but rather, a reflection of availability and physiological exchange rates during periods of growth (Dermott and Lum 1986).

In summary, mussels are useful indicators of ecological integrity, but, as with their use as sentinel organisms, further work is needed to increase their utility in this application. The following areas of research are recommended, expanding upon some of the research needs identified in the previous section:

1. Continued research on defining and quantifying biological (including fish host influences) and environmental (including physical habitat) factors that confound interpretation of mussel measurements.
2. Continued method development, particularly for short-term, physiological response measures, shell layer analysis and interpretation techniques, and nondestructive monitoring techniques.
3. Continued refinement of traditional mussel monitoring techniques (i.e., growth and contaminant body burdens).
4. Investigation of methods for assessing impact based on body burden data, or a combined approach of body burden and effects monitoring.

REFERENCES

Adams, W. F. and Alderman, J. M. Reviewing the status of your state's molluskan fauna: The case for a systematic approach, In *Conservation and Management of Freshwater Mussels. Proceedings of a UMRCC Symposium*, 1992 Oct 12–14, St. Louis, MO, Cummings, K. S., Buchanan, A. C., and Koch, L. M., Eds., pp. 83–88, 1993.

Adams, T. G., Atchison, G. J., and Vetter, R. J., The use of the three-ridge clam (*Amblema perplicata*) to monitor trace metal contamination, *Hydrobiologia*, 83, 67–72, 1981.

[AEPSC/APCO] American Electric Power Service Corporation/Appalachian Power Company, *Mussel Transplant Studies on the Clinch River near Appalachian Power Company's Clinch River Plant*, Report submitted to State Water Control Board, Richmond, VA, 1987.

Aldridge, D. W., Payne, B. S., and Miller, A. C., The effects of intermittent exposure to suspended solids and turbulence on three species of freshwater mussels, *Environ. Pollut.*, 45, 17–28, 1987.

Allen, W. R., Studies of the biology of freshwater mussels. II. The nature and degree of response to certain physical and chemical stimuli, *Ohio J. Sci.*, 23, 57–82, 1923.

Allran, J., Newton, T., O'Donnell, J., Bartsch, M., and Richardson, B., Delivery system for ammonia in sediment toxicity tests with juvenile unionids [poster], *Society of Environmental Toxicology and Chemistry 23rd Annual Meeting*, Nov 16–20, Salt Lake City, UT, 2002.

Amyot, J. P. and Downing, J. A., Endo and epibenthic distribution of the unionid mollusk *Elliptio complanata*, *J. N. Am. Benthol. Soc.*, 10, 280–285, 1991.

Anderson, R. V., Concentration of cadmium, copper, lead, and zinc in six species of freshwater clams, *Bull. Environ. Contam. Toxicol.*, 18, 492–496, 1977.

Anderson, R. V., Romano, M. A., and Pederson, T. R., Use of regression analysis to characterize mussel communities [abstract], In *Conservation and Management of Freshwater Mussels. Proceedings of a UMRCC Symposium*, 1992 Oct 12–14, St. Louis, MO, Cummings, K. S., Buchanan, A. C., and Koch, L. M., Eds., p. 177, 1993.

Arbuckle, K. E. and Downing, J. A., Freshwater mussel abundance and species richness: GIS relationships with watershed land use and geology, *Can. J. Fish. Aqat. Sci.*, 59, 310–316, 2002.

Arter, H. E., Effect of eutrophication on species composition and growth of freshwater mussels (Molluska, Unionidae) in Lake Hallwil (Aargau, Switzerland), *Aquat. Sci.*, 51/2, 87–99, 1989.

[ASTM] American Society for Testing and Materials, Standard guide for preparation of biological samples for inorganic chemical analysis, D 4638-95a, *2002 Annual Book of ASTM Standards, Vol. 11.01*, ASTM, Philadelphia, PA, 1999.

[ASTM] American Society for Testing and Materials, Standard guide for conducting in-situ field bioassays with caged bivalves, E 2122-02, *2003 Annual Book of ASTM Standards, Vol. 11.05*, ASTM, Philadelphia, PA, 2001.

Bailey, R. C., Correlations between species richness and exposure: Freshwater mollusks and macrophytes, *Hydrobiologia*, 162, 183–191, 1988.

Baker, F. C., The effects of sewage and other pollution on animal life of rivers and streams, *Trans. Ill. State Acad. Sci.*, 13, 271–279, 1920.

Baker, S. M. and Hornbach, D. J., Acute physiological effects of zebra mussel (*Dreissena polymorpha*) infestation on two unionid mussels, *Actinonaias ligamentina* and *Amblema plicata*, *Can. J. Fish. Aquat. Sci.*, 54, 512–519, 1997.

Balogh, K. V., Comparison of mussels and crustacean plankton to monitor heavy metal pollution, *Water Air Soil Pollut.*, 37, 281–292, 1988.

Balogh, K. V. and Mastala, Z., The influence of size and glochidia bearing upon the heavy metal accumulation in gills of *Anodonta piscinalis* (Nilss.), *Chemosphere*, 28, 1539–1550, 1994.

Balogh, K. V. and Salanki, J., The dynamics of mercury and cadmium uptake into different organs of *Anodonta cygnea* L, *Water Res.*, 18, 1381–1387, 1984.

Barfield, M. L., Clem, S. A., and Farris, J. L., Comparison of acute testing methods of *Utterbackia imbecillis* (Say) with variation on age, diet, and *in vitro* transformation method, In *Conservation and Management of Freshwater Mussels II. Initiatives for the Future. UMRCC Symposium Proceedings*, 1995 Oct 16–18, St. Louis, MO, Cummings, K. S., Buchanan, A. C., Mayer, C. A., and Naimo, T. J., Eds., pp. 257–264, 1997.

Bartsch, M. R., Waller, D. L., Cope, W. G., and Gutreuter, S., Emersion and thermal tolerances of three species of unionid mussels: survival and behavioral effects, *J. Shellfish Res.*, 19, 233–240, 2000.

Bates, J. M., The impact of impoundment on the mussel fauna of Kentucky Reservoir, Tennessee River, *Am. Midl. Nat.*, 68, 232–236, 1962.

Becker, K., Merlini, L., de Bertrand, N., de Alencastro, L. F., and Tarradellas, J., Elevated levels of organotins in Lake Geneva: Bivalves as sentinel organism, *Bull. Environ. Contam. Toxicol.*, 48, 37–44, 1992.

Beckvar, N., Salazar, S., Salazar, M., and Finkelstein, K., An in situ assessment of mercury contamination in the Sudbury River, Massachusetts, using transplanted freshwater mussels (*Elliptio complanata*), *Can. J. Fish. Aquat. Sci.*, 57, 1103–1112, 2000.

Bedford, J. W., Roelofs, E. W., and Zabik, M. J., The freshwater mussel as a biological monitor of pesticide concentration in a lotic environment, *Limnol. Oceanogr.*, 13, 118–126, 1968.

Berg, D. J., Haag, W. R., Guttman, S. I., and Sickel, J. B., Mantle biopsy: A technique for nondestructive tissue-sampling of freshwater mussels, *J. N. Am. Benthol. Soc.*, 14, 577–581, 1995.

Birdsall, K., Kukor, J. J., and Cheney, M. A., Uptake of polycyclic aromatic hydrocarbon compounds by the gills of the bivalve mollusk *Elliptio complanata*, *Environ. Toxicol. Chem.*, 20, 309–316, 2001.

Black, M. C., Ferrell, J. R., Horning, R. C., and Martin, L. K., DNA strand breakage in freshwater mussels (*Anodonta grandis*) exposed to lead in the laboratory and field, *Environ. Toxicol. Chem.*, 15, 802–808, 1996.

Blaise, C., Trottier, S., Gagne, F., Lallement, C., and Hansen, P. D., Immunocompetence of bivalve hemocytes as evaluated by a miniaturized phagocytosis assay, *Environ. Toxicol.*, 17, 160–169, 2002.

Blaise, C., Gagne, F., Salazar, M., Salazar, S., Trottier, S., and Hansen, P.-D., Experimentally-induced feminisation of freshwater mussels after long term exposure to a municipal effluent, *Fresenius Environ. Bull.*, 12(8), 865–870, 2003.

Brungs, W. A., *Distribution of Cobalt 60, Zinc 65, Strontium 85, and Cesium 137 in a Freshwater Pond*, U.S. Public Health Service, 999-RH-24, Washington, DC, p. 52, 1967.

Cairns, J. and Smith, E. P., The statistical validity of biomonitoring data, In *Biological Monitoring of Aquatic Systems*, Loeb, S. L. and Spacie, A., Eds., Lewis Publishing, Boca Raton, FL, pp. 49–68, 1994.

Cairns, J., McCormick, P. V., and Niederlehner, B. R., A proposed framework for developing indicators of ecosystem health, *Hydrobiologia*, 263, 1–44, 1993.

Campbell, J. H. and Evans, R. D., Inorganic and organic ligand binding of lead and cadmium and resultant implication for bioavailability, *Sci. Total Environ.*, 62, 219–227, 1987.

Campbell, J. H. and Evans, R. D., Cadmium concentrations in the freshwater mussel (*Elliptio complanata*) and their relationship to water chemistry, *Arch. Environ. Contam. Toxicol.*, 20, 125–131, 1991.

Carell, B., Forberg, S., Grundelius, E., Henrikson, L., Johnels, A., Lindh, U., Mutvei, H., Olsson, M., Svardstrom, K., and Westermark, T., Can mussel shells reveal environmental history?, *Ambio*, 16, 2–10, 1987.

Carell, B., Dunca, E., Gardenfors, U., Kulakowski, E., Lindh, U., Mutvei, H., Nystrom, J., Seire, A., Slepukhina, T., Timm, H. et al., Biomonitoring of pollutants in a historic perspective, emphasis on mussel and snail shell methodology, *Ann. Chim.*, 85, 353–370, 1995.

Cassini, A., Tallandini, L., Favero, N., and Albergoni, V., Cadmium bioaccumulation studies in the freshwater molluscs *Anodonta cygnea* and *Unio elongatulus, Comp. Biochem. Physiol.*, 84C, 35–41, 1986.

Cawley, E. T., Sampling adequacy in population studies of freshwater mussels, In *Conservation and Management of Freshwater Mussels. Proceedings of a UMRCC Symposium*, 1992 Oct 12–14, St. Louis, MO, Cummings, K. S., Buchanan, A. C., and Koch, L. M., Eds., pp. 168–172, 1993.

Chandler, J. H., Toxicity of rotenone to selected aquatic invertebrates and frog larvae, *Prog. Fish-Cult.*, 44, 78–80, 1982.

Chen, L. Y., Heath, A. G., and Neves, R., An evaluation of air and water transport of freshwater mussels (Bivalvia: Unionidae), *Am. Malacol. Bull.*, 16, 147–154, 2001.

Cheney, M. A. and Criddle, R. S., Heavy metal effects on the metabolic activity of *Elliptio complanata*: a calorimetric method, *J. Environ. Qual.*, 25, 235–240, 1996.

Cherry, D. S., Farris, J. L., and Belanger, S. E., *Use of Ecological, Biochemical, and Toxicological Procedures for Identifying and Quantifying the Extent of Stressful Constituents in Power Plant Effluents*, Report prepared for American Electric Power Company, Virginia Polytech. Inst. State Univ., Biol. Dept., Univ. Center Environ. Stud., Blacksburg, VA, p. 339, 1988.

Cherry, D. S., Farris, J. L., and Neves, R. J., *Laboratory and Field Ecotoxicological Studies at the Clinch River Plant, Virginia*, Report prepared for American Electric Power Company, Virginia Polytech. Inst. State Univ., Biol. Dept., Univ. Center Environ. Stud., Blacksburg, VA, p. 228, 1991.

Cherry, D. S., Scheller, J. L., Cooper, N. L., and Bidwell, J. R., Potential effects of Asian clam (*Corbicula fluminea*) die-offs on native freshwater mussels (Unionidae) I: water-column ammonia levels and ammonia toxicity, *J. N. Am. Benthol. Soc.*, 24, 369–380, 2005.

Couillard, Y., Campbell, P. G. C., and Tessier, A., Response of metallothionein concentrations in a freshwater bivalve (*Anodonta grandis*) along an environmental cadmium gradient, *Limnol. Oceanogr.*, 38, 299–313, 1993.

Couillard, Y., Campbell, P. G. C., Tessier, A., Pellerin-Massicotte, J., and Auclair, J. C., Field transplantation of a freshwater bivalve, *Pyganodon grandis*, across a metal contamination gradient. I. Temporal changes in metallothionein and metal (Cd, Cu, and Zn) concentrations in soft tissues, *Can. J. Fish. Aquat. Sci.*, 52, 690–702, 1995a.

Couillard, Y., Campbell, P. G. C., Pellerin-Massicotte, J., and Auclair, J. C., Field transplantation of a freshwater bivalve, *Pyganodon grandis*, across a metal contamination gradient. II. Metallothionein response to Cd and Zn exposure, evidence for cytotoxicity, and links to effects at higher levels of biological organization, *Can. J. Fish. Aquat. Sci.*, 52, 703–715, 1995b.

Couillard, Y., Giguere, A., Campbell, P. G. C., Perceval, O., Pinel-Alloul, B., and Hare, L., Field evaluation of the use of metallothionein as a biomarker for metal contamination and toxic effects in the freshwater bivalve *Pyganodon grandis*: Subcellular metal partitioning [poster], *Society of Environmental Toxicology and Chemistry 20th Annual Meeting*, Nov 14–18, Philadelphia, PA, 1999.

Curry, C. A., The freshwater clam (*Elliptio complanata*), a practical tool for monitoring water quality, *Water Pollut. Res. Can.*, 13, 45–52, 1977.

Czarnezki, J. M., Use of the pocketbook mussel, *Lampsilis ventricosa*, for monitoring heavy metal pollution in an Ozark stream, *Bull. Environ. Contam. Toxicol.*, 38, 641–646, 1987.

Davies, W. D., Some effects of condenser discharge water on the growth and distribution of the mussel population of a large river, MSc Thesis, Ohio State Univ., Columbus, OH, 1963.

Day, K. E., Metcalfe, J. L., and Batchelor, S. P., Changes in intracellular free amino acids in tissues of the caged mussel, *Elliptio complanata*, exposed to contaminated environments, *Arch. Environ. Contam. Toxicol.*, 19, 816–827, 1990.

Dermott, R. M. and Lum, K. R., Metal concentrations in the annual shell layers of the bivalve *Elliptio complanata, Environ. Pollut.*, 12, 131–143, 1986.

Diamond, J. M. and Serveiss, V. B., Identifying sources of stress to native aquatic fauna using a watershed ecological risk assessment framework, *Environ. Sci. Technol.*, 35, 4711–4718, 2001.

Dimock, R. V. and Wright, A. H., Sensitivity of juvenile freshwater mussels to hypoxic, thermal and acid stress, *J. Elisha Mitchell Sci. Soc.*, 109, 183–192, 1993.

Dobrowolski, R. and Skowronska, M., Concentration and discrimination of selected trace metals by freshwater mollusks, *Bull. Environ. Contam. Toxicol.*, 69, 509–515, 2002.

Doherty, F. G., The asiatic clam, *Corbicula* spp., as a biological monitor in freshwater environments, *Environ. Monit. Assess.*, 15, 143–181, 1990.

Doran, W. J., Cope, W. G., Rada, R. G., and Sandheinrich, M. B., Acetylcholinesterase inhibition in the threeridge mussel (*Amblema plicata*) by chlorpyrifos: implications for biomonitoring, *Ecotoxicol. Environ. Saf.*, 49, 91–98, 2001.

Doyotte, A., Cossu, C., Jacquin, M.-C., Babut, M., and Vasseur, P., Antioxidant enzymes, glutathione and lipid peroxidation as relevant biomarkers of experimental or field exposure in the gills and the digestive gland of the freshwater bivalve *Unio tumidus*, *Aquat. Toxicol.*, 39, 93–110, 1997.

Dunn, H. L., Development of strategies for sampling freshwater mussels (Bivalvia: Unionidae), In *Freshwater Mollusk Symposia Proceedings, Part II*, 1999 Mar, Chattanooga, TN, Johnson, P. D., Butler, R.S., Eds., pp. 161–167, 2000.

Elder, J. F. and Collins, J. J., Freshwater molluscs as indicators of bioavailability and toxicity of metals in surface-water systems, *Rev. Environ. Contam. Toxicol.*, 122, 37–79, 1991.

Ellis, M. M., Erosion silt as a factor in aquatic environments, *Ecology*, 17, 29–42, 1936.

Englund, V. P. M. and Heino, M. P., Valve movement of the freshwater mussel *Anodonta anatina*: A reciprocal transplant experiment between lake and river, *Hydrobiologia*, 328, 49–56, 1996.

Farris, J. L., Belanger, S. E., Cherry, D. S., and Cairns, J., Cellulolytic activity as a novel approach to assess long-term zinc stress to *Corbicula*, *Water Res.*, 23, 1275–1283, 1989.

Farris, J. L., Cherry, D. S., and Neves, R. J., *Validation of Copper Concentrations in Laboratory Testing for Site-Specific Copper Criteria in the Clinch River*, Report prepared for American Electric Power Company, Virginia Polytech. Inst. State Univ., Biol. Dept., Univ. Center Environ. Haz. Mat. Stud., Blacksburg, VA, p. 54, 1991.

Fikes, M. H. and Tubb, R. A., Dieldrin uptake in the three-ridge naiad, *J. Wildl. Manage.*, 36, 802–809, 1972.

Fischer, J. C., Farris, J. L., Neves, R. J., and Cherry, D. S. An examination of growth rates of three freshwater mussel species in the Clinch River, Virginia to Tennessee [abstract], In *Conservation and Management of Freshwater Mussels. Proceedings of a UMRCC Symposium*, 1992 Oct 12–14, St. Louis, MO, Cummings, K. S., Buchanan, A. C., and Koch, L. M., Eds., p. 180, 1993.

Fleming, W. J., Augspurger, T. P., and Alderman, J. A., Freshwater mussel die-off attributed to anticholinesterase poisoning, *Environ. Toxicol. Chem.*, 14, 877–879, 1995.

Foster, R. B. and Bates, J. M., Use of freshwater mussels to monitor point source industrial discharges, *Environ. Sci. Technol.*, 12, 958–962, 1978.

Gaglione, P. and Ravera, O., Manganese-54 concentration in fall-out, water, and *Unio* mussels of Lake Maggiore, 1960–1963, *Nature*, 204, 1215–1216, 1964.

Gagne, F., Blaise, C., Lachance, B., Sunahara, G. I., and Sabik, H., Evidence of coprostanol estrogenicity to the freshwater mussel *Elliptio complanata*, *Environ. Pollut.*, 115, 97–106, 2001a.

Gagne, F., Blaise, C., Salazar, M., Salazar, S., and Hansen, P. D., Evaluation of estrogenic effects of municipal effluents to the freshwater mussel *Elliptio complanata*, *Comp. Biochem. Physiol.*, 128C, 213–225, 2001b.

Gagne, F., Marcogliese, D. J., Blaise, C., and Gendron, A. D., Occurrence of compounds estrogenic to freshwater mussels in surface waters in an urban area, *Environ. Toxicol.*, 16, 260–268, 2001c.

Gagne, F., Blaise, C., Aoyama, I., Luo, R., Gagnon, C., Couillard, Y., Campbell, P. G. C., and Salazar, M., Biomarker study of a municipal effluent dispersion plume in two species of freshwater mussels, *Environ. Toxicol.*, 17, 149–159, 2002.

Gardner, K. and Skulberg, O., Radionuclide accumulation by *Anodonta piscinalis* Nilsson (Lamellibranchiata) in a continuous flow system, *Hydrobiologia*, 26, 151–169, 1965.

Gardner, W. S., Miller, W. H., and Imlay, M. J., Free amino acids in mantle tissues of the bivalve *Amblema plicata*: possible relation to environmental stress, *Bull. Environ. Contam. Toxicol.*, 26, 157–162, 1981.

Giguere, A., Couillard, Y., Perceval, O., Campbell, P. G. C., Pellerin, J., Hare, L., and Pinel-Alloul, B., Field evaluation of the use of metallothionein as a biomarker for metal contamination and toxic effects in the freshwater bivalve *Pyganodon grandis*: responses at the organism level [poster], *Society of Environmental Toxicology and Chemistry 20th Annual Meeting*, Nov 14–18, Philadelphia, PA, 1999.

Goldberg, E. D., The mussel watch—a first step in global marine monitoring, *Mar. Pollut. Bull.*, 6, 111, 1975.

Goudreau, S. E., Neves, R. J., and Sheehan, R. J., Effects of wastewater treatment plant effluents on freshwater mollusks in the upper Clinch River, Virginia, USA, *Hydrobiologia*, 252, 211–230, 1993.

Granmo, A. and Varanka, I., Effects of heavy metals and a surfactant on the activity of fresh-water mussel larvae, *Vatten*, 35, 283–290, 1979.

Green, R. H., *Sampling Design and Statistical Methods for Environmental Biologists*, Wiley, New York, 1979.

Green, R. H., Singh, S. M., and Bailey, R. C., Bivalve mollusks as response systems for modeling spatial and temporal environmental patterns, *Sci. Total Environ.*, 46, 147–169, 1985.

Green, R. H., Bailey, R. C., Hinch, S. G., Metcalfe, J. L., and Young, V. H., Use of freshwater mussels (Bivalvia: Unionidae) to monitor the nearshore environment of lakes, *J. Great Lakes Res.*, 15, 635–644, 1989.

Greseth, S. L., Cope, W. G., Rada, R. G., Waller, D. L., and Bartsch, M. R., Biochemical composition of three species of unionid mussels after emersion, *J. Molluscan Stud.*, 69, 101–106, 2003.

Grier, N. M., Morphological features of certain mussel-shells found in Lake Erie, compared with those of the corresponding species found in the drainage of the upper Ohio, *Ann. Carnegie Mus.*, 13, 145–182, 1920.

Haag, W. R., Berg, D. J., Garton, D. W., and Farris, J. L., Reduced survival and fitness in native bivalves in response to fouling by the introduced zebra mussel (*Dreissena polymorpha*) in western Lake Erie, *Can. J. Fish. Aquat. Sci.*, 50, 13–19, 1993.

Hameed, P. S. and Raj, A. I. M., Effect of copper, cadmium and mercury on crystalline style of the freshwater mussel *Lamellidens marginalis* (Lamarck), *Indian J. Environ. Health*, 31, 131–136, 1989.

Hammer, U. T., Merkowsky, A. T., and Huang, P. M., Effects of oxygen concentrations on release of mercury from sediments and accumulation by *Ceratophyllum demersum* and *Anodonta grandis*, *Arch. Environ. Contam. Toxicol.*, 17, 257–262, 1988.

Hanna, M., Effect of sample storage conditions on body burden analysis of organic contaminants in unionid mussels, *Water Pollut. Res. J. Can.*, 27, 833–843, 1992.

Hansten, C., Heino, M., and Pynnonen, K., Viability of glochidia of *Anodonta anatina* (Unionidae) exposed to selected metals and chelating agents, *Aquat. Toxicol.*, 34, 1–12, 1996.

Harris, J. L., Rust, P., Chordas, III S. W., and Harp, G. L., Distribution and population structure of commercial mussel beds in the Black River, Arkansas [abstract], In *Conservation and Management of Freshwater Mussels. Proceedings of a UMRCC Symposium*, 1992 Oct 12–14, St. Louis, MO, Cummings, K. S., Buchanan, A. C., and Koch, L. M., Eds., p. 181, 1993.

Haukioja, E. and Hakala, T., Measuring growth from shell rings in populations of *Anodonta piscinalis* (Pelecypoda, Unionidae), *Ann. Zool. Fenn.*, 15, 60–65, 1978.

Havlik, M. E. and Marking, L. L., *Effects of Contaminants on Naiad Mollusks (Unionidae): a Review*, Resource Publ. 164, U.S. Department of the Interior, Fish and Wildlife Service, Washington, DC, p. 20, 1987.

Hayer, F. and Pihan, J. C., Accumulation of extractable organic halogens (EOX) by the freshwater mussel, *Anodonta cygnea* L., exposed to chlorine bleached pulp and paper mill effluents, *Chemosphere*, 32, 791–803, 1996.

Hayer, F., Wagner, P., and Pihan, J. C., Monitoring of extrac organic halogens (EOX) in chlorine bleached pulp and paper mill effluents using four species of transplanted aquatic mollusks, *Chemosphere*, 33, 2321–2334, 1996.

Hemelraad, J. and Herwig, H. J., Cadmium kinetics in freshwater clams. IV. Histochemical localization of cadmium in *Anodonta cygnea* and *Anodonta anatina*, exposed to cadmium chloride, *Arch. Environ. Contam. Toxicol.*, 17, 333–343, 1988.

Hemelraad, J., Holwerda, D. A., and Zandee, D. I., Cadmium kinetics in freshwater clams. III. The pattern of cadmium accumulation in *Anodonta cygnea*, *Arch. Environ. Contam. Toxicol.*, 15, 1–7, 1986a.

Hemelraad, J., Holwerda, D. A., Teerds, K. J., Herwig, H. J., and Zandee, D. I., Cadmium kinetics in freshwater clams. II. A comparative study of cadmium uptake and cellular distribution in the Unionidae *Anodonta cygnea*, *Anodonta anatina*, and *Unio pictorum*, *Arch. Environ. Contam. Toxicol.*, 15, 9–21, 1986b.

Hemelraad, J., Herwig, H. J., van Donselaar, E. G., Holwerda, D. A., and Zandee, D. I., Effects of cadmium in freshwater clams. II. Ultrastructural changes in the renal system of *Anodonta cygnea*, *Arch. Environ. Contam. Toxicol.*, 19, 691–698, 1990a.

Hemelraad, J., Holwerda, D. A., Herwig, H. J., and Zandee, D. I., Effects of cadmium in freshwater clams. III. Interaction with energy metabolism in *Anodonta cygnea*, *Arch. Environ. Contam. Toxicol.*, 19, 699–703, 1990b.

Hemelraad, J., Holwerda, D. A., Wijnne, H. J. A., and Zandee, D. I., Effects of cadmium in freshwater clams. I. Interaction with essential elements in *Anodonta cygnea*, *Arch. Environ. Contam. Toxicol.*, 19, 686–690, 1990c.

Henry, R. P. and Saintsing, D. G., Carbonic anhydrase activity and ion regulation in three species of osmoregulating bivalve mollusks, *Physiol. Zool.*, 56, 274–280, 1983.

Herve, S., Monitoring of organochlorine compounds in Finnish inland waters polluted by pulp and paper effluents using the mussel incubation method, *Water Sci. Technol.*, 24, 397–402, 1991.

Herve, S., Heinonen, P., Paukku, R., Knuutila, M., Koistinen, J., and Paasivirta, J., Mussel incubation method for monitoring organochlorine pollutants in watercourses. Four-year application in Finland, *Chemosphere*, 17, 1945–1961, 1988.

Herve, S., Paukku, R., Paasivirta, J., Heinonen, P., and Sodergren, A., Uptake of organochlorines from lake water by hexane-filled dialysis membranes and by mussels, *Chemosphere*, 22, 997–1001, 1991.

Hinch, S. G. and Green, R. H., The effects of source and destination on growth and metal uptake in freshwater clams reciprocally transplanted among south central Ontario lakes, *Can. J. Zool.*, 67, 855–863, 1989.

Hinch, S. G. and Stephenson, L. A., Size- and age-specific patterns of trace metal concentrations in freshwater clams from an acid-sensitive and a circumneutral lake, *Can. J. Zool.*, 65, 2436–2442, 1987.

Hinch, S. G., Bailey, R. C., and Green, R. H., Growth of *Lampsilis radiata* (Bivalvia: Unionidae) in sand and mud: a reciprocal transplant experiment, *Can. J. Fish. Aquat. Sci.*, 43, 548–552, 1986.

Holland-Bartels, L. E., Physical factors and their influence on the mussel fauna of a main channel border habitat of the upper Mississippi River, *J. N. Am. Benthol. Soc.*, 9, 327–335, 1990.

Holwerda, D. A., Cadmium kinetics in freshwater clams. V. Cadmium–copper interaction in metal accumulation by *Anodonta cygnea* and characterization of the metal binding protein, *Arch. Environ. Contam. Toxicol.*, 21, 432–437, 1991.

Holwerda, D. A. and Herwig, H. J., Accumulation and metabolic effects of di-n-butyltin dichloride in the freshwater clam, *Anodonta anatina*, *Bull. Environ. Contam. Toxicol.*, 36, 756–762, 1986.

Holwerda, D. A., Hemelraad, J., Veenhof, P. R., and Zandee, D. I., Cadmium accumulation and depuration in *Anodonta anatina* exposed to cadmium chloride or cadmium–EDTA complex, *Bull. Environ. Contam. Toxicol.*, 40, 373–380, 1988.

Hornbach, D. J. and Deneka, T., A comparison of a qualitative and a quantitative collection method for examining freshwater mussel assemblages, *J. N. Am. Benthol. Soc.*, 15, 587–596, 1996.

Hornbach, D. J., Deneka, T., Payne, B. S., and Miller, A. C., Shell morphometry and tissue condition of *Amblema plicata* (Say, 1817) from the upper Mississippi River, *J. Freshwat. Ecol.*, 11, 233–240, 1996.

Houslet, B. S. and Layzer, J. B., Difference in growth between two populations of Villosa taeniata in Horse Lick Creek, Kentucky, In *Conservation and Management of Freshwater Mussels II. Initiatives for the Future. UMRCC Symposium Proceedings*, 1995 Oct 16–18, St. Louis, MO, Cummings, K. S., Buchanan, A. C., Mayer, C. A., and Naimo, T. J., Eds., pp. 37–44, 1997.

Hudson, R. G. and Isom, B. G., Rearing juveniles of the freshwater mussels (Unionidae) in a laboratory setting, *Nautilus*, 98(4), 129–137, 1984.

Huebner, J. D. and Pynnonen, K. S., Viability of glochidia of two species of *Anodonta* exposed to low pH and selected metals, *Can. J. Zool.*, 70, 2348–2355, 1992.

Imlay, M. J., Effects of potassium on survival and distribution of freshwater mussels, *Malacologia*, 12, 97–111, 1973.

Imlay, M. J., Use of shells of freshwater mussels in monitoring heavy metals and environmental stresses: a review, *Malacol. Rev.*, 15, 1–14, 1982.

Ingram, W. M., Use and value of biological indicators of pollution: freshwater clams and snails, In *Biological Problems in Water Pollution*, Tarzwell, C. M., Ed., U.S. Department of Health Education and Welfare, Public Health Service, Washington, DC, pp. 94–135, 1957.

Isom, B. G. and Gooch, C., Rationale and sampling designs for freshwater mussels Unionidae in streams, large rivers, impoundments, and lakes, ASTM STP 894, In *Rationale for Sampling and Interpretation of Ecological Data in the Assessment of Freshwater Ecosystems*, Isom, B. G., Ed., American Soc. for Testing and Materials, Philadelphia, PA, pp. 46–59, 1986.

Jacobson, P. J., Sensitivity of early life stages of freshwater mussels (Bivalvia: Unionidae) to copper, MSc Thesis, Virginia Polytech. Inst. State Univ., Blacksburg, VA, 1990.

Jacobson, P. J., Farris, J. L., Cherry, D. S., and Neves, R. J., Juvenile freshwater mussel (Bivalvia: Unionidae) responses to acute toxicity testing with copper, *Environ. Toxicol. Chem.*, 12, 879–883, 1993.

Jacobson, P. J., Neves, R. J., Cherry, D. S., and Farris, J. L., Sensitivity of glochidial stages of freshwater mussels (Bivalvia: Unionidae) to copper, *Environ. Toxicol. Chem.*, 16, 2384–2392, 1997.

Jeffree, R. A. and Brown, P. L., A mechanistic and predictive model of metal accumulation by the tissue of the Australian freshwater mussel *Velesunio angasi*, *Sci. Total Environ.*, 125, 85–95, 1992.

Jeffree, R. A., Markich, S. J., and Brown, P. L., Comparative accumulation of alkaline-earth metals by two freshwater mussel species from the Nepean River, Australia: consistencies and a resolved paradox, *Aust. J. Mar. Freshwat. Res.*, 44, 609–634, 1993.

Jenner, H. A., Hemelraad, J., Marquenie, J. M., and Noppert, F., Cadmium kinetics in freshwater clams (Unionidae) under field and laboratory conditions, *Sci. Total Environ.*, 108, 205–214, 1991.

Johnson, I. C., Zam, S. G. and Keller, A. E., *Proposed Guide for Conducting Acute Toxicity Tests with the Early Life Stages of Freshwater Mussels*, Report submitted to U.S. Environmental Protection Agency. KBN Eng. Appl. Sci., Inc., Gainesville, FL, 1990.

Jones, W. G. and Walker, K. F., Accumulation of iron, manganese, zinc, and cadmium by the Australian freshwater mussel *Velesunio ambiguus* (Phillipi) and its potential as a biological monitor, *Aust. J. Mar. Freshwat. Res.*, 30, 741–751, 1979.

Kabeer Ahmad, I., Sethuraman, M., Begum, M. R., and Ramana Rao, K. V., Effect of malathion on ciliary activity of freshwater mussel, *Lamellidens marginalis* (Lamarck), *Comp. Physiol. Ecol.*, 4, 71–73, 1979.

Kauss, P. B. and Hamdy, Y. S., Biological monitoring of organochlorine contaminants in the St. Clair and Detroit Rivers using introduced clams, *Elliptio complanatus*, *J. Great Lakes Res.*, 11, 247–263, 1985.

Kauss, P. B. and Hamdy, Y. S., Polycyclic aromatic hydrocarbons in surficial sediments and caged mussels of the St. Marys River, 1985, *Hydrobiologia*, 219, 37–62, 1991.

Keller, A. E., Acute toxicity of several pesticides, organic compounds, and a wastewater effluent to the freshwater mussel, *Anodonta imbecilis, Ceriodaphnia dubia*, and *Pimephales promelas*, *Bull. Environ. Contam. Toxicol.*, 51, 696–702, 1993.

Keller, A. E. and Pepin, R., Sensitivities of unionid mussels to aquatic contaminants relative to standard test organisms [poster], *Society of Environmental Toxicology and Chemistry 20th Annual Meeting*, Nov 14–18, Philadelphia, PA, 1999.

Keller, A. E. and Ruessler, D. S., The toxicity of malathion to unionid mussels: Relationship to expected environmental concentrations, *Environ. Toxicol. Chem.*, 16, 1028–1033, 1997.

Keller, A. E. and Zam, S. G., Simplification of in vitro culture techniques for freshwater mussels, *Environ. Toxicol. Chem.*, 9, 1291–1296, 1990.

Keller, A. E. and Zam, S. G., The acute toxicity of selected metals to the freshwater mussel, *Anodonta imbecilis*, *Environ. Toxicol. Chem.*, 10, 539–546, 1991.

Klaine, S. J. and Williams, L. A., In situ bioassay of storm-water in the basins [abstract], In *Clinch-Powell Rivers Bi-State Conference: Program and Abstracts*, 1992 May 28–30; Cumberland Gap, TN, Hoadley, K. L., Ed., p. 22, 1992.

Klemm, D. J., Lewis, P. A., Fulk, F., and Lazorchak, J. M., *Macroinvertebrate Field and Laboratory Methods for Evaluating the Biological Integrity of Surface Waters*, EPA/600/4-90/030, U.S. Environmental Protection Agency, Cincinnati, OH, p. 256, 1990.

Klose, P. N. and Potera, G. T., *Biological Impact Studies of the Faulkner Ash Site*, Report submitted to Maryland Department of Natural Resources, PPSP MP-56, Power Plant Siting Program, Annapolis, p. 61, 1984.

Koenig, B. G. and Metcalfe, C. D., The distribution of PCB congeners in bivalves, *Elliptio complanata*, introduced into the Otonabee River, Peterborough, Ontario, *Chemosphere*, 21, 1441–1449, 1990.

Koppar, B. J., Kulkarni, R. S., and Venkatachari, S. A. T., Application of static bioassay procedure in determining the comparative relative toxicity of pesticide methyl parathion on two freshwater mussels, *J. Environ. Biol.*, 14, 183–193, 1993.

Kovalak, W. F., Dennis, S. D., and Bates, J. M., Sampling effort required to find rare species of freshwater mussels, ASTM STP 894, In *Rationale for Sampling and Interpretation of Ecological Data In the Assessment of Freshwater Ecosystems*, Isom, B. G., Ed., American Soc. for Testing and Materials, Philadelphia, PA, pp. 34–45, 1986.

Lagerspetz, K. Y. H., Anneli Korhonen, I., and Tuska, A. J., Heat shock response and thermal acclimation effects in the gills of *Anodonta cygnea*: Ciliary activity, stress proteins and membrane fluidity, *J. Thermal Biol.*, 20, 43–48, 1995.

Layzer, J. B. and Madison, L. M., Dry tissue weight as an indicator of mussel condition—a cautionary note, In *Conservation and Management of Freshwater Mussels II. Initiatives for the Future. UMRCC Symposium Proceedings*, 1995 Oct 16–18, St. Louis, MO, Cummings, K. S., Buchanan, A. C., Mayer, C. A., and Naimo, T. J., Eds., pp. 170–175, 1997.

Leard, R. L., Grantham, B. J., and Pessoney, G. F., Use of selected freshwater bivalves for monitoring organochlorine pesticide residues in major Mississippi stream systems, 1972–1973, *Pestic. Monit. J.*, 14, 47–52, 1980.

Lingard, S. M., Evans, R. D., and Bourgoin, B. P., Method for the estimation of organic-bound and crystal-bound metal concentration in bivalve shells, *Bull. Environ. Contam. Toxicol.*, 48, 179–184, 1992.

Lopes, J. L. C., Casanova, I. C., Garcia de Figueireido, M. C., Nather, F. C., and Avelar, W. E. P., *Anodontites trapesialis*: A biological monitor of organochlorine pesticides, *Arch. Environ. Contam. Toxicol.*, 23, 351–354, 1992.

Makela, P., Freshwater mussel monitoring: the duck mussel *Anodonta anatina* in water quality assessment, PhD Thesis, *Univ. Joensuu Publ. Sci.*, 34, 7–47, 1995.

Makela, P. and Oikari, A. O. J., Uptake and body distribution of chlorinated phenolics in the freshwater mussel, *Anodonta anatina* L, *Ecotoxicol. Environ. Saf.*, 20, 354–362, 1990.

Makela, T. P. and Oikari, A. O. J., The effects of low water pH on the ionic balance in freshwater mussel *Anodonta anatina* L, *Ann. Zool. Fenn.*, 29, 169–175, 1992.

Makela, T. P. and Oikari, A. O. J., Pentachlorophenol accumulation in the freshwater mussels *Anodonta anatina* and *Pseudanodonta complanata*, and some physiological consequences of laboratory maintenance, *Chemosphere*, 31, 3651–3662, 1995.

Makela, T. P., Petanen, T., Kukkonen, J., and Oikari, A. O. J., Accumulation and depuration of chlorinated phenolics in the freshwater mussel (*Anodonta anatina* L.), *Ecotoxicol. Environ. Saf.*, 22, 153–163, 1991.

Makela, T. P., Lindstrom-Seppa, P., and Oikari, A. O. J., Organochlorine residues and physiological condition of the freshwater mussel *Anodonta anatina* caged in River Pielinen, eastern Finland, receiving pulp mill effluent, *Aqua Fenn.*, 22, 49–58, 1992.

Malley, D. F., Huebner, J. D., and Donkersloot, K., Effects on ionic composition of blood and tissues of *Anodonta grandis grandis* (Bivalvia) of an addition of aluminum and acid to a lake, *Arch. Environ. Contam. Toxicol.*, 17, 479–491, 1988.

Malley, D. F., Chang, P. S. S., and Hesslein, R. H., Whole lake addition of cadmium-109: radiotracer accumulation in the mussel population in the first season, *Sci. Total Environ.*, 87/88, 397–417, 1989.

Malley, D. F., Stewart, A. R., and Hall, B. D., Uptake of methyl mercury by the floater mussel, *Pyganodon grandis* (Bivalvia, Unionidae), caged in a flooded wetland, *Environ. Toxicol. Chem.*, 15, 928–936, 1996.

Manly, R. and George, W. O., The occurrence of some heavy metals in populations of the freshwater mussel *Anodonta anatina* (L.) from the River Thames, *Environ. Pollut.*, 14, 139–154, 1977.

Marquenie, J. M., Bioavailability of micropollutants, *Environ. Technol. Lett.*, 6, 351–358, 1985.

Masnado, R. G., Geis, S. W., and Sonzogni, W. C., Comparative acute toxicity of a synthetic mine effluent to *Ceriodaphnia dubia*, larval fathead minnow and the freshwater mussel *Anodonta imbecilis*, *Environ. Toxicol. Chem.*, 14, 1913–1920, 1995.

Mathis, B. J. and Cummings, T. F., Selected metals in sediments, water, and biota in the Illinois River, *J. Water Pollut. Contr. Fed.*, 45, 1573–1583, 1973.

McCuaig, J. M. and Green, R. H., Unionid growth curves derived from annual rings: A baseline model for Long Point Bay, Lake Erie, *Can. J. Fish. Aquat. Sci.*, 40, 436–442, 1983.

McKinney, A. D. and Wade, D. C., Comparative response of *Ceriodaphnia dubia* and juvenile *Anodonta imbecillis* to pulp and paper mill effluents discharged to the Tennessee River and its tributaries, *Environ. Toxicol. Chem.*, 15, 514–517, 1996.

McLeese, D. W., Zitko, V., Metcalfe, C. D., and Sergeant, D. B., Lethality of aminocarb and the components of the aminocarb formulation to juvenile Atlantic salmon, marine invertebrates, and a freshwater clam, *Chemosphere*, 9, 79–82, 1980.

McMahon, R. F., Ecology of an invasive pest bivalve, *Corbicula*, In *The Mollusca, Vol. 6, Ecology*, Russell-Hunter, W. D., Ed., Academic Press, Orlando, FL, pp. 505–561, 1983.

Metcalfe, J. L. and Charlton, M. N., Freshwater mussels as biomonitors for organic industrial contaminants and pesticides in the St. Lawrence River, *Sci. Total Environ.*, 97/98, 595–615, 1990.

Metcalfe, J. L. and Hayton, A., Comparison of leeches and mussels as biomonitors for chlorophenol pollution, *J. Great Lakes Res.*, 15, 654–668, 1989.

Metcalfe-Smith, J. L., Influence of species and sex on metal residues in freshwater mussels (Family Unionidae) from the St. Lawrence River, with implications for biomonitoring programs, *Environ. Toxicol. Chem.*, 13, 1433–1443, 1994.

Metcalfe-Smith, J. L. and Green, R. H., Aging studies on three species of freshwater mussels from a metal-polluted watershed in Nova Scotia, Canada, *Can. J. Zool.*, 70, 1284–1291, 1992.

Metcalfe-Smith, J. L., Di Maio, J., Staton, S. K., and Mackie, G. L., Effect of sampling effort on the efficiency of the timed search method for sampling freshwater mussel communities, *J. N. Am. Benthol. Soc.*, 19, 725–732, 2000.

Milam, C. D. and Farris, J. L., Risk identification associated with iron-dominated mine discharges and their effect upon freshwater bivalves, *Environ. Toxicol. Chem.*, 17, 1611–1619, 1998.

Miller, E. J., Evaluation of Verdigris River, Kansas, freshwater mussel refuge. In *Conservation and Management of Freshwater Mussels. Proceedings of a UMRCC Symposium*, 1992 Oct 12–14, St. Louis, MO, Cummings, K. S., Buchanan, A. C., and Koch, L. M., Eds., pp. 56–60, 1993.

Miller, A. C. and Nelson, D. A., *An Instruction Report on Freshwater Mussels*, Instruction Report EL-83-2, U.S. Army Eng. Waterways Exp. Sta., Vicksburg, Environmental Laboratory, Vicksburg, MS, 1983.

Miller, A. C. and Payne, B. S., The need for quantitative sampling to characterize size demography and density of freshwater mussel communities, *Am. Malacol. Bull.*, 6, 49–54, 1988.

Miller, A. C. and Payne, B. S., Qualitative versus quantitative sampling to evaluate population and community characteristics at a large river mussel bed, *Am. Midl. Nat.*, 130, 133–145, 1993.

Miller, A. C., Whiting, R., and Wilcox, D. B., An evaluation of a skimmer dredge for collecting freshwater mussels, *J. Freshwat. Ecol.*, 5, 151–154, 1989.

Miller, A. C., Payne, B. S., Schafer, D. J., and Neill, L. T., Techniques for monitoring freshwater bivalve communities and populations in large rivers. In *Conservation and Management of Freshwater Mussels. Proceedings of a UMRCC Symposium*, 1992 Oct 12–14, St. Louis, MO, Cummings, K. S., Buchanan, A. C., and Koch, L. M., Eds., pp. 147–158, 1993.

Mix, M. C., Polycyclic aromatic hydrocarbons in the aquatic environment: occurrence and biological monitoring, In *Reviews in Environmental Toxicology*, Hodgson, E., Ed., Vol. 1, Elsevier, Amsterdam, The Netherlands, pp. 51–102, 1984.

Mix, M. C., Cancerous diseases in aquatic animals and their association with environmental pollutants: a critical literature review, *Mar. Environ. Res.*, 20(1/2), 1–141, 1986.

Moulton, C. A., Fleming, W. J., and Purnell, C. E., Effects of two cholinesterase-inhibiting pesticides on freshwater mussels, *Environ. Toxicol. Chem.*, 15, 131–137, 1996.

Muncaster, B. W., Hebert, P. D. N., and Lazar, R., Biological and physical factors affecting the body burden of organic contaminants in freshwater mussels, *Arch. Environ. Contam. Toxicol.*, 19, 25–34, 1990.

Muncaster, B. W., Innes, D. J., Hebert, P. D. N., and Haffner, G. D., Patterns of organic contaminant accumulation by freshwater mussels in the St. Clair River, Ontario, *J. Great Lakes Res.*, 15, 645–653, 1989.

Naimo, T. J., Heavy metals in the threeridge mussel *Amblema plicata plicata* (Say, 1817) in the upper Mississippi River [abstract]. In *Conservation and Management of Freshwater Mussels. Proceedings of a UMRCC Symposium*, 1992 Oct 12–14, St. Louis, MO, Cummings, K. S., Buchanan, A. C., and Koch, L. M., Eds., p. 183, 1993.

Naimo, T. J., A review of the effects of heavy metals on freshwater mussels, *Ecotoxicology*, 4, 341–362, 1995.

Naimo, T. J. and Monroe, E. M., Variation in glycogen concentrations within mantle and foot tissue in *Amblema plicata plicata*: implications for tissue biopsy sampling, *Am. Malacol. Bull.*, 15, 51–56, 1999.

Naimo, T. J., Atchison, G. J., and Holland-Bartels, L. E., Sublethal effects of cadmium on physiological responses in the pocketbook mussel, *Lampsilis ventricosa*, *Environ. Toxicol. Chem.*, 11, 1013–1021, 1992.

Naimo, T. J., Waller, D. L., and Holland-Bartels, L. E., Heavy metals in the threeridge mussel *Amblema plicata plicata* (Say, 1817) in the upper Mississippi River, *J. Freshwat. Ecol.*, 7, 209–217, 1992.

Naimo, T. J., Damschen, E. D., Rada, R. G., and Monroe, E. M., Nonlethal evaluation of the physiological health of unionid mussels: methods for biopsy and glycogen analysis, *J. N. Am. Benthol. Soc.*, 17, 121–128, 1998.

Negus, C. L., A quantitative study of growth and production of unionid mussels in the River Thames at Reading, *J. Anim. Ecol.*, 35, 513–532, 1966.

Nelson, D. J., Sampling for mussels, In *Report of Freshwater Mussel Workshop*, 1981, May 19–20, Miller, A. C., Ed., U.S. Army Engineering Waterways Exp. Station, Environmental Lab., Vicksburg, MS, pp. 41–60, 1982.

Neves, R. J. and Moyer, S. N., Evaluation of techniques for age determination of freshwater mussels (Unionidae), *Am. Malacol. Bull.*, 6, 179–188, 1988.

Newton, T., Allran, J., O'Donnell, J., Bartsch, M., and Richardson, B., Effects of ammonia on juvenile unionids in laboratory sediment toxicity tests [poster], *Society of Environmental Toxicology and Chemistry 23rd Annual Meeting*, Nov 16–20, Salt Lake City, UT, 2002.

Ortmann, A. E., The destruction of the freshwater fauna in western Pennsylvania, *Proc. Am. Philos. Soc.*, 48, 90–110, 1909.

Payne, B. S. and Miller, A. C., Quantitative sampling to characterize freshwater mussel communities and populations, In *Proceedings of the Workshop on Die-Offs of Freshwater Mussels in the United States*, 1986 Jun 23–25, Davenport, IA, Neves, R. J., Ed., pp. 55–65, 1987a.

Payne, B. S. and Miller, A. C., Effects of current velocity on the freshwater bivalve *Fusconaia ebena*, *Am. Malacol. Bull.*, 5, 177–179, 1987b.

Payne, B. S., Miller, A. C., and Lei, J., Palp to gill area ratio of bivalves: a sensitive indicator of elevated suspended solids, *Reg. Riv.: Res. Manage.*, 11, 193–200, 1995.

Pekkarinen, M., Reproduction and condition of unionid mussels in the Vantaa River, South Finland, *Arch. Hydrobiol.*, 127, 357–375, 1993.

Pellinen, J., Kukkonen, J., Herb, A., Makela, P., and Oikari, A., Bioaccumulation of pulp mill effluent-related compounds in aquatic animals, *Sci. Total Environ. Suppl.*, 1993, 499–510, 1993.

Perceval, O., Couillard, Y., Pinel-Alloul, B., Methot, G., Giguere, A., Campbell, P. G. C., and Hare, L., Cadmium accumulation and metallothionein synthesis in freshwater bivalves (*Pyganodon grandis*): relative influence of the metal exposure gradient versus limnological variability, *Environ. Pollut.*, 118, 5–17, 2002.

Phillips, D. J. H., *Quantitative Aquatic Biological Indicators. Their Use to Monitor Trace Metal and Organochlorine Pollution*, Applied Science Publ., London, UK, p. 455, 1980.

Price, R. E. and Knight, L. A., Mercury, cadmium, lead, and arsenic in sediments, plankton, and clams from Lake Washington and Sardis Reservoir, Mississippi, October 1975–May 1976, *Pesticides Monit. J.*, 11, 182–189, 1978.

Pugsley, C. W., Hebert, P. D. N., and McQuarrie, P. M., Distribution of contaminants in clams and sediments from the Huron-Erie corridor. II. Lead and cadmium, *J. Great Lakes Res.*, 14, 356–368, 1988.

Pugsley, C. W., Hebert, P. D. N., Wood, G. W., Brotea, G., and Obal, T. W., Distribution of contaminants in clams and sediments from the Huron-Erie corridor. I PCBs and octachlorostyrene, *J. Great Lakes Res.*, 11, 275–289, 1985.

Pynnonen, K., Physiological responses to severe acid stress in four species of freshwater clams (Unionidae), *Arch. Environ. Contam. Toxicol.*, 19, 471–478, 1990.

Pynnonen, K., Hemolymph gases, acid–base status, and electrolyte concentration in the freshwater clams *Anodonta anatina* and *Unio tumidus* during exposure to and recovery from acidic conditions, *Physiol. Zool.*, 67, 1544–1559, 1994.

Pynnonen, K., Effect of pH, hardness and maternal pre-exposure on the toxicity of Cd, Cu and Zn to the glochidial larvae of a freshwater clam *Anodonta cygnea*, *Water Res.*, 29, 247–254, 1995.

Pynnonen, K., Changes in acid–base status, gases and electrolytes in the hemolymph of freshwater unionids during continuous and intermittent exposure to acid water, *Ann. Zool. Fenn.*, 32, 355–363, 1995.

Pynnonen, K., Holwerda, D. A., and Zandee, D. I., Occurrence of calcium concretions in various tissues of freshwater mussels, and their capacity for cadmium sequestration, *Aquat. Toxicol.*, 10, 101–114, 1987.

Rajalekshmi, P. and Mohandas, A., Effect of heavy metals on tissue glycogen levels in the freshwater mussel, *Lamellidens corrianus* (Lea), *Sci. Total Environ. Suppl.*, 1993, 617–630, 1993.

Rajalekshmi, P. and Mohandas, A., Effect of heavy metals on the rate of oxygen uptake in the freshwater mussel, *Lamellidens corrianus* (Lea), *Pollut. Res.*, 14, 1–9, 1995.

Ravinder Reddy, T., Effect of mercuric chloride on protein metabolism of a freshwater mussel, *Parreysia rugosa* (Gmelin), *J. Ecotoxicol. Environ. Monit.*, 1, 230–233, 1991.

Renaud, C. B., Kaiser, K. L. E., Comba, M. E., and Metcalfe-Smith, J. L., Comparison between lamprey ammocoetes and bivalve molluscs as biomonitors of organochlorine contaminants, *Can. J. Fish. Aquat. Sci.*, 52, 276–282, 1995.

Renzoni, A. and Bacci, E., Bodily distribution, accumulation and excretion of mercury in a fresh-water mussel, *Bull. Environ. Contam. Toxicol.*, 15, 366–373, 1976.

Roper, D. S. and Hickey, C. W., Population structure, shell morphology, age and condition of the freshwater mussel *Hyridella menziesi* (Unionacea: Hyriidae) from seven lake and river sites in the Waikato River system, *Hydrobiologia*, 284, 205–217, 1994.

Rothwell, R. J., The effect of elevated temperature on the growth of freshwater clams (Mollusca: Pelecypoda), MSc Thesis, Edinboro State College, Edinboro, PA, 1979.

Russell, R. W. and Gobas, F. A. P. C., Calibration of the freshwater mussel, *Elliptio complanata*, for quantitative biomonitoring of hexachlorobenzene and octachlorostyrene in aquatic systems, *Bull. Environ. Contam. Toxicol.*, 43, 576–582, 1989.

Russell-Hunter, W. D. and Buckley, D. E., Actuarial bioenergetics of nonmarine molluscan productivity, In *The Mollusca, Vol. 6, Ecology*, Russell-Hunter, W. D., Ed., Academic Press, Orlando, FL, pp. 463–503, 1983.

Salanki, J., New avenues in the biological indication of environmental pollution, *Acta Biol. Hung.*, 40, 295–328, 1989.

Salanki, J. and Balogh, K. V., Physiological background for using freshwater mussels in monitoring copper and lead pollution, *Hydrobiologia*, 188/189, 445–454, 1989.

Salanki, J. and Hiripi, L., Effect of heavy metals on the serotonin and dopamine systems in the central nervous system of the freshwater mussel (*Anodonta cygnea* L.), *Comp. Biochem. Physiol.*, 95C, 301–305, 1990.

Salazar, M., Critical evaluation of bivalve molluscs as a biomonitoring tool for the mining industry in Canada, In *Technical Evaluation of Molluscs as a Biomonitoring Tool for the Canadian Mining Industry*, Report prepared by Freshwater Institute, Winnipeg, Manitoba, Canada, Stewart, R. and Malley D. F., Eds., pp. 163–248, 1997.

Salazar, M. H., Salazar, S. M., Gagne, F., Blaise, C., and Trottier, S., Developing a benthic cage for long-term, in-situ tests with freshwater and marine bivalves, In *29th Annual Aquatic Toxicity Workshop Proceedings*, Oct 20–23, Whistler, British Columbia, Canada, 2002, Canadian Technical Report of Fisheries and Aquatic Sciences 2438, pp. 34–42.

Salmon, A. and Green, R. H., Environmental determinants of unionid clam distribution in the Middle Thames River, Ontario, *Can. J. Zool.*, 61, 832–838, 1983.

Sauve, S., Brousseau, P., Pellerin, J., Morin, Y., Senecal, L., Goudreau, P., and Fournier, M., Phagocytic activity of marine and freshwater bivalves: in vitro exposure of hemocytes to metals (Ag, Cd, Hg and Zn), *Aquat. Toxicol.*, 58, 189–200, 2002.

Scarpato, R., Migliore, L., and Barale, R., The micronucleus assay in *Anodonta cygnea* for the detection of drinking water mutagenicity, *Mutat. Res.*, 245, 231–237, 1990.

Schmitt, C. J., Finger, S. E., May, T. W., and Kaiser, M. S., Bioavailability of lead and cadmium from mine tailings to the pocketbook mussel (*Lampsilis ventricosa*), In *Proceedings of the Workshop on Die-Offs of Freshwater Mussels in the United States*, 1986 Jun 23–25, Davenport, IA, Neves, R. J., Ed., pp. 115–142, 1987.

Scott, J. C., Population demographics of six freshwater mussel species (Bivalvia: Unionidae) in the upper Clinch River, Virginia and Tennessee, MSc Thesis, Virginia Polytech. Inst. State Univ., Blacksburg, VA, 1994.

Servos, M. R., Malley, D. F., Mackie, G. L., and LaZerte, B. D., Lack of bioaccumulation of metals by *Elliptio complanata* (Bivalvia) during acidic snowmelt in three south-central Ontario streams, *Bull. Environ. Contam. Toxicol.*, 38, 762–768, 1987.

Shimek, B., The effect of pollution on mollusks in Iowa, *Nautilus*, 48, 109–111, 1935.

Sickel, J. B., Chandler, C. C., and Pharris, G. L., Unionid distribution and abundance relative to habitat characteristics, In *Report of Freshwater Mussels Workshop*, Miller, A. C., Ed., U.S. Army Engineering Waterways Exp. Sta., Environ. Lab., Vicksburg, MS, 1982 Oct 26–27, St. Louis, MO, pp. 169–183, 1983.

Silverman, H., McNeil, J. W., and Dietz, T. H., Interaction of trace metals Zn, Cd, and Mn, with Ca concretions in the gills of freshwater unionid mussels, *Can. J. Zool.*, 65, 828–832, 1987.

Simkiss, K. and Mason, A. Z., Metal ions: metabolic and toxic effects, In *The Mollusca, Vol. 2, Environmental Biochemistry and Physiology*, Hochachka, P. W., Ed., Academic Press, New York, pp. 101–164, 1983.

Simmons, G. M. and Reed, J. R., Mussels as indicators of biological recovery zone, *J. Water Pollut. Contr. Fed.*, 45, 2480–2492, 1973.

Sivaramakrishna, B., Radhakrishnaiah, K., and Suresh, A., Assessment of mercury toxicity by the changes in oxygen consumption and ion levels in the freshwater snail, *Pila globosa*, and the mussel, *Lamellidens marginalis*, *Bull. Environ. Contam. Toxicol.*, 46, 913–920, 1991.

Smith, A. L., Green, R. H., and Lutz, A., Uptake of mercury by freshwater clams (Family Unionidae), *J. Fish. Res. Board. Can.*, 32, 1297–1303, 1975.

Smith, D. R., Villella, R. F., Lemarie, D. P. and von Oettingen, S., How much excavation is needed to monitor freshwater mussels?, In *Freshwater mollusk Symposia Proceedings, Part II*, 1999 Mar, Chattanooga TN, Johnson, P. D., Butler, R. S., Eds., pp. 203–218, 2000.

Smith, D. R., Villella, R. F., and Lemarie, D. P., Application of adaptive cluster sampling to low-density populations of freshwater mussels, *Environ. Ecol. Stat.*, 10, 7–15, 2003.

Smolders, R., Bervoets, L., Wepener, V., and Blust, R., A conceptual framework for using mussels as biomonitors in whole effluent toxicity, *Human Ecol. Risk Assess.*, 9, 741–760, 2003.

Sparks, R. E. and Blodgett, K. D., Effect of fleeting on mussels, *1986 Annual Progress Report*, Illinois Natural History Survey, Havana, IL, Project nr 3-373-R. p. 15, 1987.

Sparks, R. E., Blodgett, K. D., Durham, L., and Horner, R., *Determination Whether the Causal Agent for Mussel Die-Offs in the Mississippi River is of Chemical or Biological Origin. Final Report*, ILENR/RE-WR 90/09, Illinois Department of Energy and Natural Resources, Springfield, IL, p. 27, 1990.

Sterrett, S. S. and Saville, L. D., A technique to separate the annual layers of a naiad shell (Mollusca, Bivalvia, Unionacea) for analysis by neutron activation, *Bull. Am. Malacol. Union*, 1974, 55–57, 1974.

Stewart, A. R., Accumulation of Cd by a freshwater mussel (*Pyganodon grandis*) is reduced in the presence of Cu, Zn, Pb, and Ni, *Can. J. Fish. Aquat. Sci.*, 56, 467–478, 1999.

Stewart, R. and Malley, D. F., *Technical Evaluation of Molluscs as a Biomonitoring Tool for the Canadian Mining Industry*, Report prepared by Freshwater Institute, Winnipeg, Manitoba, Canada, p. 162, 1997.

Storey, A. W. and Edward, D. H. D., The freshwater mussel, *Westralunio carteri* Iredale, as a biological monitor of organochlorine pesticides, *Aust. J. Mar. Fresh. Res.*, 40, 587–593, 1989.

Strayer, D. L., Smith, D. R., *A Guide to Sampling Freshwater Mussel Populations*, American Fisheries Society, Monograph 8, 2003.

Streit, B. and Winter, S., Cadmium uptake and compartmental time characteristics in the freshwater mussel *Anodonta anatina*, *Chemosphere*, 26, 1479–1490, 1993.

Sumathi, V. P., Ciliary activity levels of a bivalve mollusc in polluted water of Tirumala Pushkarani Tank—A case of biopurification, *Environ. Ecol.*, 9, 405–407, 1991.

Tessier, A., Campbell, P. G. C., Auclair, J. C., and Bisson, M., Relationships between the partitioning of trace metals in sediments and their accumulation in the tissues of the freshwater mollusc *Elliptio complanata* in a mining area, *Can. J. Fish. Aquat. Sci.*, 41, 1463–1472, 1984.

Tessier, A., Couillard, Y., Campbell, P. G. C., and Auclair, J. C., Modeling Cd partitioning in oxic lake sediments and Cd concentrations in the freshwater bivalve *Anodonta grandis*, *Limnol. Oceanogr.*, 38, 1–17, 1993.

Tessier, A., Vaillancourt, G., and Pazdernik, L., Temperature effects on cadmium and mercury kinetics in freshwater molluscs under laboratory conditions, *Arch. Environ. Contam. Toxicol.*, 26, 179–184, 1994.

Tessier, L., Vaillancourt, G., and Pazdernik, L., Comparative study of the cadmium and mercury kinetics between the short-lived gastropod *Viviparus georgianus* (Lea) and pelecypod *Elliptio complanata* (Lightfoot), under laboratory conditions, *Environ. Pollut.*, 85, 271–282, 1994.

Tessier, L., Vaillancourt, G., and Pazdernik, L., Laboratory study of Cd and Hg uptake by two freshwater molluscs in relation to concentration, age and exposure time, *Water Air Soil Pollut.*, 86, 347–357, 1996.

Tevesz, M. J. S., Matisoff, G., Frank, S. A., and McCall, P. L., Interspecific differences in manganese levels in freshwater bivalves, *Water Air Soil Pollut.*, 47, 65–70, 1989.

Turick, C. E., Sexstone, A. J., and Bissonnette, G. K., Freshwater mussels as monitors of bacteriological water quality, *Water Air Soil Pollut.*, 40, 449–460, 1988.

Varanka, I., The effect of some pesticides on the rhythmic adductor muscle on freshwater mussel larvae, *Acta Biol.*, 28, 317–332, 1977.

Varanka, I., Effect of some pesticides on the rhythmic adductor muscle activity of the freshwater mussel larvae, *Acta Biol.*, 29, 43–55, 1978.

Varanka, I., Effect of some pesticides on the rhythmic adductor muscle activity of fresh-water mussel larvae, In *Human Impacts on Life in Fresh Waters*, Salanki, J. and Biro, P., Eds., Akademiai Kiado, Symp. Biol. Hung., Budapest, Hungary, pp. 177–196, 1979.

Vaughn, C. C., Taylor, C. M., and Eberhard, K. J., A comparison of the effectiveness of timed searches vs. quadrat sampling in mussel surveys. In *Conservation and Management of Freshwater Mussels II. Initiatives for the Future. UMRCC Symposium Proceedings*, 1995 Oct 16–18, St. Louis, MO, Cummings, K. S., Buchanan, A.C., Mayer, C.A., and Naimo, T. J., Eds., pp. 157–162, 1997.

Wade, D. C., Hudson, R. G., and McKinney, A. D., Comparative response of *Ceriodaphnia dubia* and juvenile *Anodonta imbecilis* to selected complex industrial whole effluents. In *Conservation and Management of Freshwater Mussels. Proceedings of a UMRCC Symposium*, 1992 Oct 12–14, St. Louis, MO Cummings, K. S., Buchanan, A. C., Koch, L. M., Eds., pp. 109–112, 1993.

Waller, D. L., Rach, J. J., Cope, W. G., and Luoma, J. A., A sampling method for conducting relocation studies with freshwater mussels, *J. Freshwat. Ecol.*, 8, 397–399, 1993a.

Waller, D. L., Rach, J. J., Cope, W. G., Marking, L. L., Fisher, S. W., and Dabrowska, H., Toxicity of candidate molluscicides to zebra mussels (*Dreissena polymorpha*) and selected nontarget organisms, *J. Great Lakes Res.*, 19, 695–702, 1993b.

Waller, D. L., Rach, J. J., Cope, W. G., and Miller, G. A., Effects of handling and aerial exposure on the survival of unionid mussels, *J. Freshwat. Ecol.*, 10, 199–207, 1995.

Waller, D. L., Rach, J. J., and Luoma, J. A., Acute toxicity and accumulation of the piscicide 3-trifluoromethyl-nitrophenol (TFM) in freshwater mussels (Bivalvia: Unionidae), *Ecotoxicology*, 7, 113–121, 1998.

Wang, D., Couillard, Y., Campbell, P. G. C., and Jolicoeur, P., Changes in subcellular metal Partitioning in the gills of freshwater bivalves (*Pyganodon grandis*) living along an environmental cadmium gradient, *Can. J. Fish. Aquat. Sci.*, 56, 774–784, 1999.

Wang, N., Greer, E., Whites, D., Ingersoll, C., Roberts, A., Dwyer, J., Augspurger, T., Kane, C., and Tibbott, C., Developing standardized guidance for conducting toxicity tests with glochidia of freshwater mussels [poster], *Freshwater Mollusk Conservation Society Symposium*, Mar 16–19, Durham NC, 2003.

Warren, L. W., Klaine, S. J., and Finley, M. T., Development of a field bioassay with juvenile mussels, *J. N. Am. Benthol. Soc.*, 14, 341–346, 1995.

Watters, G. T., Unionids, fishes, and the species-area curve, *J. Biogeogr.*, 19, 481–490, 1992.

Watters, G.T., Sampling freshwater mussel populations: the bias of muskrat middens, *Walkerana*, 7, 63–69, 1993–1994.

Weinstein, J. E., Characterization of the acute toxicity of photoactivated fluoranthene to glochidia of the freshwater mussel, *Utterbackia imbecillis, Environ. Toxicol. Chem.*, 20, 412–419, 2001.

Widdows, J. and Donkin, P., Mussels and environmental contaminants: bioaccumulation and physiological aspects, In *The Mussel Mytilus: Ecology, Physiology, Genetics and Culture*, Gosling, E., Ed., Elsevier Science Publishers, Amsterdam, pp. 383–424, 1992.

Wilcox, D. B., Anderson, D. D., and Miller, A. C., Survey procedures and decision criteria for estimating the likelihood that *Lampsilis higginsi* is present in areas within the Upper Mississippi River System. In *Conservation and Management of Freshwater Mussels. Proceedings of a UMRCC Symposium*, 1992 Oct 12–14, St. Louis, MO, Cummings, K. S., Buchanan, A. C., Koch, L. M., Eds., pp. 163–167, 1993.

Winter, S., Cadmium uptake kinetics by freshwater mollusc soft body under hard and soft water conditions, *Chemosphere*, 32, 1937–1948, 1996.

Wurtz, C. B., Fresh-water mollusks and stream pollution, *Nautilus*, 69, 96–100, 1956.

Yokley, P., The effect of gravel dredging on mussel production, *Bull. Am. Malacol. Union*, 3, 1976.

3 A Brief Look at Freshwater Mussel (Unionacea) Biology

G. Thomas Watters

A BRIEF HISTORY

As with all aspects of natural history, the early study of freshwater mussels passed through some interesting times. After the Renaissance, a renewed interest in the sciences in general made the study of even God's lowliest creatures an acceptable avocation. Arm-chair naturalists, often relying on information unchanged from Aristotle, reported on the creation of pearls from dew swallowed by swimming mussels (Boetius in Rennie 1829), the ability of molluscs to voluntarily leave their shell (Wood 1815), and the infection of mussels by mange and gangrene (Poupart 1706). But the "enlightened" study of freshwater mussel biology begins with the Dutch haberdasher, Leeuwenhoek, who turned the fledgling hobby of microscopy towards mussels. He removed eggs and glochidia from the marsupia of an *Anodonta*, describing them in 1695 and illustrating them in 1697. He clearly believed that glochidia were larval mussels, referring to them as "oysters not yet born." But a century later Rathke (1797) stated that glochidia were parasites infesting the gills of mussels, despite Leeuwenhoek's claims to the contrary (see Heard 1999b). Rathke named the presumed parasites *Glochidium parasiticum*, from which we derive the name for these larvae. A debate ensued over their true nature. To resolve the matter, the Academie de Sciences Naturelles of Paris formed a committee to investigate the matter. In 1828, the committee reported that glochidia were indeed larvae rather than parasites, although they arrived at this conclusion in a round-about manner (Blainville 1828). In 1832, Carus carefully followed the development of unionid eggs and finally, conclusively demonstrated that glochidia were larval mussels.

The study of mussels had begun to mature. In the spirit of the age, scientists began to study mussels for mussels' sake. Prévost (1826) in Europe and Kirkland (1834) in the United States experimentally determined that most mussels had separate male and female sexes. Louis Agassiz turned his considerable talent and ego towards mussels, observing annular growth rings on shells, noting (but without realizing the importance of) the correlation between fish and mussel distributions, and bemoaning the fact that the science was conducted by "amateurs" (Agassiz 1862a, 1862b). One such "amateur" was de Quatrefages (1836), who carefully documented the existence of internal organs in glochidia (heart, stomach, intestines)—organs that did not exist. De Quatrefages was eventually debunked by Schmidt (1856).

At this time, the fish-mussel connection was still unsuspected. In 1862, the British clergyman, Houghton, reported glochidia attached to fish and artificially infested fish with the parasites (Heard 1999a). This was the first hint that glochidia might be parasites on fish. In 1866, Leydig noted (as a footnote to the dissertation of Noll, his student) that glochidia were found attached to fish, apparently unaware of Houghton's work. The same year another worker, Forel (1879), confirmed this observation. Again unaware of Houghton's previous experiment, Braun and Schierholz in the winter of 1877–1878 independently attempted to artificially infest fish with glochidia, but

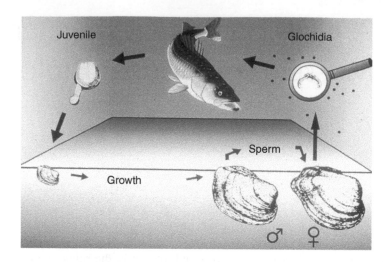

FIGURE 3.1 A typical freshwater mussel life cycle.

Schierholz's experiment failed (Schierholz 1889). Braun succeeded in infesting fish with *Anodonta* (1878a, 1878b), thus confirming Houghton's results. Glochidia had come full-circle: from larvae to parasites to parasitic larvae (Figure 3.1).

ECOLOGY

Unlike most infaunal marine bivalves, North American freshwater mussels lack true siphons or tubes for water intake and release. Because of this, most species are confined to burrowing only to the posterior edge of the shell. This is important because, for most freshwater mussels, burial depth does not become a buffer from chemicals, temperature extremes, or predation. However, a few species, such as *Pleurobema clava*, may spend much of their life buried several centimeters beneath the surface, relying on water to percolate between the substrate particles for food and oxygen. In temperate regions, mussels may burrow deeper into the substrate during winter as well.

The distribution of any given mussel species depends on three factors (Brim Box, Dorazio, and Liddell 2002), which correspond to different, increasingly finer levels of organization: (1) the overlying distribution of hosts; (2) the distribution of mussels within a river reach; and (3) the distribution of mussels on a microhabitat scale. Obviously, a mussel may not exist without its host but the converse is not true. Although there is a strong correlation between fish and mussel species richness for a drainage (Watters 1992), most mussels occupy only a portion of the overall range of their presumed hosts. Apparently, there is more to the story than just the distribution of the host. At the second level of organization, the within-stream distribution, the host's distribution probably is still important. The so-called, "big river" mussel species occur there because of the requirements of their hosts; there is nothing to prevent these mussels from living in small creeks beyond host availability. It is at the third level, microhabitats, that mussel distribution becomes confusing. Intuitively, we would suppose that microhabitats would be the eventual determinant of where mussels live, but time and again no clear-cut cause and effect is evident. For example, Brim Box, Dorazio, and Liddell (2002) found the distribution of only one of five species to be related to substrate composition. Strayer and Ralley (1993) found no strong relationship between the distribution of six mussel species and microhabitat variables except for water current speed and variability. Other studies also point out the weak association between mussel distribution and microhabitats (Tevesz and McCall 1979; Strayer 1981; Holland-Bartels 1990). It appears that stream hydro- and geomorphology variables, such as sheer stress and flow refugia, are the most important factors in mussel distribution on a fine scale (Strayer 1993; Vaughn 1997; Strayer 1999).

Mussels feed by filtering out material from the water with their extensive gills, which are much larger than is needed for respiration. The gills have a fine mesh-size indicative of a preference for minute food items. The natural components of mussel food have not been completely identified. Whereas Allen (1914, 1921), Churchill and Lewis (1924), and Fikes (1972) found the gut to contain mostly diatoms and other algae, Imlay and Paige (1972) believed that mussels fed on bacteria and protozoans. Bisbee (1984) found different proportions of algal species in the guts of two mussel species, suggesting preferences between species. The comprehensive study of Nichols and Garling (1998) demonstrated that mussels were omnivores, feeding on detritus and zooplankton, as well as algae and bacteria.

We now know that adults and juveniles do not feed upon the same material. Newly-metamorphosed juveniles do not filterfeed with their gills (which are mere buds at this stage) but rather feed on interstitial nutrients using cilia on their foot and mantle. Eventually, functional gills are formed, and there is a change to a filterfeeding mode (Tankersley, Hart, and Weiber 1997). Again, the exact food items of juveniles are the subject of debate. Yeager, Cherry, and Neves (1993) believed food for juveniles consisted of interstitial bacteria, whereas an algal/silt mix was suggested by Humphrey and Simpson (1985) and Gatenby, Neves, and Parker (1993). Small amounts of silt have been found to enhance survivorship in cultured mussels, both adults and juveniles (Hudson and Isom 1984; Humphrey 1987; Hove and Neves 1991), probably by introducing bacteria, zooplankton, and micronutrients. Juveniles grow best and have a higher survivorship when fed a diet high in lipids (Gatenby, Neves, and Parker 1997).

REPRODUCTION

Freshwater mussels typically are dioecious, but hermaphrodites have been found in many species (Poupart 1706; Fischerstrom 1761; van der Schalie 1966, 1970; Heard 1979). Some species are believed to be wholly hermaphroditic, such as *Toxolasma parvus, Lasmigona compressa,* and *Utterbackia imbecillis* (Ortmann 1912; Utterback 1916). The relative proportion of hermaphrodites among otherwise dioecious species may increase under low population densities as a means of augmenting declining population numbers (Kat 1983; Bauer 1987b). As hermaphrodites, these species may be best suited as colonizing forms capable of establishing themselves under low initial population densities or in headwater or otherwise isolated situations.

Spawning, the release of gametes, occurs at different times and frequencies depending on the species and latitude. Sperm may be released as "sperm balls" or discs—apparently hollow structures composed of sperm, flagellae facing outwards, which propel themselves through the water to a limited extent. These spheres disassociate to fertilize eggs (Barnhart and Roberts 1997). Spawning takes place in the spring for most amblemines and in the summer for most anodontines and lampsilines. These spawning patterns are described below in more detail. Mussels may migrate horizontally to congregate during spawning, presumably to increase spawning success by increasing the density of individuals (Amyot and Downing 1998). Spawning also is associated with vertical migration in the substrate. In a study of the vertical movement of eight mussel species composed of both amblemine and lampsiline taxa, Watters, O'Dee, and Chordas (2001) found that all studied taxa migrated to the surface in April–May, presumably to spawn. This migration was remarkably synchronized across the eight species and apparently was triggered by spring water temperatures. Although most North American mussels have a single spawning season per year, there is evidence that some species have multiple broods (Howard 1915; Ortmann 1919; Gordon and Smith 1990; Howells 2000).

Eggs are fertilized in the suprabranchial chambers of the gills, and developing embryos are moved to the marsupial regions of the gill where they are "brooded" until released. This marsupial region physiologically isolates the developing larvae (Kays, Silverman, and Dietz 1990) and acts as a source of maternal calcium for the construction of the glochidial shell (Silverman, Steffens, and Dietz 1985). During the gravid period, this region of the gill may not function as a site of respiration (Richard, Dietz, and Silverman 1991), or it functions in a much limited capacity (Allen 1921; Tankersley and Dimock 1992). The marsupial region may remain nonrespiratory even during the nonbreeding season (Richard, Dietz, and Silverman 1991).

Glochidia are a type of veliger larvae; freshwater mussels lack the trochophore precursor stage found in many other molluscs. Although glochidia have been recovered from Quaternary sediment cores (Brodniewicz 1968, 1969), we have no way of knowing when or how larval mussels became parasitic during their 250+ million year evolution. Watters (2001) outlined the early phylogenetic history of North American mussels, suggesting that glochidia may have been parasitic at least as early as the Cretaceous and that the explosion in mussel diversity seen during the late Mesozoic may be correlated with the rise of teleost fishes, their hosts. The earliest mussels may have lacked planktonic larvae, having instead brooded their larvae. Other freshwater bivalves, such as sphaeriids, also brood their larvae. When expelled, these larvae may have fortuitously attached to passing fish. Over time, the adaptive significance of the increased dispersal powers of these glochidia led to a phoretic relationship. As glochidia developed more efficient means of remaining attached to fish (e.g., hooks, etc.), a true parasitic symbiosis arose. This may have begun when glochidial attachment caused physical harm to the fish, eliciting a wound reaction and triggering immune system responses. These early mussels probably brooded relatively few glochidia. The earliest adaptations to their new parasitic life involved increased fecundity (as with most parasites) and the development of efficient glochidial delivery systems, such as conglutinates. The earliest parasitic mussels were conglutinate-producing forms related to recent amblemines. They produced relatively few glochidia.

North American glochidia typically occur in three morphological forms. The triangular, hooked glochidia of most anodontines and some amblemines is the most prevalent. Most lampsilines have non-hooked, D-shaped glochidia. The third morphological type, the "pick-ax" glochidium, occurs in *Potamilus*. These glochidia may be divided into two functional groups. The first, the gill parasites, comprising most of the lampsilines, lack macroscopic attachment devices. The inner rims of the glochidial shells are set with numerous fine points that enable the glochidia to grip the gill filaments. The second type, glochidia specializing in the attachment to the outside of their host, the fins, barbels, or skin have better-developed attachment structures—simple or toothed hooks, particularly developed in anodontines and *Potamilus*. These glochidia apparently require greater "gripability" due to the fact that they are more prone to being dislodged than are gill parasites, which are protected to some degree by the opercles of the host. Mechanically, glochidia fall into two groups based upon their shell-musculature design (Hoggarth and Gaunt 1988). Some are adapted for maximum glochidial shell sweep, increasing the potential area of attachment. Others emphasize closure force to minimize dislodgement. Some glochidia apparently can detect the presence of hosts through substances in host's mucus or blood (Shadoan and Dimock 2000). Henley and Neves (2001) identified fibrinogen as one such substance. Once released, glochidia may persist for several weeks and be carried by currents for considerable distances (Fisher and Dimock 2000; Zimmerman and Neves 2002), but they are heavily preyed upon by planarians, hydras, aquatic insects, etc.

Glochidia tend to be overdispersed within a host population. That is, the parasite burden is carried by a relatively small portion of the available host population. For example, of 3441 fish examined in a study of Lake Otsego, only six carried glochidia (Weir 1977). Only 14 percent of the 4800 fish examined in a study of the North Fork Holston River had glochidia (Neves and Widlak 1988). This is due to the mechanisms that mussels utilize to parasitize hosts: lures and conglutinates that tend to heavily parasitize only the few fish that actually attack these structures whereas the majority of the host population never comes into contact with them.

Once encapsulated, glochidia are true parasites on the host, feeding on host tissue enclosed between the glochidial shells and on their own larval adductor muscle (Arey 1924b, 1932a; Blystad 1924; Fisher and Dimock 2002a). The glochidial shell is perforated (Rand and Wiles 1982; Jeong, Min, and Chung 1993; Kwon et al. 1993), but whether nutrients move through these pores is uncertain. Rather, nutrient uptake is probably through the microvillae of the mantle (Wächtler, Dreher-Mansur, and Richter 2001) or the mushroom body (Fisher and Dimock 2002b). Most glochidia do not increase in size while encapsulated, margaritiferids being a notable exception.

Glochidia apparently do little harm to their hosts. When parasitized in a laboratory setting, fish often are lethargic for several hours to a day post infestation but return to normal within several days with no obvious side effects. Reports of host mortality from overinfestation are rare and are usually based on artificial situations in which enormous numbers of mussels come into close, constant contact with captive fish. These occurrences tend to involve fingerling or young-of-year fish kept in hatcheries where mussels have colonized the ponds (Moles 1983). Even so, infestations as high as 736 glochidia per fish have been reported from hatchery situations with no adverse effects (*Margaritifera margaritifera* on Atlantic salmon) (Bruno, McVicar, and Waddell 1988).

Potential hosts may possess one of two types of immunity to attached glochidia. Natural immunity occurs in unsuitable hosts, which have tissue responses against the glochidia (Howard 1914; Bauer and Vogel 1987). Acquired immunity occurs when a suitable host has been previously parasitized and has built up a temporary immunity. The number of exposures needed to achieve acquired immunity depends on the degree of prior infestations and the duration between them (Lefevre and Curtis 1910; Surber 1913; Reuling 1919; Arey 1924a; Bauer 1987a). For example, for largemouth bass exposed to *Lampsilis cardium*, three to four exposures over 30-day intervals are required to elicit complete immunity (Watters 1996). Acquired immunity to one unionid species was thought to give the host immunity to others (Reuling 1919), but this has not been substantiated. Indeed, largemouth bass possessing a complete immunity to *L. cardium* were successfully infected by glochidia of *U. imbecillis* (Watters 1996) and even its congener *Lampsilis fasciola* (Watters and O'Dee 1997). Although acquired immunity may be demonstrated in the laboratory, acquired immunity in the wildcaught fishes has been observed only once, and its overall prevalence in wild fishes is unknown (Watters and O'Dee 1996). In both natural and acquired immunity, encysted glochidia are killed by the host and either sloughed off or absorbed (Arey 1932b; Fustish and Millemann 1978; Zale and Neves 1982; Waller and Mitchell 1989). Acquired immunity apparently may be lost if no subsequent reinfestation occurs within a certain time period, and the fish may become susceptible to parasitization again (O'Dee 2000). However, the amount of time needed to lose acquired immunity is not precisely known.

Metamorphosis from glochidium to juvenile mussel takes place within the capsule in two stages (Fisher and Dimock 2002b, for *U. imbecillis*). First, during the first four days, the larval adductor muscle is digested by a region of the glochidia called the mushroom body. Second, during the last four days, the juvenile anatomical structures appear. The triggers for metamorphosis and the duration of the parasitic phase vary with mussel species and host species. Metamorphosis on different host species infested with the same mussel at the same time may be delayed by weeks depending on the host. Metamorphosis is triggered and regulated by water temperature. Heinricher and Layzer (1999) and Watters and O'Dee (1999) demonstrated that a temperature threshold existed below which metamorphosis was significantly delayed, perhaps indefinitely. Glochidia remained encapsulated until the threshold was surpassed, at which time metamorphosis took place. Conversely, the duration of metamorphosis decreases with increasing temperature (Barnhart and Roberts 1997) until an upper threshold is reached. At this point, glochidia may break free of the capsule, fail to metamorphose, and die (Dudgeon and Morton 1984). Most North American mussels have a parasitic duration of 2–6 weeks, but European species may persist for ten months (Wächtler, Dreher-Mansur, and Richter 2001).

Once free of the host, the newly metamorphosed juvenile assumes a life style quite unlike those of adults. Realization of this fact has been slow in coming. These juveniles may burrow to several centimeters beneath the surface where they rely on water percolating between the interstices of substrate particles for food and oxygen. New juveniles feed by ciliary currents on their foot and mantle—gills are present at this stage only as buds and have no filtering abilities (Yeager, Cherry, and Neves 1994). They are feeding on detritus and other interstitial material. Thus, juveniles and adults probably are utilizing different food types and living in different micro-environments and are susceptible to different changes in habitat and water chemistry. Efforts to protect the habitat of adult mussels may therefore be inadequate to protect juveniles.

REPRODUCTIVE STRATEGIES

Freshwater mussel reproductive strategies represent two mutually exclusive, adaptive choices, each choice favoring a certain aspect of the mussel life cycle at the expense of other aspects. These choices have far-reaching behavioral and phylogenetic implications. Both choices have evolved to increase the efficiency of completing the life cycle, either by specializing on a small subset of potential hosts or by generalizing on a wide variety.

Parasites in general compensate for their improbable life histories by increasing fecundity—increasing the chances of a larvae surviving to reproduce by making so many offspring that even the most complex life cycle will be completed simply by the sheer number of attempts. Freshwater mussels, once they entered into their parasitic symbiosis with fish, also increased their fecundity as a hedge against their new life cycle. It has been estimated that as few as ten in one million larvae successfully attach to a host (Bauer 2001b). Such an increase in offspring numbers carries heavy physiological burdens: calcium reserves for glochidial shells, maintenance of the surrounding marsupial medium, loss of respiratory function in the marsupial gills, increase in overall size dictated by the needs of the marsupium, etc. Any means of reducing this burden would be positively adaptive. Because the offspring number is driven by the success of completing the life cycle, perhaps the most obvious way to reduce the offspring number is to increase the efficiency with which a glochidium successfully "finds" a host. If more glochidia were more efficient parasites, then fewer offspring would be needed to attain the same level of life cycle success.

Historically mussels were considered broadcasters—simply releasing vast quantities of larvae into the water, chancing that the right host would be in the right place at the right time. We now know that true broadcasters are quite rare. Nearly all mussels investigated have evolved means of luring the correct host to their glochidia. Two mechanisms are apparent: lures and "conglutinates" (Haag, Butler, and Hartfield 1995).

Lures are the hallmark of lampsiline mussels and consist of highly specialized portions of the female mussel's mantle that mimic, in one way or another, host–prey items (Haag, Warren, and Shillingsford 1999; Haag and Warren 1999, 2000). Lures may be quite mimic-model specific. For instance, many *Lampsilis* species have mantle flaps resembling small "minnows," complete with eye spots, fins, and swimming motions. Others are less model specific, consisting of synchronous movements of papillae (some *Villosa*), writhing caruncles (*Toxolasma*), or other displays that are not recognizable (to humans at least) as specific mimics. Lures function by drawing the host to the female mussel, fooling the would-be predator into striking at a food item. Upon striking, the mussel releases a cloud of glochidia, parasitizing the host. In at least some species, the female closes her shells upon the extended marsupia, causing them to rupture.

"Conglutinates" is a collective term for structures fabricated by the female mussel, containing glochidia, that mimic host prey items. These are eaten by hosts that would normally feed on the conglutinate model. As with lures, conglutinates may be generalized or mimic specific. *Fusconaia* and *Pleurobema* release packets of glochidia that resemble "worms." *Strophitus* releases maggot-like conglutinates. The most complex conglutinates yet seen are fashioned by *Ptychobranchus*, where the conglutinates bear striking resemblances to fish eggs, fish fry (Barnhart and Roberts 1997; Watters 1999), insect larvae (Hartfield and Hartfield 1996), or simulid pupal cases. Several species assemble individual conglutinates into a single "superconglutinate" that may be played out at the end of mucus strands by the female (Haag, Butler, and Hartfield 1995; Hartfield and Butler 1997; O'Brien and Brim Box 1999). Hosts are infested when they attempt to ingest these structures. During ingestion, the host often flushes glochidia over its gills, where they attach. Although most of the glochidia are lost or ingested, enough successfully attach to make this strategy worthwhile.

Conglutinates typically are composed of glochidia embedded in a mucus matrix (most anodontines) or glochidia attached to each other by the adhesive properties of their egg membranes (most amblemines). In some groups, not all eggs are fertilized; these become structural cells, giving the conglutinate, in which the glochidia are embedded or attached, form and color (Barnhart 1997). How these eggs are "turned off" is of great interest. In some groups (*Strophitus*, for example) the glochidia are tethered to the conglutinate by a glochidial thread.

Because lures and conglutinates are more efficient at reaching a host, glochidial numbers may be much fewer than needed for a comparable broadcaster. Even though there is a physiological cost to producing conglutinates or lures, it is compensated for by the decrease in numbers of glochidia required to parasitize a host by broadcasting. Some mussels make less than 100 conglutinates, each containing less than 20 glochidia; some presumed broadcasters may produce several million glochidia. Furthermore, the glochidial stage is thought to be the life stage having the greatest mortality; this would be especially important for broadcasting species (Jansen, Bauer, and Zahner-Meike 2001).

The development of lures and conglutinates has enormous consequences for the biology and evolution of freshwater mussels, setting them on apparently irreversible and diverging courses (Bauer 2001a). These developments have conservation consequences as well. These adaptive choices are host specialization and host generalization. They are driven by the very adaptations just discussed: lures, conglutinates, and broadcasting.

A mantle lure the size, color, and shape of a small minnow represents a large suite of "can" and "cannot" consequences for the mussel—so does a conglutinate fashioned after a simulid larva. These devices play to a small, select audience. Minnow lures attract large, predatory fish, not darters, sculpins, sticklebacks, or lampreys. If one artificially parasitizes such a mussel with these unlikely attackers, more often than not, the attackers do not act as "good" hosts. This is because the mussel and its true hosts have already entered into the host-parasite Arms Race. In a teleological sense, the mussel is continuously designing a better mouse trap, a better and more efficient means of luring the correct host to its glochidia and ensuring that its larvae can withstand the host's biological defenses. The host on the other hand is continuously tweaking its immune system to ward off the parasite. The race is on—but to the exclusion of other potential hosts. By customizing the lure or conglutinate and its glochidia to a specific host, the mussel loses the ability to use other hosts. The lure so effective for bass has no charm for darters. The conglutinate shaped like a tiny fish egg holds no magic sway over walleye. Furthermore, the fine tuning that allows the glochidia to survive the immunological onslaught from its "preferred" host no longer works against other hosts. By specializing in particular hosts, the mussel is set on an irreversible path. This specialization may differ even between closely related mussel taxa (Riusech and Barnhart 2000). On the other hand, some mussels clearly seem to be host generalists; they are able to successfully parasitize a wide range of hosts. Many anodontines fall into this category.

There are obvious trade-offs between host specialists and host generalists. Specialists are more efficient at contacting their hosts and so require fewer glochidia; however, they may only use a small subset of the once available host pool. Therefore, they are susceptible to extirpation should their hosts disappear from the immediate vicinity. Generalists usually are not efficient at contacting hosts (some are broadcasters) and require large numbers of glochidia, but may successfully parasitize those hosts they do manage to contact. They are less susceptible to loss of any specific host species. However, simulations show that specialists are less affected by changes in overall host abundance than are generalists; this is due to the efficiency with which specialists can use lower numbers of hosts (Watters 1997). Generalists have evolved to exploit new habitats and new hosts; specialists have evolved to persist under low host densities.

Because of the stochastic elements of the mussel-host relationship, the greater the number of hosts available to the mussel the better. But once the host pool abundance drops below a critical threshold the mussel population may be extirpated simply by probabilistic effects. That is, mussels, whether specialists or generalists, may decline even though their hosts are present, if the host pool drops below this threshold (Watters 1997).

REPRODUCTIVE PATTERNS

Sterki (1895) noted that North American mussels could be divided into two behavioral groups based upon the duration that glochidia are held in the marsupia. These groups came to be known as tachytictic or short-term breeders and bradytictic or long-term breeders. Tachytictic breeders spawn in the spring or summer and release their glochidia later the same year, usually by July or August. Bradytictic breeders spawn in the summer or early autumn, form glochidia, and typically hold these larvae in the marsupium, overwintering them until the following spring or summer. Anodontines and lampsilines tend to be bradytictic; amblemines tend to be tachytictic. Some bradytictic forms apparently metamorphose the same year without overwintering, but there is evidence that these glochidia experience more mortality once on the host than those glochidia that overwinter in the marsupium (Corwin 1921; Higgins 1930; Tedla and Fernando 1969; Zale and Neves 1982).

But some mussels release glochidia in autumn or winter to overwinter on their hosts (Watters and O'Dee 1999, 2000), where they remain dormant until a threshold temperature is reached the following spring, at which time they metamorphose and excyst. This third reproductive pattern, termed host overwintering, plays a prominent role in some species, such as *Pyganodon grandis* and *Leptodea fragilis*. Overwintering of glochidia on hosts increases the dispersal of the species by allowing the glochidium to remain attached to its mobile hosts for a greater duration than would occur with tachytictic or bradytictic species. Host overwintering may confer greater fitness as well on the newly metamorphosed juveniles. If survival is correlated to the duration of the first year's growth before winter, then host overwintering juveniles have the longest growing season. In bradytictic species, rising spring water temperatures result in glochidial release—metamorphosis is not until several weeks later. But in host overwintering forms, spring temperatures result in metamorphosis—and an increased growth period the first year. Tachytictic species metamorphose later than any other group and have the shortest growing season; tachytictic forms are probably the most primitive.

As with reproductive strategies, there are tradeoffs between reproductive patterns as well. Although no studies have addressed the issue, it is likely that glochidia are more at risk while attached to the hosts than when brooded in the female's marsupium. While on the hosts, they may be damaged or knocked off, or the host may die. By this reasoning, host overwintering would be the most risky. The tradeoff is between the increased risk of mortality vs. the increased dispersal and lengthened growing season. The least risky is tachyticty, where the glochidial stage accounts for the shortest portion of the mussel's life in comparison with bradyticty or host overwintering, but also has the shortest growing season. The middle ground is bradyticty, which ensures a longer growing season than tachytictic forms and an equal amount of dispersal. Finally, there is mounting evidence that some temperate species have multiple broods (Watters and O'Dee 2000). In these cases, a species may have both tachytictic (in the summer) and bradytictic (over winter) reproduction. It remains to be seen how widespread is such a "hedging" of patterns.

There is relatively little information on the precise timing of glochidial release or the triggers causing their release. Watters and O'Dee (2000) found that *Lampsilis radiata luteola* released glochidia year-round as a function of water temperature—the higher the temperature the greater the number of glochidia released. Such a pattern is difficult to explain by the models of tachyticty, bradyticty, or host overwintering. Yet in the same study, *Amblema plicata* released a single, very short-lived burst of glochidia in July. Clearly there are two very different modes of glochidial release as evidenced by these two species; in one, constant glochidial release tracked water temperature, in the other, glochidial release occurred as a single event triggered by a threshold water temperature. These represent fundamental differences and deserve more study.

PARTING COMMENTS

The effects of pollutants on this complex life cycle are covered in detail elsewhere in this volume, but several important points need to be made here. *More than one animal is involved.* Efforts to conserve and manage any given mussel species are futile if the host(s) is not conserved and managed as well. Fish and mussels are very different creatures with very different life styles, requirements, and tolerances. Ecotoxicological concerns cannot concentrate on one without the other. *Mussels are parasites.* Mussels cannot be treated as free-living organisms, although they are commonly considered as such. Conservationists and researchers need to be aware of the unique aspects of the parasitic life cycle, including fecundity and host-parasite interactions. *Mussels have different life stages.* Like most invertebrates, mussels have larval and juvenile stages that are ecologically and physiologically different from their adult forms. What may be only marginally harmful to an adult may be lethal to a juvenile. *Not all mussels are created equal.* It is perhaps human nature to regard other animals as a single entity. What harms one animal harms them all. Mussels are often thought of in this way; we speak of a pollutant affecting a mussel bed or a river reach as if all mussels respond the same way. But we know this to be wrong. Mussels run the gamut in their tolerances and susceptibilities like any other kind of animal. Derailing the mussel's life cycle is dangerously simple precisely because of its complex nature.

REFERENCES

Agassiz, L., Observations on the rate of increase and other characters of fresh-water shells, unios, *Proc. Boston Soc. Nat. Hist.*, 8, 166–167, 1862a.

Agassiz, L., On reversed bivalve shells, *Proc. Boston Soc. Nat. Hist.*, 8, 100–101, 1862b.

Allen, W. R., The food and feeding habits of freshwater mussels, *Biol. Bull.*, 27, 127–147, 1914.

Allen, W. R., Studies of the biology of freshwater mussels, *Biol. Bull.*, 40, 210–241, 1921.

Amyot, J.-P. and Downing, J. A., Locomotion in *Elliptio complanata* (Mollusca: Unionidae): A reproductive function?, *Freshwat. Biol.*, 39, 351–358, 1998.

Arey, L. B., Observations on an acquired immunity to a metazoan parasite, *J. Exp. Zool.*, 38, 377–381, 1924a.

Arey, L. B., Glochidial cuticulae, teeth, and the mechanics of attachment, *J. Morphol. Phys.*, 39, 323–335, 1924b.

Arey, L. B., The nutrition of glochidia during metamorphosis, *J. Morphol.*, 53, 1–221, 1932a.

Arey, L. B., A microscopical study of glochidial immunity, *J. Morphol.*, 53, 367–379, 1932b.

Barnhart, M. C., Sterile eggs in unionid mussels and their roles in conglutinate function, *Bull. N. Am. Benthol. Soc.*, 14, 56, 1997.

Barnhart, M. C., and Roberts, A., Reproduction and fish hosts of unionids from the Ozark uplifts, In *Conservation and Management of Freshwater Mussels II*, Cummings, K. S. et al., Eds., *Proceedings of a UMRCC Symposium*, October 1995, St. Louis, MO, pp. 16–20, 1997.

Bauer, G., The parasitic stage of the freshwater pearl mussel (*Margaritifera margaritifera* L.). II. Susceptibility of brown trout, *Arch. Hydrobiol. Suppl.*, 76, 403–412, 1987a.

Bauer, G., Reproductive strategy of the freshwater pearl mussel *Margaritifera margaritifera*, *J. Anim. Ecol.*, 56, 691–704, 1987b.

Bauer, G., Life-history variation on different taxonomic levels of naiads, In *Ecology and Evolution of the Freshwater Mussels Unionoida. Ecological Studies.* Vol. 145, Bauer, G. and Wächtler, K., Eds., Springer, Berlin, pp. 83–91, 2001.

Bauer, G., Factors affecting naiad occurrence and abundance, In *Ecology and Evolution of the Freshwater Mussels Unionoida. Ecological studies*, Vol. 145, Bauer, G. and Wächtler, K., Eds., Springer, Berlin, pp. 155–162, 2001.

Bauer, G. and Vogel, C., The parasitic stage of the freshwater pearl mussel (*Margaritifera margaritifera* L.). I. Host response to glochidiosis, *Arch. Hydrobiol. Suppl.*, 76, 393–402, 1987.

Bisbee, G. D., Ingestion of phytoplankton by two species of freshwater mussels, the black sandshell, *Ligumia recta*, and the three ridger, *Amblema plicata*, from the Wisconsin River in Oneida County, Wisconsin, *Bios.*, 55, 219–225, 1984.

Blainville, H.-M. D. de, Rapport sur un Mémoire de M. Jacobson, ayant pour titre: Observations sur le développment prétendu des oeufs des moulettes ou unios et des anodontes dans leurs brachies, *Ann. Sci. Nat. Paris*, 14, 22–62, 1828.

Blystad, C. N., Significance of larval mantle of fresh-water mussels during parasitism, with notes on a new mantle condition exhibited by *Lampsilis luteola*, *Bull. U.S. Bur. Fish.*, 39, 203–219, 1924.

Braun, M., Ueber die postembryonale Entwicklung unserer Suesswassermuscheln, *Zool. Anz.*, 1, 7–10, 1878a.

Braun, M., Mittheilungen aus dem zoologischen Institut in Wuerzburg. I. Die postembryonale Entwicklung der Suesswassermuscheln, *Jahrb. deutschen Malak. Ges.*, 5, 307–319, 1878b.

Brim Box, J., Dorazio, R. M., and Liddell, W. D., Relationships between streambed substrate characteristics and freshwater mussels (Bivalvia: Unionidae) in Costal Plain streams, *J. N. Am. Benthol. Soc.*, 21, 253–260, 2002.

Brodniewicz, I., On glochidia of the genera *Unio* and *Anodonta* from the Quaternary freshwater sediments of Poland, *Acta Palaeontol. Polon.*, 13, 619–628, 1968.

Brodniewicz, I., Les glochidia et leur importance pour l'étude des depots quaternaries d'eau douce, *Mitt. Int. Ver. Theor. Angew. Limnol.*, 17, 315–318, 1969.

Bruno, D. W., McVicar, A. H., and Waddell, I. F., Natural infestation of farmed Atlantic salmon, *Salmo salar* L., parr by glochidia of the freshwater pearl mussel, *Margaritifera margaritifera* L., *Bull. Eur. Assoc. Fish. Pathol.*, 8, 23–26, 1988.

Burch, J. B., Freshwater unionacean clams (Mollusca: Pelecypoda) of North America, *EPA Biota Freshwat. Ecosyst. Ident. Manual*, 11, 176, 1973.

Carus, C. G., Neue Untersuchungen über die Entwicklungsgeschichte unserer Flussmuschel, *Verh. Kaiserl. Leopoldinisch-Carolinischen Akad. Naturf.*, 16, 1–87, 1832.

Churchill, E. P. and Lewis, S. I., Food and feeding in freshwater mussels, *Bull. U.S. Bur. Fish. [Doc. 963]*, 39, 439–471, 1924.

Corwin, R. S., Further notes on raising freshwater mussels in enclosures, *Trans. Am. Fish. Soc.*, 50, 307–311, 1921.

Dudgeon, D. and Morton, B., Site selection and attachment duration of *Anodonta woodiana* (Bivalvia: Unionacea) glochidia on fish hosts, *J. Zool. London*, 204, 355–362, 1984.

Fikes, M. H., Maintenance of the naiad *Amblema plicata* (Say, 1817) in an artificial system, *Bull. Am. Malacol. Union for 1972*, 35, 1972.

Fischerstrom, I., De concharum margaritiferarum natura, *Comm. Rebus Sci. Nat. Med. Gestis*, 10, 204–205, 1761.

Fisher, G. R. and Dimock, R. V., Viability of glochidia of *Utterbackia imbecillis* (Bivalvia: Unionidae) following their removal from the parental mussel, In *Proceedings of the Conservation, Captive Care, and Propagation of Freshwater Mussels Symposium*, Tankersley, R. A. et al., Eds., Ohio Biological Survey Special Publication, Columbus, OH, pp. 185–188, 2000.

Fisher, G. R. and Dimock, R. V., Ultrastructure of the mushroom body: Digestion during metamorphosis of *Utterbackia imbecillis* (Bivalvia: Unionidae), *Invert. Biol.*, 121, 126–135, 2002a.

Fisher, G. R. and Dimock, R. V., Morphological and molecular changes during metamorphosis in *Utterbackia imbecillis* (Bivalvia: Unionidae), *J. Molluscan Stud.*, 68, 159–164, 2002b.

Forel, F. A., Einige Beobachtungen über die Entwicklung des zelligen Muskelgewebes; Beiträge zur Entwicklungsgeschichte der Najaden, In *Inaugural-Abhandlung der medizinischen Facultät zu Würzburg.* p. 40, 1879.

Fustish, C. A. and Millemann, R. E., Glochidiosis of salmonid fishes. II. Comparison of tissue response of Coho and Chinook Salmon to experimental infection with *Margaritifera margaritifera* (L.) (Pelecypoda: Margaritanidae), *J. Parasitol.*, 64, 155–157, 1978.

Gatenby, C. M., Neves, R. J., and Parker, B. C., Preliminary observations from a study to culture recently metamorphosed mussels, *Bull. N. Am. Benthol. Soc.*, 10, 128, 1993.

Gatenby, C. M., Parker, B. C., and Neves, R. J., Growth and survival of juvenile rainbow mussels, *Villosa iris* (Lea 1829) (Bivalvia: Unionidae), reared on algal diets and sediment, *Am. Malacol. Bull.*, 14, 57–66, 1997.

Gordon, M. E. and Smith, D. G., Autumnal reproduction in *Cumberlandia monodonta* (Unionoidea: Margaritifera), *Trans. Am. Microsc. Soc.*, 109, 407–411, 1990.

Haag, W. R. and Warren, M. L., Mantle displays of freshwater mussels elicit attacks from fish, *Freshwat. Biol.*, 42, 35–40, 1999.

Haag, W. R. and Warren, M. L., Effects of light and presence of fish on lure display and larval release behaviors in two species of freshwater mussels, *Anim. Behav.*, 60, 879–886, 2000.

Haag, W. R., Butler, R. S., and Hartfield, P. D., An extraordinary reproductive strategy in freshwater bivalves: Prey mimicry to facilitate larval dispersal, *Freshwat. Biol.*, 34, 471–476, 1995.

Haag, W. R., Warren, M. L., and Shillingsford, M., Host fishes and host-attracting behavior of *Lampsilis altilis* and *Villosa vibex* (Bivalvia: Unionidae), *Am. Midl. Nat.*, 141, 149–157, 1999.

Hartfield, P. and Butler, R., Observations on the release of superconglutinates by *Lampsilis perovalis* (Conrad 1834), In *Conservation and Management of Freshwater Mussels II*, Cummings, K. S. et al., Eds., *Proceedings of a UMRCC Symposium*, October 1995, St. Louis, MO, pp. 11–13, 1997.

Hartfield, P. D. and Hartfield, E., Observations on the conglutinates of *Ptychobranchus greeni* (Conrad 1834), *Am. Midl. Nat.*, 135, 370–375, 1996.

Heard, W. H., Hermaphroditism in *Elliptio* (Pelecypoda: Unionidae), *Malacol. Rev.*, 12, 21–28, 1979.

Heard, W. H., Glochidial larvae of freshwater mussels and their fish hosts: Early discoveries and interpretations of the association, *Malacol. Rev., Suppl.*, 8, 83–88, 1999a.

Heard, W. H., A history of the controversy about *Glochidium parasiticum* Rathke, 1797 (Palaeoheterodonta: Unionoida: Unionidae), *Malacol. Rev., Suppl.*, 8, 89–106, 1999b.

Heinricher, J. R. and Layzer, J. B., Reproduction by individuals of a nonreproducing population of *Magalonaias nervosa* (Mollusca: Unionidae) following translocation, *Am. Midl. Nat.*, 141, 140–148, 1999.

Henley, W. F. and Neves, R. J., Behavioral responses of glochidia of freshwater mussels (Bivalvia: Unionidae) to chemical cues of fish, *Am. Malacol. Bull.*, 16, 131–135, 2001.

Higgins, E., Freshwater mussel investigations, *Rep. Comm. Fish. 1929, App.*, 10, 670–673, 1930.

Hoggarth, M. A. and Gaunt, A. S., Mechanics of glochidial attachment (Mollusca: Bivalvia: Unionidae), *J. Morphol.*, 198, 71–81, 1988.

Holland-Bartels, L. E., Physical factors and their influence on the mussel fauna of a main channel border habitat of the upper Mississippi River, *J. N. Am. Benthol. Soc.*, 9, 327–335, 1990.

Houghton, W., On the parasitic nature of the fry of *Anodonta cygnea*, *Quart. J. Microsc. Sci., n.s.*, 2, 162–168, 1862.

Hove, M. and Neves, R., Distribution and life history of the James River spinymussel, *Endangered Species Tech. Bull.*, 16, 9, 1991.

Howard, A. D., Experiments in propagation of fresh-water mussels of the *Quadrula* group, *Rep. U.S. Comm. Fish. 1913. App.*, 4, 1–52, 1914.

Howard, A. D., Some exceptional cases of breeding among the Unionidae, *Nautilus*, 29, 4–11, 1915.

Howells, R. G., Reproductive seasonality of freshwater mussels (Unionidae) in Texas, In *Proceedings of the Conservation, Captive Care, and Propagation of Freshwater Mussels Symposium*, Tankersley, R. A. et al., Eds., Ohio Biological Survey Special Publication, Columbus, OH, pp. 35–48, 2000.

Hudson, R. G. and Isom, B. G., Rearing juveniles of the freshwater mussels (Unionidae) in a laboratory setting, *Nautilus*, 98, 129–135, 1984.

Humphrey, C., Freshwater Mussels *(Velesunio angasi)*, In *Alligator Rivers Region Research Institute, Annual Research Summary for 1986–1987*, Australian Government Publishing Service, Canberra, pp. 107-108, 1987.

Humphrey, C. L. and Simpson, R. D., The biology and ecology of *Velesunio angasi* (Bivalvia: Hyriidae) in the Malaga Creek, Northern Territory, *Report to the Office of Supervising Scientist for the Alligator Rivers Region, Open File Record*, 2 appendices, 38, p. 476, 1985.

Imlay, M. J. and Paige, M. L., Laboratory growth of freshwater sponges, unionid mussels, and sphaeriid clams, *Prog. Fish-Cult.*, 34, 210–216, 1972.

Jansen, W., Bauer, G., and Zahner-Meike, E., Glochidial mortality in freshwater mussels, In *Ecology and Evolution of the Freshwater Mussels Unionoida. Ecological Studies*, Vol. 145, Bauer, G. and Wächtler, K., Eds., Springer, Berlin, pp. 183–211, 2001.

Jeong, K.-H., Min, B.-J., and Chung, P.-R., An anatomical study of the glochidium of *Anodonta arcaeformis*, *Malacol. Rev.*, 26, 71–79, 1993.

Kat, P. W., Sexual selection and simultaneous hermaphroditism among the Unionidae (Bivalvia: Mollusca), *J. Zool. London*, 201, 395–416, 1983.

Kays, W. T., Silverman, H., and Dietz, T. H., Water channels and water canals in the gill of the freshwater mussel, *Ligumia subrostrata*: Ultrastructure and histochemistry, *J. Exp. Zool.*, 254, 256–269, 1990.

Kirtland, J. P., Observations on the sexual characters of the animals belonging to Lamarck's family of naiads, *Am. J. Sci. Arts*, 26, 117–120, 1834.

Kwon, O.-K., Park, G.-M., Lee, J.-S., and Song, H.-B., Scanning electron microscope studies of the minute shell structure of glochidia of three species of Unionidae (Bivalvia) from Korea, *Malacol. Rev.*, 26, 63–70, 1993.

Leeuwenhoek, A. van, Arcana Naturæ Detecta, Henricum a Kroonevelt, p. 568 + index, 1695.

Leeuwenhoek, A. van, Continuatio Arcanorum Naturæ Detectorum, Henricum-a-Kroonevelt, p. 192 + index, 1697.

Lefevre, G. and Curtis, W. C., Reproduction and parasitism in the Unionidae, *J. Exp. Zool.*, 9, 79–115, 1910.

Leydig, F. [in F.C. Noll], Mittheilung über den Parasitismus junger Unioniden an Fischen, Tübingen, Inaugural-Dissertation, Frankfort-am-Main, pp. 46–47, 1866.

Moles, A., Effect of parasitism by mussel glochidia on growth of coho salmon, *Trans. Am. Fish. Soc.*, 112, 201–204, 1983.

Neves, R. J. and Widlak, J. C., Occurrence of glochidia in stream drift and on fishes of the Upper North Fork Holston River, Virginia, *Am. Midl. Nat.*, 119, 111–120, 1988.

Nichols, S. J. and Garling, D., Food web dynamics of Unionidae in a canopied river and a non-canopied lake, In *Program & Abstracts, Freshwater Mussel Symposium, Columbus, OH*, pp. 28–29, 1998.

O'Brien, C. A. and Brim Box, J., Reproductive biology and juvenile revruitment of the shinyrayed pocketbook, *Lampsilis subangulata* (Bivalvia: Unionidae) in the Gulf Coastal Plain, *Am. Midl. Nat.*, 142, 129–140, 1999.

O'Dee, S. H., Immune response of largemouth bass, *Micropterus salmoides*, to the fatmucket, *Lampsilis radiata luteola* (Unionidae), over repetitive infestations and over-wintering, Unpublished Masters Thesis, Ohio State University, p. 134, 2000.

Ortmann, A. E., Notes upon the families and genera of the najades, *Ann. Carnegie Mus.*, 8(2), 222–365, 1912.

Ortmann, A. E., A monograph of the naiads of Pennsylvania. Part III. Systematic account of the genera and species, *Mem. Carnegie Mus.*, 8, 1–385, 1919.

Poupart, M, Remarques sur les coquillages à deux coquilles, & premierement sur les moules, *Hist. l'Acad. Roy. Sci. Paris*, *1706*, 51–61, 1706.

Prévost, I., De la generation chez la mouile de peintres (*Unio Pictorum*), *Ann. Sci. Nat. Paris*, 7, 447–454, 1826.

Quatrefages, A. de, Mémoire sur la vie intra-branchiale des petites anodontes, *Ann. Sci. Nat. Paris*, 5(2), 321–336, 1836.

Rand, T. G. and Wiles, M., Species differentiation of the glochidia of *Anodonta cataracta* Say, 1817 and *Anodonta implicata* Say, 1829 (Mollusca: Unionidae) by scanning electron microscopy, *Can J. Zool.*, 60, 1722–1727, 1982.

Rathke, J., Om Dammuslingen, *Naturhistoire Selskabets Skrifter (Kjöbenhavn)*, 4, 139–179, 1797.

Rennie, J., British pearls, *Mag. Nat. Hist. J. Zool. Bot. Miner. Geol. Meteorol.*, 2, 461–462, 1829.

Reuling, F. H., Acquired immunity to an animal parasite, *J. Infect. Dis.*, 24, 337–346, 1919.

Richard, P. E., Dietz, T. H., and Silverman, H., Structure of the gill during reproduction in the unionids *Anodonta grandis, Ligumia subrostrata*, and *Carunculina parva texasensis*, *Can. J. Zool.*, 69, 1744–1754, 1991.

Riusech, F. A. and Barnhart, M. C., Host suitability and utilization in *Venustaconcha ellipsiformis* and *Venustaconcha pleasii* (Bivalvia: Unionidae) from the Ozark Plateaus, In *Proceedings of the Conservation, Captive Care, and Propagation of Freshwater Mussels Symposium*, Tankersley, R. A. et al., Eds., Ohio Biological Survey Special Publication, Columbus, OH, pp. 83–91, 2000.

Schierholz, C., Über die Entwicklung der Unioniden, *Denks. Kaiserl. Akad. Wissen. Wien. Math.-Naturw. Classe*, 55, 183–214, 1889.

Schmidt, O., Zur Entwicklungsgeschichte der Najaden, *Sitz. Math.-Naturw. Classe Kaiserl. Akad. Wissen. Wien*, 19, 183–194, 1856.

Shadoan, M. K. and Dimock, R.V., Differential sensitivity of hooked (*Utterbackia imbecillis*) and hookless (*Megalonaias nervosa*) glochidia to chemical and mechanical stimuli (Bivalvia: Unionidae), In *Proceedings of the Conservation, Captive Care, and Propagation of Freshwater Mussels Symposium*, Tankersley, R. A. et al., Eds., Ohio Biological Survey Special Publication, Columbus, OH, pp. 93–102, 2000.

Silverman, H., Steffens, W. L., and Dietz, T. H., Calcium from extracellular concretions in the gills of freshwater unionid mussels is mobilized during reproduction, *J. Exp. Zool.*, 236, 137–147, 1985.

Sterki, V., Some notes on the genital organs of Unionidae, with reference to systematics, *Nautilus*, 9, 91–94, 1895.

Strayer, D., Notes on the microhabitats of unionid mussels in some Michigan streams, *Am. Midl. Nat.*, 106, 411–415, 1981.

Strayer, D., Macrohabitats of freshwater mussels (Bivalvia: Unionacea) in streams of the northern Atlantic Slope, *J. N. Am. Benthol. Soc.*, 12, 236–246, 1993.

Strayer, D. L., Use of flow refuges by unionid mussels in rivers, *J. N. Am. Benthol. Soc.*, 18, 468–476, 1999.

Strayer, D. and Ralley, J., Microhabitat use by an assemblage of stream-dwelling unionaceans (Bivalvia), including two rare species of *Alasmidonta*, *J. N. Am. Benthol. Soc.*, 12, 247–258, 1993.

Surber, T., Notes on the natural hosts of fresh-water mussels, *Bull. Bur. Fish.*, 32(1912), 103–116, 1913.

Tankersley, R. A. and Dimock, R. V., Quantitative analysis of the structure of the marsupial gills of the freshwater mussel *Anodonta cataracta*, *Biol. Bull.*, 182, 145–154, 1992.

Tankersley, R. A., Hart, J. J., and Weiber, M. G., Developmental shifts in feeding biodynamics of juvenile *Utterbackia imbecillis* (Mollusca: Bivalvia), In *Conservation and Management of Freshwater Mussels II: Initiatives for the Future*, Cummings, K. S., Buchanan, A. C., Mayer, C. A., and Naimo, T. J., Eds., Proceedings of a UMRCC symposium, St. Louis, MO, Upper Mississippi River Conservation Committee, Rock Island, IL, pp. 282–283, 1997.

Tedla, S. and Fernando, C. H., Observations on the glochidia of *Lampsilis radiata* (Gmelin) infesting yellow Perch, *Perca flavescens* (Mitchill) in the Bay of Quinte, Lake Ontario, *Can. J. Zool.*, 47, 705–712, 1969.

Tevesz, M. J. S. and McCall, P. L., Evolution of substratum preference in bivalves (Mollusca), *J. Paleontol*, 53, 112–120, 1979.

Utterback, W. I., The naiades of Missouri, *Am. Midl. Nat.*, 4, 311–327, (see also pp. 339–354, 387–400, 432–464), 1916.

van der Schalie, H., Hermaphroditism among North American freshwater mussels, *Malacologia*, 5, 77–78, 1966.

van der Schalie, H., Hermaphroditism among North American freshwater mussels, *Malacologia*, 10, 93–112, 1970.

Vaughn, C. C., Regional patterns of mussel species distributions in North American rivers, *Ecography*, 20, 107–115, 1997.

Wächtler, K., Dreher-Mansur, M. C., and Richter, T., Larval types and early postlarval biology in naiads (Unionoida), In *Ecology and Evolution of the Freshwater Mussels Unionoida. Ecological Studies*, Vol. 145, Bauer, G. and Wächtler, K., Eds., Springer, Berlin, pp. 93–126, 2001.

Waller, D. L. and Mitchell, L. G., Gill tissue reactions in walleye *Stizostedion vitreum vitreum* and common carp *Cyprinus carpio* to glochidia of the freshwater mussel *Lampsilis radiata siliquoidea*, *Dis. Aquat. Org.*, 6, 81–87, 1989.

Watters, G. T., Unionids, fishes, and the species-area curve, *J. Biogeogr.*, 19, 481–490, 1992.

Watters, G. T., And immunity for all…, *Triannu. Unionid Rep.*, 10, 14, 1996.

Watters, G. T., Individual-based models of mussel-fish interactions: A cautionary study, In *Conservation and Management of Freshwater Mussels II: Initiatives for the Future*, Cummings, K. S., Buchanan, A. C., Mayer, C. A., and Naimo, T. J., Eds., Proceedings of a UMRCC symposium, St. Louis, MO, Upper Mississippi River Conservation Committee, Rock Island, IL, pp. 45–62, 1997.

Watters, G. T., Morphology of the conglutinate of the kidneyshell freshwater mussel, *Ptychobranchus fasciolaris*, *Invert. Biol.*, 118, 289–295, 1999.

Watters, G. T., The evolution of the Unionacea in North America, and its implications for the worldwide fauna, In *Ecology and Evolution of the Freshwater Mussels Unionoida. Ecological Studies*, Vol. 145, Bauer, G. and Wächtler, K., Eds., Springer, Berlin, pp. 281–307, 2001.

Watters, G. T. and O'Dee, S. H., Shedding of untransformed glochidia by fishes parasitized by *Lampsilis fasciola* Rafinesque, 1820 (Mollusca: Bivalvia: Unionidae): Evidence of acquired immunity in the field?, *J. Freshwat. Ecol.*, 11, 383–389, 1996.

Watters, G. T. and O'Dee, S. H., No acquired immunity against congener, *Triannu. Unionid Rep.*, 13, 39, 1997.

Watters, G. T. and O'Dee, S. H., Glochidia of the freshwater mussel *Lampsilis* overwintering on fish hosts, *J. Molluscan Stud.*, 65, 453–459, 1999.

Watters, G. T., O'Dee, S. H., Glochidial release as a function of water temperature: Beyond bradyticty and tachyticty, In *Proceedings of the Conservation, Captive Care, and Propagation of Freshwater Mussels Symposium*, Tankersley, R. A. et al., Eds., Ohio Biological Survey Special Publications, Columbus, OH, pp. 135–140, 2000.

Watters, G. T., O'Dee, S. H., and Chordas, S., Patterns of vertical migration in freshwater mussels (Bivalvia: Unionoida), *J. Freshwat. Ecol.*, 16, 541–549, 2001.

Weir, G. P., An ecology of the Unionidae in Otsego Lake with special references to immature stages, *Occ. Pap. SUNY Oneonta Biol. Field Sta. Copperstown, NY*, 4, 1–108, 1977.

Wood, W., *General Conchology*, J. Booth, London, p. 246, 1815.

Yeager, B. L., Cherry, D. S., and Neves, R. J., Interstitial feeding behavior of juvenile unionid mussels, *Assoc. SE Biol. Bull.*, 40, 113, 1993.

Yeager, B. L., Cherry, D. S., and Neves, R. J., Feeding and burrowing behaviors of juvenile rainbow mussels, *Villosa iris* (Bivalvia: Unionidae), *J. N. Am. Benthol. Soc.*, 13, 217–222, 1994.

Zale, A. V. and Neves, R. J., Fish hosts of four species of lampsiline mussels (Mollusca: Unionidae) in Big Moccasin Creek, Virginia, *Can. J. Zool.*, 60, 2535–2542, 1982.

Zimmerman, L. L. and Neves, R. J., Effects of temperature on duration of viability for glochidia of freshwater mussels (Bivalvia: Unionidae), *Am. Malacol. Bull.*, 17, 31–36, 2002.

4 Propagation and Culture of Freshwater Mussels

Cristi D. Bishop, Robert Hudson, and Jerry L. Farris

INTRODUCTION

The propagation and recovery of federally listed species has been supported by written policy from the U.S. Fish and Wildlife Service (USFWS) and National Marine Fisheries Service (NMFS). This policy outlines specific guidelines for the use of propagation as an essential tool for recovery and conservation of a declining population through support from the Endangered Species Act of 1973. Historical propagation efforts of other invertebrates, vertebrates, and plant species have indicated that this approach has circumvented the decline of certain species. Furthermore, one of the goals of species restoration is to develop sound policies based on best available technology, subsequently, the use of propagation in freshwater mussel recovery plans may allow for alternative techniques beyond conventional fish host encystments. Implementing species recovery as outlined in the Endangered Species Act, requires scientists to not only provide the technology for viable populations (i.e., surviving and reproducing individuals), but also to account for their habitat and life history requirements so that populations can maintain enough genetic variability to adapt to changes in the environment.

Unionid propagation in the United States began in the early 1900s employing techniques very similar to those used today, which provided scientists a basis for understanding at least the limited requirements of this complex group of invertebrates. Understanding the life history of this unique group of invertebrates also forces us to examine the life history, physiology, biochemistry, immunochemistry, and bioaccumulation of specific host fish. The specific morphology of the fish gill as well as the blood supply and gas exchange that it provides has been well published in the literature. This understanding of host-fish physiology was a prerequisite for the development of artificial culture media (in vitro technique) in the early 1980s, which provided information about specific requirements for nutrients that supported development and growth of exposed glochidia. Propagation techniques that support juvenile transformation continue to be used with declining populations, including federally threatened and endangered (T&E) species identified in federal recovery plans. Techniques that include the use of host fish (in vivo) and in vitro methods have generated viable juveniles for various objectives including toxicity testing, in situ monitoring, and reintroduction efforts into recovering streams (Jacobson et al. 1993; Yeager, Cherry, and Neves 1994; Morgan, Welker, and Layzer 1997). Monitoring juvenile responses and survival has furnished evidence of differential sensitivity to various contaminants, holding conditions, and feeding regimes. Effective propagation has supported determination of possible impacts on early life stages (glochidia and juvenile), where LC50s can be calculated and compared to existing water quality criteria and habitat assessments. Such information helps clarify reasons for declining populations and allows more accurate and effective evaluation of mitigation efforts of sensitive invertebrates. Juvenile propagation techniques have been employed for various uses and are generally used to reintroduce individuals into recovery streams, provide insight into the necessary requirements of a relatively unknown group of species, and assess impact by measuring endpoints that demonstrate the relative sensitivities of this early life stage to various contaminants. The following sections will address these areas of research that have provided some indication for the reasons of continued decline for this group of unique aquatic invertebrates.

BIVALVE LIFE HISTORY: UNDERSTANDING EARLY LIFE STAGE LIMITATIONS

The critical evaluation of early life stage organisms is imperative to understand requirements for survival, growth, and reproduction of a species. It provides insight into the requirements for growth, effects of environmental perturbations, and sensitivities to contaminants. Scientists from the U.S. Environmental Protection Agency (USEPA) that were developing guidance documents in the mid 1970s for toxicity test methods realized that early life stage testing was perhaps more critical to the protection of a population than older-age individuals; therefore, comparing sensitivities among different age groups is important for the determination of protection with water quality criteria. Toxicity test methods for adults, juveniles, and glochidia are outlined in Chapter 5.

Measuring lethal and sublethal effects in early life stage individuals (e.g., glochidia or juveniles) provides an understanding of the range of sensitivities to various pollutants that are known to impact the survival and reproduction of a species. The need for propagation in testing arises from the difficulty in finding naturally propagated juveniles in aquatic systems. The use of artificial techniques in culturing juveniles can offer additional opportunities for determining cause-effect relationships through laboratory or in situ toxicity tests. This will provide support for the much-needed protective measures (e.g., implementation of best management practices to reduce suspended solids, consideration of sensitive species in effluent discharge permits, and changes in the pesticide registration process to include more sensitive, bivalve species) for this declining group of invertebrates. This decline prompted the U.S. Congress to amend the Endangered Species Act in 1988 to require federal agencies (e.g., USFWS) to provide implementation plans for the recovery of federally listed species. Recovery plans generally include protective measures to prevent further population declines and support bivalve conservation measures, which are achieved by improving the habitat, translocating the species, increasing the quality of captive propagation programs, and acquiring land adjacent to recovering streams. Recovery plans for freshwater mussels generally include habitat restoration, fish host suitability, and propagation of juveniles for reintroduction. Propagation efforts are currently limited to fish host techniques as a viable alternative to natural recruitment within an aquatic system. However, the utilization of artificial culture media may be the only way that some mussels will survive extinction (personal communication, S Ahlstedt 2003). Very often, artificial culture techniques can produce significantly more juveniles than host-fish techniques. For example, in studies that used propagation techniques for fish hosts and artificial culture, there were nearly ten and twenty times more juveniles cultured from various culture media. The uncertainties inherent in artificial culture should not restrict federal policy by limiting the approach of propagation to fish host techniques only.

THE NEED FOR ARTIFICIAL PROPAGATION

The development of fish host techniques in the early 1900s, as well as the improvement of culture media for propagating mussels using artificial media (in vitro), provides a unique opportunity for researchers to identify life history requirements and survival mechanisms for this group of declining bivalves. Artificial propagation can be employed in response to environmental spills and invasive exotic species (e.g., zebra mussels and black carp), as well as help stabilize declining populations through species recovery plans. Juvenile transformation has been reported for many species using both fish host (in vivo) and artificial media (in vitro). However, some species pose a significant challenge when using either technique, including *Elliptodeus sloatianus*, which produce lanceolate conglutinates, *Lampsilis fragilis*, and *Pleurobema* spp. (Milam et al. 2000; personal communication, P Johnson 2003).

To date, there are only two propagation techniques reported in the literature, one that includes host fish (Lefevre and Curtis 1912) and another using artificial culture media (Isom and Hudson 1982). This section will provide some insight into the current development of these techniques as well as include arguments for and against each one when determining which propagation method is most appropriate (Table 4.1).

Fish Host Techniques (in Vivo)

Techniques for the determination of fish hosts have been reported and utilized by many researchers for decades (Howard 1916; Coker et al. 1921; Penn 1939; Cope 1959; Giusti et al. 1975; Jenkinson 1982; Hove and Neves 1991), while some unconventional hosts (e.g., amphibians) have also been identified as supporting juvenile metamorphosis (Seshaiya 1969). Some species have shown relatively good success using fish host techniques, while others have proven to be more difficult. Common species as well as state and federally listed species are often difficult to transform due to the lack of knowledge of life history complexities and requirements (personal communication, J Jones 2004). Two federally endangered species that are, at this point, particularly difficult are *Dromus dromas* and *Cyprogenia stegaria*.

Techniques for determining fish host suitability include the use of aeration tanks, direct gill placement, and the use of anesthetics (i.e., MS222, which is tricaine methanesulfonate and trade name Finquel) to reduce handling stress on the fish (Zale and Neves 1982). While modifications of these have been reported from various researchers, the fundamental approach is the same. Aeration tanks are often used when there are viable glochidia with several fish species and cohorts. However, if glochidia are limited and/or the fish are smaller in size or have small gill rakers, direct gill placement using pipettes is a viable alternative to aeration techniques for attachment onto the gill. Anesthetizing fish prior to encystment is undecided, in that the possible effects of the anesthetic may inadvertently impair glochidia attachment and subsequent metamorphosis. The use of MS222 is not necessary if the pipetting onto the gill is done quickly, with little stress to the fish.

Glochidial attachment can range from several days to several months depending on the mussel species, fish health (i.e., whether the individual is stressed or diseased due to other environmental factors), water temperature, and perhaps other variables presently unknown. Alternatively, fish survival can be jeopardized by excessive glochidial infestation as a result of limiting gas exchange across the gill lamellae. While 50–100 glochidia/gill for fish that are 15–25 cm in length have been reported as adequate (Hove et al. 2000), others have directly infested host fish with several thousand and achieved successful transformation and maintained fish viability once removed from the tanks (Milam et al. 2000; Winterringer 2004).

Hove et al. (2000) described considerations for conducting fish suitability tests and included several issues that are important for the successful transformation of juveniles. Fish maintenance and holding prior to and during the encystment is fundamental to the fish host technique. The use of glass or high-grade polycarbonate, flow-through tanks with adequate aeration and temperature control devices are suggested to reduce fluctuations in water quality and quantity. Providing some nutrition to host fish during the encystment period has been noted several ways: feeding fish throughout the entire period and feeding fish only through the first half of the encystment period and eliminating food thereafter. The latter method reduces the amount of fish feces in the bottom of the tank, increasing the ability to selectively isolate juveniles more efficiently. Juveniles should be siphoned from the tank bottom and collected using a sieve series for isolation. A polarized lens, which is attached to the objective lens of a dissecting microscope, can be used to reflect, through understage lighting, only prismatic objects and block out sediment and feces that often eclipse the juvenile identification and counting process (Watters 1996). High quality foods (e.g., brine shrimp or live minnows) should be promoted in the maintenance of fish hosts; however, a critical step in

TABLE 4.1
Summary of Successful Juvenile Transformation Efforts Using Various Unionid Species: 1982–2003

Species	Technique	Purpose	Reference
A. plicata	Fish host	Reintroduction	Hubbs (2000)
	Media	Culture development	Personal communication, B Hudson and M Barfield (1993)
Anodonta suborbiculata	Fish host	Host suitability	Barnhart and Roberts (1997) (blue)
Anodontoides ferussacianus	Fish host	Host suitability	Hove et al. (1997) (blue)
Toxilasma cylindrellus	Fish host	Unknown	Hudson and Isom 1984
Cyclonaias tuberculata	Fish host	Host suitability	Hove et al. (1997) (blue)
E. angustata	Media	Toxicity testing Reintroduction	Hudson, Barfield, and McKinney (1996)
E. complanata	Media	Culture development	Hudson, Barfield, and McKinney (1996)
E. crassidens	Media	Unknown	Personal communication, D Simbeck (2003)
E. icenterina	Fish host	Toxicity testing	Keller and Ruessler (1997)
Fusconaia ebena	Media	Culture development	Isom and Hudson (1982)
Fusconaia flava	Media	Reintroduction	Milam et al. (2000)
Lampsilis cardium	Fish host	Toxicity testing	Keller and Ruessler (1997)
	Media		Milam et al. (2000) Myers-Kinzie 2000
L. fasciola	Fish host	Reintroduction	Morgan, Welker, and Layzer (1997) (blue)
	Media		
L. ovata	Fish host	Culture development	Isom and Hudson (1982)
L. rafinesqueana	Fish host	Host suitability	Barnhart and Roberts (1997) and (blue) Shiver (2002)
L. reeveiana	Fish host	Host suitability	Barnhart and Roberts (1997) (blue)
L. siliquoidea	Media	Reintroduction	Milam et al. (2000)
	Media	Survival and growth	Myers-Kinzie (2000)
L. streckeri	Fish host	Host suitability and reintroduction	Winterringer (2003)
L. subangulata	Fish host	Host suitability	Personal Communication, C Echevarria (2004)
L. teres	Media	Unknown	Keller and Zam (1990)
L. ventricosa	Media	Unknown	Milam et al. (2000)
Ligumia recta	Media	Culture development	Isom and Hudson (1982), Milam et al. (2000)
M. conradicus	Fish host	Reintroduction	Morgan, Welker, and Layzer (1997) (blue)
Megalonaias gigantia	Media	Unknown	Personal Communication, B Isom and D Simbeck (2003)
M. nervosa	Fish host	Reintroduction	Hubbs (2000)
Pleurobema coccineum	Fish host	Host suitability	Hove et al. (1997) (blue)
P. cordatum	Media	Culture development	Hudson and Isom (1984)
Ptychobranchus occidentalis	Fish host	Host suitability	Barnhart and Roberts (1997) (blue)
P. cataracta	Media	Unknown	Dimmock and Wright (1993)
P. grandis	Fish host	Toxicity testing	Keller and Ruessler (1997)
	Fish host	Reintroduction	Milam et al. (2000)
	Media		Personal Communication, B Isom (2003)
S. undulatus	Fish host	Host suitability	Hove et al. (1997) (blue)
U. imbecillis	Fish host	Toxicity testing	Keller and Zam (1991)

(continued)

TABLE 4.1 *(Continued)*

Species	Technique	Purpose	Reference
			Warren (1996)
			Clem (1998)
U. imbecillis	Media	Culture development	Isom and Hudson (1982)
			Barfield et al. (1997) (blue)
		Toxicity testing	Hudson and Shelbourne (1990)
			Wade et al. (1989)
	Fish host	Physiological effects	Dimmock and Wright (1993)
	Fish host	Viability	Fisher and Dimmock (2000)
	Media	Unknown	Keller and Zam (1990)
Venustaconcha ellipsiformis	Fish host	Host suitability	Riusech and Barnhart (2000)
V. pleasii	Fish host	Host suitability	Riusech and Barnhart (2000)
V. iris	Fish host	Toxicity testing	Jacobson et al. (1993)
	Fish host	Behavior	Yeager et al. (1994)
	Media	Unknown	Pers. comm. D Simbeck
V. liensosa	Fish host	Toxicity testing	Keller and Ruessler (1997)
	Fish host	Host suitability	Pers. comm. C Echevarria
	Media	Unknown	Keller and Zam (1990)
V. taeniata	Fish host	Reintroduction	Morgan, Welker, and Layzer (1997) (blue)
V. vibex	Fish host	Host suitability	Pers. comm. C Echevarria

reducing the possibility of fungal and bacterial growths includes the prompt removal of uneaten food. Hove et al. (2000) stated that for bottom-feeding fish (e.g., minnows and other catostomids) that will feed on newly metamorphosed juveniles and sloughed glochidia, it is important to separate the fish from the bottom of the tank using a plastic net with a small mesh size (1.6 mm) secured to the tank, which allows the juveniles to fall through the mesh but keeps fish from bottom feeding.

Collection of wild host fish or even commercially spawned fish species requires some attention to detail, including the acclimation of fish once they have been brought back to the propagation facility. We suggest that fish be allowed to acclimate for several days prior to isolating for glochidia infestation. Conducting fish suitability trials should include multiple attempts using several individuals of the same fish species with glochidia from different females to assure that metamorphosis occurs in at least two different test trials (Haag 2002).

The range of fish species that co-occur with mussel populations is important to understand from a management perspective. Long-term restoration goals should include the successful recruitment of mussel fauna as well as viable host-fish populations in a river reach. Subsequently, instream habitats should accommodate both mussel and fish life history requirements to ensure that sustainable populations are being supported.

A viable mechanism for supporting juvenile reintroductions is to release infested fish into the waterbody with known glochidia species. Several reports have indicated that the release of infected fish can support the efforts of a recovery plan for both common and listed species (Milam et al. 2000; Genoa NFH, Chapter 5, Methods for Conducting Toxicity Tests Using *Corbicula Fluminea* as Surrogate Species).

Dependence on Fish Hosts—An Obligate Trait?

Identification of fish hosts for unionid species has been reported in the literature since the turn of the twentieth century when Connor (1905) identified *Lepomis gibbosus* (pumpkinseed) as a successful fish host for *Anodonta cataracta* (now *Pyganodon cataracta*, Hoeh 1990). Watters (1994) reviewed published fish host and unionid relationships in North America with approximately 95 unionid species and over 150 fish species. Since then, there have been numerous publications that provide updates to reported host-fish requirements and include new fish species that are successful candidates to support juvenile transformation (Barnhart and Roberts 1997; Dee and Watters 1998; Hove et al. 2000; Winterringer 2004). A summary of propagation techniques (e.g., fish host or artificial media), since Watters' review was reported in 1994, has been provided as an update (Table 4.2).

While the research suggests that most species require a host fish as an obligate trait of the bivalve life history, a few do not. Unionids are also typically identified as either a generalist, where its glochidia can transform on a variety of fish species, or a specialist, where only one or two host fish have been identified that aid in the successful metamorphosis of glochidia to the juvenile stage. Additionally, with unionid populations that are deemed specialists, some of these are also species listed as threatened or endangered by federal and state governments. Current debate, however, may suggest that declines in unionid populations are not necessarily due to a specific host-fish requirement but rather to some other factor that inhibits survival, growth, and reproduction post-transformation. Many recovery projects for T&E species have identified various host fish, which

TABLE 4.2
Advantages and Disadvantages of Host Fish (in Vivo) and Culture Media (in Vitro) Propagation Techniques

Use	Fish Host (in Vivo)		Culture Media (in Vitro)	
	Advantage	Disadvantage	Advantage	Disadvantage
	Ability to use infested fish to place directly into a recovering system is an easy method for novelists	With species whose host fish are unknown, a considerable effort may be needed to determine this prior to any juvenile reintroduction	Ability to obtain considerably more juveniles per unit effort	Costs can be significantly higher
Reintroduction/ monitoring	Efforts for recovering streams can secure future recruitment by knowing and assuring that fish hosts are also residing in the stream segment	Use of basin-specific host fish may be limited to the propagation effort	Method can provide viable juveniles in lieu of unreported fish hosts	Fungal and microbial infestations can eliminate transforming glochidia if not closely monitored
Host suitability	Determination of required host fish for each species	Unhealthy fish may limit the production of transformed juveniles Previously infested fish are reported to be immune to a glochidia infestation (Arey 1932)	Not applicable	Not applicable
Toxicity exposures	With some contaminants, juveniles transformed in vivo can be more sensitive		With some contaminants, juveniles transformed in vitro can be more sensitive	

support successful and viable juveniles. Listed species such as *Lampsilis streckeri* (Winterringer 2004) and *Lampsilis powelli* have been identified as generalists.

Independence from Fish Hosts

Not all unionid species require fish hosts for the transformation of juveniles. *Strophitus undulatus* and *Utterbackia imbecillis* are the two most reported species that were found to transform from glochidia to juvenile inside the marsupial pouch (Lefevre and Curtis 1912; Howard 1916; Allen 1924; personal communication, M Barfield 2003), while *Obliquaria* species may also be (Lefevre and Curtis 1912) a nonparasitic unionid. Adult females of these species apparently provide essential nutrients for metamorphosis to take place; however, quantity and quality of these critical nutrients are unknown.

ARTIFICIAL MEDIA CULTURE (IN VITRO)

History

Freshwater mussel glochidia (i.e., larvae) are naturally transformed into juveniles, which include development of the internal organs necessary for self-sustained existence as a benthic organism. This is accomplished via encystment in fish tissue (on the gill or fin). Interest in enhancing production of mussels through the use of artificial culture was seen early in the twentieth century when Ellis and Ellis (1926) reported the first successful culture of glochidia to transformation following the excision of these from the gill tissue of their fish host. Unfortunately, the details of their solutions were never published, and the host-fish attachment prior to culture initiation may have provided the stimulating factor(s), which allowed this developmental process prior to culture initiation. Much later, in the early 1980s, interest in the artificial culture of glochidia was revived by Isom. Isom, who was familiar with the transformation work of Ellis and Ellis (1926), contracted Bob Hudson to help attempt this artificial culture, based on his work with the cell culture of catfish (Hudson, Pardue, and Roberts 1980). Isom and Hudson (1982) reported success in the transformation of several species without the use of a fish host at any time in the process, a distinct improvement over the Ellis and Ellis report of 1926. This technique, which began as a modification of modern cell-culture techniques, made use of a mixture of amino acids, vitamins, and glucose in a Unionid Ringers solution (Ellis, Merrick, and Ellis 1930), along with the addition of fish plasma as a source of protein, growth stimulants, hormones, etc. Although this work began by mixing these components from scratch (using the concentrations of each found in fish plasma as guidelines), Isom and Hudson (1982) also reported success using pre-mixed, commercially available cell culture media (Eagles essential and non-essential amino acids and Medium 199), which contains nearly all of these amino acids in concentrations as high or higher than those found in fish plasma.

Even though this mixture has been used to produce thousands of juvenile mussels for toxicology research (Wade, Hudson, and McKinney 1989; Johnson, Keller, and Zam 1993; Hudson et al. 1994; Hudson, Barfield, and McKinney 1996; Barfield, Clem, and Farris 1997; Clem 1998), it has been less than convenient for other labs to use because of the requirement that there be a ready supply of fish from which blood can be drained and separated into plasma and non-plasma components. Because of this inconvenience and because the use of fish plasma introduces more variation in the results, research was initiated in 1990 to try to develop an alternative medium that was either a serum/plasma free medium or a medium using commercially available serum in minimal concentrations (Hudson and Shelbourne 1990). Keller and Zam (1990) first addressed the modification of the glochidial culture, demonstrating that other sera could be substituted for the fish plasma, with horse serum producing their best results. Hudson and Shelborne (1990) began a massive study for Don Wade of the Tennessee Valley Authority, at Presbyterian College, testing a total of 64 different medium combinations. Later, Johnson, Keller, and Zam (1993) describe culture methods in their acute toxicity testing procedure; however, these are nearly identical to those described by Isom and Hudson (1982), using fish plasma as the protein source, and the same culture constituents including identical antibiotics without any real improvements in technique. Other advances come as a result of modification of the Hudson and Shelbourne (1990) work by Barfield, Clem, and Farris (1997), Milam et al. (2000).

Culture Media Techniques

Glochidia are removed as described by Isom and Hudson (1982) or preferably by using a syringe full of control water to flush them from the female marsupia. However, differing from prior publications, glochidia are rinsed three to four times in autoclaved river water or reconstituted water rather than deionized water, and a final rinse with Unionid Ringers solution or Hank's Balanced Salt solution. Approximately 300–900 glochidia are seeded in a 3-mL total medium in each 60-mm diameter, cell-culture dish. These dishes are incubated at 21–24°C in an incubator having 4.6–5% CO_2 to maintain a pH of about 7.3 by use of a bicarbonate buffer (Isom and Hudson 1982; Milam et al. 2000).

Modification of the Media

The original medium was comprised of Eagles essential and non-essential amino acids in Unionid Ringers containing $NaHCO_3$ for pH control, vitamins, antibiotics, and glucose as the artificial portion, and fish plasma as the natural protein source, in a final ratio of two-thirds artificial medium to one-third plasma (Isom and Hudson 1982). As previously mentioned, other media were tested in 1990 (Hudson and Shelbourne 1990) in a major effort to improve results in more species, resulting in comparisons of 64 different media combinations. For the last decade, David McKinney, Chief of Environmental Services at Tennessee Wildlife Resources Agency (TWRA), has maintained interest in the culture and use of juveniles for toxicity testing, funding research that has been presented in several reports that have added to the original modification by Hudson and Shelbourne in 1990. Although the primary work has involved only one species, *U. imbecillis*, other species have also been transformed using artificial media (Table 4.3). The following sections discuss specific artificial media components and their variations.

Ionic Balance. Initially, Isom and Hudson (1982) used a modified Unionid Ringers described by Ellis, Merrick, and Ellis (1930); however, further tests show that prepared, balanced salt solutions such as Earle's or Hank's balanced salt solution (Sigma) are useful in rinsing glochidia as well as in their artificial transformation, even though the yield may be slightly lower.

Sera. Fish plasma was reported by Isom and Hudson (1982) as the choice protein additive; however, rabbit serum performs as well or nearly as well as the fish plasma when transforming *U. imbecillis*, and rabbit performance is better than porcine, horse, sheep, chicken, and fetal bovine sera (Hudson and Shelbourne 1990). These results differ from Keller and Zam (1990) who reported that horse serum was their most productive protein additive. Combinations of the above sera and plasma resulted in the highest production being found in a medium containing a fish plasma/rabbit serum combination (usually equal to or better than fish plasma alone), but fish/porcine and rabbit/porcine combinations were not significantly different from fish plasma. Fish/fetal bovine and fish/horse were significantly lower (Hudson and Shelbourne 1990). Since some laboratories do not have access to fish plasma, and since infection rates are lower when using sterile sera obtained from biochemical supply companies, rabbit serum is considered to be the best alternative. In one case, *Elliptio angustata* cultured in rabbit serum significantly outperformed fish plasma in observed transformation rates (Hudson and Shelbourne 1990).

Serum Replacements. Rabbit serum performs better than other commercially available sera (Hudson and Shelbourne 1990); however, the goal to eliminate sera or plasma altogether resulted in the testing of several serum replacements. Hudson and Shelbourne (1990) evaluated cultures

TABLE 4.3
Mussel Species Transformed Using Artificial Media Culture

Species (Subfamily)	Media Type	Transformation Time Required	Reference
(Anodontinae)			
U. imbecillis	Rabbit/TCH/TCM; Horse serum; Fish plasma (pl.) and all combinations	7 days	Isom and Hudson (1982), Wade, Hudson, and McKinney (1989), Hudson and Shelbourne(1990), Keller and Zam (1990), Dimock and Wright (1993), Barfield, Clem, and Farris (1997) and many others.
P. cataracta	Rabbit/TCH/TCM	7 days	Dimmock and Wright (1993)
P. grandis	Fish pl.	7 days	Personal communication, B Isom (2003)
(Lampsilinae)			
L. ovata	Fish pl.	22 days	Isom and Hudson (1982)
L. fasciola	Fish pl.	Unknown	Personal communication, D Simbeck (2003)
Ligumia recta	Fish pl.	15 days	Isom and Hudson (1982)
L. siliquoidea	Rabbit; Fish pl. and all combinations; Rabbit/TCH/TCM	11-20 days	Milam and Farris (1998), Milam et al. (2000), Myers-Kinzie (2000)
L. teres	Horse serum	Unknown	Keller and Zam (1990)
L. ventricosa	Rabbit; Fish pl. and all combinations	12 days	Milam et al. (2000)
V. iris	Fish pl.	Unknown	Personal communication, D Simbeck (2003)
V. lienosa	Horse serum	Unknown	Keller and Zam (1990)
Toxolasma cylindrellus	Fish pl.	20 days	Hudson and Isom (1984)
(Ambleminae)			
A. plicata	Fish pl.	12–13 days	Hudson, Barfield, and McKinney (1996)
Fusconaia ebena	Fish pl.	18 days	Isom and Hudson (1982)
Fusconaia flava	Rabbit; Fish pl. and all combinations	9–11 days	Milam et al. (2000)
Megalonias gigantia	Fish pl.	16 days	Personal communication, B Isom, and D Simbeck (2003)
Pleurobema cordatum	Fish pl.	15 days	Hudson and Isom (1984)
(Unioninae)			
E. angustata	Rabbit/TCH/TCM and Fish pl.	7 days	Hudson and Shelbourne (1990) Hudson, Barfield, and McKinney (1996)
Elliptio crassidens	Fish pl.	Unknown	Personal communication, D Simbeck (2003)
E. complanata	Rabbit/TCH/TCM and Fish pl.	15 days	Hudson, Barfield, and McKinney (1996)

Initiated using six serum replacements and found that none produced even 20% of the yield of fish plasma cultures. These replacements were then tested in combination with fish plasma, and the results showed that the CPSR (Sigma), TCM, and TCH (Protide) in combination with fish plasma produced a higher yield than the medium containing only fish plasma without any additives. These and other serum replacements were then tested in combination with other sera, and the resulting data indicated that rabbit serum in combination with TCM and TCH outperformed other sera/serum replacement combinations. The rabbit/TCH/TCM portion was one-third of the total medium at a ratio of 1:1:1 each using a 12% stock solution of TCH and TCM.

Antibiotics/Antimycotics. Multiple rinses of glochidia washed from the marsupia, as described earlier, are essential to lower the rate of infections from bacteria and fungi. To also help insure a low rate of infection, antibiotics (e.g., penicillin and streptomycin) and an antimycotic (e.g., amphotericin B) that are usually found in most cell cultures were the first to be used in developing the glochidial medium (Isom and Hudson 1982). Improvements of these techniques were done by isolating and culturing bacteria from glochidial cultures and measuring zones of inhibition from approximately one dozen commercially available antibiotics (Isom and Hudson 1982). The three antibiotics (i.e., carbenicillin, gentamicin sulfate, and rifampin) with the best inhibitory effects were reported by Isom and Hudson (1982) and are currently those used in all laboratories that have reported glochidial cultures. Much later, bacteria that were isolated from swabs of the gills of *Amblema plicata, P. cataracta*, and *U. imbecillis*, were identified and measured for inhibiting effects (e.g., growth) by fifteen antibiotics (Loveless et al. 1999). The assumption behind this work is that contamination of artificial glochidial cultures most likely comes from the parent's gill tissue, which houses these glochidia. The most effective antibiotics against the dozen bacteria isolated were neomycin, ciprofloxacin, and polymyxin B. Since these have never been evaluated for effect on actual transformation success, a mixture of antibiotics including two of these new antibiotics was tested along with a control of standard antibiotics (Isom and Hudson 1982) for their effect on mussel transformation. This mixture contained a penicillin-streptomycin-neomycin solution (5000 IU, 10 mg, and 5 mg respectively) and was added to cultures at a rate of 30 µL (low), 60 µL (medium), and 120 µL (high) per 3-mL culture dish along with polymyxin B at a rate of 4 mg/mL, 10 mg/mL, and 20 mg/mL concentrations. The penicillin and streptomycin are not in the new antibiotic best performer list (Loveless et al. 1999); however, this mixture is commercially available (Sigma Aldrich) for laboratory use. Between 600 and 800 juveniles were evaluated on day six of the glochidial culture and again on the second day after each dish had been placed in water. Each of these was compared with the control antibiotics (Isom and Hudson 1982); however, none outperformed the original control set in transformation success (Figure 4.1). These same concentrations of new antibiotics were used in combination with the control antibiotics and the results were similar (Figure 4.1). All treatments in Figure 4.1 are significantly different from the control with the exception of the treatment having medium concentrations of the new solutions mixed in combination with the control antibiotics (contingency chi-square $= 0.498$, 1 df, $p = 0.48$). One advantage of using this increased number of antibiotics may be that a wider spectrum of bacteria could be controlled better than by use of just the original antibiotics described by Isom and Hudson (1982).

Often, fungal infection will appear in one or more dishes during the culture process. When this happens, fungal mycelia are removed using sterile forceps, and 75 µL of nystatin (Sigma) is added to each culture dish. If a fungal or bacterial infection is massive, the developing glochidia are poured into a cylinder with a 112-µm mesh Nytex® screen bottom, resuspended in a sterile 200-mL beaker, and rinsed several times in Earle's or Hank's balanced salt solution. Following the rinsing, the glochidia are resuspended in a dish of balanced salt solution, aspirated out, and placed into a culture dish with fresh, complete media.

Other Medium Components. Most other medium components remain the same as first described by Isom and Hudson (1982). Eagles essential and non-essential amino acids are mixed with the Unionid Ringers with the addition of taurine and ornithine. Cultures seem to transform well without the addition of these last two amino acids, but since these are found in fish plasma and since some species of mussels may require these two amino acids, they are still used in most cultures. L-glutamine must be added weekly due to its inability to remain stable in solution. Other media (e.g., Medium 199) have also proved successful in transformation, even though their composition varies slightly from the above-modified Eagles Medium (Isom and Hudson 1982; Keller and Zam 1990). Glochidia developing artificially seem to be lower in lipid content than those developing

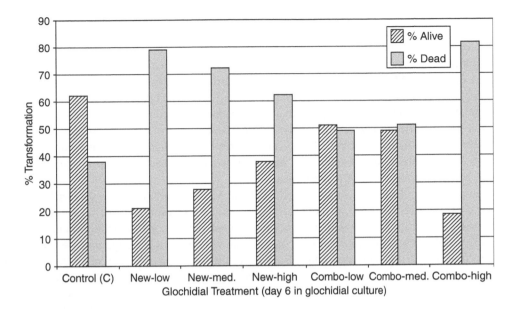

FIGURE 4.1 Results of antibiotic treatments for glochidial transformation success. (From Loveless, R., Raymond, S., Inman, J., Hudson, R., and Fairfax, M. Isolation, identification and control of bacteria which threaten in vitro cultures of freshwater mussel glochidia [presentation]. 1st Symposium of the Freshwater Mollusk Conservation Society, March 14–17, Chattanooga, TN, 1999.)

from fish hosts. Tankersley (2000) indicated that lipid levels in glochidia and juveniles vary with each brood stock, being influenced by the parent mussel. Furthermore, he indicated that the type of culture medium influenced the lipid content of the developing glochidia and subsequent juveniles. Fisher (2002) demonstrated that glochidia developing in an artificial medium (i.e., the rabbit serum mix) were lower in lipid content than in those transformed using fish hosts. Consequently, Hudson compared transformation rates with media containing squid oil (Artemate, Argent Chemical Laboratories) and cod liver oil, each added at 50 μL per culture dish. While the rates did vary, the cod liver oil seemed to be better than the squid oil and also resulted in lower fungal infection rates. It is thought that the addition of this oil enhances the lipid storage of the glochidia (Tankersley 2000).

Success of the in Vitro Cultured Juveniles

Lipid, triglycerides, cholesterol, glycogen, and protein concentrations in artificially produced juveniles are lower than those transformed on fish (Fisher 2002). However, juvenile testing with sediments from the Pigeon, Conasauga, and Telico Rivers, and Lake Monticello (Hudson and Roberts 1997; Hudson and McKissic 1999) show a 50–60% survival of juveniles held in a simple static culture for less than 90 days. These juveniles were held in cylinders with a screened bottom (112-μm mesh) and placed in sediment in a culture bowl and fed concentrated plankton two-to-three times per week. *U. imbecillis* have been grown in this system for over a year with a resulting mean size of over 9 mm. Starkey et al. (2000) used a Partitioned Aquaculture System (PAS) at Clemson University to test growth of about 100 juvenile *U. imbecillis*. This system uses planktonic algae as a treatment for fish wastes, where the wastes serve as nutrients for the plankton as they flow with the water leaving the fish holding area through a slow-moving pond, which is partitioned into lanes to allow time for blooming. In an effort to control the algae bloom, Starkey et al. (2000) used adult *Elliptio complanata* as one of the algal harvesters. They also placed 100 juvenile *U. imbecillis* in this concentrated, continuous flow of algae to determine their response to this high plankton load. The juveniles responded with very high growth, from an average of 400 μm in initial length when placed in the partitioned pond to 18.9 mm (47× increase) in just 117 days. Their most rapid growth occurred between days 47 and 61 with an average of 0.26 mm per day. Unfortunately, their survival rate was lower than desired, with only 7% survival at the end of the 117 days of the test. Many other factors may have contributed to this mortality not associated with the algae, since Hudson and Roberts (1997); Hudson and McKissic (1999) had 50–60% in 100-day-old juveniles with a lower feeding regime in a simple laboratory static culture. One factor may have been that the juveniles tested by Starkey et al. (2000) came from static cultures and already may have been malnourished. Barfield, Clem, and Farris (1997) showed that once diet-deprived juveniles were returned to a regular diet, they still had greater sensitivity to copper than those that had never been deprived of nourishment.

The media described in the section titled Artificial Mediaculture (in Vitro) vary little in amino acids and other artificial components. The mixture used by Hudson and Shelbourne (1990) and others is described in Table 4.4. The basic non-protein portion of the medium consists of the following stock with no antibiotics/mycotics and can be stored in a refrigerator (less than 4°C) for up to four weeks if glutamine is added fresh every time you make the final medium with serum and/or serum replacement. A bacterial streak on a nutrient agar plate should be made weekly to ensure sterility of this medium. One can avoid making the above mixture by using premixed Medium 199 (Sigma); however, the development rate may be slower (Isom and Hudson 1982). A protein source (i.e., serum or plasma) must be added to the above basic medium mixture. The variety of sera, plasma, and serum replacements is discussed in Modification of the Media (Table 4.5), with amounts stated in units needed to prepare one 3-mL culture dish. Every dish will receive the same volumes of the basic medium components, with the variation of each medium consisting mostly of different protein additives and/or their replacements.

Table 4.5 shows the basic non-protein, non-antibiotic/mycotic, non-lipid medium mixture, which can be combined with different plasma or serum to produce all of the variations used in most laboratories to date. Although Keller and Zam (1990) reported success using horse serum, this was not listed since others report rabbit serum or fish plasma as their protein of choice (Isom and Hudson 1982; Dimmock and Wright 1993; Milam et al. 2000). The basic medium shown in Table 4.3 may be mixed and stored for four or more weeks at 0–4°C. The serum, serum replacements, or plasma, the antibiotic/antimycotic solutions, and L-glutamine must be added fresh as the medium is being prepared for use. The above components may be mixed in batch quantities,

TABLE 4.4
Preparation of Stock Medium (MEM) for Glochidial Cultures

	Volume (mL)	Directions/Comments	Source
Unionid ringers	960	Ringers solution has as 2.2 g NaHCO₃ or 25 mM Hepes buffer. Store in refrigerator	
Essential amino acids	20	50X stock with glutamine	ICN
Non-essential amino acids	10	100X stock	Sigma
Vitamins	10	100X stock	Sigma
Dextrose solution	5	20% dextrose	Sigma
Phenol red solution	1	1% phenol red	Sigma
Taurine stock	31	Stock = 1 g/L	Sigma
L-ornithine stock	10	Stock = 1 g/L	Sigma
L-glutamine stock	0.01	Stock = 200 mM[a]	Sigma

[a] Glutamine will remain stable for only three weeks—hence, the medium must be re-made or fresh glutamine must be added every three weeks, even if stored in the refrigerator.

TABLE 4.5
Variations in Culture Media Used for Unionid Transformations

Each 60 by 15 mm Tissue Culture Dish Should Receive the Following:

Medium	All Dishes	Fish (F)	Rabbit (R)	Fish/Rabbit (F/R)	Fish/Rabbit/TCH/TCM (F/R/T)	Combo 1 (C1)	Combo 2 (C2)	Travel
Components (Ml)	1 dish	1 dish	1 dish	1 dish	1 dish	1 dish	1 dish	1 dish
MEM[a]	2	2	2	2	2	2	2	2
Carbenicillin[b]	0.15	0.15	0.15	0.15	0.15	0.15	0.15	0.15
Gentimycin[a]	0.1	0.1	0.1	0.1	0.1	0.1	0.1	0.1
Rifampicin[a]	0.003	0.003	0.003	0.003	0.003	0.003	0.003	0.003
Amphotericin B[a]	0.075	0.075	0.075	0.075	0.075	0.075	0.075	0.075
L-Glutamine[a]	0.03	0.03	0.03	0.03	0.03	0.03	0.03	0.03
Fish plasma[a]	—	1	—	0.5	0.25	—	0.2	—
TCM (2X)[a,c]	—	—	0.25	—	0.25	0.25	0.2	—
TCH (2X)[a,c]	—	—	0.25	—	0.25	0.25	0.2	—
Rabbit serum[a]	—	—	0.25	0.5	0.25	0.25	0.2	1
CPSR	—	—	—	—	—	0.25	0.2	—
Nystatin[d]	0.75	0.75	0.75	0.75	0.75	0.75	0.75	0.75
Cod Liver Oil[e]	0.5	0.5	0.5	0.5	0.5	0.5	0.5	0.5

Note: Stock Solution Concentrations per mL: carbenicillin (2 mg), gentimycin (3 mg), rifampicin (50 mg), and amphotericin (0.2 mg). TCM and TCH are used in a concentration between 12 and 24%.

[a] Change weekly.

[b] Add daily.

[c] The TCM and TCH will arrive in two two 10-mL vials each. Store in a refrigerator, not a freezer. Using a 1-mL syringe, remove 0.6 mL of TCM and TCH and add it to a 4.4 mL stock medium (MEM). Label these diluted mixtures working TCM and TCH, respectively, and store in the refrigerator.

[d] Add as needed.

[e] Add to each dish individually.

Source: (From Hudson, R. G. and Shelbourne, C. W., Improved *in vitro* culture of parasitic freshwater mussel glochidia, Report to the Tennessee Valley Authority, p. 25, 1990. Including later changes).

whereas the cod liver oil must be added to each dish separately since it will not mix with the solution. Nystatin is added as needed for fungal infections. All of the above remain fairly constant in all medium mixtures, with the exception of the plasma or serum components.

As seen in Table 4.3, most of the early transformation research was conducted using just fish plasma as a protein additive. Although Hudson and Shelbourne (1990) found a mix of rabbit serum and fish plasma to be most effective, transformation has been carried out without the addition of fish blood in most laboratories in order to reduce variability and for convenience. Some of the Anodontinae have been successfully transformed with the simplest of media using just rabbit serum and TCH and TCM additives, as have some *Elliptio* species (Table 4.3). Successful attempts to use the combination media listed in Table 4.4 have been limited to *U. imbecillis* due to shipping problems with other species. Results from Hudson and Shelbourne (1990) would indicate that a medium of fish plasma and rabbit serum may be the best beginning medium when first attempting to culture a new species. More work is needed to understand the nutritional needs of all mussels, especially in the area of lipid and protein additives.

Shipping Mussels and Glochidia

Specific techniques of shipping mussels are often not reported in the literature, and details of both successful and unsuccessful shipping are often noted from personal communications with researchers who are actively involved in mussel transport. Shipping gravid mussels is often necessary because mussels are not in the area where the propagation laboratory is located. Additionally, researchers have found that the short-term brooders (i.e., Ambleminae) are prone to aborting their embryos or glochidia following shipping (personal communication, J Jones 2004), although others have found this to be less than 5% (personal communication, P Johnson 2003), resulting in partial demibranch release during transportation. If Ambleminae are transported in wet towels in an ice chest, they often abort when returned to water. *Quadrula* species seem to be especially prone to aborting glochidia when disturbed (pers. comm. S Ahlstedt). This presents a problem in that abortion of glochidia and their deterioration results in poor success when attempting artificial culture.

Transporting adults and glochidia should include chilling individuals using ice packs (e.g., Blue Ice or frozen distilled water). The use of loose ice has the potential to melt during shipping and increases the chance of fouling by specimens if they have been allowed to come in direct contact with the ice/water. Specimens can also be placed in burlap sacks soaked with river water and shipped without a substantial amount of water (personal communication, S Ahlstedt 2003). Under these conditions, mussels will be less likely to abort during transport but can abort after they have been warmed and placed into holding systems. Others have found that long-term brooders (i.e., Lampsilines and Anodontines) hold their glochidia during shipping and handling. Hubbs and Johnson have found that mussels carried in wet towels or mesh bags in containers with ice packs, but not touching the packs, travel best. Some researchers (D. Simbeck of Tennessee Valley Authority and R. Hudson of Presbyterian College) have noticed that initiating artificial cultures seems to be more difficult from shipped parents than those that have been captured and carried in water directly to the laboratory, even considerable distances. Cultures also seem better when parents are held a day or two in the lab and then cultured, a problem in the Ambleminae since they tend to abort when held in water.

Zimmerman and Neves (2002) compared glochidia from two species over time in different temperature regimes and found that glochidia in the cooler temperatures (0–10°C) did remain viable longer than those at 25°C (75% survival at 7.5 days for *Villosa iris* and at 14.4 days for *Actinonaias pectorosa*) and were able to be transformed on fish following this time period. Comparisons of shipping glochidia in culture medium have been conducted using various packing methods with ice, limited studies indicted that the highest survival of glochidia were those that were shipped on a tray suspended on ice; however, these results are difficult to repeat, and more work needs to be done in this area.

Glochidia have been shipped free from the marsupia in river water or synthetic water, and excised, gravid marsupia have also been shipped for use in propagation efforts. Both types of shipping resulted in snap-closure response to sodium chloride upon arrival in the laboratory. However, the most appropriate way to ship glochidia is free from the marsupia because it assures that the female was not sacrificed. Alternatively, cold storage (less than 4°C) of inflated marsupia has been shown to be effective in maintaining the condition of encapsulated glochidia for toxicity testing. Acute bioassays can be successfully carried out using excised glochidia that have been stored for up to four days.

ADULT HOLDING

Holding and maintaining adult mussels in laboratory conditions is necessary to allow for transport acclimation, glochidia development, and in some cases, for reproduction to occur. *Villosa sp.* and *Lampsilis sp.* are particularly easy to maintain in the laboratory when given adequate food quantity and quality. Maintenance of these species results in relatively low mortality and considerable measured growth, indicating that these individuals are in reasonably good condition. Very often, females of *Villosa*, *Pyganodon*, *Utterbackia*, *Tritogonia*, *Elliptio*, and *Pleurobema* species have repeatedly become gravid in holding conditions (personal communication, P Johnson 2003). Additionally, these laboratory conditions have provided refugia for numerous species for more than four years (Farris, Milam, and Harris 1998). Although reproduction was not measured in these studies, it remains an outstanding issue for laboratory maintenance of adult mussels and requires additional research to determine the effects (i.e., survival and reproduction) of long-term holding facilities.

PHYSIOLOGICAL TRANSFORMATION—PHASES AND COMPARATIVE SENSITIVITIES OF DEVELOPMENT

Transformation of glochidia to juveniles on the fish gill and in media may range from 7 days to greater than 110 days, depending on mussel species, water temperature, and hostfish condition. Phases of development include the egg, glochidia, and juvenile stage. While it is thought that glochidia and egg phases are relatively protected from waterquality changes, juveniles may be more susceptible to exposures once excysted from the fish gill or fin. Several reports suggest that comparisons of glochidia and juvenile survival may be drastically different given a wide range of contaminant exposures (outlined in Chapter 5). Exposures to developing glochidia (i.e., already attached to live fish gills) include one study that measured impacts from aqueous-bound copper on hostfish. Results of this study indicate that encysted glochidia are relatively protected from acute copper exposures and will successfully metamorphose into juveniles (Jacobson 1990). Since only a single study with hostfish exposures has been reported, more work needs to be conducted to support existing literature and provide additional exposure results with a variety of priority pollutants (e.g., metals, organics, TSS, bacteria, etc.).

Comparisons of physiological conditions of juveniles transformed in vitro and in vivo indicate that, in general, those individuals that metamorphose on fish hosts tend to be healthier than those transformed in culture media (Fisher 2002). These comparisons were conducted using *U. imbecillis* where fish-reared individuals were found to absorb lipids and glycogen deposits from the fish, which provided a better survival and growth rate through several weeks. Additionally, juveniles reared from fish hosts had measurably higher concentrations of triglycerides, cholesterol, glycogen, and protein than juveniles cultured in media. Overall, juveniles that transformed on fish hosts were relatively less sensitive to thermal and hypoxic stresses following metamorphosis than those juveniles transformed in vitro.

Acute toxicity exposures of juvenile *U. imbecillis* to potassium chloride (reference solution) indicated that one-day-old individuals transformed via host fish (in vivo), were equally sensitive to juveniles transformed in culture media (in vitro) (Clem 1998). However, older-aged (e.g. greater than four days old) juvenile tests indicate that individuals transformed in vivo (an average of LC50 230 µg CuSO$_4$/L) were consistently more tolerant than those transformed in vitro (an average of LC50 94 µg CuSO$_4$/L). Similar exposures to copper sulfate, however, indicated that even one-day-old individuals that were transformed in vitro were nearly three times more sensitive than in vivo juveniles. Hudson et al. (2003) found that juveniles transformed in vitro were more sensitive to exposures with drilling compounds than juveniles transformed in vivo. Additionally, repeated tests indicated that measured sensitivity was also a function of individual size and that juveniles of similar sizes also had similar LC50 values. Further comparisons need to be made to support these findings. Other studies have shown that when diet deprived, older juveniles (7d) are more sensitive to copper concentrations than younger juveniles ($\leq$1d), which is somewhat contrary to existing sensitivity comparisons with invertebrates (Barfield, Clem, and Farris 1997). These data all seem to indicate that juvenile variation in toxic sensitivity is due to age, diet, and size. The mode of transformation appears to be coupled with diet, since artificially transformed juveniles have been shown to have lower food reserves (Fisher 2002). Furthermore, the variation in standard toxicity levels of *U. imbecillis* to copper (Barfield, Clem, and Farris 1997) in different laboratories would also seem to be diet induced. The need to provide a diverse algal food supply is imperative when growing and testing all juveniles, especially those cultured with an artificial medium. The need to add sediment for maximum growth (Isom and Hudson 1982; Gatenby et al. 1997) should be considered when growing juveniles and in any toxicity testing attempting to mimic natural conditions. While scientists do not fully understand the complex requirements that promote growth and reproduction for these early life stages, some studies have identified food quality and quantity, water quality, and habitat conditions for certain species (Dimmock and Wright 1993; Yeager, Cherry, and Neves 1994; Dunn and Layzer 1997; Mummert 2001). Early comparisons of toxic response of in vitro- vs. in vivo-transformed juveniles indicated that those transformed in an artificial medium were more sensitive to copper than the fish-transformed juveniles (Warren and Klaine 1994). These mussels were only 24, 48, and 96 h old and were not fed during the test exposure. This is consistent with Fisher's (2002) observations in that lipid levels of newly transformed juveniles are generally lower in those transformed in artificial media than those utilizing a fish host. Hudson et al. (2003) found the same phenomenon in juveniles of the same age from the two kinds of transformations; however, allowing the in vitro juveniles to age a few more days and build up some food reserves (even though their algae supply for this project was weak in that it was very much a monoculture) indicated that the size of the juveniles was more important than the kind of transformation that they experienced ($r = 0.87$ for increasing size and increasing LC50 values to a drilling compound). More comparisons, including different feeding regimes and ages of juveniles, need to be made. See Chapter 2 for details on other biological indicators including lipid concentrations.

Dimmock and Wright (1993) defined oxygen, pH, and temperature requirements for juvenile *U. imbecillis* and *P. cataracta* and found that 7–10-day-old juveniles could not survive 24 hour in an anoxic condition (see Chapter 2 for supporting evidence of ammonia and hydrogen sulfide buildup with concurrent decreases in oxygen and pH). Furthermore, temperatures above 30°C proved lethal, having a 96-hour LT50 value of 31.5°C for *U. imbecillis* and 33°C for *P. cataracta*. Slight acidity was tolerated with greater than 70% survival in all groups above a pH value of 5.0; however, they found a LC50 value of pH 4.5 for both species. Although these lower pH values are tolerated, Hudson, Barfield, and McKinney (1996) demonstrated that the sensitivity of *U. imbecillis* to aqueous copper with sediment was much greater at low acidic conditions (LC50 = 1.34 mg Cu/L at a pH of 5.5) than at higher pH values (LC50 = 2.75 mg Cu/L at a pH of 7.2).

FITNESS BEYOND THE "DROP ZONE"

Little is known about the survival, growth, and reproduction of naturally propagated unionids once they excyst from the fish gill or fin and are carried (in lotic systems) by hydrologic processes to downstream river reaches. Juvenile growth during the first year is variable among species, and consequently, collection from the wild and assessment of these young individuals is often difficult. Certain species of juveniles may only grow a few millimeters to centimeters in a typical year (O'Beirn, Neves, and Steg 1998; Payne and Miller 2000; Mummert 2001). Additionally, unionid age-to-maturity can not only vary from one to seven years depending on species (Baird 2000; Obermeyer 2000; Haag and Staton 2003) but is also relatively long when compared to other freshwater invertebrates; therefore few, if any, studies have included this endpoint as a physiological indicator of stress.

The percentage of juvenile survival that results in reproductively-viable adults is currently unknown. Artificial propagation, however, has shed some light on the ability of juveniles to grow and maintain homeostasis in a laboratory/hatchery setting for several growing seasons and has proven to be successful for certain species (Hanlon 2000). Juveniles of three species were propagated by fish-host techniques and reared for three years, after which these species matured and spawned in hatchery raceways. (Raceways are typically flow-through systems that pump hatchery source water through the concrete, glass, or plastic holding tanks, which can hold various water volumes from 100 to 10,000 L). Survival, growth, and reproduction over a three-year period provide good evidence that raceways are reasonable mimics of the real world. Hudson and Isom (1984) demonstrated conditions that allowed for an 18-times growth increase in juveniles of *U. imbecillis* (then *Anodonta imbecillis*, changed by Hoeh in 1990) in a period of 74 days. This was accomplished by using river water supplemented with sediment and plankton in a static water system. Mussels were cultured at 23–30°C with only a slight increase in growth rate at 30°C. More recently, Hudson and McKissick (1999) raised artificially transformed juveniles in a simple static water system for 93 days and found them to have increased in growth from an average of 320 to 3,257 μm (a 10-times increase) in sediment from the Conasauga River, TN. Although static cultures do indicate that juveniles can survive and grow in this type of holding, it is important to renew overlying water to reduce the potential for toxic concentrations of various materials (e.g., ammonia, bacteria, metals, etc.) to accumulate to toxic levels (Layzer, Gordon, and Anderson 1993; Michaelson and Neves 1995). Wade, Hudson, and McKinney 1989 showed that water is changed daily during acute toxicity tests to provide consistent levels of contaminants for measurable effects; however, during chronic tests or during the culture of juveniles, water renewal may be as infrequent as 7–10 days.

Juvenile U. imbecillis raised in static systems in sediment and fed concentrated plankton at varying rates have demonstrated moderately good survival for approximately 90 days as shown by mean 87-day survival rates ranging from 59 to 76% for three tests using sediments from Conasauga and Telico Rivers, TN and Lake Monticello, SC (Hudson and McKissick 1999). However, the best mean survival (76%) in the groups dropped by day 93 to only 53%, indicating some unidentified stress factor, probably food availability. Similar studies using Pigeon River, TN, sediment yielded 60% survival in 99 days (Hudson and Roberts 1997). Growth in the first three sediments above resulted in an increase in shell length from 290 mm at transformation to group averages ranging from 3718 to 3992 mm. The earlier test made with Pigeon River sediment had an average length of only 1470 mm after 99 days of culture. Of the remaining juveniles which were kept for over 1 year, shell length ranged from 6 mm (Pigeon River sediment) to just over 9 mm (Monticello). A better food supply and flowing water should increase size and survival as Starkey et al. (2000) have shown in their PAS at Clemson University (discussed in the section titled The Need for Artificial Propagation).

The addition of sediment fines in the water medium has been shown to increase growth rates by Hudson and Isom (1984) and Gatenby et al. (1997). Bivalves can utilize the organic carbon in the organic fines that coat sediment particles, particularly the smaller ones with a higher surface-to-volume ratio. Sediment is usually screened to remove larger particles by most researchers, and autoclaving this material is also practiced in an effort to eliminate fungal growth that may prove fatal to juveniles. Even though sediment is necessary for better growth and survival, Hudson et al. (2003) report that sediment pretreated with bentonite clay or EZ mud® at low levels clears the suspension of the finest clay particles, resulting in better survival of the juveniles. This may be an indication that the finest particles actually may impair gill function.

Nutrition in juveniles and adults is important for the survival, growth, and reproduction of unionid populations; however, very little is known about the quantity or quality of food source that provides minimal conditions for sustaining populations in the wild and in laboratory/hatchery conditions. A diversity of algae is found to be most effective in promoting the growth of juveniles (Hudson and Isom 1984; Gatenby et al. 1997). Gatenby et al. (1997) found that the use of algae containing higher levels of lipids (e.g., *Neochloris oleoabundans*) promoted the best growth, whereas the species of algae within each group seemed insignificant. Jerry Nichols of the USGS Great Lakes Science Center (abstract) has qualified the nutrient requirements of one unionid species (*Pyganodon grandis*) based on the gut contents identified from species located in an algal-dominated lake system and a detritus-dominated river. Determinations of gut contents in unionid species indicated that the main food source was fine-particulate organic matter, specifically detritus, algae, plankton, bacteria, and fungus. Additionally, all individuals used a carbon source and vitamin B12 from indigenous bacteria populations rather than algal communities, despite the high concentrations of diatoms and green algae found in the gut and mantle cavity of the mussels. Since this time, many researchers have improved on the technique in a variety of ways, from flowing hatchery raceways to in-river holding.

JUVENILE CULTURE: IN-RIVER HOLDING

Howard (1914, 1916) reported on transforming glochidia into juveniles via host fish contained in floating crates in a river. The definition of a food source and other water quality parameters were not provided. Although Howard reported the inability to raise juveniles in aquaria, he did place some juveniles from the crates into aquaria for three weeks and noted that those in the aquaria grew at a rate of only one-third of those in the crates. Following their successful transformation of juveniles by use of artificial media, Isom and Hudson (1982) further modified the marine bivalve culture techniques to those of Castagna and Kraeuter (1981) to achieve successful culture of the juvenile mussels in a laboratory setting (Hudson and Isom 1984). These juveniles grew in an environment of lake water along with sediment collected from the benthic area at a suspended concentration of 800 mg/L and the addition of concentrated plankton for food.

JUVENILE CULTURE: HATCHERY CONDITIONS

Juvenile culture in hatchery conditions has provided some insight into the requirements of this early life stage. Researchers have outlined several techniques that include water quality and feeding requirements, which aid in the survival and growth of juveniles prior to reintroducing them into natural systems. Water hardness concentrations ranging from 250 to 350 mg $CaCO_3$/L have been shown to support the long-term maintenance of juveniles (personal communication, J Jones 2004). Others have found that water hardness concentrations as low as 180 mg $CaCO_3$/L provide adequate levels of calcium and magnesium to support juvenile and adult survival (Farris, Milam, and Harris 1998). A feeding regime is perhaps one of the more significant variables in the successful maintenance of juveniles, and for some hatcheries, a daily feeding of *N. oleoabundans* or *Nannochloropsis oculata* (~30,000 cells/mL), which are two relatively small-celled, high-lipid algal species, provides adequate nutrition for survival and growth through several weeks to months. Holding juveniles in recirculating tanks provides several benefits including the continuous assortment of fine sediments that juveniles often use as a food source as well as maintaining a more stable environment (i.e., temperature, pH, and DO are more consistent in a flow-through system). Juveniles held under these conditions for several weeks have been reported to range in length from 700 to 1200 μm (a 7-to-12-fold increase in size from the newly-transformed juvenile) at the time of their release into streams (Milam et al. 2000).

Juvenile *Lampsilis fasciola* were cultured and held in hatchery raceways for 90 days to simulate what happens upon release of newly metamorphosed juveniles in riverine habitats (Hanlon 2000). Raceways were designed for maintaining high water quality through recirculating systems, and sediment fines were added to the concrete raceways for juvenile substrate. Survival (82% following 90 days) and growth of juveniles were monitored for "release" during various seasonal conditions (e.g., summer, fall, and winter water temperatures). Results of these trials indicated that late spring is the most ideal time for juvenile release when temperatures are warming from winter freezes to levels greater than 15°C and when water volumes are sufficient to avoid drought conditions. This study also provided information that the most appropriate age to release juveniles (*L. fasciola*) into recovering streams is when juveniles are approximately one month old. Maintaining juveniles in hatchery conditions has been shown to support growth and survival; however the unconfined release of newly metamorphosed juveniles results in greater growth than juveniles that are often held in culture dishes.

RESTORING DECLINING POPULATIONS: CASE STUDIES

Several case studies have been published outlining efforts to restore native unionids in watersheds across North America. Some studies have identified well-known causes of mollusk declines such as industrial spills and dam construction, while others are less apparent in describing causal factors (Table 4.6).

PIGEON RIVER, TENNESSEE

During the 1990s, there was an effort to restore water quality in the Pigeon River, TN. Personnel from TWRA became interested in rebuilding the missing mollusk fauna, which had been eliminated from this river for more than 50 years (personal communication, D McKinney 2003). Initial assessments using chronic sediment tests with artificially transformed (in vitro) juvenile *U. imbecillis* (Hudson and Roberts 1997) were employed to determine whether the existing river sediment was clean enough to support growth and survival of propagated juveniles. Results of these tests indicated that there was evidence that sediments from the Pigeon River would promote juvenile reintroduction. As a result, the process of restoring the diverse mollusk faunal group began in 1994 with the restocking of two common snail species (*Leptoxis praerosa* and *Lithasia verrucosa*) at fixed riffle sites (personal communication, S Ahlstedt and M Fagg 2003). Approximately 100,000–200,000 snails were removed from the Nolichuckey River, TN, over a four-to-five-year period and placed into the Pigeon River. By the fifth year, reproduction was reported in the Pigeon River through identification of young snails for both species.

The Spiny River Snail (*Io fluvialis*), which has been identified by the state for management, was introduced into the Pigeon River in the 1990s. Although mostly translocated from the Nolichuckey River, individuals from other rivers were also used. These snail populations from each river system were initially subjected to a genetic analysis for homogeneity and determined to be very homozygous. The spiny river snail showed reproduction almost immediately, and as a result, it was also reintroduced into the Holston, Lower French Broad, and Upper Tennessee Rivers.

Adults of six unioinid species (*Elliptio dilitata*, *Lampsilis ovata*, *S. undulatus*, *Alasmidonta marginata*, *L. fasciola*, and *V. iris*) were translocated into the Pigeon River in 2000 in a riffle area (personal communication, S Ahlstedt and M Fagg 2003). Individuals were monitored for survival

TABLE 4.6
Number of Individuals Reintroduced into North American Watersheds to Meet Federal Restoration Goals

Species	Morgan, Welker, and Layzer (1997)	Milam et al. (2000)	Jones, Mair, and Neves (2002)	Jones and Neves (1999)	Jones and Neves (1999, 2000)	Jones and Neves (1999, 2000) and Jones, Mair, and Neves 2002	Winterringer (2003)
L. streckeri[FE]	—	—	—	—	—	—	2,100
P. grandis	—	1,000	—	—	—	—	—
Lampsilis ventricosa	—	1,700	—	—	—	—	—
L. siliquoidea	—	2,500	—	—	—	—	—
L. fasciola[SE]	2,500	—	—	—	—	—	—
M. conradicus[SE]	15,000	—	—	—	—	—	—
V. taeniata[FE]	3,500	—	—	—	—	—	—
Epioblasma brevidens[FE]	—	—	36,400	—	14,800	5,000	—
Epioblasma capsaeformis[FE]	—	—	—	—	118,000	70,000	—
Epioblasma triquetra[FE]	—	—	—	—	1,900	1,800	—
Epioblasma walkeri[FE]	—	—	4,650	7,000	—	25,500	—
C. stegaria[FE]	—	—	—	—	—	11,400	—
Villosa perpupurea[FE]	—	—	1,100	—	—	11,200	—
Villosa trabalis[FE]	—	—	1,100	—	—	—	—
Dromus dromas[FE]	—	—	—	—	—	850	—
Pagias fabula[FE]	—	—	570	—	—	—	—

following three weeks, and all were found to be present and alive several months later in 2001. A final assessment in 2002 took place on these same individuals, and one mussel was recorded as dead. Additionally, no measurement of reproduction or growth has been attempted on this group to date. Finally, to assure that suitable host fish are present, restoration of the full complement of fish is now taking place, with many darters and other perciform fish being placed back into the river (personal communication, M Fagg and S Ahlstedt 2003).

SHOAL CREEK, ALABAMA AND TENNESSEE

Following two decades of faunal extirpation from industrial and municipal discharges, Shoal Creek was selected for mussel relocation and reintroduction activities to help recover a historically diverse mollusk community (Morgan, Welker, and Layzer 1997). A combination of translocation from a similar drainage and propagation (using host-fish techniques) of three species was included in the effort. Approximately 4,600 adult mussels, comprising 22 species, were translocated, and 21,000 artificially propagated juveniles (*Lampsilis teres*, *Medionidus conradicus*, and *Villosa taeniata*) were introduced into the creek from 1992 to 1994.While the fish fauna in the creek has recovered, successful reproduction of translocated adults has not been measured to date.

LEADING CREEK, OHIO

The goal for restoration and recovery of a creek in southeastern Ohio was identified in the mid-1990s with its focus to reintroduce mussel fauna that had been impacted by a mine effluent two years before. Three species, *Lampsilis siliquoidea*, *L. ventricosa*, and *P. grandis*, were transformed using host-fish propagation and artificial media (Milam et al. 2000). Over 6,000 juveniles of the three species were introduced into Leading Creek in the fall of 1997. Additional efforts included the translocation of adults from the watershed and on-site host-fish inoculation. Several adults were found alive following one year, and juveniles were also found alive following the two years of post-effluent release (personal communication, J Van Hassel 2004).

MIDDLE FORK LITTLE RED RIVER, ARKANSAS

Lampsilis streckeri was federally listed in 1989 by the USFWS. A recovery plan was approved two years later and included host-fish identification, juvenile propagation, and reintroduction into its reported range as some of the objectives (USFWS 1991a). Specific goals of the recovery plan included the reclassificaion of this species from endangered to threatened. Only when several criteria are met could this occur: (1) populations are discovered or reestablished, (2) populations are viable and the habitat is fully protected, and (3) population levels are deemed viable for at least 20 years. During the initial phase of the project, which included the determination of acceptable host fish, *L. streckeri* was identified as a generalist (Winterringer 2004). In fall 2001, over 2,000 juveniles were introduced (via host fish) into the Middle Fork Little Red River, AR, at two separate locations where known populations had been previously reported.

CLINCH, POWELL, HOLSTON, AND CUMBERLAND RIVERS, VIRGINIA

Several riverine sites in Virginia were identified to help determine the suitability of these water-bodies to support survival and growth of artificially propagated (via host fish) juveniles mussels.

Clinch River

A recovery plan for *C. stegaria* had been approved in 1991 and outlined the current status and need to de-list it from federally endangered status. However, the required type and amount of protected habitat needed for successful recovery is limited, and subsequently, the establishment of viable

C. stegaria populations may be unlikely (USFWS 1991b). Approximately 11,400 juveniles (*C. stegaria*) were introduced into the Clinch River in the early-to-mid-1990s. In 1999 through 2002, six other endangered mussels were artificially propagated via host fish and cultured in hatchery conditions for approximately two weeks until release into the Clinch River: *Epioblasma brevidens*—5,000; *Epioblasma capsaeformis*—70,000; *Epioblasma triquetra*—1,800; *Epioblasma walkeri*—25,500; *Villosa perpupurea*—11,200; and *D. dromas*—850 (Jones and Neves 1999; Jones and Neves 2000; Jones, Mair, and Neves 2002).

Powell River

From 1999 through 2000, three endangered mussels were artificially propagated via host fish and cultured in hatchery conditions for approximately two weeks until release into the Powell River: *E. brevidens*—14,800; *E. capsaeformis*—118,000; and *E. triquetra*—1,900 (Jones and Neves 1999; Jones and Neves 2000).

Holston River

The federal listing of *Villosa perpurpurea* was proposed in 1994 due to significant reductions in population range, where it currently survives in a few river reaches in Tennessee and Virginia (specifically in the upper Clinch River, VA, Copper Creek, VA, Obed River, TN, Emory River, TN, and Beech Creek, TN). Suggested impacts to the decline of this species include the construction of reservoirs on the Clinch, Powell, and Holston rivers, non-point source pollution from industrial development, and inadequate land use practices (Federal register Vol. 59, No. 134). In 2002, *V. perpurpurea* was artificially propagated via host fish and cultured in hatchery conditions for approximately two weeks until release into the Holston River, VA. Approximately 1,100 individuals were released during this event (Jones, Mair, and Neves 2002).

Cumberland River

An approved recovery plan for *Pegias fabula* in 1989 provided support for the reclassification of this species from endangered to threatened status (USFWS 1989). Goals and objectives of the plan that were implemented from ecological studies have indicated that population density and size (i.e., the increase in river reach inhabited by species) are increasing. In 2002, four endangered mussels, including *P. fabula*, were artificially propagated via host fish and cultured in hatchery conditions for approximately two weeks until release into the Cumberland River: *E. brevidens*—36,400; *Epioblasma florentina walkeri*—4,650; *Villosa trabalis*—1,100; and *P. fabula*—570 (Jones, Mair, and Neves 2002). *Epioblasma f. walkeri* was listed as endangered by the USFWS in 1977 (Federal register Vol. 42, No. 163). A recovery plan for this species was proposed in the federal register for the species and associated habitat.

Hiwassee River, Tennessee

The Hiwassee River in Tennessee was selected to help determine the suitability of this waterbody to support survival and growth of artificially propagated (via host fish) juvenile mussels. The federally endangered species, *Epioblasma f. walkeri*, was propagated in hatchery conditions and cultured for approximately one week followed by their reintroduction into the Hiwassee River, TN. Approximately 7,000 *Epioblasma florentina* juveniles were released in 2000, and monitoring of survival and growth will be conducted in the future to determine the success of this reintroduction effort (Jones and Neves 2000).

FEDERAL HATCHERY GOALS

A relatively new approach to supporting propagation and refugia efforts includes the use of federal fish hatcheries as research facilities. The federal hatchery system through USFWS was changing in the early nineties to enlarge the scope of federal hatcheries from a conventional stocking and management program for commercial fisheries to include conservation recovery and restoration goals for other biological species including those listed by the USFWS as threatened or endangered (personal communication, J Geiger 2004). This new scope that identifies National Fish Hatcheries (NFH) as providing support for unionid (as one example) research allowed many of the hatcheries in the early nineties to avoid being shut down as a result of federal cuts. Several hatcheries around North America have been utilized as research laboratories and have been shown to successfully support host-fish identification, T&E species recovery, refugium for listed and non-listed species, and juvenile culturing.

MAMMOTH SPRING NFH, ARKANSAS

Refugia. Over 2,500 individuals comprising 28 species of native unionids from the White and Ouachita Rivers in Arkansas have been held in refugia at the Mammoth Spring National Fish Hatchery since 1995.This refugium was designed to hold adult mussels in response to a significant zebra mussel infestation predicted by personnel at the state game and fish commission. Species were held and monitored for survival and physiological condition (cellulolytic enzyme activity), using surrogate species, for four years (however, some species are still surviving in the hatchery raceways nearly seven years after initial collection). Survival from year one (90%) to year four (60%) was measured and indicated that the hatchery provided suitable conditions (e.g., high water quality, adequate food source, and continuous water temperatures throughout the year) for short-and long-term holding of native mussels.

Propagation

Since 1994, this hatchery has supported freshwater mussel propagation for recovery and restoration projects in Arkansas and Ohio (personal communication, R Shelton 2004). Six species, of which two are federally endangered, have been successfully propagated using a combination of host fish and artificial media for the production of juveniles (*L. streckeri, Arkansia wheeleri, P. grandis, L. siliquoidea, L. ventricosa, Fusconaia flava,* and *U. imbecillis*). Approximately 10,000 juveniles (listed above) were maintained in recirculating streams for several weeks and reintroduced into watersheds to support the restoration goals of the USFWS.

LOST VALLEY SFH, MISSOURI

Since 2002, personnel at the Lost Valley State Fish Hatchery have successfully propagated, via host fish, approximately 5,000 *E. triquetra* and 40,000 *L. teres* juveniles (personal communication, S Bruenderman 2004). *E. triquetra* is considered rare by the state of Missouri and is currently listed as a candidate species by USFWS. The release of these juveniles in the summer of 2002 occurred in the Big and Bourbeuse Rivers, MO. No determination of survival and growth has been measured for these individuals.

WARM SPRINGS NFH, GEORGIA

Refugia

Due to drought conditions that were occurring in a small tributary of the Flint River, GA, 1,500 individual mussels were transported to the Warms Springs National Fish Hatchery in the late 1990s (personal communication, C Echevarria 2004). Two species were federally listed as endangered: *Lampsilis subangulata* and *Pleurobema pyriforme*. Most of the specimens recovered from the dry tributary were maintained at the hatchery in recirculating tanks for approximately one year. Bi-annual glycogen concentration is being proposed as a measurement of unionid condition for specimens held in refugia to determine whether hatchery conditions are suitable for long-term holding. Approximately 1,100 mussels were returned to the creek after normal stream flows had returned.

Propagation

Propagation efforts at the hatchery began in 2000 using a variety of host fish. During the summer of 2001, hatchery managers reported the successful transformation of *Villosa vibex*, *V. lienosa*, and *L. subangulata* juveniles. *L. subangulata* is listed as endangered by the federal government, and consideration of this listing has prompted hatchery personnel to focus efforts on propagating this and other species in the region. From these three species, nearly 8,000 juveniles were released into Spring Creek, GA. An additional 20,000 juveniles have been maintained in laboratory conditions and are being monitored for growth and survival of viable juveniles in these hatchery conditions.

White Sulphur Springs NFH, West Virginia

Refugia

In response to an emergency salvage order, White Sulfur Springs was involved in the collection and holding of various mussels species from the Ohio River in 1995. This action led to the development of a refugium and propagation program on hatchery grounds. While high mortality occurred in mussels held in less than 5 cm of substrate during winter months, the following years yielded a much higher survival of mussels held in containers with at least 20 cm of substrate.

Propagation

The propagation of two common mussels, *L. fasciola* and *V. iris*, indicated that conditions at the hatchery may be limiting for the successful transformation of other species. While juveniles were successfully propagated using fish host techniques, mean survival of *V. iris* and *L. fasciola* juveniles following three months was 50% and 6%, respectively.

Genoa NFH, Wisconsin

Propagation

The Genoa hatchery is focusing its efforts specifically on the propagation and reintroduction of juvenile *Lampsilis higginsi*. Various propagation techniques are being implemented including hatchery propagation (using host fish) and the holding of juveniles for survival and growth. Over two years, approximately 180,000 juveniles were released into watersheds known to maintain existing populations. While survival and growth in hatchery holding trials dropped, researchers determined that freshwater hydrozoans and flatworms were a significant factor in juvenile survival. Other propagation techniques include the free release and cage release of infested host fish. Nearly 3,500 host fish were released in 2001, and results indicate that for cage releases, 250 juveniles (5–10 mm in length) have been transformed and are surviving riverine conditions following 90 days.

CRITICAL USES OF EARLY LIFESTAGE UNIONIDS FOR MONITORING

The life history of most unionids includes critical steps between internal development in the adult female to the release of viable glochidia (varying in number from thousands to millions) into the water column for attachment onto specific fish hosts. This post-release stage, in which the glochidia are exposed to the aquatic environment, can be the most critical (Huebner and Pynnonen 1992; Neves 1997). High levels of mortality occur during this passage due to a low incidence of fish-host contact, although once attached to the gill, glochidia are relatively protected from in situ exposure to water-column metal mixtures (Jacobson 1990). Newly released juveniles may also be particularly sensitive to water- and sediment-associated contaminants due to their siphoning and burrowing activities once transformation occurs.

The continued improvement of artificial propagation techniques has bought scientists time in an effort to restore declining populations. It seems that identification of the source and magnitude of impairments that cause measurable declines in certain species are more critical at this point for the long-term management of unionids. Rather than provide a temporary fix to the problem, perhaps resources should continue to be utilized to determine effect levels of various contaminants from point and non-point sources (Belanger et al. 1993; Farris et al. 1994). Not only have unionids been used to expand the research knowledge of propagation, including the species-specific requirements, but have also been used in monitoring (Humphrey, Bishop, and Brown 1990) and determining sensitivities among species to various contaminants.

There have been numerous studies that include the determination of threshold limits (e.g., LC50, EC_{50}, IC_{25}, etc.) using early life stages of freshwater bivalves (Wade, Hudson, and McKinney 1989; Keller and Zam 1991; Huebner and Pynnonen 1992; Jacobson et al. 1993; Wade, Hudson, and McKinney 1993; Warren 1996; Jacobson et al. 1997; Milam and Farris 1998; Hudson et al. 2003); however, these assays are generally restricted to single-metal (more specifically as total recoverable) exposures using *U. imbecillis*, a lentic pond mussel. Comparisons using early life stages and subsequent threshold responses to pesticides are less abundant in the literature (Johnson, Keller, and Zam 1993; Keller 1993; Hudson et al. 1994; Keller and Ruessler 1997; Heinonen, Kukkonen, and Holopainen 2001; Milam et al. 2004). While reported acute responses (LC50) provide information on bivalve effects and relative sensitivities to a suite of chemicals, there is a need to evaluate additional unionid species that require different habitats. Since a majority of the federally T&E species inhabit lotic systems, inclusion of stream-dwelling unionids in developing sensitivity comparisons to contaminants should offer a wider range of protection for this group of unionids (Milam et al. 2004). The only reported study making such a comparison involved the species *U. imbecillis* and *E. angustata* (Hudson, Barfield, and McKinney 1996). In this study, the LC50 of the lentic species (*U. imbecillis*) was similar in sensitivity to the lotic species (*E. angustata*), 2.75 and 2.81 µg Cu/L, respectively.

Beyond early lifestage testing and monitoring, researchers have evaluated adult unionids to determine threshold effects (e.g., LC50, EC_{50}, IC_{25}, etc.) that now provide protective, site-specific criteria for freshwater mussels (Dobbs et al. 1994; Cherry et al. 2002). Valuable data gathered from monitoring efforts can lead to broader investigations than just using specimens to estimate acute and chronic toxicity endpoints (Farris et al. 1988; Jacobson et al. 1993). These endpoints, such as survival, growth, and reproduction, are used to predict concentrations where effects begin to occur and establish guidelines for compliance. The endpoints are used to estimate critical concentrations, effects thresholds, or guidelines for compliance. Glochidia from the same marsupia used for propagation/restoration efforts can also be used to determine contaminant effects that will provide the much-needed acute toxicity endpoints for various contaminants, all the while satisfying multi-agency objectives. There is much room for collaboration between taxonomists, conservationists, ecologists, and toxicologists that, if all are working toward the same goals, could offer significant contributions to our understanding of mussels. Stopping the decline will undoubtedly take interdisciplinary approaches that we are only beginning to understand.

REFERENCES

Allen, E., The existence of a short reproductive cycle in *Anodonta imbecillis*, *Biol. Bull.*, 46, 88–94, 1924.

Baird, M. S., Life history of the spectaclecase, *Cumberlandia monodonta* Say, 1892 (Bivalvia, Unionoidea, Margaritiferidae), MS Thesis, Southwest Missouri State University, Springfield, MO, 2000.

Barnhart, M.C. and Roberts, A., Reproduction and fish hosts of unionids from the Ozark Uplifts, In *Conservation and Management of Freshwater Mussels II: Initiatives for the Future. Proceedings of a UMRCC Symposium*, 16–18 October 1995, St. Louis, MO, pp. 16–20, Cummings, K. S., Buchanan, A. C., Mayer C. A., Naimo, T. J., Eds., Upper Mississippi River Conservation Committee Rock Island, IL 293, 1997.

Barfield, M. L., Clem, S. A., and Farris, J. L., Comparison of acute testing methods of *Utterbackia imbecillis* (Say) with variation on age, diet, and in vitro transformation method, In *Conservation and Management of Freshwater Mussels II: Initiatives for the Future. Proceedings of a UMRCC Symposium*, 16–18 Oct, 1995, St. Louis, MO, Cummings, K. S., Buchanan, A. C., Mayer, C. A., and Naimo, T. J., Eds., UMRCC, Rock Island, IL, p. 293, 1997.

Belanger, S. E., Davidson, D. H., Farris, J. L., Reed, D. L., and Cherry, D. S., Effects of cationic surfactant exposure in stream mesocosms to a bivalve mollusc, *Environ. Toxicol. Chem.*, Special journal issue, Risk assessment of surfactants, 12, 1789–1802, 1993.

Castagna, M. and Kraeuter, J. N., Manual for growing the hard clam Mercenaria, Special Report No. 249, Virginia Institute of Marine Science, p. 110, 1981.

Cherry, D. S., Van Hassel, J. H., Farris, J. L., Soucek, D. J., and Neves, R. J., Site-specific derivation of the acute copper criteria for the Clinch River, Virginia, *Hum. Evol. Risk Assess.*, 8, 591–601, 2002.

Clem, S. A., Complexities in testing early lifestage responses of freshwater mussels to metals and contaminated sediments, MS Thesis, Arkansas State University, Jonesboro, AR, 1998.

Coker, R. E., Shira, A. F., Clar, H. W., and Howard, A. D., Natural history and propagation of fresh-water mussels, *Bull. US. Bur. Fish.*, 37, 75–181, 1921.

Connor, C. H., Glochidia of *Unio* on fishes, *Nautilus*, 18, 142–143, 1905.

Cope, O. B., New parasite records from stickleback and salmon in an Alaskan stream, *Trans. Am. Microsc. Soc.*, 789, 157–162, 1959.

Dee, S. H. and Watters, G. T., New or confirmed host identifications for ten freshwater mussels, Freshwater Mollusk Symposia Proceedings, Tankersley, R. A., Warmolts, D. I., Watters, G. T., Armitage, B. J., Johnson, P. D., and Butler, R. S., Eds., Ohio Biological Survey, Columbus, OH, pp. 77–82, see also xxi + 274, 1998.

Dimmock, R. V. and Wright, A. H., Sensitivity of juvenile freshwater mussels to hypoxic thermal and acid stress, *Elish. Mitchell. Soc.*, 109, 183–192, 1993.

Dobbs, M. G., Farris, J. L., Reash, R. J., Cherry, D. S., and Cairns, J., Evaluation of the resident species procedure for developing site-specific water quality criteria for copper in Blaine Creek, Kentucky, *Environ. Toxicol. Chem.*, 13, 963–971, 1994.

Dunn, C. S. and Layzer, J. B., Evaluation of various holding facilities for maintaining freshwater mussels in captivity, In *Conservation and Management of Freshwater Mussels II: Initiatives for the Future, Proceedings of a UMRCC Symposium*, 16–18 October 1995, St. Louis, MO, pp. 205–213, Cummings, K. S., Buchanan, A. C., Mayer, C. A., and Naimo, T. J., Eds., Upper Mississippi River Conservation Committee, Rock Island, IL, p. 293, 1997.

Ellis, M. M. and Ellis, M. D., Growth and transformation of parasitic glochidia in physiological nutrient solutions, *Science*, 54(1667), 579–580, 1926.

Ellis, M. M., Merrick, A. D., and Ellis, M. D., The blood of North American fresh-water mussels under normal and adverse conditions, *Bull. Bur. Fish.*, 56, Document 1097, pp. 509–542, 1930.

Farris, J. L., Van Hassel, J. H., Belanger, S. E., Cherry, D., and Cairns, J., Application of cellulolytic activity of *Corbicula* to in-stream monitoring of power plant effluents, *Environ. Toxicol. Chem.*, 7, 701–713, 1988.

Farris, J. L., Grudzien, J. L., Belanger, S. E., Cherry, D. S., and Cairns, J., Molluscan cellulolytic activity responses to zinc exposure in laboratory and field stream comparisons, *Hydrobiologia*, 287, 161–178, 1994.

Farris, J. L., Milam, C. D., and Harris, J. L., Zebra mussel impacts on freshwater mussels in Arkansas, Final report to the Arkansas Game and Fish Commission, Little Rock, AR, p. 68, 1998.

Fisher, G., Physiological condition of in vitro- and in vivo-reared juvenile mussels. Abstract, FMCS Workshop, Shepherdstown, WV, Mar. 14–15, p. 11, 2002.

Fisher, G. R. and Dimmock, R. V. Jr., Viability of glochidia of Utterbackia imbecillis (Bivalvia: Unionidae) following their removal from the parental mussel, In Freshwater Mollusk Symposia Proceedings, Tankersley, R. A., Warmolts, D. I., Watters, G. T., Armitage, B. J., Johnson, P. D., and Butler, R. S., Eds., Ohio Biological Survey, Columbus, OH, p. xxi+, pp. 185–188, 274, 2000.

Gatenby, C. M., Parker, B. C., and Neves, R. J., Growth and survival of juvenile rainbow mussels, Villosa iris (Lea, 1829) (Bivalvia: Unionidae), reared on algal diets and sediment, Am. Malacol. Bull., 14, 57–66, 1997.

Giusti, F., Castagnolo, L., Moretti, F., Renzoni, L., and Renzoni, A., The reproductive cycle and the glochidium of Anodonta cygnea from Lago Trasimeno (Central Italy), Monitore Zoologico Italiano, 9, 99–118, 1975.

Haag, W. R., Fish host determination. Abstracts, Freshwater Mussel Workshop, Southwest Missouri State University, Springfield, MO, 2002.

Haag, W. R. and Staton, J. L., Variation in fecundity and other reproductive traits in freshwater mussels, Freshwat. Biol., 48, 2118–2130, 2003.

Hanlon, S. D., Release of Juvenile Mussels into a Fish Hatchery Raceway: Comparison of Techniques, MS Thesis, Virginia Polytechnic and State University, Blacksburg, VA, 2000.

Heinonen, J., Kukkonen, V. K., and Holopainen, I. J., Temperature- and parasite- induced changes in toxicity and lethal body burdens of PCP in the freshwater clam Pisidium amnicum, Environ. Toxicol. Chem., 20, 2778–2784, 2001.

Hoeh, W. R., Phylogenetic relationships among Eastern North American Anodonta (Bivalvia: Unionidae), Mal. Rev., 23, 63–82, 1990.

Hove, M. C. and Neves, R., Distribution and life history of the James River spinymussel, Endangered Species Tech. Bull., 16, 9, 1991.

Hove, M. C., Engelking, R. A., Peteler, M. E., Peterson, E. M., Kapusciniski, A. R., Sovell, L. A., Evers, E. R., Suitable fish hosts for glochidia of four freshwater mussels, In Conservation and Management of Freshwater Mussels II: Initiatives for the Future, Proceedings of a UMRCC Symposium, 16–18 October 1995, St. Louis, MO, pp. 21–25, Cummings, K. S., Buchanan, A. C., Mayer, C. A., Naimo, T. J., Eds., Upper Mississippi River Conservation Committee Rock Island, IL 293, 1997.

Hove, M. C., Hillegass, K. R., Kurth, J. E., Pepi, V. E., Lee, C. J., Knudsen, K. A., Kapuscinski, A. R., Mahoney, P. A., Bomier, M. M., Considerations for conducting host suitability studies, In Freshwater Mollusk Symposia Proceedings, pp. 27–34, Tankersley, R. A., Warmolts, D. I., Watters, G. T., Armitage, B. J., Johnson, P. D., Butler, R. S., Eds., Ohio Biological Survey Columbus, OH p. xxi+274, 2000.

Howard, A. D., A new record in rearing fresh-water pearl mussels, Trans. Am. Fish. Soc., 44, 45–47, 1914.

Howard, A. D., A second generation of artificially reared fresh-water mussels, Trans. Am. Fish. Soc., 46, 39–49, 1916.

Hubbs, D., Augmentation of natural reproduction by freshwater mussels to sustain shell harvest, pp. 49–52, In Freshwater Mollusk Symposia Proceedings, Tankersley, R. A., Warmolts, D. I., Watters, G. T., Armitage, B. J., Johnson, P. D., and Butler, R. S., Eds., Ohio Biological Survey, Columbus, OH, p. xxi+274, 2000.

Hudson, R. G. and Isom, B. G., Rearing juveniles of the freshwater mussels (Unionidae) in the laboratory setting, Nautilus, 98, 140–147, 1984.

Hudson, R. G. and McKissick, R. Q., Chronic toxicity effects of Conasauga and Teleco River sediments when used to culture freshwater juvenile mussels, A report submitted to the Tennessee Wildlife Resources Division, p. 49, 1999.

Hudson, R. G. and Roberts, C. E., Chronic toxicity effects of Pigeon River sediment when used to culture freshwater juvenile mussels, A report submitted to the Tennessee Wildlife Resources Division, p. 61, 1997.

Hudson, R. G. and Shelbourne, C. W., Improved in vitro culture of parasitic freshwater mussel glochidia, Report to the Tennessee Valley Authority, p. 25, 1990.

Hudson, R. G., Pardue, G. B., and Roberts, J. F., Ictalurid chromosome preparation by cell culture and squash methods, Prog. Fish-Cult., 42, 43–45, 1980.

Hudson, R. G., Wade, D. C., McKinney, A. D., and Martin, R. D., Comparative response of selected non-target aquatic organisms to commercial molluscacide, *Abstracts*, 15th Annual Meeting, Society of Environmental Toxicology and Chemistry, Denver, CO, October 30–November 3, p. 186, 1994.

Hudson, R. G., Barfield, M. L., and McKinney, A. D., Species specificity and effect of pH on the response of freshwater mussel juveniles to acute copper toxicity, Poster presentation at the 88th Annual National Shellfish Association Meeting, Baltimore, April 14–18, 1996.

Hudson, R. G., McKinney, D. A., Wetzel, J. T., Griner, J. G., Brinson, A. L., and Hinesley, J. H., Effects of drilling agents on the growth and survival of juvenile mussels, *Abstracts*, 3rd Biennial Symposium of the Freshwater Mollusk Conservation Society, Durham, March 16–19, p. 44, 2003.

Huebner, J. D. and Pynnonen, K. S., Viability of glochidia of two species of *Anodonta* exposed to low pH and selected metals, *Can. J. Zool.*, 70, 2348–2355, 1992.

Humphrey, C. L., Bishop, K. A., and Brown, V. M., Use of biological monitoring in the assessment of effects of mining wastes on aquatic ecosystems of the Alligator Rivers region, tropical northern Australia, *Environ. Monit. Assess.*, 14, 139–181, 1990.

Isom, B. G. and Hudson, R. G., In vitro culture of parasitic freshwater mussel glochidia, *Nautilus*, 96, 147–151, 1982.

Jacobson, P. J., Sensitivity of early lifestages of freshwater mussels (Bivalvia: Unionidae) to copper, MS thesis, Virginia Polytechnic Institute and State University, Blacksburg, VA, 1990.

Jacobson, P. J., Farris, J. L., Cherry, D. S., and Neves, R. J., Juvenile freshwater mussel (Bivalvia: Unionidae) responses to acute toxicity testing with copper, *Environ. Toxicol. Chem.*, 12, 879–883, 1993.

Jacobson, P. J., Neves, R. J., Cherry, D. S., and Farris, J. L., Sensitivity of glochidial stages of freshwater mussels (Bivalvia: Unionidae) to copper, *Environ. Toxicol. Chem.*, 16, 2384–2392, 1997.

Jenkinson, J., Cumberlandian mollusk conservation program, pp. 95–103, In *Report of Freshwater Mollusks Workshop*, Miller, A. C., Ed., U.S. Army Engineer Waterways Experiment Station, Environmental Laboratory, Vicksburg, MS, p. 185, 1982.

Johnson, I. S., Keller, A. E., Zam, S. G., A method for conducting acute toxicity tests with the early life stages of freshwater mussels. Standard Technical Publication 1179, ASTM, Philadelphia, PA, 1993.

Jones, J. W. and Neves, R. J., Life history and artificial culture of endangered mussels, Annual Progress Report for Tennessee Wildlife Resources Agency, Nashville, TN, p. 52, 1999.

Jones, J. W. and Neves, R. J., Life history and artificial culture of endangered mussels, Annual Progress Report for Tennessee Wildlife Resources Agency, Nashville, TN, p. 62, 2000.

Jones, J. W., Mair, R., and Neves, R. J., Life history and artificial culture of endangered mussels, Annual Progress Report for Tennessee Wildlife Resources Agency, Nashville, TN, p. 80, 2002.

Keller, A. E., Acute toxicity of several pesticides, organic compounds, and wastewater effluent to the freshwater mussel, *Anodonta imbecillis*, *Ceriodaphnia dubia*, and *Pimephales promelas*, *Bull. Environ. Contam. Toxicol.*, 51, 696–702, 1993.

Keller, A. E. and Ruessler, D. S., The toxicity of malathion to unionid mussels: Relationship to expected environmental concentrations, *Environ. Toxicol. Chem.*, 16, 1028–1033, 1997.

Keller, A. E. and Zam, S. G., Simplification of in vitro culture techniques for freshwater mussels, *Environ. Toxicol. Chem.*, 9, 1291–1296, 1990.

Keller, A. E. and Zam, S. G., The acute toxicity of selected metals to the freshwater mussel, *Utterbackia* (*formally Anodonta*) *imbecillis*, *Environ. Toxicol. Chem.*, 10, 539–546, 1991.

Layzer, J. B., Gordon, M. E., and Anderson, R. M., Mussels: The forgotten fauna of regulated rivers. A case study of the Caney Fork River, *Regul. Rivers: Res. Manage.*, 8, 63–71, 1993.

Lefevre, G. and Curtis, W. C., Studies on the reproduction and artificial propagation of freshwater mussels, *Bull. US. Bur. Fish.*, 30, 105–201, 1912.

Loveless, R., Raymond, S., Inman, J., Hudson, R., and Fairfax, M., Isolation, identification and control of bacteria which threaten in vitro cultures of freshwater mussel glochidia, *Abstracts*, 1st Symposium of the Freshwater Mollusk Conservation Society, Chattanooga, TN, March, p. 79, 1999.

Michaelson, D. L. and Neves, R. J., Life history and habitat of the endangered dwarf wedgemussel *Alasmidonta heterodon* (Bivalvia: Unionidae), *J. N. Am. Benthol. Soc.*, 14(2), 324–340, 1995.

Milam, C. D. and Farris, J. L., Risk identification associated with iron-dominated mine discharges and their effect upon freshwater bivalves, *Environ. Toxicol. Chem.*, 17, 1611–1619, 1998.

Milam, C. D., Farris, J. L., Van Hassel, J., and Barfield, M. L., Reintroduction of native freshwater mussels using in vivo and in vitro propagation techniques, pp. 53–61, In *Freshwater Mollusk Symposia Proceedings*, Tankersley, R. A., Warmolts, D. I., Watters, G. T., Armitage, B. J., Johnson, P.D, and Butler, R. S., Eds., Ohio Biological Survey, Columbus, OH, p. xxi+274, 2000.

Milam, C. D., Farris, J. L., Dwyer, F. J., and Hardesty, D. K., Acute toxicity of six freshwater mussel species (glochidia) to six chemicals: Implications for daphnids and *Utterbackia imbecillis* as surrogates for protection of freshwater mussels (Unionidae), *Arch. Environ. Contam. Toxicol.*, 48, 166–173, 2004.

Morgan, A., Welker, N. J., and Layzer, J. B., Feasibility of reintroducing threatened and endangered mussels into Shoal Creek in Alabama and Tennessee, In *Conservation and Management of Freshwater Mussels II: Initiatives for the Future, Proceedings of a UMRCC Symposium*, October 1995, St. Louis, MO, p. 196, Cummings, K. S., Buchanan, A. C., Mayer, C. A., and Naimo, T. J., Eds., Upper Mississippi River Conservation Committee, Rock Island, IL, p. 293, 1997.

Mummert, A. K., Evaluating the Feasibility of Rearing Juvenile Freshwater Mussels in a Flow-Through Pond System at White Sulphur Springs National Fish Hatchery, MS Thesis, Virginia Polytechnic Institute and State University, Blacksburg, VA, p. 164, 2001.

Myers-Kinzie, M., In vitro transformation of *Lampsilis siliquoidea*: A suitable species for glochidial and juvenile research, pp. 63–66, In *Freshwater Mollusk Symposia Proceedings*, Tankersley, R. A., Warmolts, D. I., Watters, G. T., Armitage, B. J., Johnson, P. D., and Butler, R. S., Eds., Ohio Biological Survey, Columbus, OH, p. xxi+274, 2000.

Neves, R. J., A state-of-the-unionids address, In *Conservation and Management of Freshwater Mussels II: Initiatives for the Future, Proceedings of a UMRCC Symposium*, October 1995, St. Louis, MO, p. 1, Cummings, K.S, Buchanan, A. C., Mayer, C. A., and Naimo, T. J., Eds., Upper Mississippi River Conservation Committee, Rock Island, IL, p. 293, 1997.

O'Beirn, F. X., Neves, R. J., and Steg, M. B., Survival and growth of juvenile freshwater mussels (Unionidae) in a recirculating aquaculture system, *Am. Malacol. Bull.*, 14, 165–171, 1998.

Obermeyer, B. K., Recovery plan for four freshwater mussels in Southeast Kansas, Final report for Kansas Department of Wildlife and Parks, Pratt, KS, p. 93, 2000.

Payne, B. S. and Miller, A. C., Recruitment of *Fusconaia ebena* (Bivalvia: Unionidae) in relating to discharge of the lower Ohio River, *Am. Midl. Nat.*, 144, 328–341, 2000.

Penn, G. H., A study of the life cycle of the fresh-water mussel, *Anodonta grandis*, in New Orleans, *Nautilus*, 52, 99–101, 1939.

Riusech, F. A. and Barnhart, M. C., Host suitability and utilization in *Venustaconcha ellipsiformis* and *Venustaconcha pleasii* (Bivalvia: Unionidae) from the Ozark region, In *Freshwater Mollusk Symposia Proceedings*, pp. 83–92, Tankersley, R.A., Warmolts, D.I., Watters, G.T., Armitage, B.J., Johnson, P.D., Butler, R.S., Eds., Ohio Biological Survey, Columbus, OH, xxi+274, 2000.

Seshaiya, R. V., Some observations on the life-histories of south Indian freshwater mussels, *Malacologia*, 9, 286–287, 1969.

Shiver, M. A., Reproduction and propagation of the neosho mucket, *Lampsilis rafinesqueana*, MS Thesis, Southwest Missouri State University, Springfield, MO, 2002.

Starkey, R. W., Eversole, A. G., and Brune, D. E., Growth and survival of juvenile and adult freshwater mussels in the Partitioned Aquaculture System, In *Freshwater Mollusk Symposia Proceedings*, Tankersley, R. A., Warmolts, D. I., Watters, G. T., Armitage, B. J., Johnson, P. D., and Butler, R. S., Eds., Ohio Biological Survey, Columbus, OH, pp. 109-114, 2000.

Tankersley, R. A., Fluorescence techniques for evaluating lipid content of larval and juvenile mussels, pp. 115–126, In *Freshwater Mollusk Symposia Proceedings*, Tankersle, R. A., Warmolts, D. I., Watters, G. T., Armitage, B. J., Johnson, P. D., and Butler, R. S., Eds., Ohio Biological Survey, Columbus, OH, p. xxi+274, 2000.

U.S. Fish and Wildlife Service, Fanshell (*C. stegaria*) Recovery Plan, U.S. Fish and Wildlife Service, Atlanta, GA, p. 37, 1991.

U.S. Fish and Wildlife Service, *Little-wing Pearly Mussel Recovery Plan*, U.S. Fish and Wildlife Service, Atlanta, Georgia, p. 29, 1989.

U.S. Fish and Wildlife Service, *Speckled Pocketbook Mussel (Lampsilis streckeri)* Recovery Plan, U.S. Fish and Wildlife Service, Jackson, Mississippi, p. 14, 1991.

Wade, D. C., Hudson, R. G., and McKinney, A. D., The use of juvenile freshwater mussels as a test species for evaluating environmental toxicity, *Abstracts*, 10th Annual Meeting, Society of Environmental Toxicology and Chemistry, Toronto, Ontario, October 28–November 2, p. 247, 1989.

Wade, D. C., Hudson, R. G., McKinney, A. D.,Comparative response of *Ceriodaphnia dubia* and juvenile *Anodonta imbecillis* to selected complex industrial whole effluents, In *Conservation and Management of Freshwater Mussels. Proceedings of a UMRCC Symposium*, Cummings, K. S., Buchanan, A. C., Koch, L. M., Upper Mississippi River Conservation Committee, Rock Island, IL, 109–112, 1993.

Warren, L. W., The use of juvenile mussels, *Utterbackia imbecillis* Say (Bivalvia: Unionidae), as a standardized toxicity testing organism, Ph.D. Dissertation, Clemson University, Pendleton, SC, p. 138, 1996.

Warren, L. W. and Klaine, S. J., Standardization of the juvenile mussel bioassay: Dietary requirements and life history, *Abstracts*, 15th Annual Meeting, Environmental Toxicology and Chemistry, 1994.

Watters, G. T., An annotated bibliography of the reproduction and propagation of the Unionoidea (Primarily of North America), Ohio Biol. Surv. Misc. Cont. No. 1, p. vi + 158, 1994.

Watters, G. T., Cross-polarized light to detect glochidia, Triannual Unionid Report, 9, 7, 1996.

Winterringer, R. L., Life history and population biology of the endangered speckled pocketbook (*Lampsilis streckeri*), MS Thesis, Arkansas State University, Jonesboro, AR, 2004.

Yeager, M. M., Cherry, D. S., and Neves, R. J., Feeding and burrowing behaviors of juvenile rainbow mussels, *Villosa iris* (Bivalvia: Unionidae), *J. N. Am. Benthol. Soc.*, 13, 217–222, 1994.

Zale, A. V. and Neves, R. J., Fish hosts of four species of Lampsiline mussels (Mollusca: Unionidae) in Big Moccasin Creek, *Virginia. Can. J. Zool.*, 60, 2535–2542, 1982.

Zimmerman, L. L. and Neves, R. J., Effects of temperature on duration of viability for glochidia of freshwater mussels (Bivalvia: Unionidae), *Am. Malacol. Bull.*, 17, 31–35, 2002.

5 Laboratory Toxicity Testing with Freshwater Mussels

Christopher G. Ingersoll, Nicola J. Kernaghan, Timothy S. Gross, Cristi D. Bishop, Ning Wang, and Andy Roberts

INTRODUCTION

Numerous laboratory toxicity studies have been conducted with freshwater mussels in an attempt to understand the role of contaminants in the decline of field populations of mussels (Chapter 7). In these studies, early life stages of mussels of several species were highly sensitive to some metals and ammonia in water exposures when compared to many of the most sensitive species of other invertebrates, fish, or amphibians that are commonly used to establish U.S. Environmental Protection Agency (USEPA) Water Quality Criteria (WQC) (Augspurger et al. 2003; USGS 2005a, 2005b). Importantly, results of these studies indicate WQC for individual chemicals established for the protection of aquatic organisms may not be adequately protective of sensitive stages of freshwater mussels. This chapter provides a summary of methods from over 75 laboratory toxicity studies conducted with freshwater mussels and also provides an overview of a standardized method for conducting water-only acute and chronic laboratory toxicity tests with glochidia and juvenile freshwater mussels (ASTM 2006a). Three life stages (glochidia, juveniles, and adults) have been used to conduct laboratory toxicity tests with mussels. Within this chapter, toxicity studies are separated according to the medium of exposure (aqueous, sediment, and host fish). Each section begins with a review of the methods used to conduct toxicity tests (e.g., obtaining organisms, duration of exposure, exposure chambers, and toxicity endpoints). Each section also discusses issues that have been identified regarding the routine application of the methods (e.g., to generate data for the derivation of WQC) and discusses research needs. The final section of this chapter reviews the use of the Asian clam (*Corbicula fluminea*) as a surrogate for assessing effects on native unionids. Finally, a summary of future research needs for improving methods used to conduct acute and chronic toxicity tests with freshwater mussels is provided.

AQUATIC TOXICITY TESTING WITH GLOCHIDIA, JUVENILE, AND ADULT LIFE STAGES OF FRESHWATER MUSSELS

METHODS FOR CONDUCTING ACUTE WATER-ONLY TOXICITY TESTS WITH GLOCHIDIA OF FRESHWATER MUSSELS

Review of Methods

Conditions that have been used to conduct acute toxicity tests with glochidia of freshwater mussels are summarized in Table 5.1 including the test conditions recommended in ASTM (2006a). The procedures outlined in Table 5.1 are consistent with acute toxicity testing methods for fish, macroinvertebrates, and amphibians (ASTM 2006c) and with acute toxicity testing methods for saltwater

TABLE 5.1
Summary of Test Conditions Used to Conduct Toxicity Tests with Glochidia of Freshwater Mussels

	Conditions	Johnson et al. (1990, 1993)	Lasee (1991)	Huebner and Pynnonen (1992)[a]	Goudreau, Neves, and Sheehan (1993)	Jacobson et al. (1997)	Keller and Ruessler (1997)	McCann (1993)	Klaine, Warren, and Summers (1997)	USGS (Unpublished Data)	Recommended Test Conditions in ASTM (2006a)
1	Species tested	Utterbackia imbecillis[b]	Lampsilis cardium[c]	Anodonta cygnea, Anodonta anatina	Villosa iris	Multiple species[d]	Multiple species[e]	V. iris	U. imbecillis	Multiple species[f]	NA[g]
2	Test type	Static	Static	Static	Renewal	Static	Static	Static	Static	Static, renewal, flow-through	Static, renewal, or flow-through (depending on chemical tested)
3	Test duration (hours)	24	48	24, 48, 72, 144	24	24, 48	4, 24, 48	24	24, 48	6, 24, 48	6, 24 (up to 48 depending on viability of glochidia)
4	Temperature (°C)	20	21	13	22	10–25	25	20	25	20	20
5	Light quality	Ambient lab light	NR[h]	NR	NR	NR	NR	NR	Ambient lab light	Ambient lab light	Ambient lab light
6	Light intensity	NR	NR	NR	NR	NR	NR	NR	NR	200 lux	100–1000 lux
7	Photoperiod	16L:8D	24D	Natural regime	16L:8D	16L:8D	12L:12D	NR	16L:8D	16L:8D	16L:8D
8	Test chamber	100-mL beaker	250-mL or 300-ml beaker	400-mL beaker	Basket of mesh netting in 4-L chamber	12-well plate	6-well plate	12-well plate	12-well plate	200-mL dish or 300-mL beaker	100-mL glass chamber (minimum)
9	Test solution volume (mL)	50	200	200	NR	3.5	NR	5	3.5	100	75 (minimum)
10	Glochidia collection	Shake piece of cut gill in water	Flush gills with syringe	Cut gills and press out glochidia using forceps	Flush gills with syringe	Cut gills and separate glochidia from marsupia	NR	Flush gills with syringe	Flush gills with syringe	Flush gills with syringe	Flush gills with syringe
11	Age of test organisms (hours)	NR	NR	3–24	NR	NR	NR	<2	NR	<2 to <24	<24
12	Number of organisms per test chamber	10	10	1000–3000	Several hundreds	50–75	50–100	40	50–100	About 1000	About 500 (1000 for repeated sampling during a toxicity test)
13	Number of replicate chambers per treatment	2	3	2, Counting 3 samples with about 100 glochidia	2, Counting 3 samples with about 100 glochidia	3	3 or 4	3	3	3, Counting a subsample with about 100 glochidia from each replicate	3, Counting a subsample with about 100 glochidia from each replicate
14	Feeding	None	None	None	None	None	None	None	None	None	None
15	Aeration	None	None	Yes	None	None	NR	NR	NR	None	None, if dissolved oxygen is maintained above acceptable concentration

	Reconstituted water, hardness 40–50 mg/L as CaCO$_3$	Hardness 150 mg/L as CaCO$_3$	Tap water	Dechlorinated effluent water	Dechlorinated tap water or Clinch River water, VA	Reconstituted water, hardness 47–76 mg/L as CaCO$_3$	Sinking Creek water, VA	Hardness 99–107 mg/L as CaCO$_3$	Reconstituted water, hardness 170 mg/L as CaCO$_3$	Depends on experimental design
16 Dilution water	Reconstituted water, hardness 40–50 mg/L as CaCO$_3$	Hardness 150 mg/L as CaCO$_3$	Tap water	Dechlorinated effluent water	Dechlorinated tap water or Clinch River water, VA	Reconstituted water, hardness 47–76 mg/L as CaCO$_3$	Sinking Creek water, VA	Hardness 99–107 mg/L as CaCO$_3$	Reconstituted water, hardness 170 mg/L as CaCO$_3$	Depends on experimental design
17 Water quality	DO, pH, hardness, alkalinity, conductivity	DO, pH, hardness, alkalinity, conductivity	pH, Ca, Cu, Zn	DO, pH, hardness, alkalinity, conductivity	DO, pH, hardness, alkalinity, conductivity	DO, pH, hardness, alkalinity, conductivity	DO, pH, hardness, alkalinity, conductivity	DO, pH, hardness, alkalinity, conductivity	DO, pH, ammonia, hardness, alkalinity, conductivity	DO, pH, ammonia, hardness, alkalinity, conductivity
18 Endpoint	Survival (valve closure with culture medium)	Survival (valve closure with NaCl)	Survival (valve closure with KCl)	Survival (valve closure with NaCl)	Survival (valve closure with NaCl)	Survival (valve closure with NaCl)	Survival (valve closure with salt solution)	Survival (valve closure with saline solution)	Survival (valve closure with NaCl)	Survival (valve closure with NaCl)
19 Control survival (%)	>95	>90	>80	80	>90	>80	>80	80	>90	>90 (must)

The Last Column Provides a Summary of Recommended Conditions That Can be Used to Conduct Toxicity Tests with Glochidia Based on ASTM (2006a)

[a] See also Pynnonen (1995); Hansten et al. (1996).

[b] Formerly *Anodonta imbecillis*. See also Weinstein (2001).

[c] Formerly *Lampsilis ventricosa*.

[d] *V. iris, A. pectorosa, Pyganodon grandis, L. fasciola, Medionidus conradius*. See also Jacobson (1990); Cherry et al. (2002).

[e] *Villosaosa lienosa, Villosa villosa, U. imbecillis, Megalonaias nervosa, Lampsilis teres, L. siliquoidea*. See also Jacobson (1990), McCann (1993), *V. iris, A. pectorosa, M. conradius*.

[f] *Actinonaias ligamentina, Alasmidonta heterodon, Epioblasma capsaefotmis, Lampsilis siliquoidea, Lampsilis fasciola, Lampsilis abrupta, L. rafinesqueana, Potamilus ohiensis, Pleurobema plenum, Quadrula quadrula, Quadrula pustulosa, Leptodea fragilis, Leptodea leptodon, Venustaconcha ellipsiformis, V. iris*.

[g] NA, not applicable; NR, not reported.

[h] USGS unpublished data, Columbia, MO.

bivalve mollusks (ASTM 2006c). Gravid female mussels are usually collected from the field and held in the laboratory before isolating glochidia to start a toxicity test (ASTM 2006a, Chapter 4). Alternatively, Zimmerman and Neves (2002) suggested glochidia of some species (including *Villosa iris* and *Actinonaias pectorosa*) could be extracted in the field from a female and transported back to the laboratory in cool water where glochidia can remain viable for several days without a reduction in ability to successfully attach to a host fish. This procedure may be particularly useful when glochidia of endangered species are extracted in the field and the female mussels are then immediately returned to their habitat. Mature glochidia are typically flushed from the marsupium of a female mussel using a syringe filled with water. Glochidia have also been isolated by cutting a section of gill from the female mussel and then teasing out the glochidia in water. (This technique is destructive to the adult female and may not be appropriate for use in isolating glochidia for conducting toxicity tests.) No studies were identified where glochidia were isolated for toxicity testing from conglutinates released into the water by female mussels.

Before starting an exposure, the viability of glochidia is typically evaluated by a response to the addition of a concentrated solution of NaCl or KCl. Mature and healthy glochidia will snap shut in response to the addition of a saline solution. Immature glochidia isolated from the marsupium of a female will often still be enclosed in the egg membrane and will be fragile and tend to fracture (Chris Barnhart, Missouri State University, Springfield, MO, personal communication). Tests are usually started if greater than 80 to greater than 90% viability of the glochidia is observed (Huebner and Pynnonen 1992; Jacobson et al. 1997; Klaine, Warren, and Summers 1997; ASTM 2006a). If immature glochidia are isolated from a female mussel, these glochidia should not be used for testing. Exposures are usually started the same day that glochidia are isolated from a female by pooling glochidia from at least three females without an extended acclimation period in the exposure water before the start of a toxicity test (ASTM 2006a). The viability of glochidia isolated from each female should be evaluated before they are pooled together. Toxicity tests can be conducted with glochidia obtained from one female (e.g., when a limited number of endangered species are available for testing); however, the results of tests conducted with a limited number of mussels should be interpreted with caution. Additional research is needed to determine the minimum number of females that should be sampled to obtain glochidia to start a toxicity test. This research might include an evaluation of the variability in sensitivity of glochidia obtained from individual females using a variety of toxicants.

ASTM (2006a) provides a list of recommended test conditions for conducting toxicity tests with glochidia isolated from female mussels. The list of recommended test conditions is based on the various methods outlined in Table 5.1 and on the conditions used to conduct an inter-laboratory toxicity test with glochidia (ASTM 2006a). ASTM (2006a) recommends that toxicity tests with glochidia should be conducted at 20°C with a 16L:8D photoperiod at an illuminance of about 100–1000 lux (Table 5.1). The endpoint measured in toxicity tests with glochidia is survival (viability) as determined by the response of organisms to the addition of a solution of NaCl. Glochidia that close their valves with the addition of a salt solution are classified as alive (viable) in a toxicity test. For most species, the duration of a toxicity test conducted with glochidia should be up to 24 hours with survival measured at 6 and 24 hours. Control survival is typically greater than 90% at the end of 24-hour toxicity tests conducted with glochidia. Longer duration toxicity tests with glochidia (e.g., 48 hours) can be conducted as long as control survival greater than 90% is achieved. However, toxicity tests conducted for greater than 24 hours with glochidia may not be as ecologically relevant given the short period of time between release of glochidia from a female mussel until encystment on a host fish (ASTM 2006a; Chapter 4). Effect concentrations are typically calculated based on the percentage of viable glochidia in the control at a particular sampling time.

ASTM (2006a) recommends the use of glass test chambers for conducting toxicity tests with glochidia. Test chambers should be a minimum volume of 100 mL containing a minimum of 75 mL of dilution water. Static, renewal, or flow-through conditions can be used depending on the chemical being tested. Glochidia are not fed during the toxicity test, and aeration of dilution water is typically not necessary. Dilution water should be a source of water that has been demonstrated to support survival of glochidia for the duration of the toxicity test. For site-specific evaluations, the characteristics of the dilution water should be as similar as possible to the site of interest. The number of replicates and concentrations tested depends in part on the significance level selected and the type of statistical analysis. ASTM (2006a) recommends a minimum of three replicates should be tested, each replicate containing about at least 500 glochidia (preferably 1,000 glochidia/replicate if survival is to be evaluated in subsamples of glochidia collected during the toxicity test). Survival can be determined throughout the toxicity test by subsampling each replicate (e.g., by subsampling about 100 glochidia at 6 and 24 hours and then placing these organisms into one well of a multi-well plate to determine survival with the addition of a salt solution).

Toxicity tests with glochidia have been conducted for up to 144 hours, but 24 and 48-hour exposures are most often used (Table 5.1). The relatively short duration of toxicity tests with glochidia is based on the relatively short duration between the release of glochidia into the water column and encystment on the host and on the relatively short survival time of glochidia after isolation from the female mussel (Table 5.2). If the life history of the glochidia for a particular species is not known (e.g., the host required for encystment or how long glochidia released from a female mussel can remain in the water column before encysting on a host), it might be appropriate to conduct toxicity tests with glochidia for longer than 24 hours as long as 90% control survival can be achieved at the end of the test (ASTM 2006a).

Issues Regarding the Use of Methods

Glochidia and juvenile mussels of several genera have been found to be highly sensitive to some metals and to ammonia in water exposures compared to many of the most sensitive genera of other invertebrates, fish, or amphibians that are commonly tested (Chapter 7, Cherry et al. 2002; Augspurger et al. 2003; USGS 2006a, 2006b). However, concerns have been expressed regarding the use of toxicity data generated with glochidia in the derivation of WQC (Charles Stephan, USEPA, Duluth, MN; personal communication). These concerns mainly include: (1) the duration of the toxicity tests, (2) the quality of the glochidia at the start of a test, and (3) the test acceptability criteria. The following section provides information that attempts to address these concerns. Areas of ongoing research or need for future research are also identified.

Duration of the Toxicity Test

1. How long should acute tests with glochidia be conducted (i.e., based on the life history of the species)?

   There are nearly 300 species of freshwater mussels in North America, and the length of time that glochidia remain viable after release from the marsupium of a female into the environment depends on the life history of the species and the temperature of the water (Chapter 3). Longevity of glochidia after release and before attachment to a host may exceed one week and may be dependent on temperature (Zimmerman and Neves 2002); however, some reports are anecdotal (Murphy 1942; Matterson 1948; Tedla and Fernando 1969). Glochidia of some species released in conglutinates remain viable for days or weeks after release into the environment (Chris Barnhart, personal communication). Glochidia of several species, including *Anodonta* spp., remain viable while free in the environment for 7–14 days (Howard and Anson 1922; Mackie 1984; Huebner and Pynnonen 1992; Pynnonen 1995).

   Table 5.2 provides a summary of laboratory studies that have evaluated survival times of glochidia after removal from the marsupium of the female or survival time based on results reported in toxicity tests conducted with glochidia. For example, Zimmerman and

TABLE 5.2
Survival Time of Glochidia after Removal from Female Unionid Mussels

Species	Temperature (°C)	Duration of Viability Day (% Survival)	Reference
Actinonaias ligamentina	20	7 (>90); 8 (>75); 9 (>50)	USGS (2004)
Actinonaias pectorosa	10	13 (>75)	Zimmerman and Neves (2002)
	25	5 (>75)	Zimmerman and Neves (2002)
	20	>2 (>90)[a]	Jacobson et al. (1997)
Alasmidonta heterodon	20	2 (>90); 2 (>75); 2 (>50)	USGS (2004)
Anodonta anatine	13	>3 (>90)	Huebner and Pynnonen (1992)
Anodonta cataracta	10	>14 (>90)	Jacobson (1990)
Anodonta cygnea	13	>3 (>90)	Huebner and Pynnonen (1992)
Anodonta grandis	10	>14 (>90)	Jacobson (1990)
Elliptio complanata	5	7 NR[b]	Matterson (1948)
	20	<1 (>90); 3 (>75)	Bringolf et al. (2005)
Elliptio dilatata	20	<1 (>90); 1 (>75); <2 (>50)	Bringolf et al. (2005)
Epioblasma capsaeformis	20	0.3 (>90)	Wang et al. (2003)
Lampsilis abrupta	20	2 (>90); 5 (>75); 7 (>50)	USGS (2004)
Lampsilis cardium	21	>2 (>90)[a]	Lasee (1991)
Lampsilis fasciola	20	6 (>90); 7 (>75); 8 (>50)	Wang et al. (2003)
	20	>2 (>90)[a]	Jacobson et al. (1997)
	20	1 (>90); 2 (>75); 3 (>50)	Bringolf et al. (2005)
	20	2 (>90; 4 (>75); 5 (>50)	Bringolf et al. (2005)
Lampsilis rafinesqueana	20	6 (>90); 6 (>75); 6 (>50)	USGS (2004)
Lampsilis siliquoidea	10	9 NR	Tedla and Fernado (1969)
	20	8 (>90); 9 (>75); 10 (>50)	Wang et al. (2003)
	25	>2 (>80)[a]	Keller and Ruessler (1997)
	20	1 (>90); 3 (>75); 4 (>50)	Bringolf et al. (2005)
Limpsilis teres	25	0.2 (>80)	Keller and Ruessler (1997)
Leptodea fragilis	20	1 (>90); 3 (>75); 4 (>50)	Wang et al. (2003)
Leptodea leptodon	20	1 (>90); 2 (>75)	Bringolf et al. (2005)
Leptodea leptodon	20	0.25 (>90); 1 (>75); 2 (>50)	USGS (2004)
Margaritifera falcate	11	11 NR	Murphy (1942)
Medionidus conradius	20	>2 (>90)[a]	Jacobson et al. (1997)
Megelonaias nervosa	25	1 (>80)[a]	Keller and Ruessler (1997)
Potamilus alatus	20	6 (>90) 6 (>75); 6 (>50)	Wang et al. (2003)
Potamilus ohiensis	20	5 (>90); 6 (>75); 7 (>50)	Wang et al. (2003)
Pyganodon grandis	20	>1 (>90)[f]	Jacobson et al. (1997)
Quadrula quadrula	20	1 (>90); 1 (>75); 2 (>50)	Wang et al. (2003)
Quadrula pustulosa	20	<1 (>90); 1 (>75); 1 (>50)	Wang et al. (2003)
Utterbackia imbecillis	21	10 (>80); 14 (>50)	Fisher and Dimock (2000)
	25	>2 (>80)[f]	Keller and Ruessler (1997)
	25	>2 (>80)[a]	Klaine, Warren, and Summers (1997)
	20	>1 (>90)[a]	Johnson, Zam, and Keller (1990, 1993)
Venustaconcha ellipsiformis	20	2 (>90); 3 (>75); 3 (>50)	Wang et al. (2003)
Villosa iris	10	8 (>75)	Zimmerman and Neves (2002)
	20	5 (>90); 5 (>75); 6 (>50)	Wang et al. (2003)
	25	2 (>75)	Zimmerman and Neves (2002)

(continued)

TABLE 5.2 *(Continued)*

Species	Temperature (°C)	Duration of Viability Day (% Survival)	Reference
	22	>1 (>80)[a]	Goudreau, Neves, and Sheehan (1993)
	20	>1 (>80)[a]	Scheller (1997)
	20	>2 (>90)[a]	Jacobson et al. (1997)
Villosaosa lienosa	25	>2 (>80)[a]	Keller and Ruessler (1997)
Villosa nebulosa	20	>2 (>90)[a]	Jacobson (1990)
Villosa villosa	25	>2 (>80)[a]	Keller and Ruessler (1997)

[a] The value based on control survival in 24- or 48-hour toxicity tests.
[b] NR, not reported.

Neves (2002) report that the viability of glochidia of *V. iris* was greater than 75% for 8 days at 10°C and 2 days at 25°C, and viability of glochidia of *A. pectorosa* was greater than 75% for 13 days at 10°C and 5 days at 25°C (Table 5.2). Similarly, glochidia of *Utterbackia imbecillis* may survive up to 19 days but exhibit 50% mortality within 13.5 days (Fisher and Dimock 2000). Survival of isolated glochidia from many species listed in Table 5.2 is typically greater than 90% after two to three days; however, the viability of glochidia for a particular species should be determined before the start of an exposure. For example, glochidia of *L. teres* and *E. capsaeformis* were viable for only four to six hours and glochidia of *M. nervosa* and *Q. quadrula* were viable for one day after removal from the marsupium of the female (Table 5.2). Therefore, 24 hours is a reasonable time period to conduct toxicity tests with glochidia of many species at 20°C, although shorter or longer tests might be needed for a particular species depending on glochidia survival time and the life history characteristics of the species (i.e., survival of glochidia in the control must be greater than 90% at the toxicity test) (ASTM 2006a).

The time between the release of glochidia from the marsupium of the female mussel to attachment of these glochidia on a host may only take a few seconds for some species, but hours are required for the gill tissue of a fish to migrate to form a cyst around the glochidia. During that time, the glochidia may be exposed to water-borne toxicants. Anodontinae species releases glochidia directly into water, which remain viable for days in order to effectively infest their host fish. Therefore, a prolonged glochidial test would have ecological relevance for these species. Other species release glochidia in mucus strands that coat the bottom or remain suspended on vegetation, waiting for their hosts to swim by, and still other species package glochidia in conglutinates that serve as a lure to host fish. Hence, glochidia of these species may also be in water for extended periods of time; however, it is not known how exposure to water-borne contaminants would be influenced by the mucus or conglutinate surrounding the glochidia. Toxicity tests conducted for 24 hours with glochidia may not be as ecologically relevant as toxicity tests conducted with juvenile mussels, but they may be useful for some purposes such as deriving concentrations of a chemical that may be protective of the species. Use of glochidia to evaluate the relative sensitivity of a particular mussel species to chemicals would be particularly useful when evaluating species where only a limited number of adult mussels are available for methods development or for producing juvenile mussels for toxicity testing. Moreover, the host fish for some species of mussels or techniques for transforming juvenile mussels in the laboratory may be unknown (Chapter 4).

The relatively short duration of toxicity tests with glochidia is based on the relatively short duration between release of glochidia into the water column and encystment on the host and on the relatively short survival time of glochidia after isolation from the female mussel. If the life history of a particular species is not known (e.g., the host required for encystment or how long glochidia released from a female mussel can remain in the water column before encysting on a host), it might be appropriate to conduct toxicity tests with glochidia for longer than 24 hours as long as 90% control survival can be achieved at the end of the test.

2. How long can glochidia survive and still be able to attach to a host?

Glochidia of some species can still attach to a host for several days after release from a female depending on temperature (Chris Barnhart, personal communication). The maximum time at which greater than 50% of *U. imbecillis* metamorphosed in a tissue culture medium was nine days after isolation from a female (Fisher and Dimock 2000). Zimmerman and Neves (2002) reported that glochidia can successfully attach to a host one to two weeks after isolation from a female. A future research project could be to conduct a series of toxicity tests to determine if there is a change in sensitivity over time after glochidia have been released into the environment. Sensitivity of *L. siliquoidea* glochidia held for 24 hours after isolation from a female was similar to newly-released glochidia in exposures to copper (Wang et al. 2003). The sensitivity of glochidia held in an extra piece of the marsupium in a refrigerator overnight was similar to the sensitivity of glochidia tested immediately after isolation from a female in toxicity tests conducted with zinc or copper (Jerry Farris, Arkansas State University, State University, AK; personal communication). Ultimately, it is more practical to base duration of exposure on survival of control organisms in the laboratory rather than on an estimate of the length of time glochidia can survive and still attach to a host (e.g., Table 5.2).

3. Are there data that indicate that effect concentrations do not change very much during the last half of a toxicity test (i.e., does the EC50 at 6, 24, 48, or 96 hours differ)?

There are limited studies with glochidia that have compared changes in toxicity over this timeframe. The toxicity of copper (Jacobson et al. 1997; Wang et al. 2003), ammonia (Wang et al. 2003), and chlorine (Wang et al. 2003) decreased over 48–96-hour exposures. In contrast, no change in the toxicity of several pesticides was observed in 24–48-hour exposures (Keller and Ruessler 1997; Bringolf et al. 2005). If glochidia for a particular species are able to survive for more than 24 hours, then a 24-hour toxicity test should be considered. Importantly, researchers are encouraged to design studies that generate toxicity data throughout the exposure period (e.g., reporting 6, 24, and 48-hour responses) (ASTM 2006b). However, generating data for a six-hour exposure period is logistically difficult in an eight-hour day.

Quality of Glochidia at the Start of a Toxicity Test

1. How should the quality of glochidia be determined at the start of a toxicity test? Is the use of a solution of NaCl (or KCl) to determine the percentage of glochidia exhibiting valve closure an appropriate method to judge the acceptability of glochidia used to start a toxicity test? Does the response of glochidia to a solution of NaCl (or KCl) relate to the ability of glochidia to attach to a host? Is there an independent way of determining if glochidia are alive or healthy at the start (or end) of a toxicity test?

Valve closure is an ecologically relevant endpoint that is critical for glochidia to successfully transform on the host. If glochidia do not snap shut, the glochidia should be considered ecologically dead (Huebner and Pynnonen 1992; Goudreau, Neves, and Sheehan 1993; McCann 1993; Jacobson et al. 1997). The response of glochidia in toxicity tests was similar when either KCl or fish plasma was used to make glochidia close at the end of an exposure (Huebner and Pynnonen 1992). Decreased response to KCl was considered an indication of reduced glochidia viability and thus reduced capability to attach to the fish host (Pynnonen 1995). A significant correlation was observed between the response of glochidia to KCl and the ability of glochidia of *U. imbecillis* to metamorphose to the juvenile life stage (Fisher and Dimock 2000). Zimmerman and Neves (2002) reported a correspondence between the response of glochidia of *V. iris* and *A. pectorosa* to NaCl and the ability to infest a host fish. Jacobson et al. (1997) reported glochidia of *V. iris* that responded to the addition of NaCl following an exposure to copper were able to attach to a host fish with no impairment of subsequent metamorphosis to juvenile mussels. Results of these studies indicate that addition of a solution of NaCl or KCl can be used to estimate the condition of glochidia. While either a solution of salt or fish plasma could be used to determine the percentage of organisms closing, it is easier to work with NaCl compared to KCl or fish plasma.

2. Should there be a holding time for glochidia after harvesting but before application of a saline solution to determine if glochidia that are initially closed might open?

   Mature glochidia are not typically closed after being isolated from a female mussel. Glochidia that are closed after isolation from a female may reopen after being held in clean water a few hours (Goudreau, Neves, and Sheehan 1993; Dick Neves, Teresa Newton, USGS, LaCrosse, WI; personal communications).

3. Will immature, stressed, or unhealthy glochidia close when exposed to a saline solution? Could glochidia be alive and successfully attach to a host but not close when exposed to a saline solution? Are broken glochidia frequently observed at the start of a test? Would the presence of broken glochidia be indicative of stress during harvesting?

   Immature glochidia that are free of the egg membrane or mature and healthy glochidia will close when exposed to a salinity challenge. However, immature glochidia are generally enclosed in an egg membrane and are fragile and tend to fracture, thus should not be used for toxicity testing. The best approach for avoiding the use of immature glochidia in toxicity testing is to sample female mussels at a time of the year when the organisms would be expected to be releasing mature glochidia (Jess Jones, US Geological Survey, Blacksburg, VA; personal communication). Stressed or unhealthy glochidia could either be opened or closed before the start of a test. If stressed or unhealthy glochidia were to close when exposed to a salinity challenge, then these individuals would be used in a toxicity test. Measurement of the viability of glochidia in the control at the end of a toxicity test would help to identify stressed or unhealthy glochidia. Results of reference-toxicant tests should also be used to evaluate the health of the glochidia used to conduct the test (ASTM 2006a). Broken glochidia have not been observed at the start of a test (Chris Barnhart, Jerry Farris, Dick Neves, Teresa Newton, Ning Wang, USGS, Columbia, MO; personal communications). The presence of broken glochidia may indicate that the glochidia are immature and should not be used for testing.

Test Acceptability Criteria for Toxicity Tests with Glochidia

1. What criteria should be used to judge acceptability of a toxicity test conducted with glochidia?

   ASTM (2006a) recommends that the age of glochidia should be less than 24 hours old at the start of the toxicity test. Viability of glochidia isolated at the beginning of a toxicity test must be greater than or equal to 80% (preferably greater than or equal to 90%). Average survival of glochidia in the control at the end of a test must be greater than or equal to 90%. ASTM (2006a) also recommends that subsamples of each batch of test organisms used in toxicity tests should be evaluated using a reference toxicant (e.g., NaCl or $CuSO_4$). Data from these reference-toxicant tests can be used to assess genetic strain or life-stage sensitivity of test organisms to select chemicals.

2. Should glochidia be rinsed before use in a toxicity test? Would rinsing glochidia before the start of a test be stressful to the organisms?

   Glochidia should be rinsed with culture or dilution water after removal from marsupia to (1) eliminate tissues or excess mucus from the excised glochidia that have a high potential for fungal growth and subsequently could affect the survival (toxicity tests) or transformation of glochidia (propagation) and (2) reduce the number of protozoans that may be present in the excised gill that could also affect glochidia survival or transformation (ASTM 2006a). Rinsed glochidia have been observed to successfully transform on fish or in artificial media and high control survival in toxicity tests has been reported using glochidia that have been rinsed (Huebner and Pynnonen 1992; Johnson, Keller, and Zam 1993; Myers-Kinzie 1998; Milam et al. 2005).

3. Should glochidia be acclimated to test conditions before the start of a toxicity test?

   Glochidia are not typically acclimated to the water-quality characteristics of the dilution water before the start of a toxicity test (Table 5.1). Most of these exposures are started the same day that glochidia are isolated from marsupia of the females. Therefore, minimal time is available to acclimate glochidia to the dilution water before the start of a test. In order to maintain organisms in good condition and avoid unnecessary stress, ASTM (2006a) recommends that organisms should not be subjected to rapid changes in temperature or water quality before the start of a test. Glochidia can be acclimated in a mixture of 50% culture water and 50% test water and gradually adjusted to the test temperature within about two hours before the start of an exposure (ASTM 2006a). Investigators have held adult mussels under test conditions before isolation of glochidia (e.g., Huebner and Pynnonen 1992), which would result in acclimating glochidia to the selected exposure temperature in the toxicity test. However, brooding glochidia in the marsupium are in contact with the hemolymph of the female that is physically isolated from direct contact with water (Silverman, McNeil, and Dietz 1987). In addition, glochidia are typically released instantaneously into the surrounding water from the marsupium of the female mussel. Therefore, holding the female mussels in the dilution water before isolating glochidia for toxicity testing would probably have a minimal influence on the ability of glochidia to acclimate to the conditions of the dilution water.

METHODS FOR CONDUCTING WATER-ONLY TOXICITY TESTS WITH JUVENILE FRESHWATER MUSSELS

Review of Methods

ASTM (2006a) provides a list of recommended test conditions for conducting toxicity tests with juvenile mussels. The list of recommended test conditions is based on the various methods outlined in Table 5.3 and on the conditions used to conduct an inter-laboratory toxicity test with juvenile mussels (ASTM 2006a). ASTM (2006a) recommends that toxicity tests with juvenile mussels be conducted at 20°C with a 16L:8D photoperiod at an illuminance of about 100–1,000 lux (Table 5.3). Toxicity tests are typically started with newly-transformed juvenile mussels less than five days after the release

TABLE 5.3

Summary of Test Conditions Used to Conduct Toxicity Tests with Juvenile Freshwater Mussels

Conditions	Johnson et al. (1990, 1993)	Jacobson (1990), Jacobson et al. (1993)	Lasee (1991)	Keller and Zam (1991)	Klaine, Warren, and Summers (1997)	Scheller (1997)	Myers-Kinzie (1998)	Dimock and Wright (1993)	Newton et al. (2003)	Lasee (1991)	Wade (1992)[a]	Jacobson (1990)	Valenti et al. (2005)	USGS (2004)	USGS (2005a, 2005b, 2005c)	Recommended Test Conditions ASTM (2006a)
1 Species tested	Utterbackia imbecillis[b]	Villosa nebulosa, Villosa iris, Anodonta grandis[c]	Lampsilis cardium[d]	Multiple species[e]	U. imbecillis	V. iris	Lampsilis siliquoidea	U. imbecillis Pyganodon cataracta	L. cardium	Lampsilis ventricosa	U. imbecillis	Villosa nebulosa	V. iris	Multiple species[f]	L. siliquoidea, Epioblasma capsaeformis, V. iris[g]	NA[h]
2 Test type	Renewal	Static	Static	Static	Static	Static	NR	Static	Flow through	Renewal	Renewal	Artificial stream	Renewal	Renewal, flow through	Flow through	Static, renewal or flow-through (depending on duration of exposure and chemical tested)
3 Test duration (days)	2	1	2	1-4	1-4	4	1, 2, 4	1-4	4, 10	7	9	14	21	2, 4, 10	28	Acute: ≤4 Chronic: 21-28
4 Temperature (°C)	20	20	21	22, 25, or 32	25	25	24	20	21	21	24	20	20	20	20	20
5 Light quality	Ambient lab light	NR[h]	NR	NR	NR	NR	NR	NR	Fluorescent	NR	NR	NR	NR	Fluorescent	Fluorescent	Ambient lab light
6 Light intensity	NR	NR	NR	NR	NR	NR	NR	NR	NR	NR	NR	NR	NR	200 lux	200 lux	100-1000 lux
7 Photo period	16L:8D	16L:8D	24 D	12L:12D or 16L:8D	16L:8D	NR	NR	NR	16L:8D	24D	24D	16L:8D	12L:12D	16L:8D	16L:8D	16L:8D
8 Test chamber	125-mL beaker	12-well plate	Covered 250-mL crystallizing dish	Petri dish	Petri dish	12-well plate	Petri dish	120-mm diam. tub with mesh bottom in 4-L chamber	132 by 90 by 130 mm chamber	Covered 250-mL crystallizing dish	50-mm diam. glass tub with mesh bottom in 250-mL Chamber	Dish covered with mesh	30-mL beakers submerged in a 1-L glass beaker	50- or 300-mL beaker	300-mL beaker	Static or renewal: 50-mL beakers (minimum); flow-through: 300-mL beakers (minimum)
9 Test solution volume (mL)	100	3.5	NR	15	15	5	10	NR	1200	NR	200	150	950	30 or 200	200	Static or renewal: 30 (minimum); flow-through: 200 (minimum)
10 Procedure for obtaining juveniles	Artificial media	Fish host	Fish host	Fish host or artificial media	Fish host or artificial media	Fish host	Artificial media	Artificial media	Fish host	Fish host	Artificial media	Fish host	Fish host	Fish host	Fish host	Fish host
11 Age of test organisms (days)	1-10	1-3	0, 7, 14	1-2	1-3	<3, 5, 9	<10	7-10	3-5	0	6-10	1-3	60	3-5, 60	60	Acute: <5 Chronic: 60-120
12 Number of organisms per test chamber	10	10	10	10-20	1	5	NR	10	20	50	15	15	5	5	10	Acute: <5 (minimum) Chronic: 10 (minimum)

(continued)

TABLE 5.3 (Continued)

Conditions	Johnson et al. (1990, 1993)	Jacobson (1990), Jacobson et al. (1993)	Lasee (1991)	Keller and Zam (1991)	Klaine, Warren, and Summers (1997)	Scheller (1997)	Myers-Kinzie (1998)	Dimock and Wright (1993)	Newton et al. (2003)	Lasee (1991)	Wade (1992)[a]	Jacobson (1990)	Valenti et al. (2005)	USGS (2004)	USGS (2005a, 2005b, 2005c)	Recommended Test Conditions ASTM (2006a)
13 Number of replicate chambers per treatment	2	2 or 3	3	2-4	10	4	NR	3	6	2	3	3	4	4	4	Acute:4 (minimum) Chronic:3 (minimum)
14 Feeding	None	None	None	None	None	None	None	None	None	Lab cultured phyto plankton	Algae and silt	Algae	Algae and sediment	None	Instant algae mixture[j]	Acute:none Chronic: Algae
15 Aeration	None	None	Yes	NR	None	NR	NR	Yes	Yes	None	None	None	Yes	None	None	None, if dissolved oxygen is maintained above acceptable concentration
16 Dilution Water	Reconstituted water, hardness 40–50 mg/L as CaCO3; Reconstituted water-hardness 170 mg/L as CaCO3	Clinch River water, VA; Depends on	Hardness 150 mg/L as CaCO3; experimental design	Reconstituted water, hardness 47–76 mg/L as CaCO3	Reconstituted water, hardness 99–107 mg/L as CaCO3	Sinking Creek water, VA	Hardness 100 or 200 mg/L as CaCO3	NR	Hardness 133 mg/L as CaCO3	Hardness 150 mg/L as CaCO3	Tennessee River water	Clinch River water, VA			Reconstituted water, hardness 100 mg/L as CaCO3	Reconstituted water, hardness 170 mg/L as CaCO3
17 Water Quality	DO, pH, hardness, alkalinity, conductivity	DO, pH, hardness, alkalinity, conductivity	DO, pH, hardness, alkalinity, conductivity	DO, pH, hardness, alkalinity, conductivity	DO, pH, hardness, alkalinity, conductivity	DO, pH, hardness, alkalinity, conductivity	pH, hardness	NR	DO, pH, hardness, alkalinity, conductivity	DO, pH, hardness, alkalinity, conductivity	DO, pH, hardness, alkalinity, conductivity	NR	NR	DO, pH, ammonia, hardness, alkalinity, conductivity	DO, pH, ammonia, hardness, alkalinity, conductivity	DO, pH, ammonia, hardness, alkalinity, conductivity
18 Endpoints	Survival (movement)	Survival (gaped valves, foot activity or stained with neutral red)	Survival (foot or ciliary movement)	Survival (activity and heartbeat)	Survival (gaped valves with foot and ciliary activity)	Survival (heart beat and ciliary action)	Survival (foot or valve movement)	Survival (foot, valve or ciliary activity, heartbeat)	Survival, growth, ratio of stressed to alive	Survival (foot or ciliary movement), growth (length and height)	Survival (Ciliary action)	Survival (extruded foot and gaping valves)	Survival, growth	Survival (foot or shell movement) and growth (shell length)	Survival (foot or shell movement and growth (shell length)	Survival (foot movement), growth (shell length)
19 Control survival (%)	>95	100	96	NR	>90	>80	99	>90	>95	97	>90	100	90	>90	>88	Acute:>90 (must) Chronic:>80 (should)

The last column provides a summary of recommended conditions that can be used to conduct toxicity tests with juvenile mussels as outlined in ASTM (2006a). In the last Column, Acute Tests are Tests Conducted for up to 96 hours and Chronic Tests are Tests Conducted for at least 21 days.

[a] See also Masnado, Geis, and Sonzogni (1995). McKinney and Wade (1996). Keller, Ruessler, and Kernaghan (1999).

[b] Formerly A. imbedllis.

[c] See also McCann (1993) for two- to four-day exposures with Villosa iris, A. pectomsa, M. conradius.

[d] Formerly L ventricosa.

[e] A. imbecillis, V. lienosa, V. villosa, U. imbecillis, Lampsilis straminea daibomensis, L. subangulata, Elliptic icterina. See also Keller (1993). Keller and Ruessler (1997).

[f] V. iris, E. capsaeformis, L. fasciola, L. siliquoidea, L. abrupta, L. rafinesqueana, L. leptodon.

[g] Bringolf et al. (2005) has adapted this method to conduct 21-day toxicity tests with four-month old juvenile A. ligamentina.

[h] NA, not applicable; NR, not reported.

[i] See USGS (2005a) for a description of the procedure used to prepare this instant algae mixture.

from the host; however, some toxicity tests have been started with two- to four-month-old juvenile mussels. Acute toxicity tests with juvenile mussels are typically conducted for 96 hours with survival measured at 48 and 96 hours. Chronic toxicity tests started with two- to four-month-old juvenile mussels have been conducted for 21–28 days with measures of survival (based on movement of the foot) and growth (based on shell length). Control survival is typically greater than 90% at the end of 96-hour toxicity tests conducted with juvenile mussels and is typically greater than 80% at the end of toxicity tests conducted for 10–28 days with juvenile mussels (Table 5.3; ASTM 2006a).

In acute static tests, glass test chambers should be a minimum volume of 50 mL containing a minimum of 30 mL of dilution water (ASTM 2006a). In chronic tests or in flow-through tests, glass chambers should be a minimum volume of 300 mL containing a minimum volume of 200 mL of dilution water. Static, renewal, or flow through conditions can be used depending on the chemical being tested. Juvenile mussels are not typically fed during acute toxicity tests. Algae have been used as a food source in toxicity tests conducted for 10–28 days (Table 5.3; ASTM 2006a).

The number of replicates and concentrations tested depends in part on the significance level selected and the type of statistical analysis. In 96-hour toxicity tests, ASTM (2006a) recommends a minimum of 20 organisms should be exposed to each concentration (e.g., four replicates each containing a minimum of five juvenile mussels). It may be desirable to test only five juvenile mussels in each replicate when a limited number of test organisms are available or when test organisms are relatively small (e.g., when juvenile mussels are small, it may be difficult to observe more than about five test organisms simultaneously in a replicate test chamber under the microscope). However, some investigators have tested 10–20 juvenile mussels in each replicate. In chronic toxicity tests, a minimum of three replicates should be tested, each replicate containing a minimum of 10 juvenile mussels.

Toxicity tests with juvenile mussels are typically started with organisms that have been transformed with a fish host (ASTM 2006a); however, artificial media has also been used to transform juvenile mussels for use in toxicity testing (Johnson, Keller, and Zam 1993; Clem 1998; Hudson et al. 2003). ASTM (2006a) recommends testing of juvenile mussels that have been transformed on a fish host due to uncertainties regarding the sensitivity of juvenile mussels transformed using artificial media. Numerous investigators have observed high mortality of juvenile mussels about four to six weeks after transformation (e.g., Anne Keller, USEPA, Athens, GA; Don Cherry, Jerry Farris, Teresa Newton; personal communications). As a result of this problem, the duration of toxicity tests started with newly-transformed juvenile mussels is less than 14 days with survival or growth measured at the end of the exposures (Table 5.3). Food (mixtures of different species of algae) and sediment have been added to exposure chambers, but some investigators have found that newly-transformed juvenile mussels will survive for at least 14 days without the addition of food (Table 5.3; ASTM 2006a). For example, USGS (2004) determined the acute toxicity of copper in 48-hour tests with juvenile *L. siliquoidea* and *Lampsillis rafinesqueana* that had been held for 10 days under control conditions (e.g., with the replacement of dilution water but without the addition of food). Similar 48-hour EC50s were observed in tests conducted with juvenile mussels held for 10 days before testing compared to tests started with newly-transformed juvenile mussels. Results of these tests indicate that the sensitivity of juvenile mussels did not change over the 10-day exposure without feeding. Hence, toxicity tests conducted for up to 10 days without feeding may provide reliable data for evaluating effects of chemicals on mussels in exposures longer than 4 days.

The high mortality of newly-transformed juvenile mussels in toxicity tests conducted for longer than 14 days is likely related to a lack of understanding of the nutritional requirements of mussels at this life stage. Newly-transformed juvenile mussels depend on pedal feeding to obtain food (cilia on the foot are used to move food into the juvenile mussel). Juvenile mussels gradually begin to use a combination of pedal and filter feeding to obtain food until the mussels eventually depend on filter feeding to obtain food by about six months in laboratory cultures supplied with a silt-clay sediment substrate. However, in the field, juvenile mussels probably depend on a combination of filter, deposit and pedal feeding in coarser substrates (Dick Neves, personal communication). Research is ongoing to improve culturing methods for propagation, holding, and feeding of newly-transformed juvenile mussels (Chapter 4; Keller and Zam 1990; Gatenby, Neves, and Parker 1996, 1997; Henley, Zimmerman, and Neves 2001; ASTM 2006a). Once developed, these culturing methods should help to refine methods for conducting chronic exposures with juvenile mussels.

Investigators have reported success in conducting toxicity tests for up to 28 days starting with two- to four-month-old juvenile mussels. Valenti et al. (2005) conducted 21-day exposures to mercury starting with two-month old juvenile *V. iris* held in a small amount of sediment and fed algae (*Neochloris*). USGS (2005a, 2005b, 2005c); Bringolf et al. (2005) conducted toxicity tests starting with two- to four-month-old juvenile *A. ligamentina, L. siliquoidea*, or *V. iris* and observed control survival greater than 88% in 21–28-day exposures to copper, lead, zinc, cadmium, ammonia, and several pesticides when a mixture of instant algae was used as a food source. The instant algae mixture was prepared from commercial Instant Algae brand non-viable microalgae concentrates (Reed Mariculture, Campbell, CA including *Nannochloropsis, Isochrysis, Pavlova, Tetraselmis*, and *Thalassiosira weissflogii*).

Issues Regarding the Use of Methods

Concerns have been expressed regarding the use toxicity data generated with glochidia in the derivation of WQC (Charles Stephan, personal communication). Charles Stephan concluded that acute methods for testing juvenile mussels (such as those outlined in Table 5.3) generally follow standard testing methods (e.g., ASTM 2006a, 2006b), and data generated from these types of studies should be useful in the derivation of WQC. However, there were concerns identified regarding toxicity tests conducted with juvenile mussels including: (1) the life stage tested, (2) the determination of death at the end of a test, and (3) test acceptability criteria. The following section provides information that attempts to address some of these concerns. Areas of ongoing research or needs for future research are also identified.

What Life Stage Should Be Used to Start Acute or Chronic Toxicity Tests with Juvenile Mussels?

Toxicity tests have been started with newly-transformed juvenile mussels that have either been transformed on a host or have been transformed with the use of an artificial medium (Table 5.3). Glochidia, newly-transformed juvenile mussels, and two- to four-month-old juvenile mussels have been successfully shipped via overnight carriers to other laboratories for use in toxicity testing (USGS 2004; Bringolf et al. 2005; ASTM 2006a). Toxicity tests have been successfully conducted for 10–14 days starting with newly-transformed juvenile mussels (Table 5.3), but exposures Started with newly-transformed juvenile mussels conducted for longer periods of time have resulted in high mortality in controls at about four to six weeks, probably due to nutritional limitations of the diet (e.g., Newton et al. 2003). Valenti et al. (2005), USGS (2005a, 2005b, 2005c) and Bringolf et al. (2005) conducted 21–28-day toxicity tests starting with two- to four-month-old juvenile mussels of a variety of species and observed control survival greater than 88% when algae was used as a food source.

How Should the Death of Juvenile Mussels Be Determined at the End of a Toxicity Test?

Lack of foot or shell movement, lack of ciliary activity on the foot, lack of a heart beat, or a wide gaped valve have been used to establish death in toxicity tests with juvenile mussels (Table 5.3). ASTM (2006a) recommends establishing death of juvenile mussels based on foot movement during a five-minute observation period under a microscope. If it is suspected that juvenile mussels are avoiding exposure to a chemical in a toxicity test, it may be desirable to place the suspected live test organisms into dilution water that does not contain any added test material for one to two days after the end of the toxicity test to determine whether these test organisms are alive or dead (ASTM 2006a).

What Criteria Should Be Used to Judge Acceptability of a Toxicity Test Conducted with Juvenile Mussels?

ASTM (2006a) recommends that average survival of juvenile mussels in the control at the end of a 96-hour test must be greater than or equal to 90%. An insufficient number of tests have been conducted with juvenile mussels for 10 or more days for ASTM (2006a) to provide specific guidance on control survival in longer-term tests. However, a limited number of toxicity tests have reported control survival greater than 80% in tests conducted with juvenile mussels for 10–28 days. Therefore, ASTM (2006a) recommends that average survival of juvenile mussels in the control at the end of a test conducted for 10–28 days should be greater than or equal to 80%. ASTM (2006a) also recommends that subsamples of each batch of test organisms used in toxicity tests should be evaluated using a reference toxicant (e.g., NaCl or $CuSO_4$). Data from these reference-toxicant tests can be used to assess genetic strain or life-stage sensitivity of test organisms to select chemicals.

METHODS FOR CONDUCTING WATER-ONLY TOXICITY TESTS WITH ADULT FRESHWATER MUSSELS

Review of Methods

Conditions that have been used to conduct toxicity tests with adult freshwater mussels are summarized in Table 5.4. Specific standardized methods have not been developed for conducting toxicity tests with adult mussels, but the procedures outlined in Table 5.4 are generally consistent with guidance for conducting laboratory toxicity tests with early life stages of freshwater mussels (ASTM 2006a). Exposures have been conducted under static (Keller, Ruessler, and Kernaghan 1999), renewal (Mane 1979; Holwerda and Herwig 1986), and flow-through (Naimo, Waller, and Holland-Bartels 1992a, 1992b; Imlay 1973; Kernaghan et al. 2003) conditions. A limited number of species have been used to conduct toxicity studies with adults, and these mussels are typically collected from the field.

Adults have been held under laboratory conditions from one day to several months before the start of toxicity testing. Toxicity tests are conducted under a wide variety of conditions, with exposure chambers ranging from 10 to 1,500 L. Due to the relatively low abundance of adult mussels, the number of replicates per test concentration and the number of mussels tested is generally low. Replication ranges from one to six chambers, each containing between 9 and 125 organisms. The tests have been conducted from 48 hours to 8 months, and a variety of endpoints have been used to evaluate toxicity. Survival, as measured by cessation of siphoning activity and inability to react to stimulation, is a common endpoint assessed in most adult toxicity studies. In addition, sublethal endpoints, such as respiration rate, condition indices, glycogen content, and other biochemical parameters, are frequently measured in toxicity tests conducted with adults. Water quality analysis, including temperature, dissolved oxygen, pH, conductivity, hardness, and alkalinity, are routinely measured during adult toxicity tests. In addition, bioaccumulation has also been determined (e.g., Holwerda and Herwig 1986; Naimo, Waller, and Holland-Bartels 1992a, 1992b). Control survival for adult toxicity studies is typically greater than 80 to greater than 90% at the end of the exposures.

Issues Regarding the Use of the Methods

Issues regarding the use of adult mussels in toxicity tests are similar to those for toxicity tests with glochidia and juvenile mussels. Some issues that have been raised in relation to glochidia or

TABLE 5.4
Summary of Test Conditions for Conducting Toxicity Tests with Adult Freshwater Mussels

	Conditions	Keller and Ruessler (1997)	Mane, Kachole, and Pawar (1979)	Naimo, Atchison, and Holland-Bartels (1992b)	Imlay (1973)	Raj and Hameed (1991), Jacobson (1990)	Holwerda and Herwig (1986)	Farris et al. (1991)	Kernaghan et al. (2003)	Nicola Kernaghan, University of Florida, FL (Unpublished Data)	Chris Ingersoll, USGS, Columbia, MO (Unpublished Data)
1	Species tested	Villosa lienosa, Elliptio icterina, Utterbackia imbecillis[a]	Indonaia caeruleus	Lampsilis ventricosa	Amblycorypha carinata, Lampsilis radiata siliquoidea, Fusconaia flava, Amblema plicata	Villosa iris, Actinonaias pectorosa, Pyganodon grandis, Lampsilis fasciola, Medionidus conradius	Anodonta anatina	Elliptio dilatata, M. conradicus, Pleurobema oviforme, Villosa iris	E. buckleyi	E. buckleyi	Amblema plicata
2	Test type	Static	Static-renewal	Flow-through diluter	Flow-through	Static-renewal	Static-renewal	Static-renewal	Flow-through	Static-renewal	Flow-through
3	Test duration	72–96 hours	48 hours	14, 28 days	36 days–8 months	96 hours, 10, 20, 30 days	7 months	30 days	56 days	7, 14, 21, 30, 60 days	4, 56 days
4	Temperature (°C)	25, 32	30–32	21	11–21	29	NR	16–20	17	20	20
5	Light quality	NR[b]	NR	NR	Fluorescent and incandescent	NR	NR	NR	NR	Fluorescent	Fluorescent
6	Light intensity	NR	NR	Subdued	16–22 foot candles	NR	NR	NR	NR	NR	250 lux
7	Photoperiod	12L:12D	NR	NR	Natural conditions	NR	NR	10L:14D	Natural conditions	16L:8D	16L:8D
8	Test chamber	23-L aquaria	Plexiglass aquaria	57-L glass acuaria	20-L stainless steel chamber	Plastic containers	NR	75-L fiberglass oval stream	1500-L plastic containers	46-L glass aquaria	40-L glass aquaria
9	Test solution volume (L)	23	NR	57	20	10	40	60	1500	30	25 L with a 5-cm layer of gravel
10	Number of organisms per test chamber	5–10	50	10	NR	10	50	6	125	30	9–10
11	Number of replicate chambers per treatment	2–4	1	2	1–6	1	1	1	1	2	4
12	Feeding	None	None	Yes	None	None	NR	Yes	None	Yes	Algae in pond water
13	Aeration	None	None	None	None	None	NR	None	Yes	Yes	Yes

(continued)

TABLE 5.4 *(Continued)*

Conditions	Keller and Ruessler (1997)	Mane, Kachole, and Pawar (1979)	Naimo, Atchison, and Holland-Bartels (1992b)	Imlay (1973)	Raj and Hameed (1991), Jacobson (1990)	Holwerda and Herwig (1986)	Farris et al. (1991)	Kernaghan et al. (2003)	Nicola Kernaghan, University of Florida, FL (Unpublished Data)	Chris Ingersoll, USGS, Columbia, MO (Unpublished Data)
14 Dilution water	Reconstituted water with a hardness of 76 mg/L as $CaCO_3$	River water	Reconstituted water with a hardness of 165 mg/L as $CaCO_3$	Dechlorinated city water and lake water	Filtered river water	Tap water	River water	Well water	Well water (hardness 260 mg/L as $CaCO_3$) and reconstituted water (hardness 80 mg/L as $CaCO_3$)	Mixture of well water (hardness 260 mg/L as $CaCO_3$) and pond water to a hardness of about 190 mg/L as $CaCO_3$)
15 Water quality	pH, hardness, conductivity	Temperature	Temperature, DO, pH, alkalinity, hardness, conductivity	Temperature, DO, pH, hardness, alkalinity	Temperature, DO, pH, salinity	DO, pH, Ca, Mg, Na, Fe, HCO_3, Cl, SO_4	Temperature, DO, pH, conductivity, hardness, alkalinity	Temperature, DO, pH, conductivity	Temperature, DO, pH, conductivity	Temperature, DO, pH, ammonia, conductivity, hardness, alkalinity
16 Endpoint	Survival (cessation of siphoning activity and inability to react to stimulation)	Survival and effects on neurosecretory cells, digestive gland and intestine	Respiration rate, food clearance rate, ammonia excretion rate, assimilation efficiency, tissue condition index, oxygen to nitrogen ratio	Survival	Survival, respiration rate and body weight	Carbohydrate and lipid content, lactate, succinate, acetate, and propionate	Survival and cellulolytic activity	Survival, body condition index, glycogen concentration, sex steroid concentration	Survival, body condition index, soft tissue index, glycogen concentration, sex steroid concentration	Survival, glycogen concentration, behavior
17 Control survival (%)	NR	80	>80	>97	NR	NR	NR	100	NR	>80

[a] Formly *A. imbecilis*.
[b] NR, not reported.

juvenile toxicity tests are also applicable to adult toxicity tests. These concerns mainly include: (1) the length of time that adults can be held in a laboratory, (2) the conditions for maintaining adults in the laboratory, (3) the evaluation of the health of adults, and (4) the similarities in sensitivity to contaminants between different populations of mussels. The following section attempts to address some of these concerns. Areas of ongoing research or needs for future research are also identified.

How Long Can Adults Be Held in the Laboratory?

Dunn and Layzer (1997) reported the results of several long-term holding experiments using fishery ponds and raceways. Survival of adults varied by species and according to holding conditions. In the raceway experiments, survival ranged from 43 to 100% after one year. Survival of mussels held in ponds appeared more variable, ranging from 0 to 100%. Other researchers report that adults can be successfully maintained in the laboratory for a period of several months (e.g., Chris Barnhart, Jerry Farris, Dick Neves, Teresa Newton; personal communications).

What Conditions Should Be Used to Maintain Adults in the Laboratory?

Holding conditions for adults vary by species and season. Temperatures tested ranged between 10 and 25°C, and mussels have been maintained in systems supplied by pond, river, or well water. Conditions that most closely mimic those in the environment from which the mussels were collected are recommended. Some research facilities have relied upon natural sources of food in the pond or river water to maintain an adequate diet for the captive mussels. Researchers at Virginia Tech have successfully developed a cultured algal diet to feed the mussels (Gatenby 2000).

How Should the Health of Adults Held in the Laboratory Be Evaluated?

Adults to be used in toxicity studies are only occasionally screened for background contamination levels (Nicola Kernaghan, University of Florida, Gainesville, FL; personal communication). The health of adults can be evaluated by making observations of activity, behavior, and orientation of mussels in a substrate (Jerry Farris and Teresa Newton, personal communications). Further health assessments may be achieved with the use of biochemical indicators, such as glycogen (Chapter 10). Several studies have used glycogen as a measure of the energetic status of mussels and as an indicator of their physiological condition after exposure to contaminants (Hemelraad and Herwig 1990; Holopainen and Lamberg 1997; Kernaghan et al. 2003).

Are There Similarities in Sensitivity to Contaminants between Populations of Mussels of the Same Species?

Some species of mussels appear to be more sensitive to certain contaminants than others, and several studies have been conducted to compare species sensitivities (Imlay 1973; Keller and Ruessler 1997). See Chapter 7 for comparison of toxicity endpoints between species. However, no studies have been conducted to date comparing the sensitivity of different populations of the same species of mussel.

METHODS FOR CONDUCTING SEDIMENT TOXICITY TESTS WITH FRESHWATER MUSSELS

REVIEW OF METHODS

Conditions used to conduct sediment toxicity tests with freshwater mussels are summarized in Table 5.5. Chapter 6 provides an overview of methods used to conduct sediment toxicity tests with mussels in the field. Only a few sediment toxicity tests with mussels have been conducted, and

TABLE 5.5
Summary of Test Conditions for Conducting Sediment Toxicity Tests with Freshwater Mussels

	Conditions	Keller, Ruessler, and Chaffee (1998)	Wade (1992)	Newton et al. (2003)	USGS (2005c), Nile Kemble, USGS, Coumbia, MO (Unpublished Data)
1	Species tested	Lampsilis siliquoidea, Lasmigona costata, Villosa villosa	Utterbackia imbecillis (formerly Anodonta imbecillis)	Lampsilis cardium	Lampsilis siliquoidea, Lampsilis rafinesqueana
2	Test type	Renewal	Renewal	Flow through	Renewal
3	Test duration (day)	1, 2	9	4, 10	28
4	Temperature (°C)	22	24	21	20
5	Light quality	NR[a]	NR	NR	Fluorescent lights
6	Light intensity	NR	NR	NR	200 lux
7	Photoperiod	12L:12D	24D	16L:8D	16L:8D
8	Test chamber	Glass cylinder (5 cm in diameter, 7.5 cm in height, closed on one end, with 100 um Nitex screen) in 5-L glass aquaria	Glass cylinder (5 cm in diameter, closed on one end, with 100 um Nitex screen) in 250-mL crystallizing dishes	Polycarbonate tube (4.5 cm in diameter, 11 cm in height, closed one end with 153 um mesh Nitex screen) in 12×8× 13 cm chamber	300-mL beaker
9	Test solution volume (mL)	NR	200	800 mL of overlying water, 3 cm of sediment, and 1.5 cm of sand	260 mL of overlying water and 15 mL of sieved sediment, two volume additions/day of overlying water
10	Age or life stage of test organisms	Glochidia and newly-transformed juveniles	Newly-transformed juveniles	Newly-transformed juveniles	Two- to four-month-old juvenile mussels
11	Number of organisms per test chamber	10, 50–100	15	20	10
12	Number of replicate chambers per treatment	3	3	3–6	4
13	Feeding	None	Yes	None	Instant algae mixture
14	Aeration	Yes	None	Yes	None
15	Dilution water	Well water, with a hardness of 250 mg/L as CaCO3	Sediment pore water	Well water, with a hardness of 123–190 mg/L as CaCO3	Well water, with a hardness of 140 mg/L as CaCO3

No.	Water quality	Endpoints	Control survival (%)
16	Temperature, DO, pH, conductivity	Survival	Glochidia >69 Juvenile >90
17	Temperature, DO, pH, hardness, alkalinity, conductivity	Survival	96
18	Temperature, DO, pH, hardness, alkalinity, conductivity, ammonia	Survival (foot movement, ciliary activity), growth (shell height)	99–100
	Temperature, DO, pH, hardness, alkalinity, conductivity, ammonia	Survival (foot movement), growth (shell length)	90–95

a NR, not reported.

specific standardized methods have not been developed. However, the procedures outlined in Table 5.5 are generally consistent with the methods for conducting toxicity tests with early life stages of freshwater mussels (ASTM 2006a) and with the methods developed for conducting sediment toxicity tests with other freshwater invertebrates (ASTM 2006d). Exposures to whole sediment and to pore water have been conducted. Both glochidia and juvenile mussels have been used in sediment toxicity studies. Mussel species for sediment toxicity studies have been selected according to several criteria including: availability of glochidia, suitability for culture in the laboratory, and similar sensitivity between the surrogate and target species. Glochidia or juvenile mussels used to start the exposures have been obtained using procedures outlined above in sections dealing with water-only toxicity tests with glochidia and with juvenile mussels.

Whole-sediment exposures with glochidia or newly-transformed juveniles have been conducted by placing screened cylinders containing test organisms into secondary chambers containing sediment. Three to six replicates per concentration are generally tested, with up to 100 glochidia or 10–20 newly-transformed juvenile mussels tested per replicate. The duration of exposures started with glochidia have been 24 and 48 hours, and the duration of exposures starting with newly-transformed juvenile mussels have ranged from 4 to 10 days. Control survival of glochidia ranged from 69 to 79%, and control survival of newly-transformed juveniles ranged from 90 to 100% (Table 5.5)

USGS (2005c) and Nile Kemble (USGS, Columbia, MO; unpublished data) conducted a 28-day whole-sediment toxicity test starting with two-month-old juvenile *L. siliquoidea* and *L. rafinesqueana*. Sediments were sieved to a less than 250-μm particle size before the start of the exposure. The sediments were sieved to obtain a particle size that could be used isolate juvenile mussels at the end of a sediment exposure. It is unlikely that this life stage of juvenile mussel would be able to consume larger sediment particles. Exposures were conducted in 300-mL beakers containing about 15 mL of sediment and 260 mL of overlying water with about 2 volume additions/day of overlying water. Juvenile mussels were fed an instant algae mixture twice daily (commercial Instant Algae brand non-viable microalgae concentrates; Reed Mariculture, Campbell, CA) and control survival was greater than or equal to 90%. Studies are ongoing to evaluate the influence of sieving sediment on the bioavailability of contaminants to juvenile mussels and to *Hyalella azteca* and *Chironomus dilutus* (i.e., sieved to a particle size of 63–250 μm).

Issues Regarding Use of Methods

Issues regarding the use of glochidia and juvenile mussels in sediment toxicity tests are similar to those for water-only toxicity tests with these life stages. These concerns include: (1) the duration of the toxicity tests, (2) the quality of the organisms at the start of a test, (3) the life stage tested, (4) the determination of death at the end of a test, and (5) test acceptability criteria. Many of these issues have already been addressed in earlier sections in this chapter. However, specific issues relating to sediment toxicity tests are summarized in the following section, and information is provided that attempts to address these concerns. Areas of ongoing research or need for future research are also identified.

Duration of the Toxicity Test

Typically, sediment toxicity tests with glochidia have been conducted for 24–48 hours, and tests with newly-transformed juvenile mussels have been conducted for up to 10 days. Species-dependent viability of glochidia needs to be considered, as previously discussed (Table 5.2). Sediment toxicity tests starting with two-month-old juvenile mussels have been conducted for 28 days. In addition, sediment tests should be conducted for a duration that will enable appropriate comparisons to be drawn between water-only and sediment toxicity tests and between mussels and other standard sediment test organisms, such as the amphipod *H. azteca* and the midge *C. dilutus* (USEPA 2000; ASTM 2006d).

Life Stage Tested

Glochidia, newly-transformed juvenile mussels, and two-month-old juvenile mussels have been used in sediment toxicity studies. Adult mussels, which filter-feed as opposed to pedal-feeding juvenile mussels, have less contact with the sediment, and it could therefore be hypothesized that they would be less sensitive than the early life stages of mussels. However, no studies of sediment toxicity on adult mussels were reviewed for this chapter, and comparisons between life stages have not been conducted.

Use of Control or Reference Sediments to Establish Test Acceptability

Typically, control sediment is a sediment that has previously been demonstrated to have no toxic effects on the species being tested and is used to evaluate the acceptability of a toxicity test (USEPA 2000; ASTM 2006d). In some instances, control sediments may have been sterilized or tested for contaminants before the start of a sediment toxicity test (Keller, Ruessler, and Chaffee 1998). Sediment material collected from a reference site, which is usually a relatively undisturbed location, has been used to evaluate test acceptability and to make comparisons to test sediments (USEPA 2000; ASTM 2006d). Survival of mussels in control or reference sediments has been reported in the literature to be acceptable at greater than 90% (Table 5.5); however, no standard methods for conducting toxicity tests with freshwater mussels have been developed to establish a definitive test acceptability criterion for survival.

METHODS FOR CONDUCTING HOST FISH EXPOSURE TOXICITY TESTS WITH FRESHWATER MUSSELS

REVIEW OF METHODS

Two studies have evaluated exposure of glochidia encysted on fish to toxicants (Jacobson 1990; Nicola Kernaghan, unpublished data). During the glochidial stage of development, mussels attach to the gills of a host fish and encyst in host tissues within 2–36 hours of attachment. While encysted, the glochidia change form and begin to resemble adults. Although the exact processes that occur while the glochidia are attached to the host are not fully understood, some in vitro experiments suggest that glochidia absorb organic molecules from fish tissues and require fish plasma for development and metamorphosis (Isom and Hudson 1982) in a true host–parasite relationship. Glochidia remain attached to the host fish from 7 to 10 days to several months, offering a significant period of time during which they may be exposed to host contaminant burdens.

Nicola Kernaghan collected gravid female mussels from a reference location that has historically supported healthy populations of diverse mussel species. Host fish were collected from captive pond populations for which contaminant body burdens were documented. These fish were implanted with time-release pellets of toxaphene, dichlorodiphenyldichloroethylene (DDE), or atrazine to establish body burdens representative of concentrations found in fish tissue in the environment. Three to five fish were prepared at each concentration, and an equal number of control fish were implanted with placebo pellets for each test.

Following inoculation with glochidia, fish were held separately in 40-L aquaria until glochidial transformation occurred. Juvenile mussels were collected and randomly assigned to small glass Petri dishes containing about 20 mL of water. Five replicates, each containing at least 10 juvenile mussels, were used for each implanted fish and test duration was nine days. As previously described for juvenile water-only and sediment toxicity tests, cessation of both activity and heartbeat was used as the measurement endpoint indicating death. Control survival of glochidia for host exposure toxicity tests was generally greater than 80% (Nicola Kernaghan, unpublished data).

Jacobson (1990) investigated the effect of copper on the early life stages of freshwater mussels, including the encysted glochidia of *V. nebulosa*, *A. pectorosa*, and *A. grandis*.

Largemouth bass, which were used as hosts, were obtained from a fish hatchery and were held in the laboratory for five to seven days before testing. Fish were encysted *en masse* in a rectangular 120-L polyethylene tank, filled with about 40–60 L of aerated, dechlorinated tap water. Glochidia of a species were then added to the tank, resulting in a relatively homogenous level of encystment on all fish. The infested fish were exposed to 0, 25, 100, and 200 µg Cu/L in 17-L polycarbonate carboys at 19–21°C and with a photoperiod of 16:8, light–dark. Three replicates, each containing 5 or 10 encysted fish were prepared for each concentration and test solutions were renewed every two to four days. At each renewal, the tank water was drained through a 100-µm sieve, and the number of juvenile mussels was recorded. Mean number of juvenile mussels transformed per fish was used as an endpoint for this study.

Issues Regarding the Use of Methods

Issues regarding host fish exposure studies are unique to freshwater mussels, as a result of the unique life cycle of freshwater mussels. However, issues raised regarding standardization of methods for conducting toxicity tests with glochidia or juvenile life stages of freshwater mussel described above are also applied to encystment tests. These concerns include: (1) the duration of the toxicity tests, (2) the confirmation of exposure concentrations, (3) determination of mortality of glochidia while attached to the host, (4) evaluating exposure of glochidia to contaminants while still in the marsupium of the adult female mussel, and (5) test acceptability criteria. These specific issues relating to host fish exposure studies are summarized in the following section and information is provided that attempts to address these concerns. Areas of ongoing research or need for future research are also identified.

Duration of the Toxicity Test

The length of host fish studies is primarily determined by the period of time that any given species will naturally remain encysted on the host fish. Subsequent survival studies of juvenile mussels should be based upon the recommendations for conducting water-only toxicity studies with juvenile mussels.

Confirmation of Chemical Concentrations in Host Fish and Exposed Mussels

Due to the relatively small size of glochidia and juvenile mussels, it is generally not practical to determine bioaccumulation of chemicals. Testing of host fish to determine indirect exposure concentrations would be useful, but further studies are needed on the relation between the host fish and the encysted glochidia to understand the significance of these concentrations.

Determination of Glochidia Mortality While Attached to a Contaminated Host Fish

Jacobson (1990) enumerated all juvenile mussels transformed in each replicate and calculated a mean number of transformed juvenile mussels per fish. Alternatively, Nicola Kernaghan tested a subsample of transformed juvenile mussels. There is no reliable method of determining glochidia mortality while attached to a contaminated fish, requiring further research.

Would Glochidia, While Still in the Marsupium of the Adult Female Mussel, Be Exposed to Waterborne Contaminants?

The brood chambers of adult mussels and the glochidia that they contain have been reported to be physically isolated from the general water circulation pattern in the rest of the gills of the adult mussels (Silverman, McNeil, and Dietz 1987). In addition, exposure of brooding adults or encysted glochidia to copper were not very sensitive, suggesting that the glochidia are functionally isolated from the ambient water conditions while they are in the brood chamber (Jacobson 1990). However, more research is needed to further evaluate this question.

METHODS FOR CONDUCTING TOXICITY TESTS USING *CORBICULA FLUMINEA* AS SURROGATE SPECIES

The Asian clam, *C. fluminea*, has been frequently used as a surrogate for other freshwater mussels and target test organism since the early 1970s, more than 30 years following their introduction into the United States. The use of *Corbicula* in toxicity testing has supported the hypothesis that this organism is a viable indicator of impairment in aquatic systems (Graney, Cherry, and Cairns 1983). Consequently, threshold responses of *Corbicula* have been used to assess relative impacts on native mussels (Moulton, Fleming, and Purnell 1996; Milam and Farris 1998). A review by Doherty and Cherry (1988) provides an assessment of measured effects on juvenile and adult *C. fluminea* in laboratory tests including a description of various exposure and recovery regimes with single chemicals in static or flow-through systems. More recent toxicity tests have included acute and chronic exposures with *Corbicula* using: (1) laboratory and in situ techniques; (2) multi-contaminant stressors, including metals, surfactants, pesticides, and industrial or municipal effluents; and (3) measurement of alternative endpoints to assess damage in exposed populations. The purpose of this section is to include additional data available since the review by Doherty and Cherry (1988) and to evaluate variability among responses observed in toxicity tests conducted with *C. fluminea*.

REVIEW OF METHODS

Specific standard methods for conducting toxicity tests with *Corbicula* have not been developed. However, standard methods have been published for conducting bioconcentration tests (ASTM 2006e) and acute toxicity tests with saltwater bivalves (ASTM 2006c) that could be adapted for use with *Corbicula*. Additionally, standard methods have been developed for in situ testing specifically using caged bivalves (primarily, estuarine and marine species) (ASTM 2005f). Techniques have been developed to minimize variability among exposures when measuring growth and tissue bioaccumulation.

A summary of toxicity test methods using *C. fluminea* since the completion of the review by Doherty and Cherry (1989) is presented in Table 5.6. Experimental designs for these studies include exposures to single-chemical or complex effluents by measuring a variety of toxic endpoints. Test methods compiled from this dataset are either derivations from USEPA guidance documents for acute and chronic testing (USEPA 1993) or site-specific methods developed specifically for bivalves (Farris et al. 1989; Belanger et al. 1991).

A substantial portion of the studies outlined in Table 5.6 tested adult *Corbicula* rather than juvenile life stages, perhaps because of the range of lethal and sublethal endpoints that can be measured with adult organisms (i.e., sublethal test endpoints have typically been developed using greater tissue mass than what is available in juvenile mussels). Laboratory toxicity tests that were conducted with juvenile and adult *Corbicula* ranged in duration from 4 (Moulton, Fleming, and Purnell 1996) to 56 days (Belanger, Meiers, and Bausch. 1995; Belanger et al. 2000). In situ tests ranged in duration from 31 (Soucek et al. 2000) to 90 days (Bouldin, Farris, and Milam 2003). Although various test chambers have been used, most of the studies used artificial streams in a laboratory. Artificial streams are flexible in design and have the ability to hold large volumes of water with varying depths and currents for conducting tests with *Corbicula*.

Aqueous Toxicity Testing

Many of the tests were conducted in the laboratory using a wide range of feeding regimes and photoperiods (Table 5.6). These tests included whole effluent from industrial and municipal

TABLE 5.6

Summary of Test Conditions for Conducting Toxicity Tests with *Corbicula fluminea*

	Conditions	Farris et al. (1989)	Belanger, Meiers, and Bausch 1991	Belanger et al. (1993)	Farris et al. (1994)	Belanger, Meiers, and Bausch. (1995)
1	Test type	Flow through	Flow through	Flow through	Static-renewal	Flow through
2	Test duration (days)	30	28	56	30	42 or 56
3	Chemical/effluent mixture	ZnSO$_4$	Cu, Cl, NH$_3$, bromine, chloramine, temperature	Cationic surfactant	Zinc sulfate	Anionic surfactants
4	Exposure medium	Aqueous	Aqueous	Aqueous	Aqueous[a]	Aqueous[b]
5	Laboratory/in situ	Laboratory	Laboratory	Laboratory	Laboratory	Laboratory
6	Temperature (°C)	20	7–30	NR[f]	20–25	15–25
7	Photoperiod	NR[f]	NR[f]	NR[f]	14L:10D	Simulated ambient
8	Volume of test chamber (L)	13	20	384	20	460
9	Volume of test solution (L)	0.005/min	20	19/min	20	166L/min
10	Test vessel	Artificial streams	Artificial streams	Artificial streams	Artificial streams	Artificial streams
11	Organism source	New River, VA	New River, VA; Catawba River, SC	Lower East Fork - Little Miami River, OH	New River, VA	NR[f]
12	Age of test organism	Adult	Adult; juvenile	Adult	Adult	Adult
13	Number of organisms/replicate	60	10	15	6	NR[f]
14	Number of replicates/treatment	NR[f]	3	8	3	5 and 9[c]
15	Feeding regime	Continuous	Daily	Continuous	Daily	Continuous
16	Dilution water	Hardness 71 mg/L; pH 8.1–8.4	Hardness 60–75 mg/L; pH 7.9–8.2	Hardness 37–42 mg/L; pH 7.3; TOC 7.0–13.7 mg/L	Hardness 60 mg/L; pH 7.3–8.1	NR[f]
17	Water quality	Yes	Yes	Yes	Yes	Yes
18	Endpoints	Growth, sublethal biomarker, bioaccumulation, mortality	Mortality, growth	Mortality, growth, sublethal biomarker, reproductive condition	Bioconcentration, sublethal biomarker, recovery	Biomass, abundance, density
19	Control survival	NR[f]	NR[f]	>90%	NR[f]	NR[f]
	Conditions	Cole (1995)	Moulton, Fleming, and Purnell (1996)	Baudrimont et al. (1997)	Milam and Farris (1998)	Belanger et al. (2000)

	Parameter					
1	Test type	Static-renewal	Static	Static-renewal	Static-renewal	Flow through
2	Test duration (days)	30	4	45	30	56
3	Chemical/effluent mixture	$CuSO_4$	Aldicarb/acephate[d]	Hg and Cd[e]	Coal mining effluent	Anionic surfactants
4	Exposure medium	Spiked sediment	Aqueous	Aqueous	Aqueous	Aqueous[b]
5	Laboratory/in situ	Laboratory	Laboratory	Laboratory	Laboratory	Laboratory
6	Temperature (°C)	22	21	20	25	NR[f]
7	Photoperiod	16L:8D	12L:12D	12L:12D	16L:8D	Simulated ambient
8	Volume of test chamber (L)	60	4	NR[f]	60-L oval streams	460
9	Volume of test solution (L)	40	2	12	40	166L/min
10	Test vessel	Artificial streams	4-L glass jars	Glass tank	Artificial streams	Artificial streams
11	Organism source	Saline River, AR	NR[f]	Cazaux-Sanguinet Lake, France	Saline River, AR	NR[f]
12	Age of test organism	Adult	Adult	Adult	Adult	Adult
13	Number of organisms/replicate	12	4	5	12	NR[f]
14	Number of replicates/treatment	3	NR[f]	3	3	NR[f]
15	Feeding regime	Daily	None	Twice weekly	Daily	Continuous
16	Dilution water	Hardness 59 mg/L; pH 7.3–7.8	NR[f]	NR[f]	Hardness 50–324 mg/L; pH 7.4–8.0	Hardness 165 mg/L; pH 8.0; TSS 13 mg/L
17	Water quality	Yes	No	Daily metal analysis	Twice daily iron analysis	Yes
18	Endpoints	Mortality, growth, bioconcentration, sublethal biomarker	Mortality, ChE activity, behavior response, recovery	Metallothionein induction	Sublethal biomarker, mortality, growth, feeding behavior, recovery	NOEC, density estimates
19	Control survival	100%	100%	>99%	NR[f]	NR[f]
	Conditions	Soucek et al. (2000)	Cataldo et al. (2001)	Tran, Boudou, and Massabuau (2002)	Versteeg and Rawlings (2003)	Bouldin, Farris, and Milam (2003)
1	Test type	Flow through	Static-renewal	Static-renewal	Flow through	Flow through
2	Test duration (days)	31	6	15	32	90
3	Chemical/effluent mixture	Acid mine drainage	Industrial effluent	Cadmium	Anionic surfactant	POTW effluent
4	Exposure medium	Aqueous	Aqueous/sediment/pore water	Aqueous	Aqueous	Aqueous

(continued)

TABLE 5.6 (Continued)

Conditions	Farris et al. (1989)	Belanger, Meiers, and Bausch 1991	Belanger et al. (1993)	Farris et al. (1994)	Belanger, Meiers, and Bausch. (1995)
5 Laboratory/in situ	in situ	Laboratory	Laboratory	Laboratory	in situ
6 Temperature (°C)	Ambient	25	15 and 25	15	18–27
7 Photoperiod		NR[f]	Ambient	NR[f]	Ambient
8 Volume of test chamber (L)	NA[f]	20 mL	9	460	NA[f]
9 Volume of test solution (L)	NA[f]	20 mL	NR[f]	167 L/min	NA[f]
10 Test vessel	NA[f]	20-mL tissue culture plate wells	9-L glass tanks	Artificial streams	NA[f]
11 Organism source	New River, VA	Parana River, Argentina	NR[f]	Lower East Fork Little Miami River, OH	Illinois River, AR
12 Age of test organism	Adult	Juvenile	Adult	Adult	Adult
13 Number of organisms/replicate	5	20	5	20	24
14 Number of replicates/treatment	5	3	4	5	3
15 Feeding regime	Continuous	NR[f]	Bi-daily	NR[f]	Continuous
16 Dilution water	pH 3.7–7.7	no data	no data	Hardness 140 mg/L; DO 8.1; pH 7.8–8.3; TOC 7.2 mg/L	Hardness 60–153 mg/L; pH 7.1–8.4
17 Water quality	Yes	Yes	Yes	Yes	Yes
18 Endpoints	Mortality	Mortality	Ventilatory flow rate, accumulation	Bioconcentration, LC50, growth	Mortality, sublethal biomarker, growth
19 Control survival	94%	>90%	NR[f]	NR[f]	>90%

[a] Exposures included the use of coarse sand as a burrowing substrate for clams.
[b] Use of cobble as a substrate for exposures.
[c] Five replicates were sampled at zero, two, and four weeks, and nine replicates were sampled after eight weeks.
[d] Two separate assays were conducted to determine effect - no mixture studies conducted.
[e] Tests conducted separately.

Note: NR not reported. NA not applied.

122

Freshwater Bivalve Ecotoxicology

processes, single chemicals including metals, surfactants, and pesticides, and chemical mixtures including ammonia, bromine, and chloramine. Conventional test methods have provided guidance for maintaining test organisms in a laboratory, including the requirements of a feeding regime and photoperiod for all exposures.

Laboratory Exposures

Most of the methods summarized in Table 5.6 were laboratory exposures. Providing a supplemental food source to *Corbicula* during tests seemed to be necessary regardless of the test duration. Laboratory-cultured algae (*Chlamydomonas, Ankistrodesmus, Scenedesmus,* and *Chlorella*) have been used as the food source. *Corbicula* tests conducted for as short as 15 days were fed twice daily (Tran, Boudou, and Massabuau 2002); whereas, Organisms in tests conducted for as long as 45 days were fed twice each week (Baudrimont et al. 1997). The range of feeding regimes, however infrequent, did not seem to affect the response of the test organisms to a stressor. However, the need to provide a food source when conducting laboratory toxicity tests should be considered to reduce the possibility of potential nutritional problems.

The presence of ambient photoperiods specific to the time of year is inherent with in situ tests conducted in the field. Laboratory toxicity tests are typically conducted using a consistent photoperiod, but it may be important to use light–dark cycles to eliminate stress on exposed individuals. The range of photoperiods in these tests (12:12–16:8, light–dark) reflects the time of year during which the laboratory tests were being conducted (Table 5.6).

In Situ Exposures

Only two tests among those summarized in Table 5.6 were conducted in situ (Soucek et al. 2000; Bouldin, Farris, and Milam 2003). The assumption is that a natural flora of algae was being provided to the test organisms because these tests were conducted in situ, although algal species were not identified in these studies. Several non-in situ tests used a flow-through system of riverine water that included a natural population of algae and phytoplankton (Farris et al. 1989; Belanger et al. 1991, 1993, 2000; Versteeg and Rawlings 2003). In situ tests allow for a natural light:dark photoperiod throughout the test.

Sediment Toxicity Testing

Relatively few toxicity tests have included the use of sediment as a source of contaminants in exposures to *C. fluminea*. Cataldo et al. (2001) assessed the impact of various industrial effluents and used sediments located near an industrial outfall as the exposure medium. Cataldo et al. (2001) reported *Corbicula* survival was reduced in sediments where concentrations of total chlorinated pesticides (as low as 1.4 ng/g), aliphatic hydrocarbons (as low as 3,200 ng/g), PAHs (as low as 244 ng/g), and metals (as low as 171 µg/g) were detected. Cole (1995) determined that *C. fluminea* were chronically sensitive to copper-spiked sediments in laboratory artificial stream exposures at levels as low as 17 µg/g in 30-day exposures. Cole (1995) reported reductions in the cellulolytic enzyme activity of clams, a digestive enzyme that has been shown to indirectly correlate to contaminant presence in lotic systems. Enzyme reductions directly affect the intake and assimilation of food, and consequently, mortality can occur due to starvation of the exposed individuals.

Overview of Conditions Used to Conduct Toxicity Tests with *Corbicula*

Age of Test Organisms

Generally, toxicity tests with *Corbicula* have used the adult life stage (greater than 10 mm in length) to measure mortality, bioaccumulation, and various biomarkers (Table 5.6). Doherty and Cherry

(1988) summarized published literature that included the use of *Corbicula* larvae to assess impact by metals, slimicides, biocides, herbicides, asbestos, and chlorinated compounds. Doherty and Cherry (1988) indicated that juvenile and larval life stages were generally more sensitive to metals, chlorinated compounds, biocides, and asbestos compared to adult *Corbicula*. Toxicity tests have been conducted with juvenile clams to measure impact from metal-dominated industrial effluent in laboratory exposures (Cataldo et al. 2001). While juvenile mussels may be a more sensitive life stage, adults are readily available and provide more tissue mass for sublethal endpoints such as bioaccumulation, lipid content, glycogen, and cellulolytic activity. Furthermore, behavioral responses (e.g., filtration rates, valve gaping, foot movement) are well documented and can provide complementary information to measures of survival or sublethal endpoints (Anderson et al. 1976; Rodgers et al. 1980; Milam and Farris 1998). While there is considerably more data for adult *Corbicula* in acute and chronic toxicity assessments, the use of either life stage is feasible, and decisions regarding life stage should be made based on the objectives of the study.

Organism Source

The sources of *Corbicula* used in the testing include locations in the eastern, midwestern, and southern United States (Table 5.6). Clams have also been collected for testing from a lake in France and a river in Argentina (Baudrimont et al. 1997; Cataldo et al. 2001). Clams from these tests were collected from reference sites or at least from minimally disturbed sites. Organism source (i.e., ambient levels of contaminants in clam tissues) is a significant concern in bioaccumulation studies particularly when study objectives include the measurement of certain contaminants in clam tissues and ambient concentrations exceed exposure concentrations. Organisms that are used for National Pollutant Discharge Elimination System (NPDES) testing of wastewater have been required by the USEPA to come from certified laboratories across the United States. This certification process includes stringent daily care for test organisms with monthly reference toxicity tests to account for variability of organism response (often to NaCl or Cu) within the specific laboratory culture. Annual inter-laboratory tests are conducted to assure that laboratories across the United States are within calculated control limits of each other. Furthermore, a recent federal register notice indicated that neonates of adult *Ceriodaphnia dubia* must be distributed among the test chambers to eliminate bias of brood effect from a single female on any one concentration or replicate (40 CFR 136, 2001). While tests outlined in Table 5.6 do not include this strict approach to organism culture, because they were not required by regulatory compliance guidelines, reassurance of background physiological status is important and (e.g., how the clams respond without exposure to selected contaminants) is often found in the literature using reference controls and background data analyses (Cole 1995; Milam and Farris 1998; Bouldin, Farris, and Milam 2003; Versteeg and Rawlings 2003).

Duration of Exposure

Most of the toxicity tests with *Corbicula* have been conducted for more than 30 days with measures of survival, growth, biomarkers, enzyme activity, and protein induction. Unless contaminant concentrations in aqueous or sediment exposures are relatively high, acute tests may be limiting since *Corbicula* often respond to elevated concentrations by an avoidance behavior (e.g., shell closure), reducing the effectiveness of short-term toxicity tests; Varanka 1986, 1987). Cole (1995) indicated that to determine a concentration range of copper for a chronic sediment exposure (30 days), acute range-finding tests indicated that sediment-bound copper concentrations that ranged from 2,000 to 4,000 µg/g were required to induce mortality following an 11-day exposure. Additionally, expected environmental concentrations of many metals and pesticides are often much lower than what is needed to elicit a lethal response in *Corbicula*.

Flow-Through Versus Static Exposures

About half of the laboratory tests used flow-through systems to reduce physiological stress on test organisms or to reduce changes in water quality during exposures (Table 5.6), which often accompany smaller, static chambers (USEPA 1993; ASTM 2006g). None of the tests that included static or renewal exposures in Table 5.6 suggested that the organisms experienced stress associated with lack of a flow-through system. This test parameter, albeit debatable, may need to be left up to the researcher for determination of appropriate test methods with regard to specific objectives.

Toxicity Endpoints

Toxicity endpoints from the recent reported literature include survival, growth, bioaccumulation, behavior, and numerous biomarkers such as glycogen concentration, DNA strand breakage, cellulolytic enzyme activity, and AChE inhibition (Table 5.6).

Survival. Measuring survival following an acute or chronic exposure is relatively simple and inexpensive and is routinely measured in toxicity tests. Although measures of survival alone may not offer insight to mitigate a system once it has been exposed, it does give a sense of magnitude of the contaminant that caused the lethal response.

Growth. Determination of shell width (i.e., measurement from the umbo to the ventral margin) has been used to evaluate growth during toxicity exposures. Rate of *Corbicula* growth has also been evaluated and indicates that young, adult clams can often grow shell material under optimal conditions in as little as 30 days (e.g., Belanger et al. 1986b). Although growth does not necessarily provide clear evidence for determining a response, growth can, if included with other calculated endpoints, provide a more holistic picture of the exposure.

Bioaccumulation. Numerous toxicity studies have been conducted using bioaccumulation as the sole endpoint for bivalves, including *Corbicula* (Chapter 10). However, links between bioaccumulation and toxic effects are often more revealing than body-burden alone for determination of cause–effect relationships (Cole 1995; Milam and Farris 1998; Farris et al. 1989, 1994). Milam and Farris (1998) indicated that metal body burdens in *Corbicula* as low as 373 mg Fe/g and 683 mg Zn/g dry weight were associated with significantly reduced biomarker measurements and subsequent physiological stress. In a long-term exposure study using zinc as the predominant contaminant, increased clam body burdens correlated with sublethal responses in cellulolytic enzyme activity (Farris et al. 1989). Following 10 days, aqueous zinc concentrations were 34 µg/L in overlying water (lower than the WQC of 47 µg/L) (USEPA 1980), body burdens in exposed clams reached 540 µg/g, and significant reductions in enzyme activity were measured. Cole (1995) determined that sediment-bound copper exposures to *Corbicula* were toxic (i.e., significantly lower cellulolytic enzyme activities) following a 20-day exposure. As these studies show, bioaccumulation analyses from tissues of exposed clams can provide valuable data in determining levels of effect when combined with other endpoints such as mortality, behavior, and sublethal endpoints. Farris et al. (1994) suggested that the use of biomarkers in conjunction with body burdens and other stress indicators can effectively quantify the effects of heavy metal discharges. However, bioaccumulation analysis alone cannot always be used to predict responses to a population.

Physiological Indicators. Several physiological enzymes, including cellulase and cholinesterase (ChE), have been measured in laboratory or in situ exposures of *Corbicula*. ChE is a natural enzyme produced to regulate synaptic responses in certain tissues, and organophosphate and carbamate pesticides are known to inhibit ChE. Moulton, Fleming, and Purnell (1996) compared ChE activity in muscle tissue of *Elliptio complanata* and *Corbicula* and determined that in a 96-hour aqueous exposure *Corbicula* were less sensitive to ChE-inhibiting pesticides than *E. complanata*. However, data generated by Moulton, Fleming, and Purnell (1996) provide support for using adductor muscle tissue in freshwater mussels for determination of ChE activity.

Cellulolytic enzyme activity has been used as a biomarker in mollusk monitoring efforts (Farris et al. 1989) and has been used to determine effects in exposures of *Corbicula* to various effluents, metal mixtures, pesticides, and surfactants (Belanger et al. 1993; Farris et al. 1994; Cole 1995; Milam and Farris 1998; Bouldin, Farris, and Milam 2003). This particular biomarker is a digestive enzyme, which responds (i.e., typically an inverse relationship to a pollutant concentration) to aqueous and sediment-bound contaminants by reducing the concentration of enzyme that is normally used to digest cellulose, thereby starving the individual.

As indicated in the previous paragraph on bioaccumulation, this particular biomarker has been utilized to measure damage in *Corbicula* from numerous contaminants. Bouldin, Farris, and Milam (2003) measured long-term (90 days) in situ exposures of *Corbicula* in streams receiving treated sewage and indicated that reductions of cellulolytic enzyme activity were associated with storm-related fluctuations of nutrients, metals, and fecal coliform bacteria instream. Sensitivity comparisons of clams and unionid mussels have also been determined using this biomarker and indicate that *Corbicula* are at times more sensitive (i.e., elicit a greater response to contaminants measured in reduced cellulolytic activity) than unionids following long-term exposures to copper (Jerry Farris, personal communication).

Measurement of elevated protein induction has often been considered a stress response in organisms exposed to metals or metal complexes. Baudrimont et al. (1997) conducted 45-day exposures of *Corbicula* to cadmium and mercury. Measured endpoints included metallothionein (MT) induction as a response to metal exposures in clams. Metallothionein is a metal-binding protein that is expressed when exposed to relatively high concentrations of some metals. Analysis of the MT protein has been shown to indicate a direct relation between concentration of metals and induction of MT in tissues (Roesijadi and Robinson 1994). Metallothionein induction provided a substantial biomarker response to cadmium exposures but was less predictive of inorganic mercury (Baudrimont et al. 1997). Despite this, MT induction provides one component to the understanding effects of metals on *Corbicula*.

Behavior.   Behavioral responses, such as ventilatory flow rates, feeding behavior, and valve movement are important indicators of effect when combined with other endpoints such as bioaccumulation and mortality. Ventilatory flow rates are important when assessing metal uptake from an aqueous phase and from algal-bound particles in order to determine route of uptake and subsequent impact on individuals (Tran, Boudou, and Massabuau 2002). Feeding behavior (i.e., quantity, quality, and rate of intake of phytoplankton and zooplankton) has been measured in laboratory tests to indicate the ability of individual clams to continue filtering during chronic, aqueous exposures (Belanger, Cherry, and Cairns 1986a; Doherty 1986; Milam and Farris 1998; Gatenby et al. 1999). Reduced valve movement can indicate the onset of starvation and ultimate mortality when observed and measured for longer periods of time (usually greater than 30 days). Hyperactive valve movement, could also indicate impact from organophosphate pesticide exposures since these compounds are known to inhibit ChE activity in muscle tissue (Galgani and Bocquene 1990; Moulton, Fleming, and Purnell 1996; Bocquene, Roig, and Fournier 1997).

ISSUES REGARDING THE USE OF THE METHODS

Advantages and disadvantages of conditions that have been used to conduct toxicity tests with *Corbicula* are summarized in Table 5.7. Static and renewal tests conducted in the laboratory are typically less costly than flow-through exposures. Although less water is often associated with laboratory tests using smaller chambers, there may be a greater potential for problems with low dissolved oxygen as well as a loss of toxicant due to adsorption or volatilization. In situ tests can offer a more realistic environmental exposure but are at risk of being subjected to flooding, drought, predation, and vandalism.

TABLE 5.7
Evaluation of Conditions Used to Conduct Toxicity Tests with *Corbicula fluminea*

Parameter	Advantages	Disadvantages
Static renewal and non-renewal	Simple and less expensive Smaller volume of water required than for flow-through tests	Increased risk of stress on test organism Decrease in DO concentration due to metabolic waste Loss of toxicant (non-renewal) due to adsorption or volatilization may occur
Flow through	DO concentrations are readily maintained throughout the test period Decrease in loss of contaminant due to various fate models, and lack of renewal is reduced Flow-through water often has indigenous algal communities, which reduces the need for daily feeding from cultured cells Can provide a more realistic evaluation of contaminant source	Costly to maintain Increase in laboratory space is often needed to properly maintain a flow-through system Large volumes of water needed
Test duration (>30 days)	Can analyze additional test endpoints other than mortality More realistic to contaminant levels in the environment	Costly, from a single endpoint experimental design—less "bang for your buck" if you have fewer measured test endpoints
Laboratory/in situ	Laboratory tests are local, with less travel and time spent going to a field site Analyst has control over exposure in the laboratory In situ tests can be more realistic than laboratory exposures	In situ tests are at greater risk from flooding, drought, theft, and predation Laboratory tests are highly controlled and may misrepresent actual environmental exposure
Toxicity endpoints	Lethal endpoints are simple and less expensive Lethal endpoints can detect impact at higher concentrations Sublethal endpoints can provide a more comprehensive impact assessment Sublethal endpoints can offer insight into other endpoints such as bioaccumulation, growth, and biomarker analysis	Lethal endpoints may only be measured from high-contaminant concentrations Sublethal endpoints are more costly
Organism age	Adult *Corbicula* are relatively easy to acquire Adults can offer more tissue for various analyses Adults are easy to maintain in the laboratory or field Juveniles/larvae are often more sensitive to contaminants	There is limited availability of juveniles or larvae There is only a small amount of tissue available for juveniles or larvae

Managing freshwater mussel populations through surrogate species is complex. The more information that the science can provide to natural resource managers, whether it be recovery determinations following contaminant exposure, toxicity thresholds for certain life stages, or comparative species or life stage sensitivities, the better prepared they will be in supporting recovery efforts for native populations. There is evidence suggesting that *Corbicula* can be used in lieu of native mussels, to determine the effects (and potential recovery) of various contaminants in aquatic

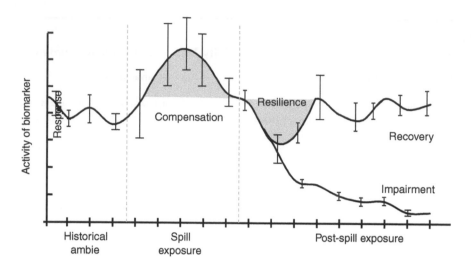

FIGURE 5.1 Theoretical sublethal biomarker response levels of *Corbicula* during a long-term exposure and recovery assessment following a spill.

ecosystems. However, additional direct comparisons between the response of *Corbicula* and native mussels to contaminants are needed. While the decline of native mussels continues, the need for toxicological research with these species grows. The use of *Corbicula* in toxicity assessments supports the demand for sensitivity thresholds of various inorganic and organic compounds while not exploiting already-threatened native mussels.

Resource managers may need to identify environmental damage of an exposed population via accidental spills or discharges downstream of the impact zone. Recovery periods following a contaminant exposure can be used to the determine resilience and elasticity of a population (Figure. 5.1). Variability among individuals in an unimpacted population (as shown by the error bars) is relatively moderate during ambient conditions and a natural increase and decrease of responses is expected. During exposure to a non-specific contaminant, this variability can increase because individuals respond differently to toxic compounds and are often compensating (e.g., elevated responses of some biomarkers) during the stress period. Following this acute or chronic exposure (resilience), measured responses can determine whether a population can ultimately recover to pre-stress conditions or if it will succumb to the exposure/impairment and eventually die. Doherty and Cherry (1988) discussed the importance of recovery in their review and concluded that these periods provide valuable information for determining time-to-impairment and time-to-recovery for body burdens in *Corbicula*.

CONCLUSIONS AND RECOMMENDATIONS

A variety of methods for conducting acute or chronic toxicity tests with freshwater mussels have been published (Table 5.1, Table 5.3 through Table 5.6). However, historic lack of standardization of these methods has limited the use of data generated from these studies. The recent development of a standard for conducting laboratory toxicity tests with glochidia and juvenile mussels will help to address this problem (ASTM 2006a). Additional effort should be placed on conducting research that will help to further standardize methods for conducting toxicity tests with freshwater mussels through a consensus-based process such as ASTM. The following is a list of suggested research that would be useful in helping to further develop standardized methods for conducting toxicity tests with freshwater mussels:

1. Further evaluate the influence of handling, holding, and acclimation on adult, juvenile, or glochidia used to conduct toxicity tests.
2. Determine the minimum number of female mussels that should be sampled to obtain glochidia or juvenile mussels used to start a toxicity test. These studies might include an evaluation of the variability in the sensitivity of glochidia or juvenile mussels obtained from individual females using a variety of chemicals with different toxic modes of action.
3. Further evaluate the influence of contaminant exposure on immature glochidia developing within the marsupium of the female mussel.
4. Further evaluate methods for conducting chronic toxicity tests with juvenile mussels, focusing on establishing the feeding requirements for newly-transformed juvenile mussels for a variety of mussel species. Ongoing research to improve culturing methods for the propagation, holding, and feeding of newly-transformed juvenile mussels will hopefully provide useful information that can be adapted to establish methods for conducting chronic toxicity tests with juvenile mussels.
5. Further evaluate the utility of conducting acute or chronic toxicity tests with juvenile mussels greater than two months old with a variety of species of mussels. These studies might include comparisons of the sensitivities of newly-release juvenile mussels and older juvenile mussels using acute toxicity tests with a variety of chemicals.
6. Further develop endpoints for establishing effects in toxicity tests with mussels (e.g., behavior, growth, and biomarkers).
7. Conduct additional round-robin (ring) toxicity tests using specified methods to evaluate inter-laboratory variability in control and toxic responses of mussels to a variety of chemicals with different toxic modes of action. ASTM (2006a) summarizes the results of acute copper round-robin tests with glochidia and juvenile mussels for five laboratories.
8. Evaluate the relative sensitivity of glochidial, juvenile, and adult life stages of mussels to a variety of different chemicals in acute or chronic toxicity tests.
9. Compare the response of various species of mussels to the response of other surrogate species (e.g., trout, cladocerans, amphipods, and *Corbicula*) in toxicity tests conducted using a variety of different chemicals.
10. Compare the response of different populations of a species collected from different geographic regions to a variety of chemicals in laboratory toxicity tests.
11. Compare the response of mussels tested in laboratory toxicity tests to the response of mussels exposed in the field (either using in situ exposure containers or in a natural habitat).

REFERENCES

American Society for Testing and Materials (ASTM), Standard guide for conducting laboratory toxicity tests with freshwater mussels (ASTM E2455-06), *ASTM Annual Book of Standards Volume 11.05*, ASTM, West Conshohocken, PA, 2006a.

ASTM, Standard guide for conducting acute toxicity tests on test materials with fishes, macroinvertebrates, and amphibians (ASTM E729-96 (2002)), *ASTM Annual Book Of Standards Volume 11.05*, ASTM, West Conshohocken, PA, 2006b.

ASTM, Standard guide for conducting static acute toxicity tests starting with embryos of four species of saltwater bivalve mollusks (ASTM E724-98 (2002)), *ASTM Annual Book of Standards Volume 11.05*, ASTM, West Conshohocken, PA, 2006c.

ASTM, Standard test method for measuring the toxicity of sediment-associated contaminants with freshwater invertebrates (ASTM E1706-05), *ASTM Annual Book of Standards Volume 11.05*, ASTM, West Conshohocken, PA, 2006d.

ASTM, Standard guide for conducting bioconcentration tests with fishes and saltwater bivalve mollusks. (E1022-94(2002)), *ASTM Annual Book of Standards Volume 11.05*, ASTM, West Conshohocken, PA, 2006e.

ASTM, Standard guide for conducting in-situ field bioassays with caged bivalves (E2122-02), *ASTM Annual Book of Standards Volume 11.05*, ASTM, West Conshohocken, PA, 2006f.

ASTM, Standard guide for conducting static and flow-through acute toxicity test with mysids from the West Coast of the United States (ASTM E1463-92 (2004)), *ASTM Annual Book of Standards Volume 11.05*, ASTM, West Conshohocken, PA, 2006g.

Anderson, K. B., Thompson, C. M., Sparks, R. E., and Paparo, A. A., Effects of potassium on adult asiatic clams, *Corbicula manilensis*, *Ill. Nat. Hist. Surv. Biol. Note*, 98, 7, 1976.

Augspurger, T., Keller, A. E., Black, M. C., Cope, G., and Dwyer, F. J., Derivation of water quality guidance for protection of freshwater mussels (Unionidae) from ammonia exposure, *Environ. Toxicol. Chem.*, 22, 2569–2575, 2003.

Baudrimont, M., Metivaud, J., Maury-Brachet, R., Ribeyre, F., and Boudou, A., Bioaccumulation and metal-lothionein response in the Asiatic clam (*Corbicula fluminea*) after experimental exposure to cadmium and inorganic mercury, *Environ. Toxicol. Chem.*, 16, 2096–2105, 1997.

Belanger, S. E., The effect of dissolved oxygen, sediment, and sewage treatment plant discharges upon growth, survival, and density of Asiatic clams, *Hydrobiologia*, 218, 113–126, 1991.

Belanger, S. E., Cherry, D. S., and Cairns, J., Uptake of chrysoltile asbestos fibers alters growth and reproduction of Asiatic clams, *Can. J. Fish. Aquat. Sci.*, 43, 43–52, 1986.

Belanger, S. E., Farris, J. L., Cherry, D. S., and Cairns, J., Growth of Asiatic clams (*Corbicula* sp.) during and after long-term zinc exposure in field-located and laboratory artificial streams, *Arch. Environ. Contam. Toxicol.*, 15, 427–434, 1986.

Belanger, S. E., Farris, J. L., Sappington, K. G., Cherry, D. S., and Cairns, J., Sensitivity of the Asiatic clam to various biocidal control agents, *J. Am. Water Works Assoc.*, 83, 79–87, 1991.

Belanger, S. E., Davidson, D. H., Farris, J. L., Reed, D., and Cherry, D. S., Effects of cationic surfactant exposure to a bivalve mollusc in stream mesocosms, *Environ. Toxicol. Chem.*, 12, 1789–1802, 1993.

Belanger, S. E., Meiers, E. M., and Bausch, R. G., Direct and indirect ecotoxicological effects of alkyl sulfate and alkyl ethoxysulfate on macroinvertebrates in stream mesocosms, *Aquat. Toxicol.*, 33, 65–87, 1995.

Belanger, S. E., Cuckert, J. B., Bowling, J. W., Begley, W. M., Davidson, E. M., LeBlanc, E. M., and Lee, D. M., Responses of aquatic communities to 25-6 alcohol ethoxylate in model stream ecosystems, *Aquat. Toxicol.*, 48, 135–150, 2000.

Bocquene, G., Roig, A., and Fournier, D., Cholinesterases from the common oyster (*Crassostrea gigas*): Evidence for the presence of a soluble acetylcholinesterase insensitive to organophosphate and carbamate inhibitors, *FEBS Lett.*, 407, 261–266, 1997.

Bouldin, J. L., Farris, J. L., and Milam, C. D., Measured responses of the Asian Clam (*Corbicula fluminea*) to runoff associated with urbanization: A multifaceted biomonitoring approach in a watershed assessment of Northwest Arkansas, Submitted to City of Rogers, Rogers, AR, p. 44, 2003.

Bringolf, R. B., Humphries, L. F., Lazaro P., Eads, C., Barnhart, M. C., Shea, D., Levine, J., and Cope W. G., Assessing the hazards of current use pesticides to early life stages of native freshwater mussels. Presented at the Freshwater Mollusk Conservation Society Symposium in St. Paul, MN, May 15–18, 2005.

Cataldo, D., Colombo, J. C., Boltovskoy, D., Bilos, C., and Landoni, P., Environmental toxicity assessment in the Paraná River delta (Argentina): Simultaneous evaluation of selected pollutants and mortality rates of *Corbicula fluminea* (Bivalvia) early juveniles, *Environ. Pollut.*, 112, 379–389, 2001.

Cherry, D. S., Van Hassel, J. H., Farris, J. L., Soucek, D. J., and Neves, R. J., Site-specific derivation of the acute copper criteria for the Clinch River, Virginia, *Human Ecol. Risk Assess.*, 8, 591–601, 2002.

Clem, S. A., Complexities in testing early life stage responses of freshwater mussels to metals and contaminated sediments, MS Thesis, Arkansas State University, Jonesboro, AR, 1998.

Code of Federal Regulations, Guidelines Establishing Test Procedures for the Analysis of Pollutants; Whole Effluent Toxicity Test Methods, Environmental Protection Agency, 40 CFR part 136, 2001.

Cole, R., Differentiation of copper uptake from sediment or overlying water by the Asian clam, *Corbicula fluminea*, and the modification of sediment characters in a 30-day artificial stream test, Honors Thesis, Arkansas State University, Jonesboro, AR, 1995.

Dimock, R. V. and Wright, A. H., Sensitivity of juvenile freshwater mussels to hypoxic, thermal and acid stress, *J. Elisha Mitchell Sci. Soc.*, 109, 183–192, 1993.

Doherty, F. G., A multidisciplinary study of the Asiatic clam, *Corbicula fluminea*, from the New River, Virginia. PhD. dissertation, Virginia Polytechnic Institute and State University, Blacksburg, VA, 1986.

Doherty, F. G. and Cherry, D. S., Tolerance of the Asiatic Clam *Corbicula* spp. to lethal levels of toxic stressors—a review, *Environ. Pollut.*, 51, 269–313, 1988.

Dunn, C. S. and Layzer, J. B., Evaluation of various holding facilities for maintaining freshwater mussels in captivity, In *Conservation and Management of Freshwater Mussels II: Initiatives for the Future*, Cummings, K. S. et al., Eds., *Proceedings of a UMRCC Symposium*, St. Louis, MO, pp. 205–213, 1997.

Farris, J. L., Belanger, S. E., Cherry, D. S., and Cairns, J., Cellulolytic activity as a novel approach to assess long-term zinc stress to *Corbicula*, *Water Resour.*, 23, 1275–1283, 1989.

Farris, J. L., Grudzien, J. L., Belanger, S. E., Cherry, D. S., and Cairns, J., Molluscan cellulolytic activity responses to zinc exposure in laboratory and field stream comparisons, *Hydrobiologia*, 287, 161–178, 1994.

Farris, J. L., Van Hassel, J. H., and Cherry, D. S., Refinement of site-specific copper criteria for the protection of mussel population, *Presented at the 12th Annual Meeting of SETAC*, Seattle, WA, November 3–7, 1991.

Fisher, G. R. and Dimock, R. V., Viability of glochidia of *Utterbackia imbecillis* (Bivalvia: Unionidae) following their removal from the parent mussel, In *Proceedings of the First Freshwater Mussel Conservation Society Symposium, 1999*, Ohio Biological Survey, pp. 185–188, 2000.

Galgani, F. and Bocquene, G., In vitro inhibition of acetylcholinesterase from four marine species by organophosphates and carbamates, *Bull. Environ. Contam. Toxicol.*, 45, 243–249, 1990.

Gatenby, C. M., A study of holding conditions, feed rations, and algal foods for the captive care of freshwater mussels, PhD Dissertation, Virginia Polytechnic Institute and State University, Blacksburg, VA, 2000.

Gatenby, C. M., Neves, R. J., and Parker, B. C., Influence of sediment and algal food on cultured juvenile freshwater mussels, *J. North Am. Benth. Soc.*, 15, 597–609, 1996.

Gatenby, C. M., Neves, R. J., and Parker, B. C., Growth and survival of juvenile rainbow mussels, *Villosa iris* (Lea 1829) (Bivalvia: Unionidae), reared on algal diets and sediment, *Am. Malacol Bull.*, 14, 57–66, 1997.

Gatenby, C. M., Patterson, M. A., Parker, B. C., and Neves R. J., Filtration rates of the unionid, *Villosa iris* (Lea 1829) fed at different algal cell concentrations and different feeding regimes. North American Benthological Society Annual meeting, Duluth, MN, 1999.

Goudreau, S. E., Neves, R. J., and Sheehan, R. J., Effects of wastewater treatment plant effluents on freshwater mollusks in the upper Cinch River, Virginia, USA, *Hydrobiologia*, 252, 211–230, 1993.

Graney, R. L., Cherry, D. S., and Cairns, J., The influence of substrate, pH, diet, and temperature upon cadmium accumulation in the Asiatic clam (*Corbicula fluminea*) in laboratory artificial streams, *Water Res.*, 18, 833–842, 1983.

Hansten, C., Heino, M., and Pynnonen, K., Viability of glochidia of *Anodonta anatina* (Unionidae) exposed to select metals and chelating agents, *Aquat. Toxicol.*, 34, 1–12, 1996.

Hemelraad, J. and Herwig, H. J., Effects of cadmium in freshwater clams. III. Interaction with energy metabolism in *Anodonta cygnea*, *Arch. Environ. Contam. Toxicol.*, 19, 699–703, 1990.

Henley, W. F., Zimmerman, L. L., and Neves, R. J., Design and evaluation of recirculating water systems for maintenance and propagation of freshwater mussels, *N. Am. J. Aquaculture*, 63, 144–155, 2001.

Holopainen, I. J., Lamberg, S., Tellervo Valtonen, E., and Rantanen, J., Effects of parasites on life history of the freshwater bivalve, *Pisidium amnicum*, in Eastern Finland, *Archiv Fur. Hydrobiologie*, 139, 461–477, 1997.

Holwerda, D. A. and Herwig, H. J., Accumulation and metabolic effects of di-n-butylin dichloride in the freshwater clam, *Anodonta anatina, Bull. Environ. Contam. Toxicol.*, 36, 756–762, 1986.

Howard, A. D. and Anson, B. J., Phases in the parasitism of the Unionidae, *J. Parasitol.*, 9, 68–82, 1922.

Hudson, R. G., McKinney, D. A., Wetzel, J. T., Griner, J. G., Brinson, A. L., and Hinesley, J. H., Effects of drilling agents on the growth and survival of juvenile mussels, Third Biennial Symposium of the Freshwater Mollusk Conservation Society, Durham, NC, 2003.

Huebner, J. D. and Pynnonen, K. S., Viability of glochidia of two species of *Anodonta* exposed to low pH and selected metals, *Can. J. Zool.*, 70, 2348–2355, 1992.

Imlay, M., Effects of potassium on survival and distribution of freshwater mussels, *Malacological*, 12, 97–113, 1973.

Isom, B. G. and Hudson, R. G., In vitro culture of parasitic freshwater mussel glochidia, *Nautilus*, 96, 147–151, 1982.

Jacobson, P. J., Sensitivity of early life stages of freshwater mussels (Bivalvia: Unionidae) to copper, Masters Thesis, Virginia Polytechnic Institute and State University, Blacksburg, VA, 1990.

Jacobson, P. J., Farris, J. L., Cherry, D. S., and Neves, R. J., Juvenile freshwater mussel (Bivalvia: Unionidae) responses to acute toxicity testing with copper, *Environ. Toxicol. Chem.*, 12, 879–883, 1993.

Jacobson, P. J., Neves, R. J., Cherry, D. S., and Farris, J. L., Sensitivity of glochidial stages of freshwater mussels (Bivalvia: Unionidae) to copper, *Environ. Toxicol. Chem.*, 11, 2384–2392, 1997.

Johnson, I. C., Zam, S. G., and Keller A. E., Proposed guide for conducting acute tests with early life stages of freshwater mussels, A report submitted by KBN Engineering and Applied Sciences to the USEPA Office of Pesticide Programs, Washington, DC, 1990.

Johnson, I. C., Keller, A. E., and Zam, S. G., A method for conducting acute toxicity tests with early life stages of freshwater mussels, In *Environmental Toxicology and Risk Assessment, ASTM STP 1179*, Landis, W. G., Hughes, J. S., and Lewis, M. A., Eds., American Society for Testing and Materials, Philadelphia, PA, pp. 381–396, 1993.

Keller, A. E., Acute toxicity of several pesticides, organic compounds, and wastewater effluent to the freshwater mussel, *Anodonta imbecillis, Ceriodaphnia dubia*, and *Pimephales promelas, Bull. Environ. Contam. Toxicol.*, 51, 696–702, 1993.

Keller, A. E. and Ruessler, D. S., The toxicity of Malathion to unionid mussels: Relationship to expected environmental concentrations, *Environ. Toxicol. Chem.*, 16, 1028–1033, 1997.

Keller, A. E. and Zam, S. G., Simplification of in vitro culture techniques for freshwater mussels, *Environ. Toxicol. Chem.*, 9, 1291–1296, 1990.

Keller, A. E. and Zam, S. G., The acute toxicity of selected metals to the freshwater mussel, *Anodonta imbecillis, Environ. Toxicol. Chem.*, 10, 539–546, 1991.

Keller, A. E., Ruessler, D. S., and Chaffee, C. M., Testing the toxicity of sediments contaminated with diesel fuel using glochidia and juvenile mussels (Bivalvia, Unionidae), *Aquat. Ecosyst. Health Manage.*, 1, 37–47, 1998.

Keller, A. E., Ruessler, D. S., and Kernaghan, N. J., Effect of test conditions on the toxicity of copper to juvenile unionid mussels, In *Environmental Toxicology and Risk Assessment, ASTM STP 1364*, Henshel, D. S., Black, M. C., and Harrass, M. C., Eds., American Society for Testing and Materials, West Conshohocken, PA, pp. 381–396, 1999.

Kernaghan, N. J., Ruessler, D. S., Holm, S. E., and Gross, T. S., An evaluation of the potential effects of paper mill effluents on freshwater mussels in Rice Creek, Florida, In *Pulp and Paper Mill Effluent Environmental Fate and Effects, 5th International Conference on Environmental Fate and Effects of Pulp and Paper Mill Effluents; 2003 June 1–4; Seattle*, Borton, D. L., Hall, T. J., Fisher, R. P., and Thomas, J. E., Eds., DEStech Publications, Lancaster, pp. 455–463, 2004.

Klaine, S. J., Warren, L. W., and Summers, J. M., Final report on the use of juvenile mussels (*Utterbackia imbecillis*, Say) as a standardized toxicity testing organism, Submitted to Anne Keller, USEPA, Athens, GA, 1997.

Lasee, B. A., Histological and ultrastructural studies of larval and juvenile *Lampsilis* from the upper Mississippi River, PhD Thesis, Iowa State University, Ames Iowa, 1991.

Mackie, G. L., Bivalves, In *The Mollusca*, Wilbur, K. M., Ed., Academic Press, New York, pp. 351–418, 1984.

Mane, U. H., Effect of pesticides and narcotants of bivalve mollusks, *Malacological*, 18, 347–360, 1979.

Masnado, R. G., Geis, S. W., and Sonzogni, W. C., Comparative toxicity of a synthetic mine effluent to Ceriodaphnia dubia, larval fathead minnow and the freshwater mussel *Anodonta imbecillis, Environ. Toxicol. Chem.*, 14, 1913–1920, 1995.

Matterson, M. P., Life history of *Elliptio complanatus* (Dillwyn 1817), *Am. Midl. Nat.*, 40, 690–723, 1948.

McCann, M. T., Toxicity of zinc, copper, and sediments to early life stages of freshwater mussels in the Powell River, Virginia, Masters thesis, Virginia Polytechnic Institute and State University, Blacksburg, VA, 1993.

McKinney, A. D. and Wade, D. C., Comparative response of *Ceriodaphnia dubia* and juvenile *Anodonta imbecillis* to pulp and paper mill effluents discharged into the Tennessee River and its tributaries, *Environ. Toxicol. Chem.*, 15, 514–517, 1996.

Milam, C. D. and Farris, J. L., Relating threshold responses of *Corbicula fluminea* to assess damage in resident mussel populations exposed to partially treated mine water, *Environ. Toxicol. Chem.*, 17, 1611–1619, 1998.

Milam, C. D., Farris, J. L., Dwyer, F. J., and Hardesty, D. K., Acute toxicity of six freshwater mussel species (glochidia) to six chemicals: Implications for daphnids and *Utterbackia imbecillis* as surrogates for protection of freshwater mussels (Unionidae), *Arch. Environ. Contam. Toxicol.*, 48, 166–173, 2005.

Moulton, C. A., Fleming, W. J., and Purnell, C. E., Effects of two cholinesterase-inhibiting pesticides on freshwater mussels, *Environ. Toxicol. Chem.*, 15, 131–137, 1996.

Murphy, G., Relationship of the freshwater mussel to trout in the Truckee River, *Calif. Fish Game*, 28, 89–102, 1942.

Myers-Kinzie, M., Factors affecting survival and recruitment of Unionid mussels in small midwestern streams, PhD Thesis, Purdue University, West Lafayette, IN, 1998.

Naimo, T. J., Waller, D. L., and Holland-Bartels, L. E., Heavy metals in the Threeridge mussel *Amblema plicata plicata* (Say, 1817) in the Upper Mississippi River, *J. Freshwat. Ecol.*, 7, 209–217, 1992.

Naimo, T. J., Atchison, G. J., and Holland-Bartels, L. E., Sublethal effects of cadmium on physiological responses in the Pocketbook mussel, *Lampsilis ventricosa*, *Environ. Toxicol. Chem.*, 11, 1013–1021, 1992.

Newton, T. J., Allran, J. W., O'Donnell, J. A., Bartsch, M. R., and Richardson, W. B., Effects of ammonia on juvenile freshwater mussels (*Lampsilis cardium*) in laboratory sediment toxicity tests, *Environ. Toxicol. Chem.*, 22, 2554–2560, 2003.

Pynnonen, K., Effect of pH, hardness and maternal pre-exposure on the toxicity of Cd, Cu and Zn to the glochidial larvae of a freshwater clam, *Anodonta cygnea*, *Water Res.*, 29, 247–254, 1995.

Raj, A. and Hameed, P. S., Effect of copper, cadmium and mercury on metabolism of freshwater mussel, Lamellidens marginalgis (Larmarck), *J. Environ. Biol.*, 12, 131–135, 1991.

Rodgers, J. H., Cherry, D. S., Clark, J. R., Dickson, K. L., and Cairns, J., Comparison of heavy metal interactions in acute and artificial stream bioassay techniques for the Asiatic clam (*Corbicula fluminea*), In *Aquatic Toxicology, ASTM STP 707*, Eaton, J. G., Parrish, P. R., and Hendricks, A. C., Eds., American Society for Testing and Materials, Philadelphia, PA, pp. 266–280, 1980.

Roesijadi, G. and Robinson, W. E., Metal regulation in aquatic animals: Mechanisms of uptake, accumulation, and release, In *Aquatic Toxicology: Molecular, Biochemical, and Cellular Perspectives*, Malins, D. C. and Ostrander, G. K., Eds., CRC Press, Boca Raton, FL, pp. 387–420, 1994.

Scheller, J. L., The effect of dieoffs of Asian clams (*Corbicula fluminea*) on native freshwater mussels (Unionidae), Masters Thesis, Virginia Polytechnic Institute and State University, Blacksburg, VA, 1997.

Silverman, H., McNeil, J. W., and Dietz, T. H., Interaction of trace metals Zn, Cd, and Mn with Ca concentrations in the gills of freshwater Unionid mussels, *Can. J. Zool.*, 65, 828–832, 1987.

Soucek, D. J., Cherry, D. S., Currie, R. J., Latimer, H. A., and Trent, G. C., Laboratory to field validation in an integrative assessment of an acid mine drainage-impacted watershed, *Environ. Toxicol. Chem.*, 19, 1036–1043, 2000.

Telda, S. and Fernando, C. H., Observations on the glochidia of *Lampsilis radiata* (Gmelin) infesting yellow perch, *Perca flavescens* (Mitchill) in the Bay Quinte, Lake Ontario, *Can. J. Zool.*, 47, 705–712, 1969.

Tran, D., Boudou, A., and Massabuau, J. C., Relationship between feeding-induced ventilatory activity and bioaccumulation of dissolved and algal-bound cadmium in the Asiatic Clam *Corbicula fluminea*, *Environ. Toxicol. Chem.*, 21, 327–333, 2002.

U.S. Environmental Protection Agency (USEPA), Ambient water quality criteria for zinc, EPA 440/5-80-079, Washington, DC, 1980.

USEPA, Methods for measuring the acute toxicity of effluents and receiving waters to freshwater and marine organisms, 4th ed., EPA/600/4-90/027F, Washington, DC, 1993.

USEPA, Methods for measuring the toxicity and bioaccumulation of sediment-associated contaminants with freshwater invertebrates, 2nd ed., EPA/600/R-99/064, Washington, DC, 2000.

U.S. Geological Survey (USGS), Quarterly report for the project entitled "Developing Water Quality Standards for Recovery of Imperiled Freshwater Mussels (Family Unionidae)," prepared for the U.S. Fish and Wildlife Service and the U.S. Environmental Protection Agency by the Columbia Environmental Research Center, Columbia, MO, Project 301812N037 (USFWS) and DW14937809-01-0 (USEPA), 2004.

USGS, Quarterly report for the project entitled "Developing Water Quality Standards for Recovery of Imperiled Freshwater Mussels (Family Unionidae)," prepared for the U.S. Fish and Wildlife Service and the U.S. Environmental Protection Agency by the Columbia Environmental Research Center, Columbia, MO, Project 301812N037 (USFWS) and DW14937809-01-0 (USEPA), 2005a.

USGS, Quarterly report for the project entitled "Developing Water Quality Standards for Recovery of Imperiled Freshwater Mussels (Family Unionidae)," prepared for the U.S. Fish and Wildlife Service and the U.S. Environmental Protection Agency by the Columbia Environmental Research Center, Columbia, MO, Project 301812N037 (USFWS) and DW14937809-01-0 (USEPA), 2005b.

USGS, Quarterly report for the project entitled "Determination of the sensitivity of Ozark mussels to lead, zinc, and cadmium in water and sediment," prepared for the U.S. Fish and Wildlife Service by the Columbia Environmental Research Center, Columbia, MO, Project 83359RX, 2005c.

Valenti, T. W., Cherry, D. S., Neves, R. J., and Schmerfeld, J., Acute and chronic toxicity of mercury to early life stages of the rainbow mussel, *Villosa iris* (Bivalve: Unionidae), *Environ. Toxicol. Chem.*, 24, 1242–1246, 2005.

Varanka, I., Toxicity of moquitocides on freshwater mussel larvae, *Acta. Biologicia Hungaria*, 37, 143–158, 1986.

Varanka, I., Effect of mosquito killer insecticides on freshwater mussels, *Comp. Biochem. Physiol.*, 86, 157–162, 1987.

Versteeg, D. J. and Rawlings, J. M., Bioconcentration and toxicity of dodcylbenzene sulfonate ($C_{12}LAS$) to aquatic organisms exposed in experimental streams, *Arch. Environ. Contam. Toxicol.*, 44, 237–246, 2003.

Wade, D. C., Definitive evaluation of Wheeler Reservoir sediment toxicity using juvenile freshwater mussels (*Anodontaimbecillis* Say) TVA/WR-92/25. Tennessee Valley Authority, Muscle Shoals, AL, USA, 1992.

Wang, N., Greer, E., Whites, D., Ingersoll, C. G., Roberts, A., Dwyer, J., Augspurger, T., Kane, C., and Tibbott, C., An evaluation of the viability of glochidia after removal from mussels, Presented at the Freshwater Mollusk Conservation Society Symposium in Durham, NC, 2003.

Weinstein, J. E., Characterization of the acute toxicity of photoactivated fluoranthene to glochidia of the freshwater mussel, *Utterbackia imbecillis*, *Environ. Toxicol. Chem.*, 20, 412–419, 2001.

Zimmerman, L. L. and Neves, R. J., Effects of temperature on the longevity and viability of glochidia of freshwater mussels (Bivalvia: Unionidae), *Am. Malacol. Bull.*, 16, 31–35, 2002.

6 In Situ Toxicity Testing of Unionids

Mindy Yeager Armstead and Jessica L. Yeager

INTRODUCTION

As more regulatory attention is given to risk-based decision making, in-stream surveys and in situ toxicity testing are rapidly becoming more important in the regulatory arena (Karr and Chu 1999). This represents a dramatic change from the more traditional approach of regulation by meeting specific criteria. Traditional toxicity testing is used to evaluate the concentrations of a given chemical and the duration of exposure required to produce the criterion effects (Rand 1995). This type of testing has been used extensively for evaluating the potential in-stream impacts of discharges and for setting discharge limits or exposure concentrations that protect organisms and communities from chemical stressors (Webber 1993). Traditionally, toxicity tests have been designed to evaluate lethal or sub-lethal endpoints that indicate impairment to aquatic organisms or communities (i.e., growth, reproductive impairment, and reduced diversity) or to measure bioaccumulation, which may or may not be associated with adverse effects (Rand 1995). Much effort has gone into the standardization of toxicity testing methodologies, from the culturing of organisms used in testing to the statistical analyses used to determine significant impacts. Standardization is important for ensuring that the stressor being evaluated is responsible for any effects identified and not for organism health, laboratory personnel error, food of poor nutrient value, substandard water, temperature stress, dirty glassware, or any number of variables that can contribute to impairment in test organism performance. Minimizing the test variability enhances the confidence in the cause and effect relationship being demonstrated between the stressor and the response.

The benefits of minimizing laboratory toxicity test variability are significant, and these tests have many applications; however, the results do not always extrapolate to in situ effects (La Point and Waller 2000). Protocols for in-stream assessments and monitoring are each different in the balance between environmental realism and reproducibility. In situ testing, exposing organisms to contaminants in the field as opposed to under laboratory conditions, incorporates the natural variability of an ecosystem into the test rather than intentionally minimizing or excluding variability as is done in traditional toxicity testing (Chappie and Burton 2000).

Biological surveys go beyond the limitations of in situ testing by indicating the conditions of indigenous organisms over their entire lifespan as opposed to an exposure period. Each test type or survey has inherent benefits and limitations. Methods for in-stream surveys and community assessments, such as the Rapid Bioassessment Protocols for benthic macroinvertebrates and the Index of Biotic Integrity for fish (Barbour et al. 1999), have been standardized for widespread use and are applicable to many aquatic assessments. The use of standardized assessment and testing methods allows for conclusions to be drawn on the biological health of an ecosystem as compared with the expected or potential communities based on historical databases, professional knowledge, or reference conditions. These surveys allow for the evaluation of: overall in-stream community conditions, the effects of specific National Pollutant Discharge Elimination System (NPDES) permitted outfalls or other point sources, non-point source impacts, and comparisons to theoretic or measured references.

BENEFITS OF IN SITU TESTING

In situ testing increases the environmental realism lacking in standard laboratory testing thereby more accurately predicting in-stream individual and community impacts from test organism responses. Using the natural water from the system of concern can profoundly affect the test outcome. Hardness values can increase or decrease metals toxicity, suspended solids can bind up contaminants and render them unavailable, alkalinity can buffer the impacts of acidic releases, and pH can increase or decrease the percentage of metals that are in the more toxic dissolved form. While site water is sometimes used in laboratory testing, this is often not the case for chronic, flow-through, and sediment tests where large volumes of water are required.

Two of the most important parameters fluctuating in the field that are not often replicated in laboratory testing are temperature and dissolved oxygen. Both tend to fluctuate diurnally as well as temporally. Increased temperatures may stress test organisms by limiting oxygen because saturation is inversely related to temperature. Higher temperatures also increase organism metabolism, which can lead to increased uptake rates of environmental contaminants and organism responses not seen in laboratory testing. Dissolved oxygen limitations, which tend to occur under low-flow, high-temperature conditions or at night, may stress organisms in the field and make them more susceptible to additional stressors. There are many other variables that are controlled in laboratory testing that can alter the predictability of laboratory testing. These include, but are not limited to: light regime, light intensity, food quality and quantity, competitive and predatory interactions with other resident taxa, habitat and substrate limitations, and flow variability.

In addition to variability in environmental conditions and the test organisms, in situ testing also incorporates the toxicant or stressor variability. Most laboratory tests are conducted with a grab sample (or a series of grabs, which is termed a composite) that represents conditions at that particular instance. This situation is analogous to the difference between the information gained from a photograph versus a videotape. In situ testing exposes organisms to changing levels of toxicants or stressors in conjunction with other environmental variations. This is particularly useful when toxicity may be intermittent and the actual impacts on organisms and community structure may be more or less severe than are indicated by laboratory bioassays and chemical analyses. For example, the toxicity associated with stormwater events is greatest during the initial rise in the hydrograph of the first flush of the stormwater. The first flush of urban stormwater often contains heavy metals, sewage inputs, hydrocarbons, pesticides, and deicing salts (Lieb and Carline 2000). Industrial stormwater also may exhibit intermittent toxicity with the runoff constituents specific to the industry. Stormwater from agricultural properties is difficult to represent in laboratory testing but has been successfully monitored by in situ testing (Crane et al. 1995). For stormwater, the characteristics of the runoff will be variable with each storm event and will depend on many factors including: time since last rainfall, activities in the drainage area since last rainfall, intensity of rainfall, duration of rainfall, and pH of rainwater.

In situ testing is also useful for predicting the affects of toxicants that volatilize quickly or demonstrate photo-induced toxicity, such as polynuclear aromatic hydrocarbons (Burton, Pitt, and Clark 2000). These methods are also preferred when multiple stressors are present and test organism exposure would be variable for the different stressors over the test period. Another benefit of in situ testing is specific to sediment testing. The physical characteristics of the sediments are known to affect toxicant bioavailability. Characteristics that affect toxicity of the sediments and porewater, such as dissolved oxygen, pH, and oxygenation/reduction potential, may be altered during collection and transportation of samples (Burton 1991). In-stream testing minimizes the alteration of chemical and physical properties of the sediment to more realistically depict in-stream effects (Chappie and Burton 2000).

LIMITATIONS OF IN SITU TESTING

There are several disadvantages or limitation of in situ testing that include the unknown effects of testing on the organisms. Factors such as acclimation to site conditions (e.g., temperature and water quality), effects of transportation and handling, and caging artifacts (e.g., food availability, flow, suspended solids, and predation) all influence test outcome (Chappie and Burton 2000). Site-related disadvantages of in situ testing include vandalism or loss of test chambers, unknown field stresses, and variability or difficulty in chamber or cage placement. Variations in field conditions often make data interpretation difficult (Pereira et al. 2000). This is exacerbated by not knowing the expected organism performance as you would in standardized laboratory testing. With the organism response not predictable, the inclusion of a background or reference station is mandated. Under this scenario, there is no confidence in comparisons between tests. For example, when determined with standard toxicity testing, if one effluent has an LC50 of 50% and another effluent has an LC50 of 25%, it can be surmised that one effluent is more toxic. Likewise, a benthic macroinvertebrate community can be determined to be healthy based on metric scores compared to regional reference conditions or published metrics scores. Many states have multimetric indices with standard performance categories that deem benthic communities as excellent, good, fair, or poor. For the performance of in situ organisms, comparisons can be made to laboratory performance or a database of other studies, but ultimately, the performance of the organisms will be specific for the conditions of each test. Due to these limitations and the novelty of many in situ test methods, in situ testing has primarily been conducted in conjunction with laboratory testing to confirm or validate the laboratory results. Some researchers also use field surveys, such as benthic macroinvertebrate community surveys, to confirm or validate the in situ testing results.

IN SITU METHODS

A number of in situ bioassays have been developed with a wide variety of organisms, endpoints, and test chambers. In situ test methods have been developed for freshwater, estuarine, and marine taxa including: cladocerans (Sasson-Brickson and Burton 1991; Ireland et al. 1996; Pereira et al. 1999; Maltby et al. 2000), mussels (Foe and Knight 1987; Gray 1989; Belanger et al. 1990; Yeager 1994; Warren, Klaine, and Finley 1995; Salazar et al. 2002), midges (Chappie and Burton 1997; Sibley et al. 1999; Crane et al. 2000), amphipods (Crane and Maltby 1991; Shaw and Manning 1996; Chappie and Burton 1997; Schulz and Liess 1999; Maltby et al. 2000; Kater et al. 2001), oligochaetes (Sibley et al. 1999), mayflies (Shaw and Manning 1996), caddisflies (Schultz and Liess 1999), and many fish species (Simonin et al. 1993). Chappie and Burton (2000) provide a thorough summary of the various organisms used in testing. Similarly, a wide range of endpoints have been employed including: mortality (Matthiessen et al. 1995), reproduction, growth, bioaccumulation (Sibley et al. 1999), mouth-part deformity (Meregalli, Vermeulen, and Ollevier 2000), feeding rate (Matthiessen et al. 1995), gape (Sloof, de Zwart, and Marquenie 1983), and valve movements (Kramer, Jenner, and Zwart 1989). Test chamber design has been variable to accommodate the organisms, test conditions, and study designs. A review of the available literature indicated that exposure chambers are usually polyvinyl chloride or Plexiglas with mesh screen covering openings that allow water and small particulates to move through. Exposure chambers are secured in a variety of ways to allow for exposure to the water column (Chappie and Burton 1997), sediment and water (Sibley et al. 1999), or sediment alone with no water-column exposure (Crane et al. 2000).

Test organisms used for in-stream testing are either laboratory-reared organisms or organisms indigenous to the system being evaluated. Laboratory-reared organisms are readily available due to standardized culturing procedures, and the documentation accompanying cultured organisms, along with reference toxicant testing, provides knowledge of a test organism's general health and condition. Indigenous organisms are directly related to the system under study and may be acclimated to the stream (if not transplanted). However, it is often difficult to obtain suitable size ranges in numbers sufficient for testing, and it may be difficult to maintain some organisms in the laboratory prior to testing. Moving field-collected organisms to another watershed for testing sometimes occurs but should be approached cautiously due to the potential for introduced species to become established, the possibility of transporting pathogens, and other reasons.

IN SITU TESTING WITH FRESHWATER MUSSELS

It has been suggested that mussels and fish were the most widely used organisms for in situ testing due to their availability from the aquaculture industry and the general public concern for these commercial organisms (Chappie and Burton 2000). This may be true as it applies to marine bivalves, but it certainly does not apply to freshwater unionid mussels. The pelagic larval stage of the juvenile marine bivalve lends itself to culturing while the parasitic glochidial stage of the freshwater unionid makes laboratory culturing difficult and cumbersome. Additionally, while much effort has gone into culturing marine species for consumption, commercial propagation of unionids is not widespread. Efforts on behalf of the unionids have been limited primarily to researching the unionids' sensitivities to various toxicants and potential use as test organisms, researching the propagation of endangered species (Buddensiek 1995), and determining the life history (Zale and Neves 1982; Neves, Weaver, and Zale 1985; Neves and Widlak 1987). Some species of adults are available from commercial suppliers where they are raised in ponds, but juveniles are primarily cultured at research facilities. The introduced Asiatic clam (*Corbicula fluminea*) and zebra mussel (*Dreissena polymorpha*) both lack the parasitic lifestage of the unionid and offer unique possibilities as test organisms in laboratory and in situ testing. This discussion will primarily focus on the use of unionid mussels, particularly the juvenile lifestage, as in situ test organisms. References to other test organisms, both marine and freshwater, will be included as they relate to in situ testing.

Freshwater mussels are uniquely adapted for use as ecological indicators primarily because they are sessile, long-lived organisms believed to be highly sensitive to ecosystem stress. One of the reasons unionids are believed to be highly sensitive is that they are declining worldwide in systems that may or may not show other signs of stress. Also, in many areas, juvenile recruitment is negligible where adult populations continue to exist (Scott 1994). The need to protect declining populations of an increasing number of federally listed threatened and endangered species of unionids also contributes to the interest in using freshwater mussels in testing. Being filter-feeders, mussels also have a propensity to accumulate contaminants from the water column, and they have limited ability to rid their bodies of the contaminants (ASTM 2001). These qualities make mussels highly desirable as test organisms.

There are significant limitations on the use of unionids for testing, which include: the difficulty in obtaining organisms, variability in response, lack of information on the general condition of test organisms, and limitations on the test endpoints. As indicated earlier, adult mussels are only available commercially from a few sources. There is limited information on these farm-raised organisms regarding their general condition, there are limitations on the size range of harvestable, farm-raised mussels (i.e., only larger sizes are available, and there may be a great variability in ages), there are little or no data on the expected responses of organisms to reference toxicants, and the organisms may have to be transported great distances, which contributes to handling and acclimation stress on the organisms. Additionally, there are significant concerns on placing transplanted mussels into stream systems where native mussels exist due to the introduction of non-native species, parasites, or other diseases. Using field-collected organisms is often not possible, as it is with other invertebrates, due to the low numbers of mussels for harvest, disturbance of threatened or endangered taxa during harvesting, and the aforementioned limitations on farm-reared organisms. While some freshwater unionids are available for field collection for testing, a suitable population for extensive sampling is the exception and not the norm, and this method of obtaining test organisms would not be recommended as a strategy for standardized testing practices.

When used in testing, the endpoints applied to adult mussels may be limited. While they are excellent for use in bioaccumulation studies, unionid growth is slow, and in many situations, it may be undesirable to sacrifice adults, such as when threatened or endangered species are involved or population densities are low. There are some techniques for harvesting tissue for biochemical analysis that do not require sacrifice of the individual (Naimo et al. 1998); however, adults are not generally useful for in situ testing with traditional endpoints such as mortality, growth, and reproduction.

When available, juvenile mussels have been found to be suitable laboratory test organisms with sensitivities equal to or greater than standard test species (Jacobson et al. 1993; Keller and Zam 1991). Methods have been developed for procuring juveniles using both artificial media transformation (Isom and Hudson 1982) and encystment of the juveniles on the appropriate fish hosts. The general condition and the sensitivity of juveniles, however, is still dependent on a number of factors such as the general condition of the female from which glochidia were harvested, the water chemistry of the system where both the adult and the juveniles are reared, the general condition of the fish host, quality and quantity of the food source provided, and a general lack of knowledge on the ecological preferences of many unionid species. Although improving quickly through continued research, the same lack of standardization and variability in organism response (between taxa and within taxa) discussed previously limits the widespread use of juvenile mussels in both laboratory and in situ testing.

Other freshwater bivalves that have been used for in situ testing are introduced Asiatic clams and zebra mussels. These organisms are not limited by many of the constraints described above. They are potentially useful test organisms because:

1. They are easily harvestable from the field.
2. They are numerous enough for use in testing.
3. They do not require sophisticated culturing techniques for harvesting juveniles (Doherty 1986) (although juveniles are limited to harvesting during reproductive seasons).
4. They provide sufficient tissue for biochemical testing.
5. There is little concern for sacrificing adults.

Extreme care must be taken, however, when using these organisms to ensure they are not being introduced into an un-invaded system and that no parasites or diseases are transferred with the test organisms (ASTM 2001).

ADULT UNIONID MUSSEL IN SITU TESTING

Adult freshwater mussels are not generally used for in situ testing, utilizing traditional endpoints of mortality, growth, or reproductive success. They are more often used for assessing the bioavailability of contaminants and indicating the bioaccumulation potential in contaminated areas, particularly with regard to metals. Field studies have included tissue sampling of field-collected adults (Tessier et al. 1984; Naimo, Waller, and Holland-Bartels 1992; Hickey, Roper, and Buckland 1995; Metcalfe-Smith, Green, and Grapentine 1996) as well as caging and transplanting studies (Adams, Atchison, and Vetter 1981; Couillard et al. 1994; Hickey, Roper, and Buckland 1995). Adult mussels have been used for the comparison of metal tissue concentrations from mussels exposed to water columns, sediment, or porewater with the exposure cage design variable, to accommodate the different exposure scenarios. While metals concentrations generally increase in organisms at the most contaminated sites, tissue concentrations may not correlate with water column, sediment, or porewater concentrations of contaminants of concern and may be dependent on many factors including the mechanism of uptake (i.e., ingestion of water, ingestion of sediments, and direct adsorption), the physiochemical forms of metals that affect bioavailability (Tessier et al. 1984), and other sediment characteristics that affect bioavailability (Tessier et al. 1984; Stewart 1999).

In situ testing has been conducted with readily available, field-collected adult individuals (sampled from study sites or transplanted from a common site) or specimens purchased from a few commercial suppliers. There are several benefits to collecting mussels from contaminated sites and using them for analysis (as opposed to transplanting them from a reference site and exposing them to a contaminated site). Given the longevity of unionids and their sessile nature, field-collected mussels have an extended exposure period and can indicate past and present in-stream exposures. This would be particularly useful for identifying low levels of exposure over time or intermittent exposures. This strategy is limited to areas where unionids exist in sufficient numbers for sampling, which often does not occur in areas of suspected contamination. When adults are transplanted in caging studies, mussels can be placed in areas where they may not have occurred previously or may have been eliminated. This allows for more control over the exposure period and for the collection of baseline data for the determination of bioaccumulation factors. However, as indicated previously, few commercial suppliers are available for adult freshwater mussels, and field collection, though sometimes possible, is not an option for widespread use in testing due to the factors mentioned previously.

Adult mussel testing methodologies have been successfully demonstrated for short exposure periods and extended study periods. Adams, Atchison, and Vetter (1981) collected *Amblema perplicata* (now *A. plicata*) from an uncontaminated site and placed them in polyethelyene cages at contaminated sites. Differences in zinc and cadmium concentrations were found in the gill tissue and digestive glands of the organisms after only one week of exposure. *Elliptio complanata* were successfully exposed for a year in a study evaluating both cage design and sex reversal as a potential test endpoint (Salazar et al. 2002). These caging studies evaluating biochemical endpoints are discussed in greater detail in Chapter 9. The caging apparatus found to be preferable in these long-term studies was a plastic tub with an internal mesh chamber, which was buried in the sediment. Survival of mussels placed in individual compartments and placed on the surface of sediment (a design used successfully in marine studies) was substantially reduced in these chambers.

The limitations of using freshwater mussels in testing, such as the limited availability of organisms, slow growth, and the ability to avoid toxicants, can often be overcome with variations in study design. Alternatives to sacrificing the mussels used in testing include the use of alternative endpoints such as biochemical indicators, valve closure, and filtering activity (Salánki and Balogh 1989), or non-lethal tissue sampling (Naimo et al. 1998).

Biochemical indicators present an early warning as they may be predictive of biological effects. They are generally specific to a contaminant or class of contaminants, they respond in a concentration-dependent manner, and they are related to the health or fitness of the organisms to be protected. The effects of other environmental or organism-specific influences on biochemical indicators should be well understood so that they can be minimized (Couillard et al. 1994). Metallothionein and glycogen have demonstrated promise as biochemical indicators of stress that can be used in lieu of more traditional measurements such as growth in short-term (relative to mussel longevity) studies. Glutathione S-transferases measured in *Anodonta cygnaea* exposed to agricultural runoff did not correlate well with traditional, sediment-testing organisms or in-stream monitoring (Crane et al. 1995).

Despite the obvious advantages of using freshwater mussels for in situ studies, there is evidence from the marine literature that factors such as species, age, size, sex, reproductive cycle, and nutritional status can influence bioaccumulation in marine species (Metcalfe-Smith et al. 1996). There is little information available to describe the influences these factors may have on contaminant uptake and accumulation in freshwater taxa; however, one study suggested that species, size, age, and possibly growth rate are significant factors that should be considered (Metcalfe-Smith et al. 1996). In studies monitoring the uptake of xenobiotics, it was found that variable filtration rates between a lake and a river altered the test organism's uptake rate (Englund and Heino 1996). Many environmental factors such as food availability, suspended solids, temperature, and other factors can alter valve closure and filtration rates, which could lead to differences in contaminant uptake. In another study, the source of test organisms was found to effect growth and metal uptake in *E. complanata* (Hinch and Green 1989) further supporting the need for caution when drawing conclusions on bioavailability of contaminants using freshwater mussels. Generally, without significant improvements in test standardization and culturing techniques, in situ testing using adult freshwater mussels will generally be limited to "upstream/downstream" comparisons with conclusions drawn for the conditions existing at the time of testing with the broader applicability to other areas and conditions largely unknown.

JUVENILE UNIONID IN SITU TESTING

As is described in Chapter 4, in the discussion of juvenile mussel culturing, obtaining juvenile unionids for testing is a time-consuming and labor-intensive process. Culture methods, using either artificial media or encystments on fish, are available for several species. Some species have also been used in toxicity tests using standardized methods (Keller and Zam 1991; Jacobson et al. 1993; Keller, Ruessler, and Kernaghan 1999), and a database on organism sensitivity, condition, and performance standards is developing rapidly for some widely distributed taxa. Research is also underway on culturing methods for several endangered species for the purpose of reintroductions. The amount of information known, regarding fish hosts, lifecycle preferences, feeding behaviors, and other critical variables for developing test methods, varies for the different unionid taxa. There are only three studies, of which we are aware, that utilized in situ testing of juvenile unionid mussels. The studies are described below.

Kentucky Lake Study

A study was conducted in Kentucky Lake (an impoundment on the Tennessee River) in which 6-week-old, media-transformed *Utterbackia imbecillis* were placed at three locations for a 7-day in situ test (Warren, Klaine, and Finley 1995). The sampling locations represented variable levels of known or suspected sediment toxicity ranging from a reference condition with 11 mussel taxa present, an intermediately impaired location (5 mussel taxa present), and an impaired site with no unionid mussels in the benthic fauna. In situ test chambers were constructed from glass vials with 105 μm Teflon™ mesh attached to the ends. The glass tubes were affixed on the bottom and middle of plastic storage crates to test sediment and water column toxicity (Figure 6.1).

At the reference site, mortality ranged from 0.00 to 100.00% in the water-column-exposed organisms and from 0.00 to 33.00% in the sediment-exposed organisms. Recovery of the mussels was reduced due to some juveniles escaping from the exposure chambers. At the reference location, recovery was 64.44% in the water-column-exposed organisms and 86.67% in the sediment-exposed organisms. The juvenile mussel responses from in situ testing were consistent with the knowledge of site contaminants and the impairment seen in adult mussel populations at the sites indicating that the in situ testing was reflective of the biological condition at the sites.

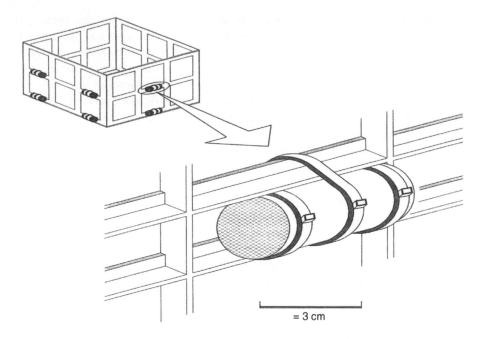

FIGURE 6.1 Test chambers deployed in Kentucky Lake (an impoundment on the Tennessee River) for 7-day in situ testing of juvenile *Utterbackia imbecillis*. (From Warren, L. W., Klaine, S. J., and Finley, M. T., *J. N. Am. Benthol. Soc.*, 14(2), 341–346, 1995. With permission.)

Clinch River Study

In situ juvenile mussel testing using *Villosa iris* was combined with traditional laboratory sediment bioassays (*Chironomus riparius* and *Daphna magna*), chemical analyses, and benthic macroinvertebrates surveys to determine the impact of point and non-point contaminants on mussels in the Clinch River, Virginia. Water chemistry analysis and laboratory sediment testing indicated variable contaminant levels and intermittent sediment toxicity over several sampling events (Yeager 1994). In situ testing and the biological survey completed the triad approach for determining if the intermittent toxicity seen in laboratory testing was reflected in the stream benthic community, and whether mussel recruitment may be inhibited by sediment-bound toxicants and the intermittent toxicity.

Twelve sites, spanning 114.3 river miles, were examined based on the availability of background information indicating intermittent sediment toxicity at these sites. Two of the sites included in this study were considered reference sites. The sites used in this research were part of a larger study and were selected due to substrate characteristics that currently or historically supported mussel populations.

Test chambers were constructed using aquarium uplift tubing that was cut away and fitted with 105 μm nylon screening to create a flow-through holding chamber (Figure 6.2). Eight 1-to-2-week-old *V. iris* were placed in each tube that was then fitted with a cotton plug and wired in place. Two sets of four tubes containing juveniles were maintained in a 2-L beaker of Clinch River water in the laboratory as a laboratory reference. These mussels were fed a detrital suspension containing silt, clay, and organic fractions of sediment from an upstream reference site on the Clinch River and a tri-algal suspension containing *Chlamydomonas*, *Ankistrodesmus*, and *Chlorella* (Foe and Knight 1986).

The test chambers were placed into test tube racks, which were secured onto bricks. On July 15 and 16, 1994, two bricks with two tubes, each containing eight juveniles, were placed at the

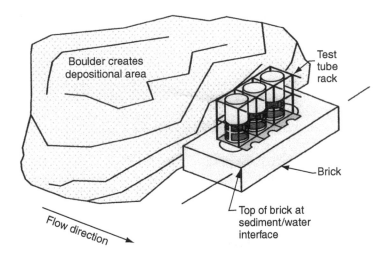

FIGURE 6.2 Test chambers deployed in the Clinch River, Virginia for 2- and 3-week in situ testing of juvenile *Villosa iris*. (From Yeager, M. M., Abiotic and biotic factors influencing the decline of native unionid mussels in the Clinch River, Virginia, PhD Dissertation, Virginia Polytechnic Institute and State University, Blacksburg, VA, 1994.)

twelve sites. Depositional areas behind large rocks were excavated, and the bricks were placed in the area of low flow where the sediments would build up around the test chambers. Due to high flow from rain events, retrieval of all of the test chambers was not possible after 2 weeks; so, some mussels were retrieved after 3 weeks of exposure. Endpoints for the testing were juvenile growth and mortality.

Four tubes (two bricks) were lost from the study sites. Overall, 92.29% of the organisms were recovered, and survival averaged 80.50% with the lost test chambers considered dead for the purpose of the analysis. Of the test chambers recovered, 98.96% of the mussels were retrieved from the chambers, and 86.5% of these were alive. Overall, survival ranged from 43.75 to 100% in the test chambers and was not significantly different between reference and potentially impacted sites. Mussels placed at the upstream reference site had reduced survival and growth as compared to the intermediately located reference site and several test sites. Subsequently, this site was not useful for comparisons to other sites. One hundred percent survival and greater than 400 µm growth in the laboratory references indicate that the organisms were suitable for use in testing.

The major findings of the study were as follows:

- The intermediately located reference site had greater than 90% juvenile mussel survival and ranked intermediately in mussel growth. This site also had a healthy benthic macroinvertebrate community, so there was agreement between the in situ testing and field surveys.
- In general, at the four most impacted sites, there was also agreement between mussel growth and the benthic macroinvertebrate community analysis. Other researchers have indicated that in situ testing most accurately predicts heavy impacts and is more variable at the intermediately impaired sites.
- At three of the potentially impaired sites, the benthic macroinvertebrate sampling and mussel testing indicated a healthy benthic macroinvertebrate community despite documented declines in the mussel communities at the sites. There are many potential sources of impairment to the mussel communities at these sites that would not be indicated by the testing described herein. For example, there may be limitations on fish host species

reducing recruitment, winter road treatments, spring pesticide runoff, siltation in depositional areas (juvenile mussel habitat), Asiatic clam predation on the juvenile mussels (Yeager, Neves, and Cherry 1999), and many other factors that may contribute to this discrepancy.

- Two sites had no agreement between the impaired benthic macroinvertebrate communities and the high growth of the mussels used in testing. This may indicate habitat limitations at these sites not affecting the caged animals or possibly organic enrichment and subsequent dissolved oxygen limitations that were not experienced under the high-flow events occurring during the test period.
- Finally, two sites, including the upstream reference site, showed significant mussel impairment that was not reflected in the benthic community. Since juvenile mussels are extremely sensitive, it is possible that this site may have impairment that is not reflected in more tolerant organisms.

This research indicates that a suit of parameters and a weight of evidence approach is necessary for assessing the biological health of a stream reach. In situ testing is found to be an integral part of assessing the in-stream condition.

St. Croix Riverway Study

The effects of porewater ammonia on juvenile unionid mussels (*Lampsilis cardium*) were evaluated in the St. Croix River in the Summer of 2000 and 2001 (Bartsch et al. 2003). Juveniles were placed in the water column and sediments at twelve sites for a 10-day exposure period (2000) and at eight sites for 4-, 10-, and 28-day (2001) exposure periods. In situ test endpoints included survival and growth, and the porewater and surface water quality monitoring included total ammonia nitrate, unionized ammonia, dissolved oxygen, pH, and temperature. Test chambers were polyvinyl chloride cylinders measuring approximately 6 in. by 1.5 in. Each chamber had two 1.5 in. holes covered with 153 μm Nitex® mesh, and the ends were covered with the same mesh. The chambers were deployed using cable ties to attach them to a plastic-coated garden stake (Figure 6.3).

During the Summer 2000 exposure period, all of the chambers that were deployed were recovered. Juvenile mussel recovery was 89% from the sediments and 90% from the water column, and survival was significantly lower in the sediments (47%) as compared to the water column (86%). In the Summer 2001 exposure period, 92% of the exposure chambers were recovered with juvenile mussel recovery decreasing over the exposure period. After 4 days, 90% of the mussels were retrieved. The recovery dropped to 86% and 71% after 10-day and 28-day exposure periods, respectively. Survival of the juveniles was variable measuring 45%, 28%, and 41% over the 4-, 10-, and 28-day exposure periods. Neither survival nor growth was consistently predicted by porewater ammonia despite unionized ammonia concentrations, which, based on laboratory testing, was expected to cause impairment. Data indicated that dissolved oxygen was positively related to growth, while temperature was positively related to growth and negatively related to survival. Caging effects were believed to be minimal. The lack of impairment to the test organisms due to unionized ammonia was attributed to variable conditions in the river environment including episodic toxicity and changing influences on the percentage of ammonia existing in the unionized form.

IN SITU TESTING WITH NONUNIONID BIVALVES

Standardized methods exist for in situ testing of bivalves, both marine and freshwater (ASTM 2001). These methods were primarily developed using marine organisms and may be more easily applied to nonunionid bivalves for the reasons discussed above.

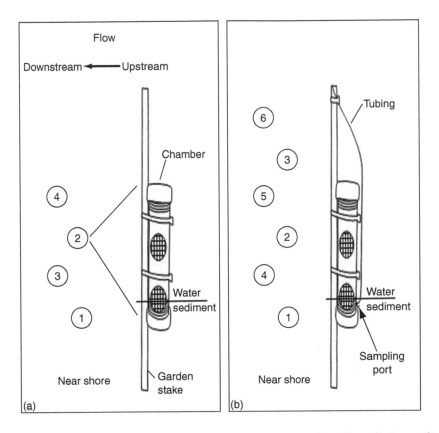

FIGURE 6.3 Test chambers deployed in the St. Croix River for in situ testing of juvenile *Lampsilis cardium.* (From Bartsch et al., *Environ. Toxicol. Chem.*, 22, 2561–2568, 2003. With permission.)

When considering the use of nonunionid bivalves in toxicity testing, it is important to note the differences between effects and impairment. While unionids are believed to be the most sensitive, mostly due to their decline in areas that continue to support other bivalve species (such as *C. fluminea* and *D. polymorpha*), they are not necessarily the first to exhibit a measurable endpoint. The previously described limitations, such as the availability of unionids for testing and the necessity to preserve live specimens, sometimes make it desirable to utilize surrogate species in testing. Several freshwater bivalve species have been used successfully for in situ testing, primarily in cages or bags, with mortality, growth, and physiological endpoints (Belanger et al. 1990; Farris et al. 1998; Salazar et al. 2002).

FRESHWATER CLAMS

The fingernail clam, *Sphaerium fabale*, was evaluated as an in situ test organism by Smith and Beauchamp (2000). The study design allowed for comparison of mortality, growth, and natality as endpoints with exposure periods ranging from 70 to 135 days. Results indicated that the locally available native clams were useful as in situ test organisms with mortality and growth showing good correlation with known site contamination levels as indicated by previous testing and the benthic macroinvertebrate community health. As other studies have indicated, bivalve testing endpoints of growth and mortality were most similar to the benthic macroinvertebrate community health in reference or severely impaired sites. Initially, the intermediately contaminated sites had discrepancies between the evaluation methods; however, after approximately 50 days of exposure, the methods began to converge. This supports the idea that the exposure periods in other studies (such as those described in Yeager 1994) may have simply been too short to represent all of the potential sources of impairment in the watersheds. The authors also found that natality did not necessarily correspond well with the growth and mortality endpoints and suggested that the age of the organisms and seasonality of collection and testing are critical when evaluating natality as an endpoint.

The fingernail clam is desirable as a test organism due to its widespread distribution and lifecycle in which fully formed offspring are released from the parent at sizes suitable for use in testing.

NONNATIVE TAXA

Asiatic clams have been found to be useful in laboratory, mesocosm, and field studies. They have been shown to respond to several environmental stressors including thermal effluent (Foe and Knight 1987), sewage treatment plants (Belanger 1991) metals (Farris et al. 1988; Belanger et al. 1990), and acid mine drainage (Soucek, Schmidt, and Cherry 2001). Endpoints employed in the in situ testing have included survival, growth, and biochemical markers. In a study assessing the sensitivity of several potential endpoints from in situ clam testing, Foe and Knight (1987) recommend monitoring juvenile clam (< 10 mm) shell and tissue growth rate, adult and juvenile clam mortality and possibly adult clam tissue burden, and juvenile clam clearance and gill histopathology. Newly released juvenile Asiatic clams (2–4 weeks old) were not found to correlate well with juvenile unionid responses in laboratory sediment testing (personal observation); however, no field studies using juvenile clams were found in the literature. Cellulolytic activity has been shown to be a particularly sensitive and efficient method for monitoring stress in clam populations (Farris et al. 1988). Both the enzyme activity assay and survival have been demonstrated to correlate well with field bioassessments (Farris et al. 1998; Soucek, Schmidt, and Cherry 2001). However, if these introduced species are to be used as surrogates for the protection of native taxa, more information is needed on the relationship of clam in situ testing endpoints with native fauna impairment.

The zebra mussels, a more recent introduction than *Corbicula*, also show potential for use in bioaccumulation studies in larger rivers and lakes where they have established populations (Johns 2001). However, much research is still needed in this area for establishing relationships between tissue concentrations of contaminants and the contaminant concentrations in streams (Roditi et al. 1999; Rutzke et al. 2000). More information is also needed to correlate the biochemical endpoints with in-stream effects of contaminants.

CONCLUSIONS

In situ testing is becoming an integral component of risk assessments and regulatory compliance monitoring. As the in situ testing methods develop, more standardization of techniques is likely to occur so that conclusions from monitoring will be broadly applicable and comparable from study to study. However, due to the nature of in situ monitoring, site-specific components will always weigh heavily in study design and data interpretation. In fact, no two tests will ever be truly comparable because exposure conditions are not replicable. Unionid mussels are an excellent choice for inclusion in field testing due to their sessile nature, longevity, and potential to show impairment earlier than many other aquatic residents. Their usefulness is limited by their limited availability, slow growth, and the lack of information on species differences pertaining to culturing and rearing, contaminant sensitivity, and ecological preferences. Other bivalves such as fingernail clams are excellent choices for test organisms, where available. The introduced species, *C. fluminea*, also shows excellent promise as a surrogate test organism for freshwater communities. However, a stronger correlation needs to be developed between organism responses to contaminants and effects in the biotic communities of streams being surveyed.

REFERENCES

Adams, T. G., Atchison, G. J., and Vetter, R. J., The use of the three-ridge clam (*Amblema perplicata*) to monitor trace metal contamination, *Hydrobiologia*, 83, 67–72, 1981.

[ASTM] American Society for Testing and Materials, Standard guide for conducting in situ field bioassays with caged bivalves, 2003, *Annual Book of ASTM Standards*, Vol. 11.05, ASTM, Philadelphia, PA, E2122–02, 2001.

Barbour, M. T., Gerritsen, J., Snyder, B. D., and Stribling, J. B., *Rapid Bioassessment Protocols for Use in Wadeable Streams and Rivers, Second Edition EPA 841-B-99-002*, U.S. Environmental Protection Agency, Office of Water, Washington, DC, 1999.

Bartsch, M. R., Newton, T. J., Allran, J. W., O'Donnell, J. A., and Richardson, W. B., Effects of pore water ammonia on in situ survival and growth of juvenile mussels (*Lampsilis cardium*) in the St. Croix Riverway, Wisconsim, USA, *Environ. Toxicol. Chem.*, 22, 2561–2568.

Belanger, S. E., The effect of dissolved oxygen, sediment, and sewage treatment plant discharges upon growth, survival and density of Asiatic clams, *Hydrobiologia*, 218, 113–126, 1991.

Belanger, S. E., Farris, J. L., Cherry, D. S., and Cairns, J., Validation of *Corbicula fluminea* growth reductions induced by copper in artificial streams and river systems, *Can. J. Fish. Aquat. Sci.*, 47, 904–914, 1990.

Buddensiek, V., The culture of juvenile freshwater pearl mussels *Margaritifera margaritifera L.* in cages: A contribution to conservation programmes and the knowledge of habitat requirements, *Biol. Conserv.*, 74(1995), 33–40, 1995.

Burton, G. A., Assessing the toxicity of freshwater sediments, *Environ. Toxicol. Chem.*, 10, 1585–1627, 1991.

Burton, G. A., Pitt, R., and Clark, S., The role of traditional and novel toxicity test methods in assessing stormwater and sediment contamination, *Crit. Rev. Environ. Sci. Technol.*, 30(4), 413–447, 2000.

Chappie, D. J. and Burton, G. A., Optimization of in situ bioassays with *Hyalella azteca* and *Chironomus tentans*, *Environ. Toxicol. Chem.*, 16, 559–564, 1997.

Chappie, D. J. and Burton, G. A., Applications of aquatic and sediment toxicity testing in situ, *Soil Sediment Contam.*, 9(3), 219–245, 2000.

Couillard, Y., Campbell, P. G. C., Tessier, A., Pellerin-Massicotte, J., and Auclair, J. C., Field transplantation of a freshwater bivalve, *Pyganodon grandis*, across a metal contamination gradient. 1. Temporal changes in metallothionein and metal (Cd, Cu, and Zn) concentrations in soft tissues, *Can. J. Fish. Aquat. Sci.*, 52, 690–702, 1994.

Crane, M. and Maltby, L., The lethal and sublethal responses of *Gammarus pulex* to stress: Sensitivity and sources of variation in an in situ bioassay, *Environ. Toxicol. Chem.*, 10, 1331–1339, 1991.

Crane, M., Delaney, P., Mainstone, C., and Clarke, S., Measurement by in situ bioassay of water quality in an agricultural catchment, *Water Resour.*, 29(11), 2441–2448, 1995.

Crane, M., Higman, M., Olsen, T., Simpson, P., Callaghan, A., Fisher, T., and Kheir, R., An in situ system for exposing aquatic invertebrates to contaminated sediments, *Environ. Toxicol. Chem.*, 19(11), 2715–2719, 2000.

Doherty, F. G., A multidisciplinary study of the asiatic clam, *Corbicula fluminea*, from the New River, Virginia, *PhD dissertation*, Virginia Polytechnic Institute and State University, Blacksburg, VA, 1986.

Englund, V. P. M. and Heino, M. P., Valve movement of the freshwater mussel *Anodonta anatine*: A reciprocal transplant experiment between lake and river, *Hydrobiologia*, 328, 49–56, 1996.

Farris, J. L., Van Hassel, J. H., Belanger, S. E., Cherry, D. S., and Cairns, J., Application of cellulolytic activity of asiatic clams (*Corbicula* sp.) to in-stream monitoring of power plant effluents, *Environ. Toxicol. Chem.*, 7, 701–713, 1988.

Foe, C. and Knight, A., Growth of *Corbicula fluminea* (bivalvia) fed artificial and algal diets, *Hydrobiologia*, 133, 155–164, 1986.

Foe, C. and Knight, A., Assessment of the biological impact of point source discharges employing asiatic clams, *Arch. Environ. Contam. Toxicol.*, 16, 39–51, 1987.

Gray, J. S., Do bioassays adequately predict ecological effects of pollutants? *Hydrobiologia*, 188/189, 397–402, 1989.

Hickey, C. W., Roper, D. S., and Buckland, S. J., Metal concentrations of resident and transplanted freshwater mussels *Hyridella menziesi* (Unionacea: Hyriidae) and sediments in the Waikato River, New Zealand, *Sci. Total Environ.*, 175(1995), 163–177, 1995.

Hinch, S. G. and Green, R. H., The effects of source and destination on growth and metal uptake in freshwater clams reciprocally transplanted among south central Ontario lakes, *Can. J. Zool.*, 67, 855–863, 1989.

Ireland, D. S., Burton, G. A. Jr. and Hess, G. G., In situ toxicity evaluations of turbidity and photoinduction of polycyclic aromatic hydrocarbons, *Environ. Toxicol. Chem.*, 15, 574–581, 1996.

Isom, B. G. and Hudson, R. G., In vitro culture of parasitic freshwater mussel glochidia, *The Nautilus*, 96, 147–151, 1982.

Jacobson, P. J., Farris, J. L., Cherry, D. S., and Neves, R. J., Juvenile freshwater mussel (Bivalvia: Unionidae) responses to acute toxicity testing with copper, *Environ. Toxicol. Chem.*, 12, 879–883, 1993.

Johns, C., Spatial distribution of total cadmium, copper, and zinc in the zebra mussel (*Dreissena polymorpha*) along the Upper St. Lawrence River, *J. Great Lakes Res.*, 27(3), 354–366, 2001.

Karr, R. J. and Chu, E. W., *Restoring Life in Running Waters: Better Biological Monitoring*, Island Press, Washington, DC, 1999.

Kater, B. J., Postma, J. F., Dubbeldam, M., and Prins, J. T. H. J., Comparison of laboratory and in situ sediment bioassays using *Corophium voluator*, *Environ. Toxicol. Chem.*, 20, 1291–1295, 2001.

Keller, A. E. and Zam, S. G., The acute toxicity of selected metals to the freshwater mussel, *Anodonta imbecilis*, *Environ. Toxicol. Chem.*, 10, 539–546, 1991.

Keller, E., Ruessler, D. S., and Kernaghan, N. J., Effect of test conditions on the toxicity of copper to juvenile unionid mussels, *Am. Soc. Test. Mater. Spec. Tech. Publ.*, 1364, 329–340, 1999.

Kramer, K. J. M., Jenner, H. A., and Zwart, D. D., The valve movement response of mussels: A tool in biological monitoring, *Hydrobiologia*, 188/189, 433–443, 1989.

La Point, T. W. and Waller, W. T., Field assessments in conjunction with whole effluent toxicity testing, *Environ. Toxicol. Chem.*, 19(1), 14–24, 2000.

Lieb, D. A. and Carline, R. F., Effects of urban runoff from a detention pond on water quality, temperature and caged *Gammarus minus* (Say) (Amphipoda) in a headwater stream, *Hydrobiologia*, 441, 107–116, 2000.

Maltby, L., Clayton, S. A., Yu, H., McLoughlin, N., Wood, R. M., and Yin, D., Using single-species toxicity tests, community-level responses, and toxicity identification evaluations to investigate effluent impacts, *Environ. Toxicol. Chem.*, 19(1), 151–157, 2000.

Matthiessen, P., Sheahan, D., Harrison, R., Kirby, M., Rycroft, R., Turnbull, A., Volkner, C., and Williams, R., Use of a *Gammarus pulex* bioassay to measure the effects of transient carbofuran runoff from farmland, *Ecotoxicol. Environ. Saf.*, 30, 111–119, 1995.

Meregalli, G., Vermeulen, A. C., and Ollevier, F., The use of chironomid deformation in an in situ test for sediment toxicity, *Ecotoxicol. Environ. Saf.*, 47, 231–238, 2000.

Metcalfe-Smith, J. L., Green, R. H., and Grapentine, L. C., Influence of biological factors on concentrations of metals in the tissues of freshwater mussels (*Elliptio complanata* and *Lampsilis radiata radiata*) from the St. Lawrence River, *Can. J. Fish. Aquat. Sci.*, 53, 205–219, 1996.

Naimo, T. J., Waller, D. L., and Holland-Bartels, L. E., Heavy metals in the threeridge mussel *Amblema plicata plicata* (Say, 1817) in the Upper Mississippi River, *J. Freshwater Ecol.*, 7(2), 209–215, 1992.

Naimo, T. J., Damschen, E. D., Rada, R. G., and Monroe, E. M., Nonlethal evaluation of the physiological health of unionid mussels: Methods for biopsy and glycogen analysis, *J. N. Am. Benthol. Soc.*, 17(1), 121–128, 1998.

Neves, R. J. and Widlak, J. C., Habitat ecology of juvenile freshwater mussels (Bivalvia: Unionidae) in a headwater stream in Virginia, *Am. Malacol. Bull.*, 5, 1–7, 1987.

Neves, R. J., Weaver, L. R., and Zale, A. V., An evaluation of host fish suitability for glochidia of *Villosa vanuxemi* and *V. nebulosa* (Pelecypoda: Unionidae), *Am. Midland Nat.*, 113, 13–19, 1985.

Pereira, A. M. M., Soares, A. M. V. D. M., Goncalves, F., and Ribeiro, R., Test chambers and test procedures for in situ toxicity testing with Zooplankton, *Environ. Toxicol. Chem.*, 18(9), 1956–1964, 1999.

Pereira, A. M. M., Soares, A. M. V. M., Goncalves, F., and Ribeiro, R., Water-column, sediment, and in situ chronic bioassays with cladocerans, *Ecotoxicol. Environ. Saf.*, 47, 27–38, 2000.

Rand, G. M., Ed., *Fundamentals of Aquatic Toxicology, Second Edition. Ecological Services, Inc*, Taylor & Francis, Washington, DC., 1995.

Roditi, H. A. and Fisher, N. S., Rates and routes of trace element uptake in zebra mussels, *Limnol. Oceanogr.*, 44(7), 1730–1749, 1999.

Rutzke, M. A., Gutenmann, W. H., Lisk, D. J., and Mills, E. L., Toxic and nutrient element concentrations in soft tissues of zebra and quagga mussels from Lakes Erie and Ontario, *Chemosphere*, 40, 1353–1356, 2000.

Salánki, J. and Balogh, K. V., Physiological background for using freshwater mussels in monitoring copper and lead pollution, *Hydrobiologia*, 188/189, 445–454, 1989.

Salazar, M. H., Salazar, S. M., Gagne, F., Blaise, C., and Trottier, S., Developing a benthic cage for long-term, in-situ tests with freshwater and marine bivalves, In *Proceedings, 29th Annual Aquatic Toxicity Workshop*, October 20–23, Fairmount Chateau Whistler, Whistler, BC, 2002.

Sasson-Brickson, G. and Burton, G.A. Jr. In situ and laboratory sediment toxicity testing with *Ceriodaphnia dubia*, *Environ. Toxicol. Chem.*, 10, 201–207, 1991.

Schulz, R. and Liess, M., Validity and ecological relevance of an active in situ bioassay using *Gammarus pulex* and *Limnephilus lunatus*, *Environ. Toxicol. Chem.*, 18, 2243–2250, 1999.

Scott, J. C., Population demographics of six freshwater mussel species (Bivalvia: Unionidae) in the Upper Clinch River, Virginia and Tennessee, *Masters Thesis*, Virginia Polytechnic Institute and State University, Blacksburg, VA, 1994.

Shaw, J. L. and Manning, J. P., Evaluating macroinvertebrate population and community level effects in outdoor microcosms: Use of in situ bioassays and multivariate analysis, *Environ. Toxicol. Chem.*, 15, 608–617, 1996.

Sibley, P. K., Benoit, D. A., Balcer, M. D., Phipps, G. L., West, C. W., Hoke, R. A., and Ankley, G. T., In situ bioassay chamber for assessment of sediment toxicity and bioaccumulation using benthic invertebrates, *Environ. Toxicol. Chem.*, 18(10), 2325–2336, 1999.

Simonin, H. A., Kretser, W. A., Bath, D. W., Olson, M., and Gallagher, J., In situ bioassays of brook trout (*Salvelinus fontinalis*) and blacknose dace (*Rhinichthys atratulus*) in Adirondack streams affected by episodic acidification, *Can. J. Fish. Aquat. Sci.*, 50, 902–912, 1993.

Sloof, W., de Zwart, D., and Marquenie, J. M., Detection limits of a biological monitoring system for chemical water pollution based on mussel activity, *Bull. Environ. Contam. Toxicol.*, 30, 400–405, 1983.

Smith, J. G. and Beauchamp, J. J., Evaluation of caging designs and a fingernail clam for use in an in situ bioassay, *Environ. Monit. Assess.*, 62, 205–230, 2000.

Soucek, D. J., Schmidt, T. S., and Cherry, D. S., In situ studies with Asian clams (*Corbicula fluminea*) detect acid mine drainage and nutrient inputs in low-order streams, *Can. J. Fish. Aquat. Sci.*, 58, 602–608, 2001.

Stewart, A. R., Accumulation of Cd by a freshwater mussel (*Pyganodon grandis*) is reduced in the presence of Cu, Zn, Pb, and Ni, *Can. J. Fish. Aquat. Sci.*, 56, 467–478, 1999.

Tessier, A., Campbell, P. G. C., Auclair, J. C., and Bisson, M., Relationships between the partitioning of trace metals in sediments and their accumulation in the tissues of the freshwater Mollusc *Elliptio complanata* in a mining area, *Can. J. Fish. Aquat. Sci.*, 41, 1463–1472, 1984.

Warren, L. W., Klaine, S. J., and Finley, M. T., Development of a field bioassay with juvenile mussels, *J. N. Am. Benthol. Soc.*, 14(2), 341–346, 1995.

Webber, C. I., Ed., *Methods for Measuring the Acute Toxicity of Effluents and Receiving Waters to Freshwater and Marine Organisms*, 4th ed., USEPA, Cincinatti, EPA/600/4-90/02, 1993.

Yeager, M. M., Abiotic and biotic factors influencing the decline of native unionid mussels in the Clinch River, Virginia, *PhD Dissertation*, Virginia Polytechnic Institute and State University. Blacksburg, VA, 1994.

Yeager, M. M., Neves, R. J., and Cherry, D. S., Competitive interactions between early life stages of *Villosa iris* (Bivalvia: Unionidae) and adult Asian clams (*Corbicula fluminea*), *Proceedings of the First Freshwater Mollusk Conservation Society Symposium*, pp. 253–259, 1999.

Zale, A. V. and Neves, R. J., Fish hosts of four species of lampsiline mussels (Mollusca: Uniondae) in Big Moccasin Creek, Virginia, *Can. J. Zool.*, 60, 2535–2542, 1982.

7 Unionid Mussel Sensitivity to Environmental Contaminants

Anne Keller, Mike Lydy, and D. Shane Ruessler

INTRODUCTION

In the 1960s and 1970s, chemical impacts on aquatic toxicity tests that lasted 2–4 days (NAS/NAE 1973; Johnson and Finley 1980; Mayer and Ellersieck 1986). These test lengths were adequate to determine short-term effects using death as the endpoint. While such relatively simple tests are not an exact measure of chemical toxicity in a stream or lake because local factors can ameliorate or exacerbate effects, they served as the basis of most early water quality criteria because they comprised the best and most abundant available data. At that time, the art and science of toxicity testing was in its infancy and chronic tests require substantially more sophistication relative to equipment, facilities and expertise. Acute tests provide repeatable results, as well as being simple, rapid, inexpensive, and provide an easily recognizable endpoint-death. Acute data were and are still used to compare toxicity among species, trophic levels, different formulations, and different compounds (Johnson and Finley 1980; Mayer and Ellersieck 1986).

US water quality criteria were originally based upon toxicity data from a suite of aquatic species that represented 95% of those tested (Stephan et al. 1985). For each chemical, a set of test data that include plant and animal species from several trophic levels, having different habitat requirements, relying on different food sources, and with different life spans were used. Taken together, responses of several taxa more adequately portray the toxicity of a chemical to the ecosystem than do toxicity data from just one or two species. Data from bioaccumulation studies, field exposures and any available chronic data were also included when available. The use of a suite of species also provided the opportunity to include data from species that are important to society because they are food sources, of recreational value, or are species of special concern because they are threatened or endangered. Included in the latter category are many species of unionid mussels. However, because the life history of unionids makes laboratory culture very difficult, no acute toxicity data were available for unionids during early criteria development.

Though fewer chronic than acute toxicity tests exposures have been conducted over the years because of the added expense and difficulty of maintaining aquatic species in the laboratory for extended periods, they are of great value in better estimating the effects of chemicals in aquatic systems. Results from such tests provide regulators a fine-tipped pen with which to establish a more realistic margin of safety than the current approach, which may rely on the use of an arbitrary value when more specific data are lacking. Chronic toxicity tests have been developed for a number of aquatic species, but test methods for unionid mussels are still being developed or refined. Roadblocks to the development of chronic tests include the difficulty of maintaining unionids in the laboratory, the fact that tests would have to be longer than for many other taxa due to lengthy unionid life spans, the need to determine appropriate endpoints, etc. These issues are being addressed in a collaborative research project now underway by the U.S. Geological Survey and the USEPA.

This chapter reviews the sensitivity of freshwater mussels to metals, pesticides, and other contaminants of freshwater systems. Data from these mollusks, in decline throughout North America and other parts of the world, were not used to establish water quality criteria for fresh waters because such information was virtually nonexistent in the early-to-mid-1980s when the USEPA began developing its water quality program (Stephan et al. 1985; Augspurger et al. 2003). However, in response to the precipitous loss of species and decreased abundance of unionid mussels in recent years, attention has turned to water pollution as a possible cause. This spurred interest among a number of researchers to establish test protocols and collect much needed data. Recently, the USEPA has begun to use these data to evaluate the protectiveness of a number of metal and other chemical criteria to unionids. It is possible that criteria for several metals and ammonia will be modified by inclusion of unionid data into the database. Current water quality criteria are included in chapter tables for comparison to data for this imperiled fauna.

METAL TOXICITY

The complicated reproductive strategy of freshwater mussels was probably the major factor limiting the availability of early life stage toxicity data (Chapter 5). Unlike Asian or fingernail clams, most unionid larvae, called glochidia, must attach to a fish host for 7–30 days or more, during which they transform and grow into juveniles, are transported to new areas, and drop off. This larval transformation process made laboratory culture of unionids difficult until new methods were developed (Isom and Hudson 1982; Keller and Zam 1990). Beginning in the late 1980s and early 1990s, several laboratories began to measure the sensitivity of unionid glochidia, juveniles and adults (Schweinforth and Wade 1990; Keller and Zam 1991; Jacobson et al. 1993). Even so, toxicity data for unionids are available for only a fraction of the contaminants that enter the aquatic systems of North America.

Several researchers have evaluated the toxicity of mining-related contaminants to unionids that inhabit nearby streams (Cherry and Farris 1991; Cherry, Farris, and Neves 1991; McCann 1993; Hansten, Heino, and Pynnonen 1996). These tests were based on the change in glochidial closing response when salt is added to water in their test chamber following exposure to a metal for a predetermined length of time. Healthy glochidia close when salted, a response that mimics their reaction to fish mucus.

The toxicity of cadmium and copper have been tested frequently, probably because they are common contaminants in industrialized areas and are very toxic to aquatic organisms (Holwerda and Herwig 1986; Hemelraad, Holwerda, and Zandee 1986a; Jacobson 1990; Farris, Van Hassel, and Cherry 1991; Keller and Zam 1991; Lasee 1991; Naimo, Waller, and Holland-Bartels 1992a; McCann 1993), but many other metals also have been evaluated. Less is known about the toxicity of organic contaminants to mussels.

ACUTE TOXICITY OF METALS

Glochidia tests are performed by exposing the larvae to a contaminant and then testing their viability after 24–48 h (McCann 1993; Hansten, Heino, and Pynnonen 1996; Keller and Ruessler 1997), determining transformation success after attachment to a host fish (Jacobson et al. 1997) or measuring activity defined as the number of valve openings and closings in a given time period (Varanka 1977). Fewer than two dozen papers evaluating the toxicity of metals to glochidia have been published (Chapter 5).

Several approaches to glochidia tests have been used. These include measuring changes in the "snapping" response (Granmo and Varanka 1979), their closing response after exposure to a contaminant (Jacobson et al. 1997; Keller and Ruessler 1997), their uptake of vital stain (Jacobson et al. 1997), and transformation success after exposure of glochidia to a chemical (Jacobson et al. 1997).

Granmo and Varanka (1979) conducted a study of copper and zinc toxicity to *Anodonta cygnea* (L.) glochidia based on how glochidial "snapping" activity was modulated after exposure to contaminants. The opening and closing of glochidia valves is an important part of their attracting host fish and in attaching to them. They determined that copper and zinc exposure reduced the snapping significantly. Since the tests were performed using concentrations up to 1000 times higher than current US water quality criteria (USEPA 1999a), results are of only limited value in determining the impact of metals on the likelihood of attachment to a host.

Since developing glochidia are held in the brood chamber (gill) of the female mussel, isolated from ambient by only a few layers of cells prior to their release, the potential impact of contaminants on glochidia while within the brood chambers is also of interest. Cherry, Farris, and Neves (1991) found that the viability of *Villosa nebulosa* developing glochidia was not impacted when the adult was exposed to copper (12–192 µg/L). These results support conclusions of other research that glochidia are isolated from the outside environment during residence in the female's gills (Silverman, McNeil, and Dietz 1987; Lasee 1991; Richard, Dietz, and Silverman 1991). In contrast, Huebner and Pynnonen (1992) and Jacobson et al. (1997), found that glochidia from gravid females exposed to metals were sometimes less viable than unexposed glochidia. So, this issue remains unresolved.

Hansten, Heino, and Pynnonen (1996) tested glochidia of *Anodonta anatina*, *Villosa iris*, *Medionidus conradicus*, *A. cygnea*, *Actinonaias pectorosa*, and *A. anatina* for sensitivity to cadmium, zinc, and copper. Toxicity was apparent at metal concentrations similar to US acute water quality criteria recommendations (Table 7.1). Not unexpectedly, humic acids, EDTA, iron, and manganese, all chelators of metals, ameliorated toxicity (Hansten, Heino, and Pynnonen 1996).

Published results of juvenile unionid mollusk toxicity tests are somewhat more numerous than for glochidia tests, and several test parameters have been evaluated for their effects on toxicity. Juvenile age, test temperature, and water hardness are known to impact the toxicity of metals (Jacobson 1990; Keller and Zam 1991; Lasee 1991; McCann 1993). Increased hardness and lower test temperature decreased toxicity, and older juveniles (14 days) were somewhat less sensitive to metals than were younger ones (0 days). These findings are generally similar to those for other aquatic species.

Keller and Zam (1991) evaluated the 48- and 96-h toxicity of zinc, copper, cadmium, mercury, chromium, and nickel to juvenile *Utterbackia imbecilis* mussels in soft (40–48 mg/L as $CaCO_3$) and moderately hard water (80–100 mg/L as $CaCO_3$). Zinc was the least toxic metal while cadmium was most toxic to these juveniles. Zinc toxicity (LC50) ranged from 268 to 438 µg/L at 96 h, depending on water hardness. McCann (1993) reported similar values—339–1,185 µg/L at 48 h. The current US criterion recommendation for zinc is 120 µg/L in water with 100 mg/L hardness (USEPA 1999a).

Jacobson et al. (1993) measured sublethal copper toxicity in exposed juvenile *V. iris* and *Villosa grandis* based on their uptake of neutral red, a vital stain. Uptake ceased at 29 µg/L Cu indicating morbidity, while the 24 h LC50 was 83 µg/L for *V. iris*. These concentrations are similar to the current acute and chronic criteria recommendations for copper in water of 100 mg/L $CaCO_3$ hardness, 13 and 9 µg/L, respectively (USEPA 1999a).

Unionid sensitivity has been compared to other aquatic species in side-by-side tests (Keller 1993; Masnado, Geis, and Sonzongi 1995). Masnado, Geis, and Sonzongi (1995) used different concentrations of metals in a series of synthetic effluents (e.g., chromium, copper, zinc, cadmium, and nickel) to determine the threat a mine effluent would pose to downstream populations of unionid mussels. Fathead minnows (*Pimephales* promelas) and *Ceriodaphia dubia* were more sensitive to the effluents than were *U. imbecilis* mussels. Keller (1993) exposed *C. dubia* and juvenile *U. imbecilis* mussels to an effluent containing 6.4 mg/L chromium. The 48-h LC50s were 61 µg Cr/L for *U. imbecilis* and 36 µg Cr/L for *C. dubia*. By 96 h, the mussel and zooplankton LC50s each had decreased by one-third. The current acute criterion recommendation for chromium in water of 100 mg/L $CaCO_3$ hardness is 74 µg Cr/L (USEPA 1999a).

TABLE 7.1
Summary of Selected Metal and Inorganic Toxicity Data for Unionid Mussels

Reference	Species	Glochidia	Juvenile	Adult	Exposure (h)	Hardness	Test (Temp.)	Chemical or Physical LC50s										
								Cu (µg/L)	Cd (µg/L)	Zn (µg/L)	Hg (µg/L)	Ni (µg/L)	Cr (µg/L)	K (µg/L)	F (mg/L)	Ammonia at pH 8.0 (mg/L)	Temperature (°C)	pH (SU)
								13[a]	2.0[b]	120[a]	1.4[a]	470[a]	570[a]	—	—	8.4[c]	NA	6.5–9.0[d]
Grammo and Varanka (1979)	A. cygnea	x			—	—	20	10,000+	—	—	—	—	—	—	—	—	—	—
Keller and Zam (1991)	U. imbecillis		x		48	60	23	171	57	355	216	240	295	—	—	—	—	—
Keller and Zam (1991)	U. imbecillis		x		96	60	23	86	9	268	147	190	39	—	—	—	—	—
Keller and Zam (1991)	U. imbecillis		x		48	80	23	388	137	588	233	471	1187	—	—	—	—	—
Keller and Zam (1991)	U. imbecillis		x		96	80	23	199	107	438	171	252	618	—	—	—	—	—
Huebner and Pynnonen (1992)	A. cygnea	x			72	—	13	6	6	150	—	—	—	—	—	—	—	—
Hansen, Heino, and Pynnonen (1996)	A. anatina	x			72	—	4	50	8	20	—	—	—	—	—	—	—	—
Imlay (1971)	Unknown			x	months	—	—	25	—	—	—	—	—	4000–7000	—	—	—	—
McCann (1993)	Medionidus conradicus	x			48	60	20	—	—	492	—	—	—	—	—	—	—	—
McCann (1993)	A. pectorosa	x			48	50	20	—	—	274	—	—	—	—	—	—	—	—
McCann (1993)	A. pectorosa	x			48	160	20	—	—	664–739	—	—	—	—	—	—	—	—
McCann (1993)	V. iris	x			48	40–50	20	—	—	577—1155	—	—	—	—	—	—	—	—
McCann (1993)	V. iris	x			48	140–160	20	—	—	836—1230	—	—	—	—	—	—	—	—
McCann (1993)	A. pectorosa		x		48	40	20	—	—	360—370	—	—	—	—	—	—	—	—
McCann (1993)	A. pectorosa		x		48	160	20	—	—	1060—1186	—	—	—	—	—	—	—	—
McCann (1993)	V. iris		x		48	50	20	—	—	339	—	—	—	—	—	—	—	—
McCann (1993)	V. iris		x		48	160	20	—	—	1122	—	—	—	—	—	—	—	—
McCann (1993)	A. pectorosa		x		48	40–50	20	52–63	—	—	—	—	—	—	—	—	—	—
McCann (1993)	A. pectorosa		x		48	140–160	20	76–156	—	—	—	—	—	—	—	—	—	—

Reference	Species																		
Dimock and Wright (1993)	*Pyganodon cataracta*	—	x	96	—	20	—	—	—	—	—	—	—	—	—	—	—	33	45
Dimock and Wright (1993)	*U. imbecillis*	—	x	—	—	—	—	—	—	—	—	—	—	—	—	—	—	—	—
Jacobson (1990)	*V. nebulosa*	x	—	48	160	20	56	—	—	—	—	—	—	—	—	—	—	—	—
Jacobson (1990)	*A. pectorosa*	x	—	48	170	20	51	—	—	—	—	—	—	—	—	—	—	—	—
Jacobson (1990)	*Anodonta grandis*	x	—	24	170	10	347	—	—	—	—	—	—	—	—	—	—	—	—
Jacobson (1990)	*A. grandis*	x	—	24	50	20	46	—	—	—	—	—	—	—	—	—	—	—	—
Jacobson (1990)	*Lampsillis fasciola*	x	—	48	170	20	40	—	—	—	—	—	—	—	—	—	—	—	—
Jacobson (1990)	*M. conradicus*	x	—	48	170	20	16	—	—	—	—	—	—	—	—	—	—	—	—
Jacobson (1990)	*V. iris*	—	x	24	190	20	83	—	—	—	—	—	—	—	—	—	—	—	—
Jacobson (1990)	*A. grandis*	—	x	24	70	20	44	—	—	—	—	—	—	—	—	—	—	—	—
Cherry, Farris, and Neves (1991)	*Ptychobranchus fasciolaris*	x	—	48	170	20	—	—	212	—	—	—	—	—	—	—	—	—	—
Cherry, Farris, and Neves (1991)	*A. pectorosa*	x	—	48	170	20	—	—	309	—	—	—	—	—	—	—	—	—	—
Cherry, Farris, and Neves (1991)	*M. conradicus*	x	—	48	170	20	—	—	570	—	—	—	—	—	—	—	—	—	—
Cherry, Farris, and Neves (1991)	*V. nebulosa*	x	—	48	170	20	—	—	656	—	—	—	—	—	—	—	—	—	—
Klaine, Warren, and Summers (1997)	*U. imbecillis*	x	—	96	100	25	67	—	—	—	—	—	—	—	—	—	—	—	—
Klaine, Warren, and Summers (1997)	*U. imbecillis*	—	x	96	100	25	—	47	—	—	—	—	—	—	—	—	—	—	—
Augspurger et al. (2003)	*Various*	x	x	24–96, ≤60	—	12–25	—	—	—	—	—	—	—	—	—	0.57–15.5	—	—	—
Keller and Augspurger (2005)	*Alasmidonta raveneliana*	x	—	24	30	25	—	—	—	—	—	—	—	—	—	288	—	—	—
Keller and Augspurger (2005)	*U. imbecillis*	x	—	24	34	25	—	—	—	—	—	—	—	—	—	234	—	—	—
Keller and Augspurger (2005)	*A. raveneliana*	—	x	96	28	25	—	—	—	—	—	—	—	—	—	303	—	—	—
Keller and Augspurger (2005)	*A. pectorosa*	x	—	96	30	25	—	—	—	—	—	—	—	—	—	178	—	—	—

(continued)

TABLE 7.1   (Continued)

Reference	Species	Glochidia	Juvenile	Adult	Exposure (h)	Hardness	Test (Temp.)	Cu (µg/L)	Cd (µg/L)	Zn (µg/L)	Hg (µg/L)	Ni (µg/L)	Cr (µg/L)	K (µg/L)	F (mg/L)	Ammonia at pH 8.0 (mg/L)	Temperature (°C)	pH (SU)
								13[a]	2.0[b]	120[a]	1.4[a]	470[a]	570[a]	—	—	8.4[c]	NA	6.5–9.0[d]
Keller and Augspurger (2005)	*U. imbecillis*	—	x		96	34	25	—	—	—	—	—	—	—	234	—	—	—
Keller and Augspurger (2005)	*L. fasciola*	—	x		96	32	25	—	—	—	—	—	—	—	172	—	—	—

Columns under heading: **Chemical or Physical LC50s**

[a] USEPA 2002, acute criterion recommendation; listed in the dissolved fraction of the metal concentration.
[b] USEPA 1999a, acute criterion recommendation; listed in the dissolved fraction of the metal concentration.
[c] USEPA 1999b, acute criterion recommendation.
[d] USEPA 1976.

Lasee (1991) conducted a histological and ultrastructural study of *Lampsilis ventricosa* in which she examined the impact of cadmium on tissues and organs. She ran toxicity tests as part of the study and calculated 48-h LC50s of 141 µg Cd/L–345 µg Cd/L at 150 mg/L hardness for juveniles at 0–14-days posttransformation. These values are similar to those seen by Keller and Zam (1991), who reported 48-h LC50s of 9 µg Cd/L–107 µg Cd/L at ~40 and 80 mg/L hardness, respectively. The current US criterion recommendation is 4.3 µg Cd/L in water with 100 mg/L hardness (USEPA 1999a).

Few adult unionid mussel toxicity tests have been reported, probably because their maintenance requirements in the laboratory have not been well characterized. Imlay (1971) described 25 µg Cu/L as "lethal" to mussels (species not identified). A set of 28-day flow-through copper toxicity tests was performed by Keller et al. (unpublished data) in 1996 using adult mussels in well water. The LC50s for *U. imbecilis* and *Elliptio buckleyi* were 69 and 4.5 µg Cu/L, respectively, at a hardness of 185 mg/L as $CaCO_3$.

SUBLETHAL TOXICITY OF METALS

Exposures of mussels to low concentrations of a metal for a long period of time (greater than seven days) permit the measurement of sublethal effects on processes such as growth (Hinch and Green 1989; Schweinforth and Wade 1990; Lasee 1991; Metcalfe-Smith and Green 1992), enzyme production (Reddy and Chari 1985), ionic balance (Malley, Huebner, and Donkersloot 1988; Pynnonen 1991; Sivaramakrishna, Radhakrishnaiah, and Suresh 1991), amino acid content of tissues (Gardner, Miller, and Imlay 1981), metallothionein production (Couillard, Campbell, and Tessier 1993; Malley et al. 1993), and others. Some of these responses may prove to be useful as indicators or biomarkers of exposure to metals and may improve the use of mussels as sentinels of ecosystem health. Virtually no information is available on the sublethal impact of metal pollution on glochidia or juvenile mussels.

A series of papers published by Jenner et al. (1991) and Hemelraad, Holwerda, and Zandee (1986a, 1986b, 1990a, 1990b) monitored tissue uptake and responses of *A. anatina* and *A. cygnea* to sublethal cadmium exposure. They found that cadmium accumulated in soft tissues linearly at low concentrations and in a biphasic manner at higher concentrations; that gills accumulated the greatest amount of cadmium; that exposure to cadmium disturbed energy metabolism; and that ionic balance of the hemolymph and tissues was disrupted.

Reports from a number of other laboratories amplify these results. Oxygen consumption, ciliary activity, and heartbeat were significantly reduced in *Lamellidens marginalis* exposed to lethal and sublethal cadmium concentrations (6 and 2 mg/L Cd) for one to ten days (Radhakrishnaiah 1988). These physiological impacts resulted from increased mucus production by the gills during cadmium exposure. In a longer study by Naimo, Waller, and Holland-Bartels (1992a) respiration rates decreased in *L. ventricosa* exposed to sublethal concentrations of cadmium for 28 days. However, ammonia excretion, mussel condition, and food assimilation efficiency were not found to change significantly, perhaps a result of the high variability among individual animals (Naimo, Waller, and Holland-Bartels 1992a). Mucus production also increased in the animals tested by Naimo, Waller, and Holland-Bartels (1992b). ATPase activity and ciliary activity of gills in *A. cygnea* was decreased following exposure to cadmium (Pirovarova, Lagerspetz, and Skulskii 1992). Similarly, Raj and Hameed (1991) determined that sublethal concentrations of copper and mercury accelerated the respiratory rate of *L. marginalis*, while cadmium depressed it. In this latter study, body weight decreased following 30-day exposures to copper. Finally, the synthesis of porphyrins (part of cytochrome, enzyme, vitamin, and myoglobin molecules) was disrupted in *Elliptio complanata* and *A. grandis* mussels exposed to low concentrations of cadmium (Chamberland et al. 1995). Digestive activity and efficiency can also be impacted by cadmium toxicity. Both Hameed and Raj (1989) and Lasee (1991) found that exposure to copper, cadmium, or mercury resulted in the dissolution of the crystalline style in *Lamellidens marginalis* and *L. ventricosa*, respectively. The style grinds ingested food before it is digested. Exposure to mercury caused the fastest dissolution of the style of the three metals tested (Hameed and Raj 1989), and resulted in the slowest recovery.

Metallothionein production was induced by exposure of several mussel species to cadmium, copper, zinc, and other metals (Couillard, Campbell, and Tessier 1993; Malley et al. 1993; Couillard et al. 1995). These proteins serve as protectors from metal toxicity and have been used as biomarkers of exposure to metals. Perhaps in attempting to relate cause and effect for declining mussel populations, biomarkers such as these would be useful.

The reported impacts of sublethal metal stress on mollusks strongly suggest that while exposure to metals may not be immediately apparent, lethality may result from eventual disruption of metabolic activities, enzyme functions, respiration, and other important processes. For the long-lived unionid mussels, repeated insults by metal pollution may be partially responsible for their continual decline. A number of mussel LC50s are similar to water quality criteria used to establish effluent concentration limits (e.g., copper and zinc). Most criteria were established based on data that lacked unionid toxicity test results because such data did not exist at the time. So, even though calculations include a built-in uncertainty factor designed to be protective (Stephan et al. 1985), the lack of unionid toxicity data in those calculations may mean that some metal criteria are not adequately protective of freshwater mussels.

ORGANIC CHEMICAL TOXICITY

ACUTE TOXICITY OF ORGANIC CONTAMINANTS

The published literature describing the impacts of acute exposure of mussels to various organic compounds is more limited than for metals. Most of the available information describes responses to pesticides; these compounds are often found in aquatic systems as an indirect result of runoff or atmospheric deposition, although spills—both intentional and unintentional—also occur (Mulla and Mian 1981). Some documents have reviewed the toxic effects of contaminants; an excellent compendium of toxicity data is included in the work of Havlik and Marking (1987) and will not be duplicated here. Acute toxicity data are vital to develop adequately protective restrictions on pesticide use in areas where they may detrimentally affect sensitive or endangered species of unionid mussels and other mollusks (Keller 1993) and to assess the risk posed by chemical spills. Most toxicity tests have found freshwater mollusks to be less sensitive to pesticides, herbicides, and other organic compounds than are the target organisms or other taxa.

Toxicity tests using glochidia have been reported for only a few organic compounds using permanent valve closure or inability to respond to stimuli as the measure of lethality. In all of the tests except one, glochidia were found to be very insensitive to tested chemicals, including atrazine, cyhalothrin, carbaryl, malathion, and several pesticides used in eastern Europe (Varanka 1979; McLeese et al. 1980; Varanka 1987; Johnson, Keller, and Zam 1993; Keller and Ruessler 1997) (Table 7.2). In contrast, Conners and Black (2004) determined that *U. imbecillis* glochidia were as sensitive or more sensitive to glyphosate and carbaryl than other aquatic invertebrates. Weinstein and Polk (2001) reported that photo-activated anthracene and pyrene were toxic to *U. imbecillis* glochidia at environmentally relevant concentrations. Anthracene was more toxic with a 24-h LC50 of 1.93 µg/L followed by pyrene at 2.63 µg/L.

Varanka (1977) investigated the effect of several pesticides on tryptamine-induced activity of the adductor muscle of *A. cygnea* glochidia but found that most effects occurred at concentrations far exceeding environmental concentrations. However, malathion caused decreased adductor activity at a concentration of 75 µg/L, which is a realistic environmental concentration. In 24-h toxicity tests, Conners and Black (2004) found that *U. imbecillis* glochidia were sensitive to copper, atrazine, glyphosate, and carbaryl, as measured by death and genotoxicity.

TABLE 7.2
Summary of Selected Toxicity Data for Organic Compounds to Unionid Mussels

Reference	Species	Glochidia	Juvenile	Adult	Exposure (h)	Test (Temp.)	TFM[a]	Chemical LC50s												
								Malathion (µg/L)	Pyrene (mg/L)	Anthracene (mg/L)	PCP (µg/L)	Toxaphene (µg/L)	Chlordane (µg/L)	Aquathol (µg/L)	Hydrothol (µg/L)	Baythroid (Mg/L)	2,4-D (Mg/L)	Atrazine (µg/L)	Cyhalothrin (mg/L)	Carbaryl (mg/L)
							—	0.1[b]	—	—	19[c]	0.73[c]	2.4[c]	—	—	—	—	1.5[d]	—	—
Johnson, Keller, and Zam (1993)	U. imbecillis	x			48	23												>60	>1	23.7
Waller et al. (1993)	Obliquaria reflexa			x	48	17	1.87									10,000				
Chandler and Marking (1975)	Elliptio sp./Plectomerus sp.			x	96	—	2–9													
Keller (1993)	U. imbecillis	x			48	23					610	740	880		4,850					
Keller and Ruessler (1997)	U. imbecillis	x			24	25		366												
	Villosa lienosa	x			24	25		54												
	Lampsilis teres	x			24	25		28												
	L. siliquoidea	x			24	25		8–54												
	Megalonaias nervosa	x			4	25		22												
	U. imbecillis		x		96	25		215												
	Elliptio icterina		x		96	25		32												
	Loxoconcha claibornensis		x		96	25		24												
	L. subangulata		x		96	25		28												
	V. lienosa		x		96	25		111												
	Viaa villosa		x		96	25		142												
Wade, Hudson, and McKinney (1989)	U. imbecillis	x			48	24								4,600			4,600			
Weinstein and Polk (2001)	U. imbecillis	x			24	25			2.63	1.93										

a TFM = 3-trifluoromethyl-4-nitrophenol.
b USEPA 2002, chronic criterion recommendation.
c USEPA 2002, acute criterion recommendation.
d 2003, acute criterion recommendation.

Neither 2,4-D, the mosquito larvicide BTI, nor the herbicide aquathol-K was toxic to juvenile *U. imbecillis* after 9-day exposures at concentrations up to twice the accepted application rate (Wade, Hudson, and McKinney 1989). Keller (1993) determined that seven of eight organic compounds (including hydrothol, toxaphene, and pentachlorophenol) were less toxic to *U. imbecillis* at 48 h than to zooplankton or fish. PCP was equally toxic to *U. imbecillis*, zooplankton, and fish (Keller 1993).

Moderate sensitivity to the lampricide TFM (3-trifluoromethyl-4-nitrophenol) was measured in adult *Elliptio* spp. and *Plectomerus* spp. (Chandler and Marking 1975), but neither of these mollusks was as sensitive as lamprey larvae or other aquatic taxa. Similar results were reported by Waller et al. (1993) and Waller, Bills, and Johnson (1997). Juvenile and adult *E. complanata* and *Anodonta cataracta* and adult *O. reflexa* were not impacted by TFM at suggested application rates. In fact, Waller et al. (1993) determined that mussels were among the least sensitive taxa to 15 organic chemicals being considered as zebra mussel control agents. Similar results were noted in tests performed by Chandler and Marking (1979) in tests with 20 fishery chemicals and Keller and Ruessler (1997) in tests with malathion and various juvenile unionids.

Warren (1992) saw no significant difference in survival between adult *E. buckleyi* controls and animals exposed to the herbicide glyphosate (Sonar) at recommended dosages in field exposures monitored for six months, and after exposure to Sonar in the laboratory for seven days at concentrations up to 100 times the recommended application rate.

Sublethal Effects of Organic Contaminants

Sublethal responses of adult mussels following exposure to pesticides and other organics include decreased enzyme activity, abnormal shell growth, changes in metabolism, heart rate, and siphoning activity, and others. Relatively fewer studies have evaluated sublethal impacts of pesticides and other organic contaminants than have been reported for metals.

Machado et al. (1990) reported abnormal shell growth in *A. cygnea* exposed to the insecticide diflubenzuron, designed to retard juvenile metamorphosis. The effective concentration was 200 mg/L, far higher than the expected environmental concentration. This could increase the mussels' vulnerability to predation or shell erosion, though even this response was elicited at a concentration much higher than was effective in crustaceans, closer relatives to insects.

Analogous changes in other tissues have been reported for mussels exposed to various organic contaminants. Mane, Akarte, and Kulkarni (1986) recorded biochemical changes in mussels exposed to fenthion, an organophosphate pesticide, including the altered distribution of protein, cholesterol, and particularly glycogen and lipids, in the mantle tissue, gills, hepatopancreas, gonad, foot, and adductor muscles of *Indonaia caeruleus*. A general decrease in glycogen content, the main energy reserve in mussels, was also observed for *I. caeruleus* (Mane, Akarte, and Kulkarni 1986; Makela, Lindstrom-Seppa, and Oikari 1992). Toxicity tests and assessments of acetylcholinesterase inhibition in *E. complanata* following exposure to aldicarb and acephate (Moulton, Fleming, and Purnell 1996) indicated that activity was inhibited after pesticide exposure and was affected by test temperature but recovered after 12 days. Considerable variability in enzyme activity was reported in control animals and may have masked the impact of the pesticides on the shell closing response that is mediated by the activity of the enzyme in the adductor muscle. However, the researchers recommended further evaluation of the assay as a measure of field exposure of mussels to agricultural chemicals.

Rao, Rao, and Rao (1983) measured the effects of malathion (40 mg/L) and methyl parathion (10 mg/L) on *L. marginalis* heart rates. Both pesticides caused a decrease in heart rate, but these responses were elicited only at concentrations higher than expected environmental concentrations (EEC). Similar inhibition was noted by Senthilmurugan et al. (1994) in the same species when exposed to phosphamidon at 0.015 µg/L, possibly leading to the disruption of metabolic processes and growth.

Adult *A. cygnea*, *A. anatine*, and *Unio pictorum* mussels displayed altered lengths of resting and active siphoning (Varanka 1987) to the extent that inadequate nutrition and slower growth might result when exposed to the mosquito insecticides (Fyfanon–malathion, K-Othrin, Unitox 7–7% dichlorvos-and Unitox 20–20% dichlorvos). These results were generated at exposure levels three-to-five orders of magnitude lower than the acute LC50, and the author maintains that these results bolster the contention that unionids, while not succumbing immediately, are sensitive to pesticides.

Another sublethal response was reported after exposure to organochlorine pesticides—the production of glutathione S-transferase (GST), a molecule that degrades chlorinated pesticides. GST activity was significantly reduced in *A. anatina* following exposure to a pulp mill effluent (Makela, Lindstrom-Seppa, and Oikari 1992).

Growth and reproduction, processes related to many of the physiological phenomena described above, are also potentially affected by exposure to organic contaminants. Makela, Lindstrom-Seppa, and Oikari (1992) found that pulp mill effluent had a positive effect on growth, contrary to expectations, and attributed these results to increased food availability due to the eutrophication of the stream caused by nutrients in the effluent. In the same study, Makela, Lindstrom-Seppa, and Oikari (1992) noted that reproductive failures had occurred in marine species exposed to organic contamination but that pulp mill effluents had no adverse effects on female unionid gill index—a measure of gill width versus length used to describe reproductive output.

In a study conducted on *I. caeruleus*, it was discovered that DDT, endrin, and thiometon exposure resulted in complete loss of lumen from digestive tubules in the hepatopancreas (Mane, Kachole, and Pawar 1979). In addition, the ciliated cells and connective tissue of the intestine became highly irregular after exposure, indicating a potential for impairment of the digestive system.

Most investigations into the impact of pesticides, herbicides, and other organic compounds on mollusks have concluded that effects occur only at concentrations far above EECs based on exposures of adults. For example, ciliary activity, an index of feeding rate, initially increased following the exposure of *L. marginalis* to 1 mg/L malathion, and then activity was inhibited. At an exposure of 8 mg/L malathion, inhibition of almost 80% was recorded (Ahamad et al. 1979). *L. marginalis* also had reduced ciliary activity after exposure to methyl parathion, but doses of at least 20 mg/L were required to elicit reductions in ciliary activity (Basha, Swami, and Pushpanjali 1988). These concentrations are two-to three orders of magnitude higher than the EEC.

Since the impacts of contaminants on freshwater mussels occur at various cellular and bio-chemical levels, the use of a lethal endpoint as the sole indicator of toxicity may be inadequate. Die-offs of mussel populations can occur as a result of exposures exceeding acutely toxic thresholds, as in the case of the organophosphate pesticide poisoning documented by Fleming, Augspurger, and Alderman (1995). However, in many cases large-scale mortalities remain unexplained. Protection from acutely lethal exposures to contaminants is an important step in conserving mollusks, but it does not guarantee protection. Although the development of acute water quality criteria recommendations based on the use of zooplankton or other surrogate organisms may protect mollusks from lethal exposures, the evaluation of sublethal impacts will require more direct research with mollusks themselves (Keller 1993).

The data reviewed in this section suggest that sublethal effects could have serious consequences for mollusk populations. Structural changes in shells and tissues could affect the ability of some species to deal with additional stresses or their general ability to function normally. Interferences with developmental processes, such as the attachment of glochidia to their hosts, could seriously endanger the sustainability of the populations. The effects of contaminants on feeding activity and digestive tissues could also limit the abilities of populations to survive in the face of other stressors such as habitat alteration.

OTHER POLLUTANTS

Even though concentrations and loads of contaminants emanating from wastewater treatment facilities, e.g., ammonia, chlorine, heat, and nutrients, have been greatly reduced by modern wastewater treatment, unionid populations are still impacted by them. Mouthon (1996a) evaluated the relationship between 11 water quality parameters and both mollusk species diversity and abundance at 96 river sites. He concluded that dissolved oxygen concentrations of less than 7 mg/L, ammonia concentrations greater than 0.6 mg/L, nitrates at greater than 20 mg/L, or phosphates at concentrations greater than 0.3 mg/L resulted in lower species diversity and abundance of mollusks. Of the 48 species of gastropod and bivalve mollusks he evaluated, unionids were the first group to be eliminated by pollution, while physid snails were the most tolerant taxa (Mouthon 1996b).

Ammonia and chlorine are common constituents of effluents emanating from many wastewater treatment plants. Though criteria may be protective of the species represented in early toxicity databases, toxicity tests conducted by Horne and McIntosh (1979), Wade (1992), Goudreau, Neves, and Sheehan (1993), Hickey and Vickers (1994), Myers-Kinzie (1998), and Keller, Ruessler, and Kernaghan (1998) indicate that unionids are more sensitive to unionized ammonia than are most aquatic taxa. Augspurger et al. (2003) analyzed ammonia toxicity data for glochidia and juvenile unionids. This included 10 species in 32 tests. They proposed that an appropriate water quality criterion would be between 0.3 and 1.0 mg/L total ammonia as N at pH 8, compared to the USEPA continuous criteria concentration recommendation of 1.24 mg/L total ammonia as N at pH 8 and 25°C (USEPA 1999b). The only reported chlorine-related toxicity data were produced by Goudreau, Neves, and Sheehan (1993) with *V. iris* glochidia. She calculated an LC50 of 0.084 mg/L for the species.

Other water quality criteria such as pH and oxygen appear to be protective of most unionids. Dimock and Wright (1993) calculated a pH LC50 of 4.5 for juvenile *U. imbecillis* and *A. cataracta*. Oxygen at a concentration of 0.1 mg/L resulted in total mortality of juvenile *U. imbecillis* mussels after 24 h (Dimock and Wright 1993). In contrast, *Amblema plicata* survived for up to 10 weeks at 0 mg/L oxygen but may need 6 mg/L dissolved oxygen for normal growth (Imlay 1971). The typical criterion for oxygen in warmwater systems is currently 5.0 mg/L.

Finally, data indicate that unionids are sensitive to potassium concentrations but not to ambient concentrations of fluoride that originate from drinking water passing through wastewater treatment plants. Based on toxicity test results, Imlay (1973) suggested that potassium concentrations might explain lack of abundance or diversity of unionid mussels in certain geographical areas. He predicted that rivers having potassium concentrations above 7 mg/L (LC_{90} at 36–52 days) would not have mussels, while rivers with potassium levels of less than 4 mg/L would. Imlay found a strong correlation between ambient potassium concentrations and mussel distributions in the Upper Midwest. Keller and Ruessler (unpublished) measured a 96-h LC50 of 24 mg/L K^+ for juvenile *Villosa vibex*. Keller and Augspurger (2005) tested several unionids and reported that fluoride at concentrations typical of North Carolina streams was not toxic to glochidia or juveniles. Growth effects occurred at concentrations as low as one-tenth of the 96-h LC50s, and one-fifth of the 9-day LC50s, based on static renewal tests. The effective concentrations were still 17 times as high as the North Carolina water quality criterion of 1.8 mg/L. However, because growth impacts were measurable after such a short time in this study and mussels live for decades, it would be worthwhile to perform more lengthy exposures at lower concentrations. Smaller adults are known to suffer greater predation effects and reduced reproductive success compared to those of normal size, and this could have a negative impact on long term population maintenance.

SUMMARY

Long-lived, sedentary, filter-feeding animals such as unionids experience repeated exposures to chemicals that can decrease their ability to withstand additional stressors. We currently have mostly acute toxicity data for unionids, and more for metals than for organic compound. But, even from these data it is appears that the decline in unionid abundance and species diversity in many American rivers may be exacerbated by heavy metal, pesticide, ammonia or other chemical pollution. For example, criteria recommendations for some metals, ammonia and PCP are similar to or higher than the acute LC50 for some unionids. The availability of more chronic toxicity test results would clarify the extent to which ambient concentrations of anthropogenic pollutants are causing the decline in mussel species throughout the world, since most available data are from acute exposures, and unionids can live for 50 or more years. Of particular interest are the effects of chronic exposure to low-metal concentrations, chemical mixtures, endocrine disruptors, tributyltin, and many herbicides. More field studies are needed to determine how habitat alterations, siltation, and toxic chemicals interact and stress mussels. Until such information is available, we cannot adequately evaluate the protectiveness of current water quality criteria recommendations, relative to this imperiled fauna.

REFERENCES

Ahamad, I. K., Sethuraman, M., Begum, R., and Ramana Rao, K. V., Effect of malathion on ciliary activity of fresh water mussel, *Lamellidens marginalis* (Lamarck), *Comp. Physiol. Ecol.*, 4, 71–73, 1979.

Augspurger, T., Keller, A. E., Black, M. C., Cope, W. G., and Dwyer, F. J., Water quality guidance for protection of freshwater mussels (Unionidae) from ammonia exposure, *Environ. Toxicol. Chem.*, 22, 2569–2575, 2003.

Basha, S. M., Swami, K. S., and Pushpanjali, A., Ciliary and cardiac activity of the freshwater mussel, *Lamellidens marginalis* (Lamarck) as an index evaluating organophosphate toxicity, *J. Environ. Biol.*, 9, 313–318, 1988.

Chamberland, G., Belanger, D., Lariviere, N., Vermette, L., Klaverkamp, J. F., and Blais, J. S., Abnormal porphyrin profile in mussels exposed to low concentrations of cadmium in an experimental Precambrian Shield lake, *Can. J. Fish. Aquat. Sci.*, 52, 1286–1293, 1995.

Chandler, J. H. and Marking, L. L., Toxicity of the lampricide 3-trifluoromethyl-4-nitrophenol (TFM) t selected aquatic invertebrates and frog larvae, *Invest. Fish. Contr.*, 62, 1–7, 1975.

Chandler, J. H. and Marking, L. L., Toxicity of fishery chemicals to the Asiatic clam, *Corbicula manilensis. Progr. Fish-Cult.*, 41, 148–151, 1979.

Cherry, D. S. and Farris, J. L., *Site-Specific Copper Criteria for the Protection of Aquatic Life of the Clinch River at Clinch River Plant, Carbo, Virginia*, Virginia Polytechnic Institute State University, Blacksburg, VA, p. 41, 1991.

Cherry, D. S., Farris, J. L., and Neves, R. J., Laboratory and field ecotoxicological studies at the clinch River Plant, Virginia Final Report to the American Electric Power Service Corporation, Virginia Polytechnic Institute State University, Columbus, OH. Blacksburg, VA, p. 228, 1991.

Conners, D. E. and Black, M. C., Evaluation of lethality and genotoxicity in the freshwater mussel *Utterbackia imbecillis* (Bivalvia: Unionidae) exposed singly and in combination to chemicals used in lawn care, *Archiv. Environ. Contam. Toxicol.*, 46, 362–371, 2004.

Couillard, Y., Campbell, P. G. C., and Tessier, A., Response of metallothionein concentrations in a freshwater bivalve (*Anodonta grandis*) along an environmental cadmium gradient, *Limnol. Oceanogr.*, 38(2), 299–313, 1993.

Couillard, Y., Campbell, P. G. C., Tessier, A., Pellerin-Massicotte, J., and Auclair, J. C., Field transplantation of a freshwater bivalve, *Pyganodon grandis*, across a metal contamination gradient. II. Metallothionein response to Cd and Zn exposure, evidence for cytotoxicity, and links to effects at higher levels of biological organization, *Can. J. Fish. Aquat. Sci.*, 52, 703–715, 1995.

Dimock, R. V. and Wright, A. H., Sensitivity of juvenile freshwater mussels to hypoxic, thermal and acid stress, *J. Elisha Mitchell Sci. Soc.*, 109, 183–192, 1993.

Farris, J. L., Van Hassel, J. H., and Cherry, D. S., Refinement of site-specific copper criteria for the protection of mussel populations [abstract], SETAC, 12th annual meeting, Seattle, WA, pp. 3–7, November 1991.

Fleming, W. J., Augspurger, T. P., and Alderman, J. A., Freshwater mussel die-off attributed to anticholinesterase poisoning, *Environ. Toxicol. Chem.*, 14, 877–879, 1995.

Gardner, W. W., Miller, W. H., and Imlay, M. J., Free amino acids in mantle tissues of the bivalve *Amblema plicata*: Possible relation to environmental stress, *Bull. Environ. Contam. Toxicol.*, 26, 157–162, 1981.

Goudreau, S. E., Neves, R. J., and Sheehan, R. J., Effects of wastewater treatment plant effluents on freshwater mollusks in the upper Clinch River, Virginia, *Hydrobiologia*, 252, 211–230, 1993.

Granmo, A. and Varanka, I., Effects of heavy metals and a surfactant on the activity of fresh-water mussel larvae, *Vatten*, 35, 283–290, 1979.

Hameed, P. S. and Raj, A. I. M., Effect of copper, cadmium and mercury on crystalline style of the freshwater mussel *Lamellidens marginalis* (Lamarck), *Indian J. Environ. Health*, 31, 131–136, 1989.

Hansten, C., Heino, M., and Pynnonen, K., Viability of glochidia of *Anodonta anatina* (Unionidae) exposed to selected metals and chelating agents, *Aquat. Toxicol.*, 34, 1–12, 1996.

Havlik, M. E. and Marking, L. L., *Effects of Contaminants on Naiad Mollusks (Unionidae): A Review*, U.S. Department of the Interior, Fish and Wildlife Service Resource Publications 164, Washington, DC, p. 20, 1987.

Hemelraad, J., Holwerda, D. A., and Zandee, D. I., Cadmium kinetics in freshwater clams (I). The pattern of cadmium accumulation in *Anodonta cygnea*, *Arch. Environ. Contam. Toxicol.*, 15, 1–7, 1986a.

Hemelraad, J., Holwerda, D. A., Teerds, K. J., Herwig, H. J., and Zandee, D. I., Cadmium kinetics in freshwater clams (II). A comparative study of cadmium uptake and cellular distribution in the unionidae *Anodonta cygnea*, *Anodonta anatina*, and *Unio pictorum*, *Arch. Environ. Contam. Toxicol.*, 15, 9–21, 1986b.

Hemelraad, J., Holwerda, D. A., Herwig, H. J., and Zandee, D. I., Effects of cadmium in freshwater clams. III. Interaction with energy metabolism in *Anodonta cygnea*, *Arch. Environ. Contam. Toxicol.*, 19, 699–703, 1990.

Hemelraad, J., Holwerda, D. A., Wijnne, H. J. A., and Zandee, D. I., Effects of cadmium in freshwater clams. I. Interaction with essential elements in *Anodonta cygnea*, *Arch. Environ. Contam. Toxicol.*, 19, 686–690, 1990.

Hickey, C. W. and Vickers, M. L., Toxicity of ammonia to nine native New Zealand freshwater invertebrate species, *Arch. Environ. Contam. Toxicol.*, 26, 292–298, 1994.

Hinch, S. G. and Green, R. H., The effects of source and destination on growth and metal uptake in freshwater clams reciprocally transplanted among south central Ontario lakes, *Can. J. Zool.*, 67, 855–863, 1989.

Holwerda, D. A. and Herwig, H. J., Accumulation and metabolic effects of di-n-butyltin dichloride in the freshwater clam, *Anodonta anatina*, *Bull. Environ. Contam. Toxicol.*, 36, 756–762, 1986.

Horne, F. R. and McIntosh, S., Factors influencing distribution of mussels in the Blanco River of central Texas, *Nautilus*, 94, 119–133, 1979.

Huebner, J. D. and Pynnonen, K. S., Viability of glochidia of two species of *Anodonta* exposed to low pH and selected metals, *Can. J. Zool.*, 70, 2348–2355, 1992.

Imlay, M. J., Bioassay tests with naiades, In *Proceeding of the Symposium on Rare and Endangered Mollusks (Naiads)*, Jorgensen, S. E. and Sharp, R. W., Eds., U.S. Fish and Wildlife Service, Bureau of Sport Fisheries and Wildlife, Twin Cities, MN, 1971.

Imlay, M. J., Effects of potassium on survival and distribution of freshwater mussels, *Malacologia*, 12, 97–111, 1973.

Isom, B. G. and Hudson, R. G., In vitro culture of parasitic mussel glochidia, *Nautilus*, 96, 147–151, 1982.

Jacobson, P. J., Sensitivity of early life stages of freshwater mussels (Bivalvia: Unionidae) to copper MSc Thesis, Virginia Polytechnic Institute and State University, Blacksburg, VA, 1990.

Jacobson, P. J., Farris, J. L., Cherry, D. S., and Neves, R. J., Juvenile freshwater mussel (Bivalvia: Unionidae) responses to acute toxicity testing with copper, *Environ. Toxicol. Chem.*, 12, 879–883, 1993.

Jacobson, P. J., Neves, R. J., Cherry, D. S., and Farris, J. L., Sensitivity of glochidial stages of freshwater mussels (Bivalvia: Unionidae) to copper, *Environ. Toxicol. Chem.*, 16, 2384–2392, 1997.

Jenner, H. A., Hemelraad, J., Marquenie, J. M., and Noppert, F., Cadmium kinetics in freshwater clams (Unionidae) under field and laboratory conditions, *Sci. Tot. Environ.*, 108, 205–214, 1991.

Johnson, W. W. and Finley, M. T., *Handbook of Acute Toxicity of Chemicals to Fish and Aquatic Invertebrates*, Resource Publication 137, p. 98, 1980.

Johnson, I. C., Keller, A. E., and Zam, S. G., A method for conducting acute toxicity tests with the early life stages of freshwater mussels, In *Environmental Toxicology and Risk Assessment*, Landis, W. G., Hughes, J. S., and Lewis, M. A., Eds., American Society for Testing and Materials, Philadelphia, PA, pp. 381–396, 1993. ASTM STP 1179.

Keller, A. E., Acute toxicity of several pesticides, organic compounds and a wastewater effluent to the freshwater mussel, *Utterbackia imbecillis, Ceriodaphnia dubia* and *Pimephales promelas, Bull. Environ. Contam. Toxicol.*, 51, 696–702, 1993.

Keller, A. E. and Augspurger, T., Toxicity of fluoride to the endangered unionid mussel, *Alasmidonta raveneliana*, and surrogate species, *Bull. Environ. Contam. Toxicol.*, 74, 242–249, 2005.

Keller, A. E. and Zam, S. G., Simplification of *in vitro* culture techniques for freshwater mussels, *Environ. Toxicol. Chem.*, 9, 1291–1296, 1990.

Keller, A. E. and Zam, S. G., The acute toxicity of selected metals to the freshwater mussel, *Anodonta (Utterbackia) imbecillis, Environ. Toxicol. Chem.*, 10, 539–546, 1991.

Keller, A. E. and Ruessler, D. S., The toxicity of malathion to unionid mussels: Relationship to expected environmental concentrations, *Environ. Toxicol. Chem.*, 16, 1028–1033, 1997.

Keller, A. E., Ruessler, S. D., and Kernaghan, N. A., Comparative sensitivity of three species of unionid mussels to a suite of toxicants: Glochidia and juveniles [abstract], SETAC, 19th Annual Meeting, Charlotte, NC, pp. 15–19, November 1998.

Klaine, S. J., Warren, L. W., and Summers, J. M., The use of juvenile mussels (*Utterbackia imbecillis*, Say) as a standardized testing organism, A report to the USEPA, Athens, GA, Clemson University, Pendleton, SC, 1997.

Lasee, B. A., Histological and ultrastructural studies of larval and juvenile *Lampsilis* (Bivalvia) from the Upper Mississippi River PhD Diss., Iowa State University, Ames, IA, 1991.

Machado, J., Coimbra, J., Castilho, F., and Sa, C., Effects of diflubenzuron on shell formation of the freshwater clam, *Anodonta cygnea, Arch. Environ. Contam. Toxicol.*, 19, 35–39, 1990.

Makela, T. P., Lindstrom-Seppa, P., and Oikari, A. O. J., Organochlorine residues and physiological condition of the freshwater mussel *Anodonta anatina* caged in River Pielinen, eastern Finland, receiving pulp and mill effluent, *Aqua Fennica*, 22, 49–58, 1992.

Malley, D. F., Huebner, J. D., and Donkersloot, K., Effects on ionic composition of blood and tissues of *Anodonta grandis grandis* (Bivalvia) of an addition of aluminum and acid to a lake, *Arch. Environ. Contam. Toxicol.*, 17, 479–491, 1988.

Malley, D. F., Klaverkamp, J. F., Brown, S. B., and Chang, P. S. S., Increase in metallothionein in freshwater mussels *Anodonta grandis grandis* exposed to cadmium in the laboratory and the field, *Water Pollut. Res. J. Can.*, 28(1), 253–273, 1993.

Mane, U. H., Kachole, M. S., and Pawar, S. S., Effect of pesticides and narcotants on bivalve mollusks, *Malacologia*, 18, 347–360, 1979.

Mane, U. H., Akarte, S. R., and Kulkarni, D. A., Acute toxicity of fenthion to freshwater lamellibranch mollusk, *Indonaia caeruleus* (Prashad 1918) from Godavari River at Paithan—A biochemical approach, *Bull. Environ. Contam. Toxicol.*, 37, 622–628, 1986.

Masnado, R. G., Geis, S. W., and Sonzongi, W. C., Comparative acute toxicity of a synthetic mine effluent to *Ceriodaphnia dubia*, larval fathead minnows and the freshwater mussel *Utterbackia imbecillis, Environ. Toxicol. Chem.*, 14, 1913–1920, 1995.

Mayer, F. L. and Ellersieck, M. R., *Manual of Acute Toxicity: Interpretation of Database for 410 Chemicals and 66 Species of Freshwater Animals*, USFWS Resource Publication 166, Washington, DC, p. 506, 1986.

McCann, M. T., Toxicity of zinc, copper, and sediments to early life stages of freshwater mussels in the Powell River, Virginia MSc Thesis, Virginia Polytechnic Institute and State University, Blacksburg, VA, 1993.

McLeese, D. W., Zitko, V., Metcalfe, C. D., and Sergeant, D. B., Lethality of aminocarb and the components of the aminocarb formulation to juvenile Atlantic salmon, marine invertebrates and a freshwater clam, *Chemosphere*, 9, 79–82, 1980.

Metcalfe-Smith, J. L. and Green, R. H., Aging studies on three species of freshwater mussels from a metal-polluted watershed in Nova Scotia, Canada, *Can. J. Zool.*, 70, 1284–1291, 1992.

Moulton, C. A., Fleming, W. J., and Purnell, C. E., Effects of two cholinesterase-inhibiting pesticides on freshwater mussels, *Environ. Toxicol. Chem.*, 15, 131–137, 1996.

Mouthon, J., Molluscs and biodegradable pollution in rivers: Proposal for a scale of sensitivity of species, *Hydrobiologia*, 317, 221–229, 1996a.

Mouthon, J., Molluscs and biodegradable pollution in rivers: Studies into the limiting values of 11 physico–chemical variables, *Hydrobiologia*, 319, 57–63, 1996b.

Mulla, M. S. and Mian, L. S., Biological and environmental impacts of the insecticides malathion and parathion on non-target biota in aquatic systems, *Residue Rev.*, 78, 101–135, 1981.

Myers-Kinzie, M., Factors affecting survival and recruitment of Unionid mussels in small midwestern streams PhD Diss., Purdue University, West Lafayette, IN, 1998.

Naimo, T. J., Waller, D. L., and Holland-Bartels, L. E., Heavy metals in the threeridge mussel *Amblema plicata plicata* (Say 1817) in the upper Mississippi River, *J. Freshwater Ecol.*, 2, 209–217, 1992a.

Naimo, T. J., Atchison, G. J., and Holland-Bartels, L. E., Sublethal effects of cadmium on physiological responses in the pocketbook mussel, *Lampsilis ventricosa*, *Environ. Toxicol. Chem.*, 11, 1013–1021, 1992b.

[NAS/NAE] National Academy of Sciences and National Academy of Engineering, *Water Quality Criteria 1972*, U.S. Government Printing Office, Washington, DC, p. 592, 1973.

Pirovarova, N. B., Lagerspetz, K. Y. H., and Skulskii, I. A., Effect of cadmium on ciliary and ATPase activity in the gills of freshwater mussel *Anodonta cygnea*, *Comp. Biochem. Physiol.*, 103C, 27–30, 1992.

Pynnonen, K., Influence of aluminum and H+ on the electrolyte homeostasis in the unionidae *Anodonta anatina* L. and *Unio pictorum* L., *Arch. Environ. Contam. Toxicol.*, 20, 218–225, 1991.

Radhakrishnaiah, K., Effect of cadmium on the freshwater mussel, *Lamellidens marginalis* (Lamarck)—A physiological approach, *J. Environ. Biol.*, 9, 73–78, 1988.

Raj, A. I. M. and Hameed, P. S., Effect of copper, cadmium and mercury on metabolism of the freshwater mussel *Lamellidens marginalis* (Lamarck), *J. Environ. Biol.*, 12, 131–135, 1991.

Rao, K. V. R., Rao, K. R. S. S., and Rao, K. S. P., Cardiac responses to malathion and methyl parathion in the mussel, *Lamellidens marginalis* (Lamarck), *J. Environ. Biol.*, 4, 65–68, 1983.

Reddy, T. R. and Chari, N., Effect of sublethal concentrations of mercury and copper on AIAT, AAT and GDH of the freshwater mussel, *Parreysia rugosa* (G), *J. Environ. Biol.*, 6, 67–70, 1985.

Richard, P. E., Dietz, T. H., and Silverman, H., Structure of the gill during reproduction in the unionids *Anodonta grandis*, *Ligumia subrostrata*, and *Caranculina parva texasensis*, *Can. J. Zool.*, 69, 1744–1754, 1991.

Schweinforth, R. L. and Wade, D. C., Effects from subchronic 90-day exposure of *in vitro*-transformed juvenile freshwater mussels (*Anodonta imbecillis*) to manganese [abstract], SETAC, 11th annual meeting, 1990.

Senthilmurugan, S., Rajasekarapandian, M., Amsath, A., and Sayeenathan, R., Effect of phosphamidon on body weight, oxygen consumption and heart beat in the freshwater mussel, *Lamellidens marginalis*, *J. Ecobiol.*, 6, 5–8, 1994.

Silverman, H., McNeil, J. W., and Dietz, T. H., Interaction of trace metals Zn, Cd, and Mn, with Ca concretions in the gills of freshwater unionid mussels, *Can. J. Zool.*, 65, 828–832, 1987.

Sivaramakrishna, B., Radhakrishnaiah, K., and Suresh, A., Assessment of mercury toxicity by the changes in oxygen consumption and ion levels in the freshwater snail *Pila globosa*, and the mussel, *Lamellidens marginalis*, *Bull. Environ. Contam. Toxicol.*, 46, 913–920, 1991.

Stephan, C. E., Mount, D. I., Hansen, D. J., Gentile, J. H., Chapman, G. A., and Brungs, W. A., *Guidelines for Deriving Numerical National Water Quality Criteria for the Protection of Aquatic Organisms and Their Uses*, U.S. Environmental Protection Agency, Duluth, MN, 1985. PB85-227049.

U.S. Environmental Protection Agency, *Quality Criteria for Water*, Office of Water, Washington, DC, 1976, PB-263.

U.S. Environmental Protection Agency, *National Water Quality Criteria Correction*, Office of Water, Washington, DC, April 1999a. EPA-822-Z-99-001.

U.S. Environmental Protection Agency, *Update of Ambient Water Quality Criteria for Ammonia*, Office of Water, Office of Science and Technology, Washington, DC, p. 153, December, 1999b, EPA 822-R-99-014.

U.S. Environmental Protection Agency, *National Recommended Water Quality Criteria*, Office of Water, Office of Science and Technology, Washington, DC, 2002. EPA-822-R-02-047.

U.S. Environmental Protection Agency, *Notice of Revised Draft Ambient Water Quality Criteria Document for Atrazine and Request for Scientific Reviews*, November 2003. EPA-822-F-03-006.

Varanka, I., The effect of some pesticides on the rhythmic activity of adductor muscle of fresh-water mussel larvae, *Acta Biol. Acad. Sci. Hung.*, 28, 317–332, 1977.

Varanka, I., Effect of some pesticides on rhythmic adductor muscle activity of fresh-water mussel larvae, In *Human Impacts on Life in Fresh Waters*, Salanki, J. and Biro, P., Eds., *Symp. Biol. Hung.*, Akademiai Kiado, Budapest, Hungary, pp. 177–196, 1979.

Varanka, I., Effect of mosquito killer insecticides on freshwater mussels, *Comp. Biochem. Physiol.*, 86C, 157–162, 1987.

Wade, D. C., Definitive evaluation of Wheeler Reservoir sediments toxicity using juvenile freshwater mussels (*Anodonta imbecillis* Say), Water Resources Division, Tennessee Valley Authority, Mussel Shoals, AL, p. 25, 1992.

Wade, D. C., Hudson, R. G., and McKinney, A. D., The use of juvenile freshwater mussels as laboratory test species for evaluating environmental toxicity [abstract], SETAC, 10th Annual Meeting, 1989.

Waller, D. L., Bills, T., and Johnson, D., Field evaluation of acute and sublethal effects of TFM to juvenile and adult freshwater mussels [abstract], SETAC, 18th annual meeting, San Francisco, CA, pp. 16–20, November 1997.

Waller, D. L., Rach, J. J., Cope, W. G., Marking, L. L., Fisher, S. W., and Dabrowska, H., Toxicity of candidate molluscicides to zebra mussels (*Dreissena polymorpha*) and selected nontarget organisms, *J. Great Lakes Res.*, 19, 695–702, 1993.

Warren, G., *Effects of Exposure to Glyphosate (Sonar) on* Elliptio buckleyi *Mussels*, Florida Game and Fresh Water Fish Commission, Lake Okeechobee, FL, 1992.

Weinstein, J. E. and Polk, K. D., Phototoxicity of anthracene and pyrene to glochidia of the freshwater mussel *U. Imbecillis*, *Environ. Toxicol. Chem.*, 20, 2021–2028, 2001.

8 Toxicokinetics of Environmental Contaminants in Freshwater Bivalves

Waverly A. Thorsen, W. Gregory Cope, and Damian Shea

INTRODUCTION

Bivalves have been used for decades as sentinel organisms to monitor pollution in the aquatic environment (Foster and Bates 1978; Farrington et al. 1983; Colombo et al. 1995; Peven, Uhler, and Querzoli 1996; Blackmore and Wang 2003). Many different classes of chemicals have been studied in this way including hydrophobic organic contaminants (HOCs), such as polycyclic aromatic hydrocarbons (PAHs), polychlorinated biphenyls (PCBs), and organochlorine (OC) pesticides, as well as inorganic contaminants such as the heavy metals cadmium (Cd), lead (Pb) and mercury (Hg) and the radionuclides plutonium (^{239,240}Pu) and cesium (^{137}Cs). The use of bivalves for biomonitoring of environmental pollution addresses difficulties associated with determining aqueous contaminant concentrations (Farrington et al. 1983). Many HOCs exhibit very low water solubilities (e.g., coronene: 1.4×10^{-4} mg/L, at 25°C), which require large sample sizes for adequate instrumental analysis. Moreover, trace metals require "ultraclean" techniques and are also frequently found in very low concentrations in the aqueous phase, sometimes at levels close to instrument detection limits (i.e., pg/L). Additionally, random water sampling may not capture real trends in pollutant concentrations over an integrated time scale.

In an attempt to overcome these obstacles, native bivalves are frequently collected worldwide, extracted, and analyzed for pollutant tissue burdens to provide preliminary information at sites suspected of contamination or to monitor chemical and waste discharge effluents. However, to effectively understand and correlate the relationship between concentrations of pollutants in the aquatic environment to concentrations in bivalve tissue and potential toxic effects, it is best to have an understanding of the kinetics involved in the uptake, distribution, and elimination of pollutants by/from mussel tissues. Additionally, this information is required to understand and predict concentrations in other environmental compartments, such as predicting aqueous or sediment exposure concentrations from bivalve tissue burdens (Neff and Burns 1996).

Traditionally, marine bivalves such as the blue mussel, *Mytilus edulis*, have been used for environmental monitoring due to concern for pollution in coastal and estuarine areas (Farrington et al. 1983; Salanki and Balogh 1989; Beliaeff et al. 2002). However, more recently (1980s) freshwater bivalves have been increasingly utilized to assess the quality of lakes, rivers, and streams of concern, not only for the protection of human health, but also to better explain recent major declines of many North American freshwater mussel populations (e.g., Keller and Zam 1991; Naimo 1995; Jacobson et al. 1997). Generally, information gleaned from freshwater bivalves has demonstrated similarities to marine bivalves; however, physiologies can vary greatly between species, age, body size, ingestion rate, reproductive state, stress, and location, among other factors (Landrum et al. 1994;

Naimo 1995; Morrison et al. 1996). Therefore, in an attempt to better evaluate pollutant fate and to effectively protect and remediate the natural environment, it would be beneficial to understand the toxicokinetics of both marine and freshwater mussels. The intent of this chapter is to present background information and to assess the toxicokinetic information available for freshwater bivalves (mussels and clams). Where data are limited, information on marine bivalves will be presented and, in some cases, will be presented in tandem with freshwater bivalve information in a comparative context. This chapter is not meant to be an exhaustive review of the literature pertaining to these issues, but rather it is meant to aid researchers, managers, and others, in understanding the bioaccumulation of organic and inorganic contaminants in freshwater bivalves.

UPTAKE AND ELIMINATION

Bivalves are exposed to and take up pollutants in tandem with their primary respiratory and feeding mechanisms; chemicals enter mussels actively and passively as they filter water through their gills for respiration and feeding (dietary exposure), or in the case of inorganic contaminants such as metals, through facilitated diffusion, active transport, or endocytosis (Marigomez et al. 2002). Additionally, some bivalve species are exposed to pollutants through pedal feeding or gut ingestion of sediment (McMahon and Bogan 2001). Therefore, chemical uptake can occur in a direct fashion when mussels draw large quantities of water (up to 11 L/mussel/day for Unionidae, Naimo 1995) into their gills or, in an indirect fashion, when ingestion of sediment occurs and chemicals desorb (passively or through facilitated desorption) from the sediment particles into the bivalve gut and become assimilated. Once chemicals enter the organism, they partition into or associate with tissues. For example, heavy metals will accumulate primarily in muscles and organ (soft) tissues (Plette et al. 1999; Markich, Brown, and Jeffree 2001; Marigomez et al. 2002) and organic pollutants will accumulate in the lipid (Farrington et al. 1983; Di Toro et al. 1991). Generally, uptake is very rapid when the bivalve is first exposed and then levels off, sometimes requiring extensive time periods for an equilibrium state to be reached (Figure 8.1a). A similar trend (Figure 8.1b) is observed for the elimination process, which may be rapid at first and then level off, some compounds never being fully eliminated (i.e., some compounds with half-lives of 20 years).

Uptake and elimination rates for both HOCs and metals can be determined through field and/or laboratory studies. One potential concern in these types of studies is the possibility that the bivalves stop siphoning. Although this is more likely to influence studies of shorter duration, it should be taken into consideration when analyzing the data. A typical uptake/elimination experiment consists of "clean" bivalves (referenced or depurated prior to commencement of the study) exposed to a constant chemical concentration in water, and sampled at increasing time intervals, to determine the chemical concentrations in tissue over time. For example, bivalves can be collected from a relatively uncontaminated field reference site, and deployed at a contaminated field site, or brought back to the laboratory for contaminant exposure. After sufficient exposure time, the organisms are removed and placed in clean water for measurement of the elimination (depuration) rate of the compounds. In the natural environment, elimination of certain chemicals might require extensive time periods. In locations where exposure levels are constant or increasing, bivalves may not eliminate the chemicals. In many instances, bivalves will accumulate contaminants to levels significantly higher than those in the water column. This can pose toxicity risks to the mussel and predatory animals or can result in biomagnification and subsequent increases in contaminant concentrations progressively up the food web.

BIOCONCENTRATION

The accumulation of contaminants from the water column by bivalves is referred to as "bioconcentration." Bioconcentration is defined as the partitioning of a contaminant from an aqueous phase into an organism and will occur when the contaminant uptake rate is greater than that for

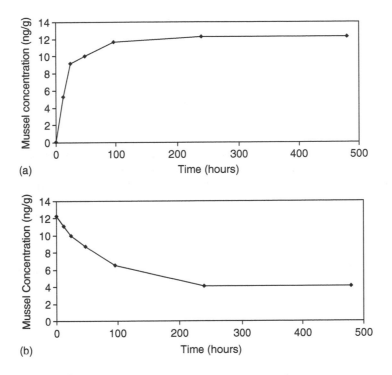

FIGURE 8.1 Hypothetical uptake (a) and elimination (b) curve in a freshwater mussel. Note in this example, the rapid uptake that initially occurs, followed by a leveling off of the concentration of the contaminant in mussel tissue. The leveling off is considered steady-state and, in this example, is reached following about 100 hours of exposure. The rate of elimination is also rapid and is essentially the reverse of the uptake curve. When placed in clean water, the mussels initially depurate the contaminant rapidly from their tissues and then reach a plateau, where no further elimination occurs on this time scale.

elimination. Typically, this leads to high concentrations of chemicals in bivalve tissues. For HOCs, partitioning generally occurs between the dissolved phase of the water and the organism lipid. The most basic example of partitioning is defined as the octanol–water partition coefficient, or K_{ow}:

$$K_{ow} = [\text{contaminant}]_{octanol}/[\text{contaminant}]_{water}$$

The K_{ow} is a measurement of a chemical's affinity for octanol versus water. In many cases, octanol is used as a surrogate for the organism lipid. A chemical with a lesser K_{ow} value (less than 100) will partition less into the lipid than a chemical with a greater K_{ow} (greater than 1,000). This type of partitioning will occur between the aqueous phase and bivalve lipid until a steady-state condition has been reached (i.e., the concentration in the organism relative to the exposure system is unchanging with time). Once steady-state or equilibrium has been reached, it is generally referred to as "equilibrium partitioning." In a simple system, equilibrium partitioning can be modeled by comparing the affinities (i.e., solubilities and fugacities) of a chemical for bivalve lipid versus water (Figure 8.2). To determine the extent of bioconcentration of a chemical in tissues, a "bioconcentration factor" or BCF can be calculated. The BCF is defined as the pollutant concentration in the bivalvel tissue (C_{tissue}) divided by the dissolved aqueous pollutant concentration (C_{water}) at steady-state:

$$BCF = C_{tissue}/C_{water}$$

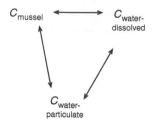

FIGURE 8.2 Diagram of the equilibrium partitioning approach. The hydrophobic organic contaminant partitions between the dissolved phase in the water column, the particulate phase in the water column, and the mussel lipid/tissues. According to Le Chaltelier's principle, when a system at equilibrium is disrupted (e.g., contaminant removed from particulate phase by a mussel), it will shift to re-establish equilibrium (e.g., system responds to change by contaminant from dissolved phase binding to particulate phase). This model assumes all rates are relatively rapid.

The BCF can also be determined by dividing the empirically derived contaminant uptake rate constant (k_1) by the empirically derived elimination rate constant (k_2):

$$BCF = k_1/k_2$$

In general, the BCF is related to the hydrophobic character of the contaminant. In this way, BCF values typically correlate in a linear fashion to K_{ow} values (Geyer et al. 1982; Mackay 1982; Hawker and Connell 1986; Pruell et al. 1986; Schuurmann and Klein 1988; Thorsen 2003) (Figure 8.3).

In many cases, a steady state bioconcentration regression equation can be developed by linearly regressing a log BCF versus a log K_{ow} plot. The resulting equation for the line takes the form of

$$\log BCF = m \log K_{ow} + b$$

where m and b are the slope and y-intercept of the line, respectively. This equation can model the bioconcentration of hydrophobic organic pollutants by bivalves and can be used to predict aqueous exposure concentrations.

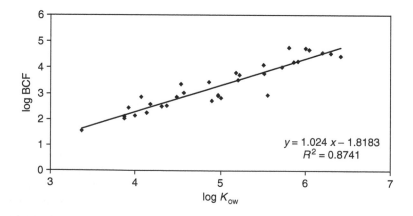

FIGURE 8.3 Example of a linear regression plot of log BCF versus log K_{ow}, based on empirical data (From Thorsen, W. A., Bioavailability of particulate-sorbed polycyclic aromatic hydrocarbons, PhD Thesis, North Carolina State Univ., Raleigh, NC, 2003). Linear regression has been performed and the resultant regression equation takes the form: log BCF = $m$log K_{ow} + b. This regression equation (through simple mathematical procedures) can be used to predict aqueous exposure concentrations based on tissue residues.

The "partitioning" of metals, however, generally refers to the adsorption of metals onto active sites in/on target tissues, such as anionic sites on bivalve gills (Kramer et al. 1997; Marigomez et al. 2002), rather than absorption into a bivalve lipid. A bioconcentration factor, though slightly less utilitarian than for HOCs due to very slow uptake rate constants, can similarly be computed by

$$BCF_{metal} = C_{tissue}/C_{water}$$

where C_{tissue} is the moles of metal per gram of soft weight tissue and C_{water} is the moles of metal dissolved per mL (or L) of water. This BCF value must also be calculated when the system has reached steady-state. More complex equations exist for predicting bioconcentration (and uptake, elimination rates) when a system is not at steady state and are discussed elsewhere (Russell and Gobas 1989; Butte 1991). The bioconcentration of metals is affected by many factors, including water pH, hardness, alkalinity, conductivity, and dissolved organic and inorganic matter, which will be discussed in following sections.

BIOACCUMULATION

While bioconcentration refers only to the uptake of chemicals directly from the water, the term bioaccumulation does not differentiate between uptake media and includes chemical accumulation into organisms from both abiotic (i.e., water and sediment) and biotic (i.e., food) sources. For example, bivalves can bioaccumulate chemicals and metals from the water column and the sediment phase in the natural environment. Typically, scientists may model this relationship by calculating either a bioaccumulation factor (BAF) or a biota-sediment accumulation factor (BSAF). The BAF includes exposure due to water and food sources, whereas the BSAF (only used for HOCs) models the partitioning/association of a chemical between the lipid phases in the organism and the sediment, where the sediment "lipid" phase is considered to be organic carbon. The BAF is represented by

$$BAF = C_{tissue}/C_{food} + C_{water} + C_{other\ exposures}$$

whereas the BSAF is mathematically defined as

$$BSAF = (C_{tissue}/\text{lipid fraction})/(C_{sediment}/\text{organic carbon fraction})$$

where the chemical concentration in the bivalve (C_{tissue}) and sediment ($C_{sediment}$) are normalized to the mass fraction of organism lipid and sediment organic carbon, respectively. Similar to the BCF calculation, a BSAF value is calculated when the chemical has reached a steady-state within the study system. Theoretically, BSAF values will equal unity or one. However, BSAF values may be less than one if the bivalve metabolizes the chemical or the system has not reached steady-state (chemicals may not be fully available to the organism due to very slow desorption or very strong binding). BSAF values can also be greater than one because organic carbon is generally less "lipid-like" than the organism lipid due to hydrophilic components of natural organic matter (Di Toro et al. 1991). The calculation of BSAF values can lend information about a particular chemical's bioavailability (see Bioavailability and Biotic Ligand Models).

Metals do not interact with organisms in the environment in the same way that HOCs do. As previously mentioned, while HOCs generally partition (*ab*sorb) into the lipid phase of a bivalve, metals *ad*sorb to the gill and other anionic sites on tissue surfaces or are actively transported via membrane pumps. For example, metals such as cadmium can enter a bivalve by binding to membrane transport ligands. Bioaccumulation of metals, including filtration of water and ingestion of food particles, in bivalves can be similarly measured through the use of a BAF:

$$BAF = C_{tissue}/C_{water,dissolved}$$

Bioaccumulation factors for metals are more difficult to interpret than for organics because the interactions between a target site (biological organism) and the metal are complicated by competition for binding sites and many more environmental variables than simply dissolved or particulate organic carbon. For all chemicals and metals, bioaccumulation is the balance between all means of chemical uptake and all means of elimination.

METABOLISM AND BIOTRANSFORMATION

For those contaminants that bivalves are capable of metabolizing, BCF, BAF, and BSAF values will be decreased. In general, the lesser metabolic capacities in bivalves makes them adequate sentinels of aquatic environmental pollution (James 1989); however, bivalves have been shown to metabolize certain classes of compounds better than others. For example, mussels possess only minimal abilities to biotransform PAHs, and therefore, are good sentinels of the accumulation of PAHs. Some marine mussels (*M. edulis*), however, have been shown to metabolize the PCB, hexachlorobiphenyl (HCBP) (Bauer, Weigelt, and Ernst 1989), and therefore, will exhibit lower BCF values. Additionally, bivalves have been shown to possess detoxification systems including low molecular weight proteins like metallothionein (MT) and lysosomal granules that make metals complex and chelate, thereby altering the metal uptake/distribution/elimination kinetics (Naimo 1995; Tessier and Blais 1996; Vesk and Byrne 1999; Byrne and Vesk 2000; Baudrimont et al. 2002).

BIOAVAILABILITY AND BIOTIC LIGAND MODELS

Underlying all of the previous concepts is the notion of bioavailability. Bioavailability can be defined as the percentage of a chemical fully available for uptake by an organism. Different chemicals and inorganic contaminants have unique bioavailabilites, which will depend on many factors including water conditions such as hardness, pH, temperature, and turbidity, as well as the physical–chemical characteristics of the compound such as water solubility, vapor pressure, and speciation (ionic state). For example, chemicals that exhibit very low water solubilities readily sorb to organic carbon phases in the water column, such as particulate or dissolved organic carbon (POC, DOC). The rate of desorption and co-occurrence of the mussel with the particle(s) partially determines the chemical's bioavailability. If the rate of desorption is rapid relative to the co-occurrence of the particle and the organism, the chemical may be fully bioavailable. However, if the rate of desorption is very slow, the chemical may not be readily available. HOCs may frequently become associated with natural organic matter in the aqueous and sediment phases, whereas metals may become complexed to various organic (DOC) and inorganic compounds present in the water such as calcium and potassium carbonates ($CaCO_3$, KCO_3).

The bioavailability of a chemical is important to understand both to ensure the protection of aquatic organisms and to implement effective and cost-efficient remediation techniques. This is particularly important because underpredictions of toxicity can result in unacceptable risks to organisms, whereas overpredictions of toxicity can require costly practices for clean-up. For instance, bivalve tissue burdens are traditionally compared directly to total aqueous or sediment-contaminant concentrations, without regard for the bioavailable fraction. This method can overpredict the actual exposure concentrations bivalves (and other aquatic organisms) receive and may result in costly, yet ineffective, remediation of a site. Moreover, sediment concentrations of total metal do not always correlate well with bivalve tissue burdens. Rather, it may be the speciation of the metal (e.g., Hg^{2+} versus CH_3Hg), or ratio of metal concentration to the amount of acid-volatile sulfate in the sediment (Di Toro et al. 1992), that best determines the metal concentration in and subsequent toxicity to the bivalve. One can see the problems that may arise when regulatory and remediation techniques are based on incorrect assessments of chemical bioavailability.

HYDROPHOBIC ORGANIC CONTAMINANTS

UPTAKE

As previously stated, HOCs primarily partition into a bivalve lipid, which is considered essentially an "infinite sink" whereby saturation of the pool does not occur. The uptake of a hydrophobic organic chemical into bivalve tissues can be defined mathematically as

$$dC_{tissue}/dt = k_1 C_{water} - k_2 C_{tissue}$$

where dC_{tissue}/dt is the change in bivalve contaminant concentration over change in time (t), k_1 is the uptake rate constant of the chemical, C_{water} is the aqueous chemical concentration, k_2 is the elimination rate (see Elimination), and C_{tissue} is the concentration of chemical in the bivalve (see Landrum, Lee, and Lydy 1992 for a review of toxicokinetic models). If the concentration of the pollutant in the water column changes, this change will be mirrored in the bivalve over several days to weeks. This process is considered first-order on a natural log (ln) basis. By integration, the above equation becomes

$$C_{tissue} = (k_1/k_2)C_{water}(1 - e^{-k_2 t}).$$

Bivalves primarily take up HOCs directly from the water column (Thomann and Komlos 1999; Birdsall, Kukor, and Cheney 2001) through their gills, although some studies have suggested additional chemical inputs from dietary exposure (Brieger and Hunter 1993; Gossiaux, Landrum, and Fisher 1996; Bjork and Gilek 1997; Raikow and Hamilton 2001), and direct sediment ingestion via pedal feeding mechanisms (McMahon and Bogan 2001; Raikow and Hamilton 2001). There is debate in the literature over the relative contribution of each of these uptake routes; however, it should be noted that once the system has attained steady-state ($dC/dt = 0$), the route of contaminant exposure is irrelevant (Di Toro et al. 1991). Because of their minimal metabolic capabilities for metabolizing the majority of HOCs (Farrington et al. 1983; James 1989), bivalves accumulate these contaminants to high levels in their lipid tissues, which can often reach many orders of magnitude greater than the corresponding concentrations in water or sediment phases. Despite the common use of freshwater bivalves for monitoring aquatic environments, relatively little information is known regarding HOC uptake rate constants, compared with that for marine bivalves. Moreover, much of the freshwater and marine data represent only a few species. For instance, the majority of the freshwater uptake studies focus on *Dreissena polymorpha*, whereas the majority of marine uptake studies use *M. edulis*.

There are various ranges in reported k_1 values for freshwater bivalves depending on species, and study variables such as temperature, exposure environment, mussel size, and lipid content (Table 8.1a,b; Table 8.2a,b for study summaries, Fisher et al. 1993; Bruner, Fisher, and Landrum 1994; Gossiaux, Landrum, and Fisher 1996; Fisher et al. 1999). However, based on the available data, most k_1 values compare well, with only a few exceptions (Table 8.1a). Many studies demonstrate initial rapid uptake during initial exposure for both freshwater and marine species (Lee, Sauerheber, and Benson 1972; Obana et al. 1983; Bjork and Gilek 1997; Birdsall, Kukor, and Cheney 2001). For example, Birdsall, Kukor, and Cheney (2001) reported rapid uptake of the PAHs naphthalene (N0), anthracene (AN), and chrysene (C0) by *Elliptio complanata* gills. Their data demonstrated that the

TABLE 8.1a
Published Uptake and Elimination Rate Constants for Various Freshwater and Marine Bivalve Species as a Function of Chemical Class and Water Solubility

Species	Chemical Class[a]	Log K_{ow}	k_1 (mL/g day)	k_2 (day^{-1})	References
Freshwater					
E. complanata	PAH (34)	3.37(N0)–7.64(CO)		0.0400(PE)–0.2600(26DMN0)	Thorsen (2003)
	PAH (38)	3.37(N0)–7.64(CO)			
	PAH (45)	3.37(N0)–7.64(CO)			
E. complanata	PAH (14)	3.92(AC)–6.75(DA)	369.0–2,133.0	0.0370(BkF)–0.0217(F0)	Gewurtz et al. (2002)
D. polymorpha	PCP	5.12		0.8600–1.5600	Fisher et al. (1999)
C. fluminea	PCP	5.12		0.3900–0.4000	Basack et al. (1997)
C. leana	Pesticides (3)	3.22(OX)–4.22(TBC)	24.2(TBC)–338.0(CNF)	0.0450(CNF)–0.0600(TBC)	Uno et al. (1997)
D. polymorpha	PAH (2)	5.18(PY)–6.04(BaP)	672.0(BaP)–32,737.0(BaP)	0.0240–0.3840(BaP)	Gossiaux, Landrum, and Fisher (1996)
D. polymorpha	PCB (2)	5.90(PCP)–6.90(HCBP)	2,280.0(PCP)–26,448.0(HCBP)	0.0240(HCBP)–0.1920(PCP)	
D. polymorpha	TCBT (8)	6.73(28)–7.54(25)	683.3(52)–848.7(80)	0.0052(27)–0.0226(21)	Van Haelst et al. (1996a)
D. polymorpha	PCB (36)	5.60(42)–7.36(180)		0.0420(183)–0.1720(64)	Morrison et al. (1995)
A. anatina	PCP	5.12			Makela and Oikari (1995)
P. complanata	PCP	5.12			
D. polymorpha	PAH (2)	5.18(PY)–6.04(BaP)	7,680.0(PY)–31,200.0(BaP)	0.1920(BaP)–0.5760(PY)	Bruner, Fisher, and Landrum (1994)
D. polymorpha	PCB (2)	5.9(TCBP)–6.9(HCBP)	9,120.0–40,320.0(HCBP)	0.1200(HCBP)–0.5040(TCBP)	
D. polymorpha	PAH (2)	5.18(PY)–6.04(BaP)	10,272.0(PY)–20,112.0(BaP)	0.0090(PY, BaP)	Fisher et al. (1993)
D. polymorpha	PCB (2)	5.90(TCBP)–6.90(HCBP)	4,008.0–25,752.0(HCBP)	0.0040(HCBP)–0.0170(TCBP)	
D. polymorpha	OC (1)	6.19(DDT)	2,976.0–17,664.0(DDT)	0.0070–0.0080(DDT)	Brieger and Hunter (1993)
	PCB (2)	6.36(77)–7.42(169)	551.0(77)–1,480.0(169)	0.0340(169)–0.0350(77)	
E. complanata	HCB, OCS	5.45(HCB)–6.29(OCS)	650.0(HCB)–1,010.0(OCS)	0.4100(HCB)–0.1600(OCS)	Russell and Gobas (1989)

Marine					
M. edulis	PCB (3)	5.67(31)–6.92(153)	2,160.0–168,000.0(153)	0.0288(153)–0.1368(31)	Bjork and Gilek (1997)
C. virginica	PAH (7)	4.57(P0)–7.0(IP)		0.0200(FL)–0.0770(BaP)	Sericano, Wade, and Brooks (1996)
M. mercenaria	PCB (9)			0.0053(149)–0.01540(110)	Tanacredi and Cardenas (1991)
	PAH (9)	3.37(N0)–6.04(BaP)		ND over 45 days	
C. virginica	PAH (14)	4.57(P0)–6.50(BghiF)	330.0(P0)–2,365.0(MPY)	0.0090(BF)–0.1180(FL)	Bender et al. (1988)
M. mercenaria	PAH (14)	4.57(P0)–6.50(BghiF)	187.0(MP0)–2,842.0(BaA)	0.0870(BaP)–0.2130(FL)	
M. edulis	PAH (6)	3.90–6.10		0.0231(FL)–0.0578(BkF)	Pruell et al. (1986)
	PCB (4)	5.00–6.60		0.0150(HCBP)–0.0420(TCBP)	
Short-necked clam	PAH (4)	4.42(D0)–5.89(D3)		0.1000(D3)–0.2400(D2)	Ogata et al. (1984)
Oyster	PAH (4)	4.42(D0)–5.89(D3)			
Mussel	PAH (4)	4.42(D0)–5.89(D3)			

Abbreviations: AC=acenaphthene; BaP=beazo[a]pyrene; BghiF=benzo[ghi]fluoranthene; BkF=benzo[k]fluoranthene; C0=chrysene; CNF=chlornitrofen; D0=dibenzothiophene; DA=dibenzanthracene; D2=dimethyldibenzothiophene; D3=trimethyldibenzothiophene; 2,6DMN0=2,6-dimethylnaphthalene; F0=fluorene; FL=fluoranthene; HCB=hexachlorobenzene; IP=indenopyrene; N0=naphthalene; OC=organochlorine; OCS=octachlorostyrene; OX=oxadiazon; PAH=polycyclic aromatic hydrocarbon; PCB=polychlorinated biphenyl (number in parentheses refers to IUPAC PCB congener); PCP=pentachlorophenol; PE=perylene; PY=pyrene; TBC=thiobencarb; TCBP=tetrachlorobiphenyl; TCBT=tetrachlorobenzyltoluene.

[a]Number in parentheses refers to total number of chemicals studied within the chemical class.

TABLE 8.1b
Published Solubility Values, Bioconcentration Factors, and Half-Lives for Various Freshwater and Marine Bivalves as a Function of Chemical Class

Species	Chemical Class	log K_{ow}	log BCF	$T_{1/2}$ (days)	References
Freshwater					
E. complanata	PAH (34)	3.37(N0)–7.64(CO)	1.54(N0)–4.66(PE)		Thorsen (2003)
	PAH (38)	3.37(N0)–7.64(CO)	1.90(N0)–5.20(CO)		
	PAH (45)	3.37(N0)–7.64(CO)	1.60(AN)–5.51(C4)		
E. complanata	PAH (14)	3.92(AC)–6.75(DA)		2.60(26DMN0)–16.50(PE)	Gewurtz et al. (2002)
D. polymorpha	PCP	5.12	2.60–3.10	3.20(F0)–18.70(BkF)	Fisher et al. (1999)
Corbicula fluminea	PCP	5.12		0.44–0.81	Basack et al. (1997)
				1.73–1.78	
C. leana	Pesticides (3)	3.22(OX)–4.22(TBC)	2.34(OX)–4.14(CNF)	11.60(TBC)–15.40(CNF)	Uno et al. (1997)
D. polymorpha	PAH (2)	5.18(PY)–6.04(BaP)	4.34(PY)–5.43(BaP)	1.75(BaP)–28.80(BaP)	Gossiaux, Landrum, and Fisher (1996)
D. polymorpha	PCB (2)	5.90(PCP)–6.90(HCBP)	4.00(PCP)–5.74(HCBP)	3.60(PCP)–28.80(HCBP)	Van Haelst et al. (1996a, 1996b)
	TCBT (8)	6.73(28)–7.54(25)	4.43(80)–5.19(27)	18.60(80)–71.80(22)	
D. polymorpha	PCB (36)	5.60(42)–7.36(180)		4.00(64)–16.50(183)	Morrison et al. (1995)
A. anatina	PCP	5.12	1.90–2.10		Makela and Oikari (1995)
P. complanata	PCP	5.12	1.80–1.90		
D. polymorpha	PAH (2)	5.18(PY)–6.04(BaP)	4.11(PY)–4.92(BaP)	1.20(PY)–3.60(BaP)	Bruner, Fisher, and Landrum (1994)
D. polymorpha	PCB (2)	5.90(TCBP)–6.90(HCBP)	4.32(TCBP)–5.38(HCBP)	1.40(TCBP)–5.80(HCBP)	Fisher et al. (1993)
	PAH (2)	5.18(PY)–6.04(BaP)	4.65(PY)–4.88(BaP)	2.60(BaP)–3.00(PY)	
	PCB (2)	5.90(TCBP)–6.90(HCBP)	4.62(HCBP)–5.43(HCBP)	1.70(TCBP)–7.20(HCBP)	
	OC (1)	6.19(DDT)	4.72–5.03(DDT)	3.60–4.30(DDT)	
D. polymorpha	PCB (2)	6.36(77)–7.42(169)	4.02(77)–4.45(169)	19.80(77)–20.40(169)	Brieger and Hunter (1993)
E. complanata	HCB, OCS	5.45(HCB)–6.29(OCS)	3.56(HCB)–4.16(OCS)	1.70(HCB)–4.30(OCS)	Russell and Gobas (1989)
Marine					
M. edulis	PCB (3)	5.67(31)–6.92(153)	4.70(49)–6.80(153)BAFs	5.00(31)–24.20(153)	Bjork and Gilek (1997)
C. virginica	PAH (7)	4.57(P0)–7.00(IP)		9.00(BaP)–26.00(FL)	Sericano, Wade, and Brooks (1996)

Species	Compound				Reference
M. edulis	PCB (9)				Bergen, Nelson, and Pruell (1996)
	PCB (21)	5.07(8)–7.42(169)	About 5.30–7.10	22.00(26)–130.00(149)	
C. virginica	PAH (14)	4.57(P0)–6.50(BghiF)	3.20(P0)–4.90(BF)	5.90(FL)–77.00(BF)	Bender et al. (1988)
M. mercenaria	PAH (14)	4.57(P0)–6.50(BghiF)	3.20(MP0)–4.40(BghiF)	3.30(FL)–8.00(BaP)	
M. edulis	PAH (6)	3.90–6.10	2.00–4.40	11.90–29.80	Pruell et al. (1986)
	PCB (4)	5.00–6.60	4.50–6.60	16.30–45.60	
Short-necked clam	PAH (4)	4.42(D0)–5.89(D3)	2.17(D0)–2.58(D3)	2.90(D2)–6.90(D3)	Ogata et al. (1984)
Oyster	PAH (4)	4.42(D0)–5.89(D3)	3.12(D0)–4.45(D3)		
Mussel	PAH (4)	4.42(D0)–5.89(D3)	2.87(D1)–3.62(D3)		

Abbreviations: AC=acenaphthene; BaP=beazo[a]pyrene; BghiF=benzo[ghi]fluoranthene; BkF=benzo[k]fluoranthene; C0=chrysene; CNF=chlornitrofen; D0=dibenzothiophene; DA=dibenzanthracene; D2=dimethyldibenzothiophene;D3=trimethyldibenzothiophene;2,6DMN0=2,6-dimethylnaphthalene; F0=fluorene; FL=fluoranthene; HCB=hexachlorobenzene; IP=indenopyrene; N0=naphthalene; OC=organochlorine; OCS=octachlorostyrene; OX=oxadiazon; PAH=polycyclic aromatic hydrocarbon; PCB=polychlorinated biphenyl (number in parentheses refers to IUPAC PCB congener); PE=perylene; PY=pyrene; TBC=thiobencarb; TCBP=tetrachlorobiphenyl; TCBT=tetrachlorobenzyltoluene.

TABLE 8.2a
Summary of Exposure and Test Duration for Toxicokinetic Studies in the Peer-Reviewed Reference with Various HOC Classes and Freshwater and Marine Bivalves

Species	Exposure	Chemical Class	Duration	References
Freshwater				
D. polymorpha				Reeders, Bij de Vaate, and Slim (1989)
E. complanata		HCB, OCS		Russel and Gobas (1989)
D. polymorpha		PAH, PCB		Fisher et al. (1993)
D. polymorpha	Water, sediment, food	PCB, 3	21- to 100-day exposure, rapid elemination	Brieger and Hunter (1993)
D. polymorpha	Water only	PAH, PCB	6-hour uptake	Bruner, Fisher, and Landrum (1994)
A. anatina, P. complanata	Water only	PCP	Steady-state reached in 16 h	Makela and Oikari (1995)
D. polymorpha	Water and field	PCB, 36	2-day exposure, 16-day elimination	Morrison et al. (1995)
D. polymorpha	Water only	PAH, PCB	6-hour uptake, 15-day elimination	Gossiaux, Landrum, and Fisher (1996)
D. polymorpha	Water only	PCBs, OCs		Chevreuil et al. (1996)
D. polymorpha	Water only	TCBTs, 8	21-day uptake, no steady-state reached	Van Halest et al. (1996a, 1996b)
C. flumina	Water only	PCP	96-hour uptake, 72-hour elimination	Basack et al. (1997)
C. leana	River water	Pesticides	14-day uptake, 15-day elimination	Uno et al. (1997)
D. polymorpha	Water only	PAH, PCB		Fisher et al. (1999)
E. complanata	Water only	PAH, pesticides	Used excised gills	Birdsall, Kukor, and Cheney (2001)
E. complanata	Water only	PAH, 14	5-day uptake, 32-day elimination	Gewurtz et al. (2002)
E. complanata	Water only and sediment	PAH, 34–48	20-day exposure, 20-day elimination	Thorsen (2003)
Marine				
Oysters		No. 2 Fuel oil	60-day uptake, 180-day elimination	Blumer, Souza, and Sass (1970)
M. edulis	Water only	PAH		Lee, Sauerheber, and Benson (1972)
Oysters		No. 2 Fuel oil	49-day uptake, 28-day elimination	Stegman and Teal (1973)
M. edulis		PAH		Clark and Findley (1975)
Mussels		PAH, BaP		Dunn and Stich (1976)
Clams		Chronic pollution	120-day elimination	Boehm and Quinn (1977)
M. edulis		PAHs		Hansen et al. (1978)

(continued)

TABLE 8.2a *(Continued)*

Species	Exposure	Chemical Class	Duration	References
Ostrea edulis	Flow-through system	PAH, N0		Riley et al. (1981)
Tapes japonica	Water and field	PAH, 9	7- to 14-day exposure	Obana et al. (1983)
Clam, oyster, mussel	Water only	PAH, D0-D3		Ogata et al. (1984)
Oysters		PAH	15-day uptake	Pittinger et al. (1985)
M. edulis	Sediment dosed	PAH, PCB	40-day uptake, 40-day elimination	Pruell et al. (1986)
M. edulis		PAHs		Broman and Ganning (1986)
Mutiple aquatic organisms		Multiple HOCs		Hawker and Connell (1986)
P. viridis	Field	PCB, 54	17-day uptake, 32-day elimination	Tanabe, Tatsukawa, and Phillips (1987)
C. virginica, M. mercenaria	Field and laboratory	PAH, 14	28-day uptake, 28-day elimination	Bender et al. (1988)
Clams		PAH	2-day uptake, 45-day elimination	Tanacredi and Cardenas (1991)
Oysters	Water only	PCB, 77		Sericano et al. (1992)
M. edulis	Field	PCBs	28-day exposure	Bergen, Nelson, and Pruell (1996)
M. edulis	Water and food	PCBs		Gilek, Bjork, and Naef (1996)
M. edulis	Water and algae	PAH, P0	20-day exposure, 14-day elimination	Bjork and Gilek (1996)
C. virginica	Field	PAH, PCB	28- to 50-day uptake, 50-day elimination	Sericano, Wade, and Brooks (1996)
M. edulis	Water and food	PBCs		Bjork and Gilek (1997)

average uptake of AN (log K_{ow} 4.54) and C0 (log K_{ow} 5.86) was similar, and both were greater than that for N0 (log K_{ow} 3.37), which was explained by its lower lipid affinity.

Differences in k_1 can be observed when comparing the same analyte among studies, as well as when comparing different analytes with similar physico-chemical parameters. However, with a few exceptions, the differences appear to be relatively small, considering the many variables that can exist between studies. For example, k_1 values measured for benzo(a)pyrene (BaP) and HCBP in both the field and laboratory over the course of three years and at different temperatures (5–24°C) in *D. polymorpha* compare well (Table 8.1a). Specifically, for BaP the range of uptake rates is from 9,960 to 32,736 mL/g day, a factor of 3 difference. The differences between highest and lowest update rate constants for HCBP, pentachlorophenol (PCP), and pyrene (PY) are even less, at factors of 2.0, 2.6, and 2.0, respectively. Data from two collection timepoints have been omitted for this comparison due to very low uptake rate constants, which the authors believed was from over-wintered mussels experiencing stress (both occurred for mussels collected at 4°C in the field; however, when mussels were fed while being acclimated to 4°C in the laboratory, these effects were not observed) (Gossiaux, Landrum, and Fisher 1996). Therefore, it is important to consider that larger differences can occur based on the physiological state of the organism. Laboratory-derived k_1s for PCP increased from 3,960 mL/g day at 4°C to 5,928 mL/g day at 15°C, whereas field-derived k_1s showed even less difference with a more dramatic temperature increase from 4 to 24°C (3,240 versus 2,640 mL/g day, respectively) (Gossiaux, Landrum, and Fisher 1996). These

TABLE 8.2b
Summary of Variables Measured and Primary Findings for Various Bivalves Published in the Peer-Reviewed References

Species	Variables Measured	Primary Findings	References
Freshwater			
D. polymorpha		No change in k_1 within a season, but change between seasons	Reeders, Bij de Vaate, and Slim (1989)
E. complanata	BCF, k_2		Russell and Gobas (1989)
D. polymorpha	k_1		Fisher et al. (1993)
D. polymorpha	k_1, k_2, BCF/BAF		Brieger and Hunter (1993)
D. polymorpha	k_1, k_2, BCF, $T_{1/2}$	k_2 depends on the lipophilicity of a chemical	Bruner, Fisher, and Landrum (1994)
A. anatina, P. complanata	BCF		Makela and Oikari (1995)
D. polymorpha	k_2, T_{95}		Morrison et al. (1995)
D. polymorpha	k_1, k_2, BCF, $T_{1/2}$	Temperature effects, monophasic elimination	Gossiaux, Landrum, and Fisher (1996)
D. polymorpha		Responses to change in aqueous OC concentrations within 7 days	Chevreuil et al. (1996)
D. polymorpha		Bivalves: have MFO but capabilities are ≪ fish	Van Haelst et al. (1996b)
D. polymorpha	k_1, k_2, BCF, $T_{1/2}$	Log K_{ow} versus k_2: independent; mussel lipid decrease over time	Van Haelst et al. (1996a)
C. flumina	k_2	No extensive phase I metabolism	Basack et al. (1997)
C. leana	k_1, k_2, BCF	Measured pesticide concentrations in mussels in rice patties	Uno et al. (1997)
D. polymorpha	k_1, k_2, BCF, $T_{1/2}$	Temperature and pH effects	Fisher et al. (1999)
		PAH uptake due to partitioning from water to animal across gill surface	Thomann and Komlos (1999)
E. complanata		Average uptake of AN = CO > N0	Birdsall, Kukor, and Cheney (2001)
E. complanata	k_2	Linear relationship between log K_{ow} and k_2	Gewurtz et al. (2002)
E. complanata	k_2, BCF, $T_{1/2}$	Stressed mussels: lower k_2s	Thorsen (2003)
Marine			
Oysters		Little elimination after 180 days	Blumer, Souza, and Sass (1970)
M. edulis		Rapid N0, BaP uptake but no metabolism; k_2 depends on lipophilicity of chemical	Lee, Sauerheber, and Benson (1972)
Oysters		Elimination nearly complete after 28 days	Stegman and Teal (1973)
M. edulis		k_2 dependent on chemical lipophilicity	Clark and Findley (1975)

(continued)

Table 8.2b *(Continued)*

Species	Variables Measured	Primary Findings	References
Mussels		k_2 dependent on chemical lipophilicity	Dunn and Stich (1976)
Clams		Slight elimination after 120 days	Boehm and Quinn (1977)
M. edulis		High lipid tissues = rapid elimination versus low lipid tissues = slower elimination: biphasic	Hansen et al. (1978)
O. edulis		Gill: primary site: uptake + accumulation	Riley et al. (1981)
T. japonica		Rapid PAH accumulation	Obana et al. (1983)
Clam, oyster, mussel	k_1, k_2, BCF		Ogata et al. (1984)
Oysters		Analytes below detection limit within 4 days of elimination	Pittinger et al. (1985)
M. edulis	k_2, BCF, $T_{1/2}$	Slow elimination observed and k_2 depends on liophilicity of chemical	Pruell et al. (1986)
M. edulis		High lipid tissues = rapid elimination versus low lipid tissues = slower elimination: biphasic	Broman and Ganning (1986)
Mutiple aquatic organisms		Log BCF vs log K_{ow} relationship; k_2 dependent on chemical lipophilicity	Hawker and Connell (1986)
P. viridis	k_2, $T_{1/2}$, T_{90}	Rapid uptake, release of lower K_{ow} PCBs	Tanabe, Tatsukawa, and Phillips (1987)
C. virginica M. mercenaria	k_1, k_2, BCF	Clams $k_2 \gg$ oyster k_2	Bender et al. (1988)
Clams		No elimination observed in 45 days	Tanacredi and Cardenas (1991)
Oysters		Equilibrium attained in 30 days	Serciano et al. (1992)
M. edulis	BCF	Coplanar PCBs reach steady-state faster (7 days) than nonplanar PCBs (14–28 days)	Bergen, Nelson, and Pruell (1996)
M. edulis	k_2	Body size affects bioaccumulation because of influences on k_1s	Gilek, Bjork, and Naef (1996)
M. edulis	k_2	k_2 unaffected by [POC], and initial uptake rapid	Bjork and Gilek (1996)
C. virginica	k_2, $T_{1/2}$		Sericano, Wade, and Brooks (1996)
M. edulis	k_1, k_2, BAF, $T_{1/2}$	Physioligically-based model of bioaccumulation, food ration affected k_1, but not k_2	Bjork and Gilek (1997)

authors noted that others (e.g., Reeders, Bij de Vaate, and Slim 1989) have documented a lack of substantial change in *D. polymorpha* filtration activity over a temperature range of 5–20°C, which helps to explain their data (Gossiaux, Landrum, and Fisher 1996). While k_1s for some of the compounds in this study increased proportionally with increasing temperature in the field (e.g., BaP and HCBP), the trend was not consistently exhibited over the three-year time frame and led the authors to suggest that uptake kinetics do not change in a proportional manner with temperature (Gossiaux, Landrum, and Fisher 1996), at least across the range tested. Although Reeders, Bij de Vaate, and Slim (1989) reported no significant change in uptake rates in *D. polymorpha* within a season, a significant change between seasons was documented.

Variations in uptake rates with *D. polymorpha* body size and lipid content were reported by Bruner, Fisher, and Landrum (1994) for HCBP, tetrachlorobiphenyl (TCBP), BaP, and PY. The average uptake rate constant for HCBP over varying mussel lipid and size was 23,680 mL/g day (Bruner, Fisher, and Landrum 1994), which compared well with k_1s reported by Gossiaux, Landrum, and Fisher (1996) for *D. polymorpha* over varying temperatures, averaging 18,624 mL/g day in the laboratory and 21,000 mL/g day in the field. When varying pH is considered in combination with changing temperatures, differences in k_1s increase but are still within a factor of less than five on average, which translates into about an order of magnitude difference in BCF values. The reported field and laboratory k_1s in *D. polymorpha* for PCP (log K_{ow} 5.12) are 2,760 and 4,120 mL/g day (Gossiaux, Landrum, and Fisher 1996), whereas those reported for varying pH (and averaged over temperature) are lower: 1,657 (pH 6.5), 1,218 (pH 7.5), and 868 (pH 8.5) (Fisher et al. 1999). The lesser k_1s may be due to the dissociable nature of PCP in the range of ambient pH ($pK_a = 4.74$) or to a combination of effects caused by changing pH and temperature on mussel filtration rates and subsequent uptake rates. When individual values are compared, rather than averages, the variation in k_1 is increased. For instance, the smaller the mussel size (measured in shell length), the faster the uptake rate (Bruner, Fisher, and Landrum 1994). Large (21 mm) zebra mussels with high lipid content (greater than 9%) had TCBP uptake rate constants of 10,080 mL/g day, whereas smaller (15 mm) but also higher lipid content mussels (greater than 9%), had TCBP uptake rate constants twice that of the larger mussels at 23,760 mL/g day (Bruner, Fisher, and Landrum 1994).

In general, uptake rates were directly proportional to compound K_{ow}; as K_{ow} increased, k_1 increased as well. For example, as log K_{ow} values increased from 5.18 for PY to 6.90 for HCBP, the average uptake rate constant increased from 10,480 to 23,680 mL/g day, respectively. An additional study reported k_1s ranging from 2,976 to 25,752 mL/g day in *D. polymorpha* for PAHs, PCBs, and OCs (DDT) spanning a similar log K_{ow} range of 5.2–6.7 (Fisher et al. 1993). This range is comparable to the other k_1s previously listed, when values for DDT are omitted (lowest values). Moreover, k_1 values reported for HCB (hexachlorobenzene) and OCS (octachlorostyrene) in *E. complanata* also increased with increasing log K_{ow}; from 650/day for HCB (log K_{ow} 5.45) to 1,010/day for OCS (log K_{ow} 6.29) (Russell and Gobas 1989). However, these values are substantially less than those reported for *D. polymorpha*.

In contrast to the linear relationship between k_1 and K_{ow} reported by some (Russell and Gobas 1989; Bruner, Fisher, and Landrum 1994; Gossiaux, Landrum, and Fisher 1996), uptake rates for eight different TCBT congeners in *D. polymorpha* were independent of K_{ow} (Van Haelst et al. 1996a). As log K_{ow} increased from 6.73 (TCBT # 28) to 7.54 (TCBT # 25), k_1s varied little, from 772 to 803 mL/g day (Van Haelst et al. 1996a), respectively. However, when all TCBT congeners were included in the log K_{ow} range, the k_1 values demonstrated larger variation and ranged from 683.3 to 848.8 mL/g day. This may be partially explained by the high K_{ow} values or the decreased ability of highly hydrophobic compounds to permeate membranes (Van Haelst et al. 1996a). Moreover, the uptake rates reported for *D. polymorpha* for TCBT congeners are lower than those for PAHs or PCBs with similar hydrophobicity (see previous values). Uptake rate constants for PCB congener 153 (Bruner, Fisher, and Landrum 1994) and TCBT (tetrachorobenzyltoluene) congener 28 (Van Haelst et al. 1996a), which have similar log K_{ow} values (6.92 and 6.73, respectively), differ by as much as a factor of 50, from as low as 771 mL/g day for TCBT congener 28 (Van Haelst et al. 1996a) to between 9,120 and 38,592 mL/g day for congener 153 (Bruner, Fisher, and Landrum 1994), both for *D. polymorpha*.

Bjork and Gilek (1997) reported k_1s for three PCB congeners (PCBs 31, 49, 153) in the marine mussel, *M. edulis*, that ranged from 2,160 (PCB153) to 168,000 mL/g day (PCB153). While the upper range is quite large, and is about four times greater than the upper range reported for *D. polymorpha*, the freshwater mussel k_1s are still within these limits. The larger k_1 values in *M. edulis* are probably due to the addition of contaminated food in the study conducted by Bjork and Gilek (1997). In contrast, Ogata et al. (1984) reported k_1s for parent and various alkylated dibenzothiophenes in a marine short-necked clam, which were significantly less ranging from 33/ day for dibenzothiophene to 66/day for dialkylated dibenzothiophene. It should be noted that some authors (e.g., Ogata et al. 1984; Russell and Gobas 1989) have reported k_1 values in reciprocal days, which is assumed to be equivalent to mL/g day (where 1 mL = 1 g). However, this assumption may not always be valid, which may explain some of the differences observed in k_1 values.

Uptake rates for various pesticides in the Asian clam *Corbicula leana* (Uno et al. 1997) are much lower than those reported in *D. polymorpha* for compounds with similar K_{ow}s. While the log K_{ow} for the pesticides thiobencarb, oxadiazon, and chlornitrofen are less than the HOCs, the uptake rate constants are more than proportionally less, ranging from 24.2 for thiobencarb to 626.0 mL/g day for chlornitrofen in the field and 140 for thiobencarb to 338 mL/g day for chlorni-trofen in the laboratory (Uno et al. 1997). The authors attributed the low uptake rate(s) for thiobencarb to a temperature decrease of 2°C over the course of a year causing slower ventilation rates in the mussels. In contrast, reports with *D. polymorpha* show that a temperature range of 20°C does not cause substantial changes in uptake rates (Reeders, Bij de Vaate, and Slim 1989; Gossiaux, Landrum, and Fisher 1996). The large differences in uptake rates for *C. leana* versus *D. polymorpha* and *M. edulis* are probably due to a combination of species and chemical differences.

In summary, uptake rate constants were remarkably similar across temperature, season, pH, chemical, and study variables, although some differences were observed, particularly when comparing chemicals of similar log K_{ow} (TCBTs versus PCBs), low versus high lipid content, and bivalves of differing size and species. Large variation in k_1 was demonstrated for stressed mussels (Gossiaux, Landrum, and Fisher 1996), suggesting that bivalve physiology must be considered when measuring empirical uptake rates or BCFs under adverse conditions such as very low temperatures. Moreover, k_1s were greater for combined food and water exposures (Bjork and Gilek 1997). The uptake rate constants reported in this chapter represent only those for a few freshwater mussel and clam species, which demonstrates the need for further research in this area. For instance, while *D. polymorpha* uptake rate constants may not vary substantially with increases or decreases in temperature (over a 20°C range) (Gossiaux, Landrum, and Fisher 1996), this may not be the case for other freshwater bivalve species (e.g., *Corbicula*) (Uno et al. 1997).

BIOCONCENTRATION

Gossiaux, Landrum, and Fisher (1996) reported bioconcentration factors in *D. polymorpha* for BaP (log K_{ow} 6.04) that ranged from 4.38 to 5.28 log bioconcentration in field exposures at temperatures from 4 to 24°C. The BaP log BCF values had a similar range in the laboratory for temperatures from 4 to 20°C (4.60 (4°C) to 5.43 (15°C), Table 8.1b). The log BCF values for PY (log K_{ow} 5.18) in both the field and laboratory ranged from 4.34 to 4.89, over a similar temperature range. However, the authors were not convinced that steady-state had been reached due to a factor of 100 difference between BCF values calculated from C_{tissue}/C_{water} and those calculated from k_1/k_2. This implies BCF values in reality would be larger than those reported or that the organisms possess some capacity for metabolism of PY. In comparison, Bruner, Fisher, and Landrum (1994) reported similar log BCF values also in *D. polymorpha* for both BaP, ranging from 4.61 to 4.92 and PY, ranging from 4.11 to 4.54, depending on mussel lipid and size. These values compare well, especially when considering the variation in temperature, lipid content, and size.

In contrast, log BCF values reported by Thorsen (2003) for *E. complanata* are lower for both PAHs, ranging from 3.50 to 4.66 for BaP, and 2.29 to 3.79 for PY, depending on exposure source (water-only versus sediment). The differences between these data may be partially explained by lipid content of *D. polymorpha* and *E. complanata*; *D. polymorpha* were generally 7–15% lipid on a dry weight basis (Gossiaux, Landrum, and Fisher 1996) whereas *E. complanata* were much lower, typically 3–4% lipid (Thorsen 2003). The contribution of a lipid can be partly confirmed by results from Bruner, Fisher, and Landrum (1994) who reported an increase in BCF values with subsequent increase in mussel lipid content. However, this effect was only observed for the higher K_{ow} compounds (HCBP and BaP) and not for the lower K_{ow} compounds (TCBP and PY). While bivalve BCF values are not traditionally normalized to lipid content, it would be helpful to report lipid values for conversion and comparison. The addition of lipid-normalized BCF values may help to explain variations in bivalve bioconcentration. Furthermore, log BCF values determined for HCB and OCS in *E. complanata* (log K_{ow}s 5.49 and 6.29, respectively) compare well with those for PAHs of similar hydrophobicity, ranging from 3.56 to 4.16 (e.g., 3.58 and 3.64 for C2-dibenzothiophenes with log K_{ow} 5.50, and 4.23 and 4.54 for benzo(e)pyrene with log K_{ow} 6.20) (Thorsen 2003). Additional variations in BCFs may be further explained by physiological differences between *E. complanata* and *D. polymorpha*, differences in study design, or a combination of environmental and physiological factors.

Makela and Oikari (1995) reported BCF values for PCP in two freshwater mussels, *Anodonta anatina* and *Pseudanodonta complanata*, which range from 1.9 to 2.1 and 1.8 to 1.9, respectively. These BCF values are much lower than those reported by Gossiaux, Landrum, and Fisher (1996) for PCP in *D. polymorpha*, which ranged from 4.00 to 5.27, depending on the study temperature. In contrast, log BCF values reported for PCP in a different study for *D. polymorpha* with varying temperature and pH are mid range between those reported for *A. anatina*, *P. complanata* (with a range of 2.60–3.13) (Fisher et al. 1999), and *D. polymorpha* (4.00–5.27) (Gossiaux, Landrum, and Fisher 1996, Table 8.1b).

The log BCF values for HCBP determined in two separate studies on *D. polymorpha* compare well, ranging from 4.79 to 5.38 in one study (Bruner, Fisher, and Landrum 1994) and from 5.24 to 5.74 in the second (Gossiaux, Landrum, and Fisher 1996). Brieger and Hunter (1993) reported log BAF values for *D. polymorpha* of 4.02 and 4.45 for two PCB congeners, 77 (log K_{ow} 6.36) and 169 (log K_{ow} 7.42), which were lower relative to their K_{ow} values than those reported for similar log K_{ow} compounds such as TCBT congener 28 (log K_{ow} 6.73, log BCF 4.83) (Van Haelst et al. 1996a, 1996b), HCBP (log K_{ow} 6.9, log BCF range 4.8–5.7) (Bruner, Fisher, and Landrum 1994; Gossiaux, Landrum, and Fisher 1996), and TCBT congener 22 (log K_{ow} 7.43, log BCF 4.71) (Van Haelst et al. 1996a, 1996b). These differences may simply suggest a lack of steady state, as BAF values would be expected to be larger than BCF values from increased exposure to contaminated food.

The values of log BCF for various pesticides including chloronitrofen, thiobencarb, and oxadiazon have been reported for *C. leana* ranging from 2.34 for oxadiazon (log K_{ow} 3.89) to 4.14 for chlornitrofen in the field, and from 3.79 for chlornitrofen to 3.45 for thiobencarb (log K_{ow} 4.22) in the laboratory (Uno et al. 1997). It should be noted that the log BCF values for oxadiazon and thiobencarb increase with corresponding increases in hydrophobicity.

Bioconcentration factors determined for PAHs in *M. edulis* (Pruell et al. 1986) and a marine short-necked clam, oyster, and mussel (Ogata et al. 1984) compare well to those for *E. complanata* (Thorsen 2003) but are less than those reported for *D. polymorpha* (see previous comparison between *E. complanata* and *D. polymorpha*). For example, across a log K_{ow} range of 3.9–6.1, log BCF values for *M. edulis* ranged from 2.0 to 4.4 (Pruell et al. 1986), whereas across a similar log K_{ow} range of 3.37–7.60 for *E. complanata*, log BCF values ranged from 1.5 to 5.2 (Thorsen 2003). Moreover, the log BCFs reported for dibenzothiophene (D0) in marine clam, oyster, and mussel were 2.17, 3.12, and 3.13, respectively (Ogata et al. 1984), which are near the range reported for *E. complanata* of

2.69–2.93 (Thorsen 2003) and similar to those reported for thiobencarb (of similar log K_{ow} to D0: 4.22 versus 4.49) in *C. leana* of 3.25–3.48 (Uno et al. 1997, Table 8.3).

Similar to uptake rate constant data, empirically derived BCF values generally increase with increasing K_{ow} of the compound (Pruell et al. 1986; Brieger and Hunter 1993; Bruner, Fisher, and Landrum 1994; Gossiaux, Landrum, and Fisher 1996; Thorsen 2003). For example, as the log K_{ow} is increased from 5.18 (PY) to 6.90 (HCBP), the average log BCF values for *D. polymorpha* increase from 4.28 to 5.14 (Bruner, Fisher, and Landrum 1994). However, exceptions to this trend have been observed. The BCFs for compounds with log K_{ow} values greater than 6–7 tend to level off due to factors such as steric hinderance (reduction of membrane permeation), lack of steady state (very long times required to reach equilibrium), and growth dilution. Van Haelst et al. (1996a, 1996b) found no correlation between log BCF values for eight TCBT congeners and log K_{ow}. They suggested that this was due to the small range of log K_{ow} (6.73–7.54) compounds used, as well as the fact that the TCBT congeners all have log K_{ow}s > 6 (i.e., may be in the linear part of the curve).

The values of log BCF for PCBs of similar hydrophobicity reported for *M. edulis* were higher than those for PAHs: ranging from about 5.0 to 5.7 for a corresponding PCB log K_{ow} range of approximately 6.0–7.0 (Pruell et al. 1986). This log BCF range fits within that reported for *D. polymorpha* (Bruner, Fisher, and Landrum 1994; Gossiaux, Landrum, and Fisher 1996) for various PCBs over the same log K_{ow} range of 4.0–6.9. However, the differences between PAH and PCBs for freshwater mussels appear to be less pronounced (Bruner, Fisher, and Landrum 1994; Gossiaux, Landrum, and Fisher 1996). Moreover, a linear relationship between log K_{ow} and log BCF was observed for both PAHs and PCBs in *M. edulis* (Pruell et al. 1986), *E. complanata* (Thorsen 2003), and *D. polymorpha* (Bruner, Fisher, and Landrum 1994). Comparisons of steady-state bioconcentration regression equations (Table 8.4) demonstrate reasonable agreement in PAH accumulation, with few exceptions. For example, Pruell et al. (1986) reported a slope of 0.965 and a y-intercept of −1.41 for *M. edulis*, whereas Thorsen (2003) reported a slope of 0.895 and a y-intercept of −1.21 ($r^2 = 0.8325$) for *E. complanata*. However, Ogata et al. (1984) reported regression equations with slopes much less than one (0.16 for short-necked clams, 0.49 for oysters, and 0.31 for mussels) and positive y-intercepts (1.54, 1.03, 1.63, respectively). The differences may be due to the fact that the regression equations of Ogata et al. (1984) were based on the parent and alkyated homologues of dibenzothiophene only, whereas those of Pruell et al. (1986) and Thorsen (2003) were based on data sets containing greater numbers of PAHs. These data suggest a good correlation between marine and freshwater BCF values, for *M. edulis*, *E. complanata*, and *Mya arenaria*.

ELIMINATION

The elimination rate constant (k_2) can be calculated from an elimination plot of the lipid normalized, natural log (ln) of the contaminant concentration in bivalve versus time. In a first-order, one-compartment kinetic model, k_2 is the absolute value of the slope of the line, based on the equation

$$\ln C_{tissue} = -k_2 t + \ln C_{tissue,0}$$

where $C_{tissue,0}$ is the tissue chemical concentration at elimination time zero (Figure 8.4).

Bivalve elimination rate constants are also fairly consistent, depending on compound, study, and species (Table 8.1a). Elimination rates for HOCs are generally much lower than their counterpart uptake rates but similarly are dependent upon the hydrophobic character of the compounds (Dunn and Stich 1976; Bruner, Fisher, and Landrum 1994; Morrison et al. 1995; Gewurtz et al. 2002; Thorsen et al. 2004a). Gewurtz et al. (2002), who calculated k_2s for nine PAHs in *E. complanata*, showed variation from 0.037/day for benzo(k)fluoranthene (BkF) to 0.217/day

TABLE 8.3

Comparison of Bioconcentration/Bioaccumulation Factors and Uptake and Elimination Rate Constants for Similar Solubility HOC Analytes in the Peer-Reviewed Reference

Analyte	Log K_{ow}	Species	Log BCF/BAF	k_1 (mL/g day)	k_2 (day^{-1})	Exposure	References
Phenanthrene	4.54	E. complanata	3.02			Water, laboratory	Thorsen (2003)
Phenanthrene	4.54	E. complanata	3.44			Sediment, laboratory	Thorsen (2003)
Phenanthrene	4.57	E. complanata	3.06			Sediment, field	Thorsen (2003)
Phenanthrene	4.57	M. edulis	2.90			Water, food	Bjork and Gilek (1997)
Phenanthrene	4.57	C. virginica	3.21	330[a]	0.206	Sediment	Bender et al. (1988)
Phenanthrene	4.57	M. mercenaria	3.29	224[a]	0.114	Sediment	Bender et al. (1988)
Dibenzothiophene	4.49	E. complanata	2.86			Water, laboratory	Thorsen (2003)
Dibenzothiophene	4.49	E. complanata	2.69			Sediment, laboratory	Thorsen (2003)
Dibenzothiophene	4.49	E. complanata	2.93			Sediment, field	Thorsen (2003)
Dibenzothiophene	4.49	Marine clam	2.17			Water	Ogata et al. (1984)
Dibenzothiophene	4.49	Marine oyster	3.12			Water	Ogata et al. (1984)
Dibenzothiophene	4.49	Marine mussel	3.13			Water	Ogata et al. (1984)
Methyldibenzothiophene	4.86	E. complanata	3.43			Water, laboratory	Thorsen (2003)
Methyldibenzothiophene	4.86	E. complanata	3.38			Sediment, laboratory	Thorsen (2003)
Methyldibenzothiophene	4.86	E. complanata	3.07			Sediment, field	Thorsen (2003)
Methyldibenzothiophene	4.86	Marine clam	2.38			Water	Ogata et al. (1984)
Methyldibenzothiophene	4.86	Marine oyster	3.40			Water	Ogata et al. (1984)
Methyldibenzothiophene	4.86	Marine mussel	3.12			Water	Ogata et al. (1984)
Pentachlorophenol	5.20	C. fluminea	2.60–3.20	8,856–51,192	0.390–0.400	Water	Basack et al. (1997)
Pentachlorophenol	5.20	D. polymorpha	4.00–4.60	2,280–5928	0.860–1.560	Water	Fisher et al. (1993)
Pentachlorophenol	5.20	D. polymorpha			0.140–0.190	Water	Gossiaux, Landrum, and Fisher (1996)
Pentachlorophenol	5.20	A. anatine	1.90–2.10			Water	Makela and Oikari (1995)

Contaminant		Species				Exposure	Reference
Pentachlorophenol	5.20	*P. complanata*	1.80–1.90			Water	Makela and Oikari (1995)
Benzo[a]pyrene	6.04	*D. polymorpha*	4.40–5.40	9,960–32,736	0.020–0.380	Water	Gossiaux, Landrum, and Fisher (1996)
Benzo[a]pyrene	6.04	*D. polymorpha*	4.60–4.90	7,920–18,240	0.190–0.410	Water	Bruner, Fisher, and Landrum (1994)
Benzo[a]pyrene	6.04	*E. complanata*	3.50–4.70			Water, sediment	Thorsen (2003)
Benzo[a]pyrene	6.04	*M. edulis*	4.50		0.045	Sediment	Pruell et al. (1986)
Benzo[a]pyrene	6.04	*C. virginica*	4.29	639[a]	0.032	Sediment	Bender et al. (1988)
Benzo[a]pyrene	6.04	*M. mercenaria*	3.62	361[a]	0.087	Sediment	Bender et al. (1988)
Hexachlorobiphenyl	6.90	*D. polymorpha*	5.20–5.70	13,536–26,448	0.024–0.960	Water	Gossiaux, Landrum, and Fisher (1996)
Hexachlorobiphenyl	6.90	*D. polymorpha*	4.80–5.40	9,120–38,592	0.120–0.680	Water	Bruner, Fisher, and Landrum (1994)
PCB 153	6.90	*M. edulis*	4.90–6.80	2,160–168,000	0.029	Water, food	Bjork and Gilek (1997)
Indenopyrene	7	*Elliptio complanata*	4.40–4.56			Water, sediment	Thorsen (2003)
Dibenzanthracene	6.75	*Elliptio complanata*	4.80–5.20			Water, sediment	Thorsen (2003)
PCB 169	7.42	*Dreissena polymorpha*	4.50			Water, food	Brieger and Hunter (1993)
TCBT 28	6.73	*Dreissena polymorpha*	4.80			Water	Van Haelst et al. (1996a)
TCBT 52	7.26	*Dreissena polymorpha*	4.50			Water	Van Haelst et al. (1996a)
PCB		*Mytilus edulis*	5.70			Sediment	Pruell et al. (1986)

[a] Units not specified in reference.

TABLE 8.4
Comparison of Steady-State Bioconcentration Regression Equations for Organic Contaminants in Freshwater and Marine Mussels

Species	Chemical Class	Exposure	Slope	y-Intercept	r^2	References
Freshwater						
E. complanata	PAH (34)	Water, laboratory	0.895	−1.21	0.83	Thorsen (2003)
E. complanata	PAH (35)	Sediment, field	0.786	−0.98	0.78	Thorsen (2003)
E. complanata	PAH (45)	Sediment, lab	0.807	−1.12	0.73	Thorsen (2003)
Marine						
M. edulis	PAH (6)	Sediment, laboratory	0.965	−1.40		Pruell et al. (1986)
M. edulis	Multiple HOCs	Water, laboratory	0.858	−0.81	0.96	Geyer et al. (1982)
M. arenaria	PAH	Water, field	1.097	−1.54	0.85	Thorsen (2003)
M. arenaria	PAH	Sediment, field	1.042	−1.28	0.85	Thorsen (2003)
Multiple marine	Multiple HOCs		0.844	−1.23	0.83	Hawker and Connell (1986)
Marine clam	PAH (4, all D0)	Water, laboratory	0.163	1.52	0.71	Ogata et al. (1984)
Marine oyster	PAH (4, all D0)	Water, laboratory	0.494	1.03	0.62	Ogata et al. (1984)
Marine mussel	PAH (4, all D0)	Water, laboratory	0.311	1.63	0.64	Ogata et al. (1984)

for fluoranthene (FL). An inverse linear relationship was observed between analyte elimination rate constant and corresponding log K_{ow}, which the authors attributed to potential passive elimination of PAHs (Gewurtz et al. 2002). This type of response is characteristic of monophasic, first-order elimination, which has also been reported in *D. polymorpha* for lower K_{ow} compounds (Gossiaux, Landrum, and Fisher 1996). Additional k_2 values reported for 45 PAHs in *E. complanata* exposed to sediment during the uptake phase ranged from 0.04 to 0.22/day (Thorsen et al. 2004a). The authors further noted that these elimination rate constants were less than those from a water-only exposure study and suggested that it may have been due to increased stress on the mussels from an unknown fungal or bacterial growth, and subsequently, increased handling (Thorsen et al. 2004a). The k_2 values for OCS and HCB in *E. complanata* ranged from 0.16 to 0.41/day and were slightly higher when compared to similar log K_{ow} PAHs (Russell and Gobas 1989).

Moreover, Gossiaux, Landrum, and Fisher (1996) demonstrated slow elimination rate constants for *D. polymorpha* in field studies, ranging from 0.024 to 0.096/day for HCBP to 0.024 to 0.384/day for BaP. For the lower hydrophobic compounds in this study (PCP and PY), elimination was rapid during the first 24 hours and then leveled off, while elimination of HCBP and BaP was minimal over the first 24 hours, increased during the following 48–168 hours, and then slowed, suggestive of a biphasic, two-compartment model. These authors, however, classified the elimination as monophasic.

Furthermore, k_2s from the studies of Gossiaux, Landrum, and Fisher (1996) and Gewurtz et al. (2002), in *D. polymorpha* and *E. complanata*, compared well among HOCs of similar log K_{ow}. For example, Gewurtz et al. (2002) reported a k_2 for PY of 0.144/day, and this is within the range also

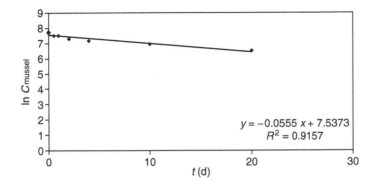

FIGURE 8.4 Example of a plot of the natural log (ln) of the lipid normalized mussel tissue concentrations versus time (Adapted from Thorsen, W. A., 2003). The absolute value of the slope of the linear regression represents the elimination rate constant, k_2, for this particular compound. The equation for the line follows the form: $\ln C_{mt} = (k_2)t + \ln C_{m0}$, where C_{mt} is the contaminant concentration in mussel tissue at time t, and C_{m0} is the contaminant concentration in mussel tissue at time $t = 0$ (initial).

reported for PY by Gossiaux, Landrum, and Fisher (1996) of 0.048–0.312/day. Moreover, a comparison of dibenzo(a,h)anthracene (DA, log K_{ow} 6.8, Thorsen 2003) and HCBP (log K_{ow} 6.9, Gossiaux, Landrum, and Fisher 1996) revealed similar k_2 values; 0.068/day for DA and 0.024–0.096/day for HCBP. The k_2 values reported by Bruner, Fisher, and Landrum (1994) were generally higher than those of Gewurtz et al. (2002) and Gossiaux, Landrum, and Fisher (1996). The k_2s for PCP in *D. polymorpha* were about two to three times less than those reported for *Corbicula fluminea*, ranging from 0.140 to 0.192/day in *D. polymorpha* (Bruner, Fisher, and Landrum 1994) to 0.39 to 0.40/day in *C. fluminea* (calculated with data from Basack et al. 1997). However, both were less than other k_2s for PCPs reported for *D. polymorpha*, which ranged from 0.86 to 1.56/day (Fisher et al. 1999), which may have been due to the combination of changing pH and temperature in that study. However, the overall consistency in k_2 values among species and studies further suggests that elimination of HOCs is highly dependent upon compound hydrophobicity, rather than other factors such as organism physiology, and can be generally described by a first order, one-compartment kinetic model (Morrison et al. 1995).

Smaller k_2 values were observed for HOCs with log K_{ow} values greater than seven. Van Haelst et al. (1996a) reported a range of 0.005–0.037/day in *D. polymorpha* for TCBT congeners ranging in log K_{ow} from 6.73 to 7.54, which were approximately two to three times less than other literature k_2s. For the same K_{ow} range, Morrison et al. (1995) demonstrated a k_2 range of 0.042–0.098/day for *D. polymorpha*, whereas Brieger and Hunter (1993) reported an elimination rate constant of 0.034/day for PCB congener 169 (log K_{ow} 7.42) in *D. polymorpha*. There seems to be a plateauing effect of k_2 values observed for HOCs with log K_{ow}s greater than about six (Morrison et al. 1995; Gewurtz et al. 2002; Thorsen et al. 2004a).

The k_2 rates reported in *C. leana* for the pesticides chlornitrofen and thiobencarb were small and varied little between laboratory and field studies, ranging from 0.045 (field) to 0.054/day (lab) for chlornitrofen and 0.049 (lab) to 0.060/day (field) for thiobencarb (Uno et al. 1997). However, these k_2 values do not follow the trend seen with K_{ow}, as the pesticide K_{ow}s are less than those for the TCBT congeners. These compounds may be metabolized to some extent by the mussels.

Elimination of HOCs in marine mussels was found to be moderately variable, ranging from rapid elimination (less than 4 days) (Pittinger et al. 1985) to no measureable depuration in 45 days (Tanacredi and Cardenas 1991). However, elimination rate constants for *M. edulis* (calculated from data in Pruell et al. 1986), a short-necked clam (species not identified) (Ogata et al. 1984) and *Crassostrea virginica* (Sericano, Wade, and Brooks 1996) exhibited similar ranges for PAHs and PCBs as those reported for freshwater bivalves. The consistency in regression equations between freshwater and marine bivalves, PAH, and PCB classes and field and laboratory studies is remarkable. A wider range in slopes and y-intercepts was observed, however, if all data for *C. virginica* (Eastern oyster) and *Mercenaria mercenaria* (Hard Clam) were included, which demonstrated negative slopes and little depuration in some instances (Tanacredi and Cardenas 1991). Further research is required for a more robust comparison of freshwater and marine bivalves and for a more complete understanding of why differences in elimination rates occur.

ATTAINMENT OF STEADY-STATE

While most bivalve uptake kinetics are generally rapid for HOCs, there is a broad range of time required to reach steady-state (Table 8.2a, b). For example, Thorsen (2003) found that steady-state was reached between the water column and mussel tissue for most PAHs within the first 24–48 hours of exposure and that all PAHs reached steady-state within 10 days. Likewise, the uptake kinetics of *D. polymorpha* are rapid (Bruner, Fisher, and Landrum 1994; Morrison et al. 1995; Gossiaux, Landrum, and Fisher 1996), which enables them to reach steady-state quickly (within a few hours, for lower hydrophobicity chemicals). In contrast, Pruell et al. (1986) reported a longer time interval to steady-state of 20 days for *M. edulis*, which may have been due to slow contaminant desorption from the sediment slurry source into the water. Moreover, Brieger and Hunter (1993) found that a steady-state for hexachlorobenzene in *D. polymorpha* required 20 days, whereas PCB congener 77 did not appear to reach steady-state, even after 30 days of exposure. The attainment of steady-state will depend on mussel lipid content, metabolic capabilities, physical– chemical characteristics of the compound (highly hydrophobic compounds may require longer time periods for equilibrium), and the availability of the chemical. In studies where sediment serves as the primary exposure media, slow chemical desorption from sediment particles may influence the time required to reach steady state (e.g., Pruell et al. 1986).

BIOACCUMULATION AND BIOAVAILABILITY

Many studies have demonstrated mussel uptake of HOCs from water, contaminated food and sediment (Augenfield et al. 1982; Brieger and Hunter 1993; Pruell et al. 1993; Bjork and Gilek 1997; Gossiaux, Landrum, and Fisher 1998). The uptake rates of PCB 77 in *D. polymorpha* increased when the exposure environment was altered to include food and sediment, in the following increasing order: water > food > sediment (Brieger and Hunter 1993). In contrast, Thorsen (2003) did not observe consistent differences in PAH BCF/BAF values determined for *E. complanata* in water-only (BCF) versus sediment (BAF) exposure studies, indicating that sediment PAH concentrations were driving the water exposure concentrations for *E. complanata*. The situation was similar when *E. complanata* were allowed to burrow into the sediment phase in both field and laboratory studies (Thorsen 2003). With the exception of benz(a)anthracene, statistically significant differences were not observed in tissue PAH burdens between *D. polymorpha* placed in the upper water column versus the sediment surface at a confined disposal facility (Roper et al. 1997). However, this may not be the situation for deposit-feeding bivalves or for bivalves exposed to PCBs, which have generally been shown to be more bioavailable than PAHs of similar physical–chemical characteristics (Lamoureux and Brownawell 1999; Kraaij et al. 2002).

While the addition of food and sediment in exposure environments may result in increased accumulation, these factors may also result in decreased bioavailability of HOCs by serving as binding agents that sequester them (Kraaij et al. 2002). Many factors can affect HOC bioavailability, including feeding and digestion mechanisms of the bivalve, as well as rates of sorption and desorption between the HOC and particle/sediment (Kraaij et al. 2002), concentrations of dissolved and particulate organic carbon and soot carbon, and aging of contaminants and sediments (Readman, Mantoura, and Rhead 1984; Schrap and Opperhuizen 1990; Gustafsson et al. 1997; Alexander 2000; Bucheli and Gustafsson 2000; Accardi-Dey and Gschwend 2002; respectively). Sediments often act as a reservoir for HOCs and modulate the corresponding water concentrations (Spacie 1994). Thus, the observation of high total-HOC sediment concentrations may not always correspond to high exposure levels to organisms because of decreased bioavailability. For instance, HOCs that are bound to sediment or particulate matter in the water column can exhibit slow desorption rates, rendering them essentially unavailable to mussels. Thorsen et al. (2004b) noted BSAF values in *E. complanata* of less than one for pyrogenic PAHs (PAHs of combustion origin). These authors observed that the greater the concentration of soot carbon, the lower the PAH bioavailability, depending on PAH source (petroleum versus combustion origin). At a creosote-contaminated site (expected to contain relatively bioavailable PAHs), BSAF values for *A. anatina* ranged from 0.79 to 1.45 for the six PAHs, acenapthene, phenanthrene, anthracene, fluorene, pyrene, and benz(a)anthracene (Hyotylainen, Karels, and Oikari 2002). Biota-sediment accumulation factors for phenanthrene were close to one, ranging from 0.80 to 0.96 (Hyotylainen, Karels, and Oikari 2002). In comparison, BSAF values have been reported for phenanthrene in *M. arenaria* (marine) and *E. complanata* that ranged from 0.17 to 1.80, depending on environmental location (PAH source) (Thorsen, Cope, and Shea 2004b).

Other studies have calculated assimilation efficiencies, rather than BSAF values to estimate the bioavailable fraction of HOCs to mussels (Morrison et al. 1996; Gossiaux, Landrum, and Fisher 1998). For example, *D. polymorpha* assimilation efficiencies (AE) were measured for PY, BaP, C0; and HCBP sorbed to algal (food) particles versus those sorbed to suspended sediment (Gossiaux, Landrum, and Fisher 1998) and demonstrated that the availability was nearly 100% assimilated from algae, but only 45–58% assimilated from suspended sediment particles. Even lower AEs from sediment particles (21%) were observed for BaP in *D. polymorpha*, and a positive correlation was noted between AE and log K_{ow} (Gossiaux, Landrum, and Fisher 1998). However, the AE from each source will vary based on factors such as algal lipid content, particulate organic carbon, and particle size. Although systems are not always at equilibrium, route of exposure is unimportant when a system is at equilibrium (or pseudo/apparent steady-state), and steady-state knowledge still serves as a simple model for predicting bioavailability of HOCs to bivalves and other benthic invertebrates. The incorporation of the many differences in species physiology and interactions with sources of contamination can quickly complicate the model. However, the difference in log BCF/BAFs of only two orders of magnitude (4.4–6.8) between HOCs with similar log K_{ow} values (Table 8.1b) among studies with exposures of water only, water and food, and the inclusion of sediment, temperature changes, mussel species, size, and lipid content changes suggests that simplification in modeling through the use of equilibrium partitioning theory is appropriate (e.g., Di Toro et al. 1991). Additional toxicological data would aid in the future comparison and summary of results.

IMPLICATIONS AND POTENTIAL FOR HOC TOXICITY

The fact that bivalves bioaccumulate HOCs from various environmental compartments necessitates the understanding of toxicokinetics to be able to assess and predict the potential for subsequent toxicity. The bioaccumulation that occurs at environmentally relevant concentrations (ng–μg/L aqueous concentrations or ng/g sediment concentrations) typically manifests in chronic, sublethal effects, rather than acute consequences such as mortality. It is often difficult to tease out specific consequences associated with contaminant exposure because other types of stresses can also induce adverse effects. However, monitoring biomarkers of exposure or effect (such as alterations of reproductive health, changes in filtration and/or ventilation rate and lipid content, reattachment success, DNA damage, lipid peroxidation, and EROD or metallothionein induction) can be extremely valuable (see Chapter 11).

METALS

UPTAKE

Metal uptake by bivalves is more complex than that for HOCs because of the increased number of interactions that can occur between metals and environmental ligands, such as dissolved organic and inorganic matter, and other complexing and competing components (Table 8.5). Metal uptake primarily involves interaction with active sites, usually on gills, and therefore follows isotherms such as the Fruendlich or Langmuir isotherms (Spacie 1994; Churchill, Walters, and Churchill 1995). However, simpler models have been used to describe metal uptake into a bivalve based on the dissolved water concentration (C_{water}) and the metal influx rate (I) (Wang 2001):

$$I = k_1 C_{water}$$

where k_1 is the uptake rate constant, and C_{water} is the metal concentration in the dissolved phase. The bioenergetic-based kinetic model is also used to assess and predict bioaccumulation of toxic metals in bivalves (and other aquatic organisms). This model requires knowledge of uptake efficiencies from water and food, as well as depuration, filtration, ingestion, assimilation, and growth rates (Roditi and Fisher 1999).

The amount of metal taken up by a bivalve will depend on characteristics of the metal, the species physiology, and the environmental conditions of the water column and sediment phase. For example, metal speciation depends on whether it is freely dissolved or bound to inorganic (OH–, Cl–) or organic (DOC, oxalate) matter, which subsequently affects its bioavailability. Additionally, competition between the metal and other cationic ligands for anionic ligands or anionic biological targets may occur. In terms of water conditions, factors such as hardness, alkalinity, pH, DOC, salinity/conductivity, total dissolved solids, and anthropogenic inputs can influence metal behavior. Metal uptake in bivalves is affected by numerous factors including the reproductive state of the organism, feeding strategy (filter feeder versus deposit feeder), length,

TABLE 8.5
Summary of Complex Interactions between Mussel, Water, and Sediment That Influence Behavior, Speciation, and Accumulation of Metals

Mussel	Water	Sediment	Metal
Gut retention time	Dissolved organic carbon concentrations	Acid-volatile sulfide concentrations	Bound versus unbound
Feeding mechanism	Particulate organic carbon concentrations	Total organic carbon	Easily labile versus not easily exchangeable
Ingestion rate	Minerals, other metals present	Total inorganic carbon	Distribution coefficients under various conditions
Assimilation rate	Total dissolved solids	Minerals, other complexing reagents such as Fe, Al, etc.	
Absorption efficiency	Hardness, alkalinity, pH		
Site saturation	Other anionic/cationic ligands present		
Miscellaneous physiology	Ionic strength/salinity		

weight and age of the organism, the source(s) of food, and assimilation and absorption efficiencies (Plette et al. 1999; Roditi and Fisher 1999; Stewart 1999; Roditi, Fisher, and Sanudo-
Wilhelmy 2000; Wang 2001).

Bivalves bioconcentrate and bioaccumulate both essential (e.g., Ca and Na) and non-essential
(e.g., Hg, Cd, and Pb) metals to high levels in their tissues. This can frequently result in concentrations that are significantly higher than the concentration present in the water column and can
cause toxicity to the bivalve as well as to predatory organisms, such as aquatic birds or terrestrial
mammals (Naimo 1995; Stuijfzand et al. 1995). Metals are taken up by bivalves via facilitated
diffusion, active transport, or endocytosis (Marigomez et al. 2002). Some non-essential metals
such as Cd can become incorporated into calcium and sodium active pump channels designed for
essential metal uptake and regulation of ionic/osmotic balance (Stewart 1999). Because these
active ion pumps are required for normal homeostasis, bivalves possess the ability to regulate
certain essential metals (e.g., Zn and Cu) to a point. This results in aqueous metal concentrations
that are not directly proportional to the corresponding tissue burden (Kraak et al. 1994b, 1994c;
Blackmore and Wang 2003). For example, Zn and Cu were not accumulated by *D. polymorpha*
exposed to low concentrations (less than 28 μg Zn/L and less than 191 μg Cu/L). The authors
suggested that this inferred a homeostatic regulation of these metals (Tessier and Blais 1996).
Homeostatic regulation can modulate metal accumulation to a threshold level. Once this threshold
level is surpassed, bioaccumulation and potential toxicity will occur.

Metal uptake takes place primarily on active sites in bivalve gills, although additional uptake
can occur in the digestive tract from ingested food particles and in the mantle, kidney, foot, and
hepatopancreas (Inza, Ribeyre, and Boudou 1998; Plette et al. 1999). The percentage of metal that
is absorbed from the total amount of water pumped through is termed the absorption efficiency (α).
The percentage of ingested metal that crosses the gut lining is termed the assimilation efficiency
(AE) (Roditi and Fisher 1999) and will depend on many factors including the length of time
required for processing, the particle type, the metal concentration associated with the particles,
and the organism's ingestion rate (Fan, Wang, and Chen 2002). For some bivalves, assimilation
efficiencies of Cd associated with different food sources have been reported to be greater than 20%
(Fan, Wang, and Chen 2002). Uptake can vary significantly based on the individual metal or
mixtures of metals studied. Frequently, metal concentrations are reported in burden (e.g., total
ng = concentration (ng/g)×tissue weight (g)) to minimize effects of size differences and/or growth
during a study and for cross-study comparisons (Naimo 1995; Beckvar et al. 2000).

TOXIC, NONESSENTIAL METALS

CADMIUM

Uptake and Accumulation

Mussels accumulate cadmium (Cd) across their gills by active uptake (binding to membrane
transported ligands or incorporation of Cd into a major ion active pump) (Stewart 1999) as well
as phagocytosis by cytosomes (Roseman et al. 1994; Marigomez et al. 2002). Roditi, Fisher, and
Sanudo-Wilhelmy (2000) reported that Cd uptake in *D. polymorpha* primarily resulted from
ingested particles rather than the dissolved phase (see Table 8.6 for a summary of Cd studies).
Moreover, Cd uptake by *Lamellidens marginalis* was shown to be primarily from the water column
through mussel gills, rather than from the sediment phase (Jana and Das 1997). Certain freshwater
mussels accumulate Cd to such an extent that some have proposed their use as biofilters in an
attempt to remediate moderately contaminated sites (Jana and Das 1997). Cadmium uptake by
freshwater mussels has been shown to be highly dependent on concentration, duration of exposure
(Das and Jana 1999), mussel length (Roseman et al. 1994), and detoxification mechanisms, and
therefore, great variation is observed. For instance, Cd accumulation in freshwater mussels ranged

TABLE 8.6
Summary of Toxicokinetic Measurements and Primary Findings by Metal for Various Freshwater and Marine Mussels in the Peer-Reviewed Reference

Metal	Species	Measurement(s)	Primary Findings	References
Cd	D. polymorpha	900 μg/g tissue after 51 days of exposure to 50 μg/g	Uptake primarily from ingested particles	Roditi, Fisher, and Sanudo-Wilhelmy (2000)
Cd	M. edulis			Everaarts (1990)
Cd	L. marginalis	5.8–600.0 μg/g tissue	Uptake primarily via gills from water column rather than from sediment phase	Jana and Das (1997)
Cd	E. complanata	65 μg/g tissue after 60 days of exposure to 50 ppb	Uptake dependent on dose, time of exposure; steady-state requires > 40 days	Tessier, Vaillancourt, and Pazdernik (1994)
Cd	L. marginalis	BCF: 35–280	10–20 ppm Cd, accumulation in liver highest, 30 ppm accumulation in gills	Das and Jana (1999)
		BAF: 1.1–5.0 (sediment exposure)		
Cd	D. polymorpha	AF: 3,000–640,000, k_i: 0.22–0.45 μg/mussel/hour (20–30 mm mussels)	Uptake dependent on mussel length, no uptake in smaller (10 mm) mussels	Roseman et al. (1994)
Cd	C. fluminea	Gill: 1,500 ng/g tissue from exposure to 68 mg/L, 30 day depuration: no loss; 120-day depuration: 25% loss (22% gills)	Highest [Cd] = visceral mass, mantle, foot; 4 days to steady-state; change in T and pH, no effect on elimination	Inza, Ribeyre, and Boudou (1998)
Cd	U. pictorum	BCF: 660	Rapid uptake; steady-state (kidney) > 3 weeks; 29-week elimination: 33% loss	Jenner et al. (1991)
Cd	E. complanata		Mussel length: no significant effect on body burdens	Tessier, Vaillancourt, and Pazdernik (1996)
Cd	D. polymorpha		Exposure concentrations: 0.30–44 μg/L; no steady-state after 27 days of exposure	Mersch, Morhain, and Mouvet (1993)
Cd	A. spp.	BCF: 981; BAF: 0.06 (sediment)	Field collected mussels; low Cd bioavailability	Dobrowoski and Skowronska (2002)
Cd	D. polymorpha			Winter (1996)
Cd	Anodonta anatina			
Cd	E. complanata	BAF: 144.5–2,089.3	Poor correlation between water and mussel concentrations	Campbell and Evans (1991)
Hg	E. complanata	Elimination Hg(II) ≫ CH_3Hg	Concentrations increase with temperature; CH_3Hg bioavailability ≫ Hg(II)	Beckvar et al. (2000)

Metal	Species	Parameters	Findings	Reference
Hg	D. polymorpha		Mussel concentrations dose and time dependent	Tessier and Blais (1996)
Hg	C. fluminea	CH_3Hg: 5,750 ng/g dry weight, all tissues; Hg: 850 ng/g dry weight all tissues	Linear uptake over 14 days in all organs except gills (steady-state 4 days). CH_3Hg uptake $\gg$ Hg(II): differences in distributions	Inza, Ribeyre, and Boudou (1998)
CH_3Hg	P. grandis		CH_3Hg highest in foot/kidney; lowest in gills/viscera	Malley, Stewart, and Hall (1996)
Hg	D. polymorpha	BAF: 14,000–250,000, depending on field location; K_ds reported	81% Hg on particles; accumulation primarily through ingested particles	Roditi, Fisher, and Sanudo-Wilhelmy (2000)
Hg	D. polymorpha	k_2: 0.043–0.056/day; $T_{1/2}$:13–16 days	4–40% assimilation from food; k_2s determined after uptake of four different food types	Roditi and Fisher (1999)
Pb	M. edulis		Slow uptake; very slow elimination: 33% eliminated after 2–3 years	Riget, Johansen, and Asmund (1997)
Pb	Anodonta spp.		Highest [Pb]:gills, across field locations; lowest [Pb]:shells	Gundacker (2000)
Pb	U. Pictorum		Low Pb bioavailability	
Pb	M. galloprovincialis	BAF: 211 (seawater); biphasic elimination	21-day exposure; $T_{1/2}$: 1.4 days, 2.5 months (biphasic); 46% distribution, shell; 49% distribution, soft tissues	Boisson, Cotret, and Fowler (1998) Marigomez et al. (2002)
Pb	M. galloprovincialis		Distribution in digestive granules	
Pb	D. polymorpha		Distribution in hemocytes	Gundacker (1999)
Pb	D. polymorpha		Significantly higher concentrations in bysall threads than in soft tissues: mechanism for detoxification	
Pb	Anodonta spp.	BCF: 250 $\gg$ BAF: 0.1 (sediment)		Jenner et al. (1991)
Ag	D. polymorpha	k_1: 3,670 mL/g day; k_2: 0.07–0.09/day, $T_{1/2}$: 8–9 days; BAF: 4.3–5.1	Variations by field location; k_1s higher than for other metals; primary accumulation from ingested particles: 64–90% accumulation from food	Roditi, Fisher, and Sanudo-Wilhelmy (2000)
Ag	M. galloprovincialis		No loss in 4 weeks of exposure; uptake: food+water $\gg$ water only $\gg$ food only	Baud, Amiard-Triquet, and Metayer (1990)

(continued)

Table 8.6   (Continued)

Metal	Species	Measurement(s)	Primary Findings	References
Ni	D. polymorpha		Body burdens increase with increasing [Ni] in water	Stuijfzand et al. (1995)
Ni	M. edulis		Body burdens increase with increasing [Ni] in water	Freidrich and Filice (1976)
Ni	L. marginalis		Accumulation occurs in hepatopancreas, mantle, adductor muscle, and foot when exposed to acute + subacute [Ni] over 15 days	Sreedevi et al. (1992)
Ni	M. edulis	BCF: 432	BCF Ni>Cr, Pb but <BCF for Cd, Zn, Cu	Lee and Kwon (1994)
Sn			Uptake greatest in gills, viscera, adductor+ mantle; uptake is concentration dependent	Moore (1991)
TBSn	D. polymorpha	k_1: 25,000/day; k_2: 0.027/day; log BCF: 5; $T_{1/2}$: 26 days	Field exposure; slow elimination (105 days); DBSn steady-state in 35 days, but accumulation ≪ TBSn	Van Slooten and Tarradellas (1994)
TBSn	M. edulis	Log BCF: 3–4; log BAF<0.3 (contaminated food); k_2: 0.17–0.36/day (6-month elimination); $T_{1/2}$: 14 days	Accumulation greatest in gills, viscera, adductor+mantle; no steady-state in 47 days (water), 30 days (food)	Laughlin, French, and Guard (1986)
TBSn	D. polymorpha	Log BCF: 4.0–4.6		Fent and Humn (1991)
Sn	M. edulis	Total Sn/TBSn: log BCF: 3.7/4.8; $T_{1/2}$: 25 days/40 days	51-day uptake and exposure	Zuolian and Jensen (1989)
TBSn	Marine mussels	Log BCF: 3.7		Shawkey and Emons (1998)

from 5.8 to 600 µg/g depending on species, exposure time, and Cd concentration (Jana and Das 1997). Specifically, *E. complanata* accumulated Cd to 65 µg/g after a 60-day exposure to 50 µg/L Cd, while *M. edulis* accumulated 900 µg Cd/g during the course of a 51-day exposure to 50 µg/g Cd (Jana and Das 1997). Moreover, a maximum uptake for *L. marginalis* exposed to low (10 and 20 mg/L) Cd concentrations occurred in the liver, whereas at high Cd concentrations (30 mg/L), the gill was the primary site of uptake (Das and Jana 1999). This is also the situation for Cd in *C. fluminea*, which exhibited gill concentrations as high as 1,500 ng/g wet weight from exposure to 68 mg Cd/L aqueous concentrations (Inza, Ribeyre, and Boudou 1998). The next highest concentrations occurred in the visceral mass, mantle, and foot, in descending order (Inza, Ribeyre, and Boudou 1998). Uptake of Cd in *Unio pictorum* was rapid, and the kidney was the primary organ of Cd accumulation (Jenner et al. 1991). Roseman et al. (1994) reported the Cd absorption rate was directly related to *D. polymorpha* and *Dreissena rostriformis bugensis* mussel length, whereas others noted that *E. complanata* mussel length had no significant effect on Cd body burdens (Tessier, Vaillancourt, and Pazdernik 1996).

Uptake rates of Cd reported in *D. polymorpha* and *D. rostriformis bugensis* (20–30 mm shell length), ranged from 0.22 to 0.45 µg/mussel/h and 0.29 to 0.49 µg/mussel/h over a 24-hour exposure period. Measureable decreases in concentrations of Cd in water were evident upon the addition of mussels to the exposure system (Roseman et al. 1994). They also noted that appreciable uptake did not occur in the smaller mussels (i.e., 10 mm) for either of these species, possibly due to inhibition of filtration by Cd, as 50% decreases in filtration rates of *D. polymorpha* (16–22 mm length) have been previously reported (Roseman et al. 1994). Cadmium uptake has been shown to decrease in the presence of other metals such as Cu, Zn, Pb, and Ni. Even though measured aqueous Cd concentrations increased, *Pyganodon grandis* accumulated less Cd (Stewart 1999) when the other metals were present. Moreover, Cd filtration times of *A. cygnea* decreased from 20 to 8 hours in the presence of 10 µg Cu/L for a 240-hour exposure and from 30 to 60 down to 7 hours with 50 µg Pb/L, also for a 240-hour exposure (Stewart 1999).

Steady-State and Bioconcentration

Various times required to reach steady-state have been noted for Cd, ranging from 4 days for *C. fluminea* (Inza, Ribeyre, and Boudou 1998) to greater than 40 days for *L. marginalis* (Das and Jana 1999). This wide range is probably due to differences in mussel species as well as exposure environment; the 4-day steady-state occurred in a water exposure, whereas the 40-day equilibrium occurred in a sediment and water exposure (Das and Jana 1999). Inza, Ribeyre, and Boudou (1998) suggested that steady state was reached within four days of Cd exposure in *C. fluminea* because of the saturation of binding sites in the clam gut, an increased elimination rate, or a physiological modification of filtration activity. In another study in which *D. polymorpha* were exposed to Cd concentrations ranging from 0.30 to 44 µg/L, steady-state was not attained after 27 days of exposure (Mersch, Morhain, and Mouvet 1993). Steady-state was achieved for Cd in the kidney of *U. pictorum* within three weeks of exposure in field and laboratory experiments with and without substratum (Jenner et al. 1991).

As expected, tissue concentration factors (BCFs) are influenced by metal concentration in the exposure media. For example, BCFs for Cd in *L. marginalis* ranged from 35 to 280, depending on Cd water concentration and tissue (i.e., the gill, liver, shell, or mantle). Concentrations varied greatly when sediment exposure was included, and BAFs were much less ranging from 1.5 to 5 (Das and Jana 1999). These values would, most likely, significantly underpredict actual equilibrium bioaccumulation because in all cases, the uptake curves were linear even after 40 days (Das and Jana 1999). In comparison, accumulation factors for Cd in *D. polymorpha* ranged from 3,000 in whole body to as high as 70,000 in the periostracum (Roseman et al. 1994) and from 150,000 to 640,000 at different field sites (Roditi, Fisher, and Sanudo-Wilhelmy 2000). However, Jenner et al. (1991) reported a much lower BCF value of approximately 660 (6,000 dry weight converted to wet weight) for *U. pictorum* kidneys. A similar BCF value (981) was reported for *Anodonta sp.* collected from the field. Bioaccumulation factors comparing *Anodonta sp.* tissue Cd concentrations to sediment Cd concentrations were significantly lower at 0.06, suggesting a decreased bioavailability of the Cd associated with the sediment phase (Dobrowoski and Skowronska 2002). Winter (1996) reported BAFs in different freshwater mussels (ranging from 1,300 to 19,000 dry weight, approximately log BAF 2.16–3.32 wet weight) and demonstrated higher Cd concentrations in *D. polymorpha* than *A. anatina*. Because of the complexity of the interactions with Cd and other metals in the environment, a comparison of accumulation from concentrations in food, sediment, or water to corresponding tissue burdens is difficult among species and studies. For instance, poor correlations between organism Cd concentrations and Cd water concentrations have been observed (Campbell and Evans 1991).

Elimination

Elimination of Cd from bivalves has also been shown to be concentration and time dependent, as well as highly variable. Following a 14-day depuration phase, *D. polymorpha* exhibited only 5% Cd loss, whereas no loss of Cd was reported in *C. fluminea* after 30 days of depuration. Even following 120 days of depuration, only 25% of the original Cd in *C. fluminea* had been eliminated (Inza, Ribeyre, and Boudou 1998), and 22% of the Cd loss occurred from the gills (Inza, Ribeyre, and Boudou 1998). Changes in temperature (12 versus 24°C) and pH (6 versus 8) had no effect on Cd elimination at the organism and individual organ levels (Inza, Ribeyre, and Boudou 1998). However, it cannot be ruled out that greater changes in temperature and pH could influence Cd elimination. After an initially rapid loss of approximately 33% of Cd in *U. pictorum*, elimination was negligible or did not occur over the course of 29 weeks (Jenner et al. 1991).

MERCURY

Uptake and Accumulation

Both methyl mercury (CH_3Hg) and inorganic mercury (Hg^{2+}) have been shown to accumulate in mussels (Malley, Stewart, and Hall 1996; Beckvar et al. 2000). Similar to Cd, exposure duration and aqueous Hg concentrations affect mussel body burdens (Tessier, Vaillancourt, and Pazdernik 1996). Additionally, Hg concentrations were shown to increase with increasing temperature in *E. complanata* (Beckvar et al. 2000), and methyl Hg was demonstrated to be more bioavailable to *E. complanata* than inorganic Hg (Beckvar et al. 2000). Cope et al. (1999) found that *D. polymorpha* accumulated significant quantities of Hg during a 143-day period in the Upper Mississippi River and that about 50% (range 30–70%) of the mean total Hg in zebra mussels was methyl Hg. Linear uptake over the course of 14 days was observed in all organs and tissues (mantle, foot and adductor, gills, and visceral mass) in *C. fluminea* for both inorganic Hg and methyl Hg with the exception of the gills, where uptake appeared to plateau following 4 days of exposure (Inza, Ribeyre, and Boudou 1998). Moreover, concentrations of methyl Hg (average 5,750 ng/g dry weight in all tissues) were nearly seven times those for inorganic Hg (average 850 ng/g dry weight in all tissues) and demonstrated differences in distribution (Inza, Ribeyre, and Boudou 1998). Concentrations were greatest in the foot, adductor muscles, and gills for methyl Hg but in the gill and visceral mass for inorganic Hg. In comparison, methyl Hg concentrations measured in *P. grandis* were greatest in the foot or kidney and least in the gills and visceral remains (Malley, Stewart, and Hall 1996). In studies comparing both water and sediment accumulation, differences in Hg phase distribution were noted. For inorganic Hg, 26% associated with the sediment, and only 2.7% accumulated in *C. fluminea* tissues, versus 14% of methyl Hg distributed to the sediment phase and 31.2% partitioned into *C. fluminea* tissues (Inza, Ribeyre, and Boudou 1998). Bioaccumulation factors for Hg in *D. polymorpha* have been reported that range from 14,000 to 250,000, depending on field location (Roditi, Fisher, and Sanudo-Wilhelmy 2000). Moreover, the calculation of particulates, such as aqueous partition coefficients (K_ds) (81% of Hg determined to be on particles), has shown that Hg is primarily accumulated in *D. polymorpha* through ingested particles (Roditi, Fisher, and Sanudo-Wilhelmy 2000), and 4–40% is assimilated from food (Roditi and Fisher 1999).

Elimination

Elimination of Hg from bivalve tissues generally appears to be slow. After 30 days of depuration, no significant loss of inorganic or methyl Hg was observed in *C. fluminea*, yet after 120 days, 30% of inorganic Hg was eliminated from mussel tissues, while 40% of methyl Hg was eliminated (Inza, Ribeyre, and Boudou 1998). Inorganic Hg has been shown to be depurated more rapidly than methyl Hg by *E. complanata* (Beckvar et al. 2000). Changes in temperature (12 versus 24°C) and pH (6 versus 8) were shown to have little to no effect on Hg elimination (Inza, Ribeyre, and Boudou 1998), but again, it should be noted that greater changes in temperature and pH could influence Hg elimination. Mercury efflux (depuration) rates determined after uptake from four different types of food (*Thalassiosira pseudonana*, *Chlorella vulgaris*, *Microcystis aeruginosa*, and a seston assemblage) ranged from 0.043 to 0.056/day, corresponding to biological half-lives of 13–16 days (Roditi and Fisher 1999).

LEAD

Uptake and Accumulation

Lead (Pb) uptake by mussels appears to be slow (Riget, Johansen, and Asmund 1997) and primarily associated with the gill, at least soon after initial exposure has occurred (Table 8.6). For example, in native *Anodonta sp.* and *U. pictorum*, the gills contained the greatest Pb concentrations among all field sites examined, whereas the least average Pb concentrations for both mussel species were found in the shells (Gundacker 2000). This author also noted that while other metal concentrations exceeded those in the environment, this was not the case for Pb, suggesting a decreased bioavailability (Gundacker 2000). This pattern is consistent with other reports demonstrating significant binding of Pb to humic acids and subsequent decreased bioavailability to *E. complanata* (Campbell and Evans 1987). In contrast to the greatest Pb accumulation occurring in the gill of freshwater mussels, Boisson, Cotret, and Fowler (1998) described nearly equivalent distribution in the marine mussel, *Mytilus galloprovincialis* with 49% of Pb in soft tissues and 46% in the shell, which they speculated meant similar uptake rates for both over the course of a 21-day exposure. Pb has been documented in digestive cells of *M. galloprovincialis* and in hemocytes of *D. polymorpha* at the electron-microscopy level (Marigomez et al. 2002). However, Gundacker (1999) reported significantly higher concentrations of Pb in *D. polymorpha* byssal threads, as opposed to soft body parts. The authors suggested the allocation of Pb to byssal threads was a mechanism for detoxification. The BCF value for Pb was much greater than the BAF value reported for *Anondonta sp.* in the field, where the BCF was 250 and the BAF (comparing *Anodonta sp.* tissue Pb concentrations to those in sediment) was 0.10 (Jenner et al. 1991). A BAF of 211 was reported for *M. galloprovincialis* in Pb based on exposure in seawater (Boisson, Cotret, and Fowler 1998). These authors noted that this BAF is two orders of magnitude lower than other BAFs for Pb, although it compared well to that reported for *Anodonta sp.* in the field and suggested it may be due to inefficient accumulation by *M. galloprovincialis* from water (Boisson, Cotret, and Fowler 1998).

Elimination

Elimination of Pb from mussels is typically very slow (Riget, Johansen, and Asmund 1997). For instance, elimination of Pb in *M. edulis* transplanted from a contaminated field site to an uncontaminated reference site eliminated only half of the original Pb contamination in two to three years, after which elimination appeared to cease (Riget, Johansen, and Asmund 1997). However, Pb elimination in *M. galloprovinicalis* was described as biphasic, with short-term compartments exhibiting a biological half-life of 1.4 days and a much slower secondary release of 2.5 months (Boisson, Cotret, and Fowler 1998). Despite the differences in these results, it is clear that Pb kinetics in mussels are slow and need to be considered when bivalves are used as sentinels of Pb pollution in the environment.

SILVER

Uptake and Accumulation

Although Ag concentrations are usually fairly low in invertebrates, it is frequently detected in organism tissues, and BCF/BAF values can be quite high (Moore 1991, Table 8.6). Roditi, Fisher, and Sanudo-Wilhelmy (2000) reported an uptake rate constant for Ag in *D. polymorpha* of 3,670 mL/g day, determined in a laboratory radiotracer study. Presumably, uptake rates would vary depending on water chemistry, species, organism differences, and other factors. The uptake rates of Ag relative to other heavy metals in this study (Cd, Cr, and Hg) were quite high (approximately 1.9, 6.1, and 1.6 times higher, respectively). In comparison, Ag accumulation and toxicity were varied for marine mussels; Ag concentrations in *M. galloprovincialis* did not reach steady-state in four weeks of exposure (Baud, Amiard-Triquet, and Metayer 1990). This study demonstrated that Ag uptake in *M. galloprovincialis* was greatest with exposure in food and water, lower with exposure in only water, and least with exposure only to food. In contrast, Roditi, Fisher, and Sanudo-Wilhelmy (2000) predicted with a bioaccumulation model that *D. polymorpha* primarily accumulated Ag from ingested particles, where anywhere from 64 to 90% of bioaccumulated Ag originated from food, depending on field characteristics and location. Obviously, differences in mussel species, feeding mechanisms, routes of exposure, and water chemistry will influence Ag accumulation. For instance, log BAFs reported for Ag in *D. polymorpha* ranged from 4.3 to 5.1, depending on field location (Roditi, Fisher, and Sanudo-Wilhelmy 2000), and in this study were lower than BAFs reported for Cd and Hg.

Elimination

Elimination of Ag from *D. polymorpha* is slow relative to uptake. For example, reported elimination rate constants range from 0.07 to 0.09/day, depending on uptake from food or water, respectively (Roditi, Fisher, and Sanudo-Wilhelmy 2000). In a separate study by the same authors, elimination rate constants in *D. polymorpha* for Ag were 0.067/day after exposure to food and 0.084/day after exposure to Ag-contaminated water. Differences in sources of Ag during uptake appear to influence the rates of subsequent metal elimination, though the significance of these differences remains unknown. Biological half-lives reported for Ag in *D. polymorpha* are 8–9 days, significantly shorter than those reported for other heavy metals such as Cd (e.g., 51 days) (Roditi, Fisher, and Sanudo-Wilhelmy 2000).

NICKEL

Although Nickel (Ni) is consistently present in the environment (Moore 1991; Stuijfzand et al. 1995), it is generally less toxic than other heavy metals such as Cd and Hg (Keller and Zam 1991). Consequently, relatively little information exists on its toxicokinetics in freshwater bivalves (Table 8.6). It has been demonstrated, however, that Ni concentrations in both freshwater (*D. polymorpha*) (Stuijfzand et al. 1995) and marine (*M. edulis*) (Friedrich and Filice 1976) mussels increase with increasing Ni concentrations in water. Nickel was reported to accumulate in various mussel tissues in *L. marginalis* including the hepatopancreas, mantle, adductor muscle, and foot, when mussels were exposed to both acute and subacute Ni concentrations over the course of 15 days (Sreedevi et al. 1992). Lee and Kwon (1994) reported a BCF value in *M. edulis* for Ni of 432, much less than that reported for Cd (2,814) and Zn (2,900), about half that reported for Cu (807), but greater than those reported for Cr (228) and Pb (127). Additional research is needed to determine uptake, distribution, and elimination of Ni in freshwater mussels to provide a thorough comparison and assessment of these values.

TIN

Uptake and Accumulation

The uptake of Tin (Sn) by bivalves is generally measured as total Sn, inorganic Sn, or organo-Sn, such as tributyl- or dibutyl-Sn (Table 8.6). The organotins are readily accumulated by mussels as a result of the hydrophobic character of the organic methyl or other alkyl group. In general, Sn concentrations in mussels are reported to be the greatest in gill, viscera, adductor, and mantle (Laughlin, French, and Guard 1986; Moore 1991), and uptake is concentration dependent (Moore 1991). Van Slooten and Tarradellas (1994) reported rapid uptake ($k_1 = 25,000$/day) of tributyltin by *D. polymorpha* in a field environment (freshwater marina), such that tissue concentrations were 63 µg/g after 35 days of exposure (steady-state). This uptake resulted in a log BCF value of approximately six based on dry weight and about five based on wet weight. Additionally, dibutyltin also reached steady-state in *D. polymorpha* after 35 days of exposure, but the accumulation was not as great as that for tributyltin (Van Slooten and Tarradellas 1994). In contrast, Laughlin, French, and Guard (1986) reported steady-state was not reached in *M. edulis* after 47 days of exposure to tributyltin in water and 30 days of exposure to tributyltin in food.

In comparison to the log BCF values reported for tributyltin previously listed (about five based on tissue wet weight), other values for *D. polymorpha* were lower, ranging from 4.0 to 4.6 based on wet weight (Fent and Hunn 1991). The study by Van Slooten and Tarradellas (1994) suggested that the BCF variations were due to lower aqueous tributyltin concentrations (e.g., the lower the water concentrations, the greater the BCF values). In comparison, log BCF values reported for organic and total tin in *M. edulis* ranged from 3.7 to 4.8 (Zuolian and Jensen 1989). These values were calculated following a 51-day exposure, but the authors noted, however, that their BCF values were much higher than others reported in similar laboratory studies (Zuolian and Jensen 1989). Additional log BCF values in marine bivalves (species not identified) for tributyltin were about 3.7 based on wet weight and 4.6 based on dry weight (Shawky and Emons 1998). The log BCF values have ranged from three to four in the laboratory and were predicted to be at least two orders of magnitude greater in the field (Laughlin, French, and Guard 1986). However, log BAF values reported for tributyltin in *M. edulis* as a result of exposure to contaminated food were less than 0.3 (Laughlin, French, and Guard 1986), suggesting the predominant source of accumulation was from the dissolved phase, although uptake of tributyltin was more rapid and resulted in increased tissue concentrations with food exposure (Laughlin, French, and Guard 1986). Salazar and Salazar (1996) reported log BCF values for juvenile *Mytlius sp.* ranging from 3.6 to 5.0, with most of the values from 4.3 to 4.6. There appears to be a wide range of BCF/BAF values documented for tributyltin, depending on species and exposure environment (e.g., aqueous Sn concentrations).

Elimination

Elimination rate constants for Sn and organotin are generally small, reported as 0.027/day for tributyltin in *D. polymorpha* (Van Slooten and Tarradellas 1994), and from 0.17 to 0.36/day for *M. edulis* (Laughlin, French, and Guard 1986). Both studies reported no return to background levels in mussel tissue over the course of 105 days to 6 months, respectively, and in the case of the 105-day depuration, *D. polymorpha* tributyltin tissue concentrations were still twice those of the background mussels (Van Slooten and Tarradellas 1994). Moreover, the elimination of tributyltin in *D. polymorpha* was faster over the first 63 days than compared to the following timepoints (Van Slooten and Tarradellas 1994). Biological half-lives reported for organotin range from 14 days (tribuytltin, *M. edulis*) (Laughlin, French, and Guard 1986) to 25.7 days (tributyltin, *D. polymorpha*) (Van Slooten and Tarradellas 1994) to 40 days (organotin, *M. edulis*) (Zuolian and Jensen 1989). A biological half-life for total Sn in *M. edulis* was reported as 25 days, similar to that for organotin (Zuolian and Jensen 1989).

METAL MIXTURES AND EFFECTS ON TOXICOKINETICS

Various reports have discussed the effects of mixtures of metals on bivalve uptake, distribution, and elimination. Understanding the effects of metal mixtures is particularly important as metals are not usually singly present in the environment, and the effects on toxicokinetics and toxicity to bivalves from a combination of metals is seldom related to those for the individual organisms (Keller and Zam 1991; Kraak 1994a; Kraak, Stuijfzand, and Admiraal 1999; Stewart 1999). Many reports focusing on metal mixtures discuss their effects on mussel filtration rates (Kraak et al. 1993). A constant theme continually demonstrated is the lack of accurate prediction of effects of individual metals from short-term tests to effects of mixtures of metals in both short- and long-term tests (Kraak et al. 1993; Kraak et al. 1994a). For example, Kraak et al. (1994a) found that the additive effects of certain metals (Cu+Zn+Cd) resulted in EC_{50} (effective concentration required to elicit some observed effect such as change in filtration rate in 50% of organisms studied, over a specified time interval) values that were below the NOECs (no observable effects concentration) for the representative individual metals. Specifically, Kraak et al. (1994a) observed less than concentration additive effects on filtration rates in *D. polymorpha* for Cu+Zn, additive effects for Zn+Cd, greater than concentration additive effects for Cu+Cd, and additive effects for the combination of all three metals. Stewart (1999) observed Cd loss from the water column was slower with the addition of a metal mixture containing Cu+Zn+Pb+Ni, such that half-lives in the water increased from 11 days (Cd only) to 22–34 days (Cd+metal mixture at different concentrations), but accumulation of Cd in *P. grandis* decreased, even with increases in aqueous Cd concentrations. Other authors discussed similar results obtained for *Anodonta cygnea* (Hemelraad et al. 1987) and *M. edulis* (Elliott, Swain, and Ritz 1986). While these data demonstrate a clear effect of metal mixtures on mussel toxicity and accumulation, more research is needed to thoroughly assess the implications. Interpreting single metal toxicity and toxicokinetic data requires caution because environmental contaminants are generally represented by mixtures rather than individual metals. Interestingly, the bulk of published research focuses on individual contaminants.

PLATINUM GROUP METALS

The environmental toxicology and chemistry of platinum (Pt) group metals have been studied little until recently. Novel research has now begun to focus on the Pt group metals, which include Pt, palladium (Pd), and rhodium (Rh), to determine whether these elements are bioavailable to bivalves and to assess their potential risk (Sures et al. 2002; Zimmermann et al. 2002). Concentrations of Pt, Pd, and Rh in *D. polymorpha* were low compared to those reported for other metals, ranging from 0.10 to 0.50 ng/g for Rh to 0.90 to 6.2 ng/g for Pd, depending on the exposure environment (Zimmermann et al. 2002). However, a relatively high bioavailability of Pd was documented for *D. polymorpha* exposed to road dust for 26 days, on par with that for Ag, Cd, and Cu, with BAF values ranging from 0.80 to 1.80 depending on water characteristics (dechlorinated tap water versus lake water containing humic material) (Zimmermann et al. 2002). Bioaccumulation factors for Pt and Rh were substantially lower, ranging from 0.60 (Pt) to between 0.04 and 0.20 for Rh (Zimmermann et al. 2002). Sures et al. (2002) found that bioaccumulation of these metals occurs in *D. polymorpha* even when water concentrations are below 0.1 ng/L. Thus, the observations that mussels bioaccumulate these metals demonstrates the need for further study in this area, especially for understanding the fate and effects of these ubiquitous and potentially hazardous contaminants in the environment.

ESSENTIAL ELEMENTS

ZINC, CALCIUM, COPPER: ENVIRONMENTAL INTERACTIONS

Zinc (Zn), calcium (Ca), and copper (Cu) toxicokinetics are unique because they represent essential elements required for biological processes including proper enzyme function, active membrane ion pumps, and normal metabolic homeostasis. Because these metals are actively taken up from the surrounding water and regulated by bivalves, they typically are not toxic at low concentrations, and therefore, uptake is not directly proportional to environmental concentrations (Camusso et al. 1994; Kraak et al. 1994a, 1994c; Mersch, Wagner, and Pihan 1996; Blackmore and Wang 2003). These metals can also interact to impede the uptake of non-essential metals such as Cd (Holwerda 1991). For example, co-exposure can result in uptake inhibition of one of the metals, competition for unique binding sites, or the physical effect of decreased filtration rate of mussels (Holwerda 1991).

Zinc reportedly can be regulated by freshwater and marine mussels at higher environmental concentrations than Cu (Kraak et al. 1994a, 1994c) whereas non-essential elements like Cd generally cannot be regulated by mussels. For example, Kraak et al. (1993) reported that *D. polymorpha* did not accumulate Zn and Cu at low (Zn $<$ 28 μg Zn/L and Cu $<$ 191 μg Cu/L) concentrations, whereas water concentrations of 9 μg Cd/L resulted in accumulation of Cd in *D. polymorpha*. Further evidence for Zn regulation in mussels is the consistency and similarity exhibited in Zn concentrations (Blackmore and Wang 2003). Moreover, in *M. edulis*, Cu uptake displaced other metals such as aluminum (Al) and molybdenum (Mo) within the course of 4 hours (Sutherland and Major 1981). Holwerda (1991) demonstrated a significant decrease (90% reduction) in the accumulation factor of Cd when *A. cygnea* was simultaneously exposed to Cu over a 6.5-week exposure period. Copper accumulation was also decreased by simultaneous Cd exposure; however, the effect was much less, decreasing the Cu uptake by only a factor of 2 (versus a factor of 12 for Cu effect on Cd uptake) (Holwerda 1991).

Certain non-essential metals may also share binding sites with essential metals. For instance, Cd is believed to compete for binding sites also used by Zn (Roditi and Fisher 1999) because of its affinity for sulfur (S) ligands (Fan, Wang, and Chen 2002), and both Ca and Mg have been shown to minimize the particulate adsorption of Cd (Roditi and Fisher 1999). Fan, Wang, and Chen (2002) reported similar findings for two marine bivalves (*Perna viridis* and *Ruditapes philippinarum*), where Cd assimilation efficiencies decreased with increasing Zn concentrations. Inverse relationships have been reported between Ca and Cd and Zn and Pb implying that bivalves will compensate for metal accumulation by actively taking up Ca and Zn from the surrounding water in an attempt to contain toxic metal concentrations in interstitial tissues (Jana and Das 1997). These observations correlate with certain freshwater mussels such as *E. complanata*, incorporating toxic metals into calcium-phosphate granules (Jana and Das 1997). [See Metal Detoxification Mechanisms.]

BIOAVAILABILITY

Metal bioavailability can depend on many different environmental factors. Based on the knowledge that simple aqueous and sediment metal measurements generally do not adequately predict metal availability and toxicity, sometimes resulting in expensive and inefficient site remediation, the concept of a water-effects ratio (WER) was developed. The WER attempts to incorporate differences in water characteristics at different locations when assessing metal toxicity (Pendergast et al. 1997). More recently, the biotic-ligand model has been used (Bianchini and Bowles 2002;

Paquin et al. 2002) in an attempt to further understand and assess the bioavailability and potential toxicity of metals in the environment. This model attempts to simplify the complex interactions between a metal and different environmental sorbents/complexing agents and directly relate metal exposure to metal toxicity. This is accomplished by measuring water characteristics including hardness (Ca), alkalinity (CO_3), dissolved organic carbon, chloride and pH. In this model, the metal can be freely dissolved in the aqueous phase, or it can be associated with multiple ligands, including biological ligands (e.g., the gill surface on a mussel), sediment ligands (e.g., acid-volatile sulfides or natural organic matter), dissolved aqueous ligands (e.g., dissolved organic matter, or dissolved inorganics, Cl^-, or other anions), and particulate ligands (e.g., particulate organic or inorganic carbon). The complexing agents can compete with each other to bind the metal, which can result in protective effects, where the metal is bound and unavailable to associate with the target ligand, or other anions can outcompete the metal for association with the biological ligand. For example, dissolved organic carbon and other organic or inorganic ligands can bind the metal, eliciting a protective effect, or cations can bind the metal such as Na^{2+}, H^+, or Ca^{2+}, thus inhibiting the metal from binding to the target tissue (Kramer et al. 1997). Therefore, metal speciation will depend on the nature of the water chemistry in different environments. However, metal bioavailability can also be influenced by the condition and physiology of the organism, including such factors as species, assimilation efficiency, clearance rate, ingestion rate, and stress level (Table 8.5). All of these factors must be considered to adequately examine metal bioavailability.

METAL DETOXIFICATION MECHANISMS

Mussels possess multiple detoxification mechanisms that can influence the uptake, distribution, and elimination of metals. These mechanisms range from simple actions such as avoidance (shell closure or decreased ventilation and filtration rate) to induction of low molecular-weight metallothionein (MT) proteins (Naimo 1995; Tessier and Blais 1996; Kadar et al. 2001; Baudrimont et al. 2002; Marigomez et al. 2002). The classification of the metals into two groups (A and B) is helpful in assessing specific detoxification mechanisms involved. Group A metals include barium (Ba), magnesium (Mg), aluminum (Al), and strontium (Sr), which exhibit high affinities towards O and P ligands, such as those found in calcium-phosphate granules (e.g., *H. depressa*) (Byrne and Vesk 2000). In contrast, Group B metals, Hg, silver (Ag), and less frequently, Cd, Cu, iron (Fe), manganese (Mn), Pb, and Zn, demonstrate affinities for S and N ligands, found in proteins like MT (Byrne and Vesk 2000). However, Group B metals will bind to calcium-phosphate granules in the absence of MTs (Byrne and Vesk 2000). Induction of MT was observed in *C. fluminea* following exposure to Cd and Zn, in which MT concentrations were induced 3.5 times those of reference levels (Baudrimont et al. 2002). During depuration of Cd, MT concentrations were shown to decrease nearly 40%, particularly early in the elimination phase (Baudrimont et al. 2002). These authors reported that 40% of accumulated Cd was bound to MT, whereas Tessier and Blais (1996) observed that 85% of Cd measured in *D. polymorpha* was bound. Thus, there may be differences among species in detoxification mechanisms. Other mechanisms that have been observed in mussels include incorporation of metals into lysosomal granules and histidine-rich glycoproteins.

IMPLICATIONS AND POTENTIAL FOR METAL TOXICITY

The ultimate effects of metal pollution in the environment on freshwater bivalves have yet to be completely assessed. While it is clear that bivalves possess the ability to accumulate metals to very high levels, it is more difficult to determine specific chronic adverse effects that emanate from the exposure. The complexity of interactions between metals and environmental conditions makes it difficult to classify behavior into simple models, however, further research in this field should include correlations between tissue burdens and physiological variables, as well as measurements of uptake and elimination rates and the combined effects of metal mixtures on toxicokinetics and bivalve health.

REFERENCES

Accardi-Dey, A. and Gschwend, P. M., Assessing the combined roles of natural organic matter and black carbon as sorbents in sediments, *Environ. Sci. Technol.*, 36, 21–29, 2002.

Alexander, M., Aging, bioavailability, and overestimation of risk from environmental pollutants, *Environ. Sci. Technol.*, 34, 4259–4265, 2000.

Augenfield, J. M., Anderson, J. W., Riley, R. G., and Thomas, B. L., The fate of polyaromatic hydrocarbons in an intertidal sediment exposure system: Bioavailability to *Macoma inquinata* (Mollusca: Pelecypoda) and *Abarenicola pacifica* (Annelida: Polychaeta), *Mar. Environ. Res.*, 7, 31–50, 1982.

Basack, S. B., Oneto, M. L., Verrengia-Guerrero, N. R., and Kesten, E. M., Accumulation and elimination of pentachlorophenol in the freshwater Bivalve *Corbicula fluminea*, *Bull. Environ. Contam. Toxicol.*, 58, 497–503, 1997.

Baud, J. P., Amiard-Triquet, C., and Metayer, C., Species-related variations of silver bioaccumulation and toxicity to three marine bivalves, *Water Res.*, 24, 995–1001, 1990.

Baudrimont, M., Andres, S., Durrieu, G., and Boudou, A., The key role of metallothioneins in the bivalve *Corbicula fluminea* during the depuration phase, after in situ exposure to Cd and Zn, *Aquat. Toxicol.*, 63, 89–102, 2002.

Bauer, I., Weigelt, S., and Ernst, W., Biotransformation of hexachlorobenzene in the blue mussel (*Mytlius edulis*), *Chemosphere*, 19, 1701–1707, 1989.

Beckvar, N., Salazar, S., Salazar, M., and Finkelstien, K., An in situ assessment of mercury contamination in the Sudbury River Massachusetts, using transplanted freshwater mussels (*Elliptio complunata*), *Can. J. Fish. Aquat. Sci.*, 57, 1103–1112, 2000.

Beliaeff, B., O'Connor, T. P., Munschy, C., Raffin, B., and Claisse, D., Comparison of polycyclic aromatic hydrocarbon levels in mussels and oysters in France and the United States, *Environ. Toxicol. Chem.*, 21, 1783–1787, 2002.

Bender, M. E., Hargis, W. J., Huggett, R. J., and Roberts, M. H., Effects of polynuclear aromatic hydrocarbons on fishes and shellfish: an overview of research in Virginia, *Mar. Environ. Res.*, 24, 237–241, 1988.

Bergen, B. J., Nelson, W. G., and Pruell, R. J., Comparison of nonplanar and coplanar PCB congener partitioning in seawater and bioaccumulation in blue mussels (*Mytilus edulis*), *Environ. Toxicol. Chem.*, 15, 1517–1523, 1996.

Bianchini, A. and Bowles, K. C., Metal sulfides in oxygenated aquatic systems: Implications for the biotic ligand model, *Comp. Biochem. Physiol.*, 133C, 51–64, 2002.

Birdsall, K., Kukor, J. J., and Cheney, M. A., Uptake of polycyclic aromatic hydrocarbons compounds by the gills of the bivalve mollusk *Elliptio complanata*, *Environ. Toxicol. Chem.*, 20, 309–316, 2001.

Bjork, M. and Gilek, M., Uptake and elimination of ^{14}C-phenanthrene by the blue mussel *Mytilus edulis* L. at different algal concentrations, *Bull. Environ. Contam. Toxicol.*, 56, 151–158, 1996.

Bjork, M. and Gilek, M., Bioaccumulation kinetics of PCB 31, 49 and 153 in the blue mussel, *Mytilus edulis* L. as a function of algal food concentration, *Aquat. Toxicol.*, 38, 101–123, 1997.

Blackmore, G. and Wang, W. X., Comparison of metal accumulation in mussels at different local and global scales, *Environ. Toxicol. Chem.*, 22, 388–395, 2003.

Blumer, M., Souza, G., and Sass, J., Hydrocarbon pollution of edible shellfish by an oil spill, *Mar. Biol.*, 5, 195–202, 1970.

Boehm, P. D. and Quinn, J. G., The persistence of chronically accumulated hydrocarbons in the hard shell clam *Mercenaria mercenaria*, *Mar. Biol.*, 44, 227–233, 1977.

Boisson, F., Cotret, O., and Fowler, S. W., Bioaccumulation and retention of lead in the mussel *Mytilus galloprovincialis* following uptake from seawater, *Sci. Total Environ.*, 222, 55–61, 1998.

Brieger, G. and Hunter, R. D., Uptake and depuration of PCB 77, PCB 169, and hexachlorobenzene by zebra mussels (*Dreissena polymorpha*), *Ecotoxicol. Environ. Saf.*, 26, 153–165, 1993.

Broman, D. and Ganning, B., Uptake and release of petroleum hydrocarbons by two brackish water bivalves, *Mytilus edulis* L. and *Macoma baltica* (L.), *Ophelia*, 25, 49–57, 1986.

Bruner, K. A., Fisher, S. W., and Landrum, P. F., The role of the zebra mussel, *Dreissena polymorpha*, in contaminant cycling: I. The effect of body size and lipid content on the bioconcentration of PCBs and PAHs, *J. Great Lakes Res.*, 20, 725–734, 1994.

Bucheli, T. D. and Gustafsson, O., Quantification of the soot-water distribution coefficient of PAHs provides mechanistic basis for enhanced sorption observations, *Environ. Sci. Technol.*, 34, 5144–5151, 2000.

Butte, W., Mathematical description of uptake, accumulation and elimination of xenobiotics in a fish/water system, In *Bioaccumulation in Aquatic Systems. Proceedings of an International Workshop, Berlin 1990*, Nagel, R. and Loskill, R., Eds., VCH, New York, NY, pp. 29–42, 1991.

Byrne, M. and Vesk, P. A., Elemental composition of mantle tissue granules in *Hyridella depressa* (Unionida) from the Hawkesbury-Nepean River system, Australia: Inferences from catchment chemistry, *Mar. Freshwat. Res.*, 51, 183–192, 2000.

Campbell, J. H. and Evans, R. D., Inorganic and organic ligand binding of lead and cadmium and resultant implications for bioavailability, *Sci. Total Environ.*, 62, 219–227, 1987.

Campbell, J. and Evans, R. D., Cadmium concentrations in the freshwater mussel (*Elliptio complanata*) and their relationship to water chemistry, *Arch. Environ. Contam. Toxicol.*, 20, 125–131, 1991.

Camusso, M., Balestrini, R., Muriano, F., and Mariani, M., Use of freshwater mussel *Dreissena polymorpha* to assess trace metal pollution in the lower river Po (Italy), *Chemosphere*, 29, 729–745, 1994.

Chevreuil, M., Blanchard, M., Teil, M. J., Carru, A. M., Testard, P., and Chesterikoff, A., Evaluation of the pollution by organochlorinated compounds (polychlorobiphenyls and pesticides) and metals (Cd, Cr, Cu, and Pb) in the water and in the zebra mussel (*Dreissena polymorpha pallas*) of the river Seine, *Water Air Soil Pollut.*, 88, 371–381, 1996.

Churchill, S. A., Walters, J. V., and Churchill, P. F., Sorption of heavy metals by prepared bacterial cell surfaces, *J. Environ. Eng.*, 121, 706–711, 1995.

Clark, R. C. and Findley, J. S., Uptake and loss of petroleum hydrocarbons by the mussel, *Mytilus edulis* in laboratory experiments, *Fish. Bull.*, 73, 508–515, 1975.

Colombo, J. C., Bilos, C., Campanaro, M., Rodriguez-Presa, M. J., and Catoggio, J. A., Bioaccumulation of polychlorinated biphenyls and chlorinated pesticides by the Asiatic clam *Corbicula fluminea*: Its use as sentinel organism in the Rio de La Plata Estuary, Argentina, *Environ. Sci. Technol.*, 29, 914–927, 1995.

Cope, W. G., Bartsch, M. R., Rada, R. G., Balogh, S. J., Rupprecht, J. E., Young, R. D., and Johnson, D. K., Bioassessment of mercury, cadmium, polychlorinated biphenyls, and pesticides in the Upper Mississippi river with zebra mussels (*Dreissena polymorpha*), *Environ. Sci. Technol.*, 33, 4385–4390, 1999.

Das, S. and Jana, B. B., Dose-dependent uptake and Eichhornia-induced elimination of cadmium in various organs of the freshwater mussel, *Lamellidens marginalis* (Linn.), *Ecol. Eng.*, 12, 207–229, 1999.

Di Toro, D. M., Zarba, C. S., Hansen, D. J., Berry, W. J., Swartz, R. C., Cowan, C. E., Pavlou, S. P., Allen, H. E., Thomas, N. A., and Paquin, P. R., Technical basis for establishing sediment quality criteria for nonionic organic chemicals using equilibrium partitioning, *Environ. Toxicol. Chem.*, 10, 1541–1583, 1991.

Di Toro, D. M., Mahony, J. D., Hansen, D. J., Scott, K. J., Carlson, A. R., and Ankley, G. T., Acid volatile sulfide predicts the acute toxicity of cadmium and nickel in sediments, *Environ. Sci. Technol.*, 26, 96–101, 1992.

Dobrowoski, R. and Skowronska, M., Concentration and discrimination of selected trace metals by freshwater mollusks, *Bull. Environ. Contam. Toxicol.*, 69, 509–515, 2002.

Dunn, B. P. and Stich, H. F., Release of the carcinogen benzo(a)pyrene from environmentally contaminated mussels, *Bull. Environ. Contam. Toxicol.*, 15, 398–401, 1976.

Elliott, N. G., Swain, R., and Ritz, D. A., Metal Interactions during accumulation by the mussel *Mytilus edulis planulatus*, *Mar. Biol.*, 93, 395–399, 1986.

Everaarts, J. M., Uptake and release of cadmium in various organs of the common mussel, *Mytilus edulis* (L.), *Bull. Environ. Contam. Toxicol.*, 45, 560–567, 1990.

Fan, W., Wang, W. X., and Chen, J., Geochemistry of Cd Cr, and Zn in highly contaminated sediments and its influences on assimilation by marine bivalves, *Environ. Sci. Technol.*, 36, 5164–5171, 2002.

Farrington, J. W., Goldberg, E. D., Risebrough, R. W., Martin, J. H., and Bower, V. T., U.S. Mussel Watch 1976–1978: An overview of the trace-metal, DDE, PCB, hydrocarbon and artificial radionuclide data, *Environ. Sci. Technol.*, 17, 490–496, 1983.

Fent, K. and Hunn, J., Phenyltins in water, sediment, and biota of freshwater marinas, *Environ. Sci. Technol.*, 25, 956–963, 1991.

Fisher, S. W., Gossiaux, D. C., Bruner, K. A., and Landrum, P. F., Investigations of the toxicokinetics of hydrophobic contaminants in the zebra mussel (*Dreissena polymorpha*), In *Zebra Mussels: Biology, Impacts and Control*, Nalepa, T. F. and Schloesser, D. W., Eds., CRC press, Boca Raton, FL, pp. 465–490, 1993.

Fisher, S. W., Hwang, H., Atanasoff, M., and Landrum, P. F., Lethal body residues for pentachlorophenol in zebra mussels (*Dreissena polymorpha*) under varying conditions of temperature and pH, *Ecotoxicol. Environ. Saf.*, 43, 274–283, 1999.

Foster, R. B. and Bates, J. M., Use of freshwater mussels to monitor point source industrial discharges, *Environ. Sci. Technol.*, 12, 958–962, 1978.

Friedrich, A. R. and Filice, F. P., Uptake and accumulation of the nickel ion by *Mytilus edulis*, *Bull. Environ. Contam. Toxicol.*, 16, 750–755, 1976.

Gewurtz, S. B., Drouillard, K. G., Lazar, R., and Haffner, G. D., Quantitative biomonitoring of PAHs using the barnes mussel (*Elliptio complanata*), *Arch. Environ. Contam. Toxicol.*, 43, 497–504, 2002.

Geyer, H., Sheehan, P., Kotzias, D., Freitag, D., and Korte, F., Prediction of ecotoxicological behavior of chemicals: Relationship between physico-chemical properties and bioaccumulation of organic chemicals in the mussel *Mytlius edulis*, *Chemosphere*, 11, 1121–1134, 1982.

Gilek, M., Bjork, M., and Naef, C., Influence of body size on the uptake, depuration and bioaccumulation of polychlorinated biphenyl congeners by Baltic sea blue mussels, *Mytilus edulis*, *Mar. Biol.*, 125(3), 499–510, 1996.

Gossiaux, D. C., Landrum, P. F., and Fisher, S. W., Effect of temperature on the accumulation kinetics of PAHs and PCBs in the zebra mussel, *Dreissena polymorpha*, *J. Great Lakes Res.*, 22, 379–388, 1996.

Gossiaux, D. C., Landrum, P. F., and Fisher, S. W., The assimilation of contaminants from suspended sediment and algae by the zebra mussel, *Dreissena polymorpha*, *Chemosphere*, 36, 3181–3197, 1998.

Gundacker, C., Tissue-specific heavy metal (Cd, Pb, Cu, and Zn) deposition in a natural population of the zebra mussel *Dreissena polymorpha* Pallas, *Chemosphere*, 38, 3339–3356, 1999.

Gundacker, C., Comparison of heavy metal bioaccumulation in freshwater molluscs of urban river habitats in Vienna, *Environ. Pollut.*, 110, 61–71, 2000.

Gustafsson, O., Haghseta, F., Chan, C., Macfarlane, J., and Gschwend, P. M., Quantification of the dilute sedimentary soot phase: implications for PAH speciation and bioavailability, *Environ. Sci. Technol.*, 31, 203–209, 1997.

Hansen, N., Jenson, V. B., Appelquist, H., and Morch, E., The uptake and release of petroleum hydrocarbons by the marine mussel *Mytilus edulis*, *Prog. Water Technol.*, 10(5–6), 351–359, 1978.

Hawker, D. W. and Connell, D. W., Bioconcentration of lipophilic compounds by some aquatic organisms, *Ecotoxicol. Environ. Saf.*, 11, 184–197, 1986.

Hemelraad, J., Kleinveld, H. A., DeRoos, A. M., Holwerda, D. A., and Zandee, D. I., Cadmium kinetics in freshwater clams. III. Effects of zinc on uptake and distribution of cadmium in *Anodonta cygnea*, *Arch. Environ. Contam. Toxicol.*, 16, 95–102, 1987.

Holwerda, D. A., Cadmium kinetics in freshwater clams. V. Cadmium–copper interaction in metal accumulation by *Anodonta cygnea* and characterization of the metal-binding protein, *Arch. Environ. Contam. Toxicol.*, 21, 432–437, 1991.

Hyotylainen, T., Karels, A., and Oikari, A., Assessment of bioavailability and effects of chemicals due to remediation actions with caging mussels (*Anodonta anatina*) at a creosote-contaminated lake sediment site, *Water Res.*, 36, 4497–4504, 2002.

Inza, B., Ribeyre, F., and Boudou, A., Dynamics of cadmium and mercury compounds (inorganic mercury or methylmercury): Uptake and depuration in *Corbicula fluminea*. Effects of temperature and pH, *Aquat. Toxicol.*, 43, 273–285, 1998.

Jacobson, P. J., Neves, R. J., Cherry, D. S., and Farris, J. L., Sensitivity of glochidial stages of freshwater mussels (Bivalvia: Unionidae) to copper, *Environ. Toxicol. Chem.*, 16, 2384–2392, 1997.

James, M. O., Biotransformation and disposition of PAH in aquatic invertebrates, In *Metabolism of Polycyclic Aromatic Hydrocarbons in the Aquatic Environment*, Varanasi, U., Ed., CRC Press, Boca Raton, FL, pp. 69–92, 1989.

Jana, B. B. and Das, S., Potential of freshwater mussel (*Lamellidens marginalis*) for cadmium clearance in a model system, *Ecol. Eng.*, 8, 179–193, 1997.

Jenner, H. A., Hemelraad, J., Marquenie, J. M., and Noppert, F., Cadmium kinetics in freshwater clams (Unionidae) under field and laboratory conditions, *Sci. Total Environ.*, 108, 205–214, 1991.

Kadar, E., Salanki, J., Jugdaohsingh, R., Powell, J. J., McCrohan, C. R., and White, K. N., Avoidance responses to aluminum in the freshwater bivalve *Anodonta cygnea*, *Aquat. Toxicol.*, 55, 137–148, 2001.

Keller, A. E. and Zam, S. G., The acute toxicity of selected metals to the freshwater mussel, *Anodonta imbecilis*, *Environ. Toxicol. Chem.*, 10, 539–546, 1991.

Kraaij, R., Seinen, W., Tolls, J., Cornelissen, G., and Belfroid, A., Direct evidence of sequestration in sediments affecting the bioavailability of hydrophobic organic chemicals to benthic deposit-feeders, *Environ. Sci. Technol.*, 36, 3525–3529, 2002.

Kraak, M. H. S., School, H., Peeters, W. H. M., and van Straalen, N. M., Chronic ecotoxicity of mixtures of Cu, Zn, and Cd to the zebra mussel *Dreissena polymorpha*, *Ecotoxicol. Environ. Saf.*, 25, 315–327, 1993.

Kraak, M. H. S., Lavy, D., Schoon, H., Toussaint, M., Peeters, W. H. M., and van Straalen, N. M., Ecotoxicity of mixtures of metals to the zebra mussel *Dreissena polymorpha*, *Environ. Toxicol. Chem.*, 13, 109–114, 1994a.

Kraak, M. H. S., Toussaint, M., Lavy, D., and Davids, C., Short-term effects of metals on the filtration rate of the zebra mussel *Dreissena polymorpha*, *Environ. Pollut.*, 84, 139–143, 1994b.

Kraak, M. H. S., Wink, Y. A., Stuijfzand, S. C., Buckert-de Jong, M. C., de Groot, C. J., and Admiraal, W., Chronic ecotoxicity of Zn and Pb to the zebra mussel Dreissena polymorpha, *Aquat. Toxicol.*, 30, 77–89, 1994c.

Kraak, M. H. S., Stuijfzand, S. C., and Admiraal, W., Short-term ecotoxicity of a mixture of five metals to the zebra mussel *Dreissena polymorpha*, *Bull. Environ. Contam. Toxicol.*, 63, 805–812, 1999.

Kramer, J. R., Allen, H. E., Davison, W., Godtfredsen, K. L., Meyer, J. S., Perdue, E. M., Tipping, E., van de Meent, D., and Westall, J. C., Chemical speciation and metal toxicity in surface freshwaters, In *Reassessment of metals criteria for aquatic life protection*, Bergman, D. and Dorward-King, E., Eds., CRC Press, Boca Raton, FL, pp. 57–70, 1997.

Lamoureux, E. and Brownawell, B. J., Chemical and biological availability of sediment-sorbed hydrophobic organic contaminants, *Environ. Toxicol. Chem.*, 18, 1733–1741, 1999.

Landrum, P. F., Lee, H., II, and Lydy, M. J., Toxicokinetics in aquatic systems: Model comparisons and use in hazard assessment, *Environ. Toxicol. Chem.*, 11, 1709–1725, 1992.

Landrum, P. F., Hayton, W. L., Lee, H., II, McCarty, L. S., Mackay, D., and McKim, J. M., Synopsis of discussion session on the kinetics behind environmental bioavailability, In *Bioavailability: Physical, Chemical, and Biological Interactions*, Hamelink, J. L., Landrum, P. F., Bergman, H. L., and Benson, W. H., Eds., Lewis Publishers, Boca Raton, FL, pp. 203–219, 1994.

Laughlin, R. B., French, W., Jr., and Guard, H. E., Accumulation of bis(tributyltin) oxide by the marine mussel *Mytilus edulis*, *Environ. Sci. Technol.*, 20, 884–890, 1986.

Lee, C. W. and Kwon, Y. T., Distribution of heavy metals in seawater, sediment and biota of Jinhaw bay, Korea, *Water Sci. Technol.*, 30, 173–177, 1994.

Lee, R. F., Sauerheber, D., and Benson, A. A., Petroleum hydrocarbons: Uptake and discharge by the marine mussel *Mytilus edulis*, *Science*, 177, 344–346, 1972.

Mackay, D., Correlation of bioconcentration factors, *Environ. Sci. Technol.*, 16, 274–278, 1982.

Makela, T. P. and Oikari, A. O. J., Pentachlorophenol accumulation in the freshwater mussels *Anodonta anatina* and *Pseudanodonta complanata*, and some physiological consequences of laboratory maintenance, *Chemosphere*, 31, 3651–3662, 1995.

Malley, D. F., Stewart, A. R., and Hall, B. D., Uptake of methyl mercury by the floater mussel, *Pyganodon grandis* (Bivalvia, Unionidae), caged in a flooded wetland, *Environ. Toxicol. Chem.*, 15, 928–936, 1996.

Marigomez, I., Soto, M., Cajaraville, M. P., Angulo, E., and Giamberini, L., Cellular and subcellular distribution of metals in molluscs, *Microsc. Res. Tech.*, 56, 358–392, 2002.

Markich, S. J., Brown, P. L., and Jeffree, R. A., Divalent metal accumulation in freshwater bivalves: An inverse relationship with metal phosphate solubility, *Sci. Total Environ.*, 275, 27–41, 2001.

McMahon, R. F. and Bogan, A. E., Mollusca: Bivalvia, In *Ecology and Classification of North American Freshwater Invertebrates* 2nd ed., Thorp, J. H. and Covich, A. P., Eds. , Academic Press, New York, pp. 331–429, 2001.

Mersch, J., Morhain, E., and Mouvet, C., Laboratory accumulation and depuration of copper and cadmium in the freshwater mussel *Dreissena polymorpha* and the aquatic moss *Rhynchostegium riparioides*, *Chemosphere*, 27, 1475–1485, 1993.

Mersch, J., Wagner, P., and Pihan, J. C., Copper in indigenous and transplanted zebra mussels in relation to changing water concentrations and body weight, *Environ. Toxicol. Chem.*, 15, 886–893, 1996.

Moore, J. W., *Inorganic Contaminants of Surface Water: Research and Monitoring Priorities*, Springer, New York, p. 334, 1991.

Morrison, H., Yankovich, T., Lazar, R., and Haffner, G. D., Elimination rate constants of 36 PCBs in zebra mussels (*Dreissena polymorpha*) and exposure dynamics in the Lake St. Clair–Lake Erie corridor, *Can. J. Fish. Aquat. Sci.*, 52, 2574–2582, 1995.

Morrison, H. A., Gobas, F. A. P. C., Lazar, R., and Haffner, G. D., Development and verification of a bioaccumulation model for organic contaminants in benthic invertebrates, *Environ. Sci. Technol.*, 30, 3377–3384, 1996.

Naimo, T. J., A review of the effects of heavy metals on freshwater mussels, *Ecotoxicology*, 4, 341–362, 1995.

Neff, J. M. and Burns, W. A., Estimation of polycyclic aromatic hydrocarbon concentrations in the water column based on tissue residues in mussels and salmon: an equilibrium partitioning approach, *Environ. Toxicol. Chem.*, 15, 2240–2253, 1996.

Obana, H., Hori, S., Nakamura, A., and Kashimoto, T., Uptake and release of polynuclear aromatic hydrocarbons by short-necked clams (*Tapes japonica*), *Water Res.*, 17, 1183–1187, 1983.

Ogata, M., Fujisawa, K., Ogino, Y., and Mano, E., Partition coefficients as a measure of bioconcentration potential of crude oil compounds in fish and shellfish, *Bull. Environ. Contam. Toxicol.*, 33, 561–567, 1984.

Paquin, P. R., Gorsuch, J. W., Apte, S., Batley, G. E., Bowles, K. C., Campbell, P. G. C., Delos, C. G., Di Toro, D. M., Dwyer, R. L., Galvez, F., Gensemer, R. W., Goss, G. G., Hogstrand, C., Janssen, C. R., McGeer, J. C., Naddy, R. B., Playl, R. C., Santor, R. C., Schneider, U., Stubblefield, W. A., Wood, C. M., and Wu, K. B., The biotic ligand model: A historical overview, *Comp. Biochem. Physiol.*, 133C, 3–35, 2002.

Pendergast, J. F., Ausley, L. W., Bro-Rasmussen, F., Cappel, C. R., Delos, C., Dorward-King, E. J., Hansen, D. J., LeBlanc, N. E., Lee, C. M., and Viteri, A., Jr., Regulatory practice for metals, In *Reassessment of Metals Criteria for Aquatic Life Protection*, Bergman, H. L. and Dorward-King, E., Eds., CRC Press, Boca Raton, FL, pp. 13–30, 1997.

Peven, C. S., Uhler, A. D., and Querzoli, F. J., Caged mussels and semipermeable membrane devices as indicators of organic contaminant uptake in Dorchester and Duxbury Bays, Massachusetts, *Environ. Toxicol. Chem.*, 15, 144–149, 1996.

Pittinger, C. A., Buikema, A. L., Hornor, S. G., and Young, R. W., Variation in tissue burdens of polycyclic aromatic hydrocarbons in indigenous and relocated oysters, *Environ. Toxicol. Chem.*, 4, 379–387, 1985.

Plette, A. C. C., Nederlof, M. M., Temminghoff, E. J. M., and van Riemsdijk, W. H., Bioavailability of heavy metals in terrestrial and aquatic systems: A quantitative approach, *Environ. Toxicol. Chem.*, 18, 1882–1890, 1999.

Pruell, R. J., Lake, J. L., Davis, W. R., and Quinn, J. G., Uptake and depuration of organic contaminants by blue mussels (*Mytilus edulis*) exposed to environmentally contaminated sediment, *Mar. Biol.*, 91, 497–507, 1986.

Pruell, R. J., Rubinstein, N. I., Taplin, B. K., LiVolsi, J. A., and Bowen, R. D., Accumulation of polychlorinated organic contaminants from sediment by three benthic marine species, *Arch. Environ. Contam. Toxicol.*, 24, 290–297, 1993.

Raikow, D. F. and Hamilton, S. K., Bivalve diets in a Midwestern U.S. stream: A stable isotope enrichment study, *Limnol. Oceanogr.*, 46, 514–522, 2001.

Readman, J. W., Mantoura, R. F. C., and Rhead, M. M., The physico-chemical speciation of polycyclic aromatic hydrocarbons (PAH) in aquatic systems, *Fresenius Z Anal. Chem.*, 319, 126–131, 1984.

Reeders, H. H., Bij de Vaate, A., and Slim, F. J., The filtration rate of Dreissena polymorpha (Bivalvia) in three Dutch lakes with reference to biological water quality management, *Freshw. Biol.*, 22, 133–141, 1989.

Riget, F., Johansen, P., and Asmund, G., Uptake and release of lead and zinc by blue mussels. Experience from transplantation experiments in Greenland, *Mar. Pollut. Bull.*, 34, 805–815, 1997.

Riley, R. T., Mix, M. C., Schaffer, R. L., and Bunting, D. L., Uptake and accumulation of naphthalene by the oyster, *Ostrea edulis*, in a flow-through system, *Mar. Biol.*, 61, 267–276, 1981.

Roditi, H. A. and Fisher, N. S., Rates and routes of trace element uptake in zebra mussels, *Limnol. Oceanogr.*, 44, 1730–1749, 1999.

Roditi, H. A., Fisher, N. S., and Sanudo-Wilhelmy, S. A., Field testing a metal bioaccumulation model for zebra mussels, *Environ. Sci. Technol.*, 34, 2817–2825, 2000.

Roper, J. M., Cherry, D. S., Simmers, J. W., and Tatem, H. E., Bioaccumulation of PAHs in the zebra mussel at Times Beach, Buffalo, New York, *Environ. Monit. Assess.*, 46, 267–277, 1997.

Roseman, E. F., Mills, E. L., Rutske, M., Gutenmann, W. H., and Lisk, D. J., Absorption of cadmium from water by North American zebra and quagga mussels (Bivalvia: Dreissenidae), *Chemosphere*, 28, 737–743, 1994.

Russell, R. W. and Gobas, F. A. P. C., Calibration of the freshwater mussel, *Elliptio complanata*, for quantitative biomonitoring of hexachlorobenzene and octachlorosytrene in aquatic systems, *Bull. Environ. Contam. Toxicol.*, 43, 576–582, 1989.

Salanki, J. and Balogh, K. V., Physiological background for using freshwater mussels in monitoring copper and lead pollution, *Hydrobiologia*, 188/189, 445–454, 1989.

Salazar, M. H. and Salazar, S. M., Mussels as bioindicators: Effects of TBT on survival, bioaccumulation and growth under natural conditions, In *Organotin*, Champ, M. A. and Seligman, P. F., Eds., Chapman and Hall, London, UK, pp. 305–330, 1996.

Schrap, S. M. and Opperhuizen, A., Relationship between bioavailability and hydrophobicity: Reduction of the uptake of organic chemicals by fish due to the sorption on particles, *Environ. Toxicol. Chem.*, 9, 715–724, 1990.

Schuurmann, G. and Klein, W., Advances in bioconcentration prediction, *Chemosphere*, 17, 1551–1574, 1988.

Sericano, J. L., Wade, T. L., El-Husseini, A. M., and Brooks, J. M., Environmental significance of the uptake and depuration of planar PCB congeners by the American oyster (*Crassostria virginica*), *Mar. Pollut. Bull.*, 24(11), 537–543, 1992.

Sericano, J. L., Wade, T. L., and Brooks, J. M., Accumulation and depuration of organic contaminants by the American oyster (*Crassostrea virginica*), *Sci. Total Environ.*, 179, 149–160, 1996.

Shawky, S. and Emons, H., Distribution pattern of organotin compounds at different trophic levels of aquatic ecosystems, *Chemosphere*, 36, 523–535, 1998.

Spacie, A., Interactions of organic pollutants with inorganic solid phases: are they important to bioavailability?, In *Bioavailability: Physical, Chemical and Biological Interactions*, Hamelink, J. L., Landrum, P. F., Bergman, H. L., and Benson, W. H., Eds., CRC Press, Boca Raton, FL, pp. 73–82, 1994.

Sreedevi, P., Suresh, A., Silvaramakrishna, B., Prabhavathi, B., and Radhakrishnaiah, K., Bioaccumulation of nickel in the organs of the freshwater fish, *Cyprinus carpio*, and the freshwater mussel, *Lamellidens marginalis*, under lethal and sublethal nickel stress, *Chemosphere*, 24, 29–36, 1992.

Stegman, J. J. and Teal, J. M., Accumulation, release and retention of petroleum hydrocarbons by the oyster *Crassostrea virginica*, *Mar. Biol.*, 44, 37–44, 1973.

Stewart, A. R., Accumulation of Cd by a freshwater mussel (*Pyganodon grandis*) is reduces in the presence of Cu, Zn, Pb, and Ni, *Can. J. Fish. Aquat. Sci.*, 56, 467–478, 1999.

Stuijfzand, S. C., Kraak, M. H. S., Wink, Y. A., and Davids, C., Short-term effects of nickel on the filtration rate of the zebra mussel *Dreissena polymorpha*, *Bull. Environ. Contam. Toxicol.*, 54, 376–381, 1995.

Sures, B., Zimmerman, S., Messerschmidt, J., and von Bohlen, A., Relevance and analysis of traffic related platinum group metals (Pt, Pd, Rh) in the aquatic biosphere, with emphasis on palladium, *Ecotoxicology*, 11, 385–392, 2002.

Sutherland, J. and Major, C. W., Internal heavy metal changes as a consequence of exposure of *Mytilus edulis*; the blue mussel, to elevated external copper(II) levels, *Comp. Biochem. Physiol.*, 168, 63–67, 1981.

Tanabe, S., Tatsukawa, R., and Phillips, D. J. H., Mussels as bioindicators of PCB pollution: A case study on uptake and release of PCB isomers and congeners in green-lipped mussels (*Perna viridis*) in Hong Kong waters, *Environ. Pollut.*, 47, 41–62, 1987.

Tanacredi, J. T. and Cardenas, R. R., Biodepuration of polynuclear aromatic hydrocarbons from a bivalve mollusc, *Mercenaria mercenaria* L., *Environ. Sci. Technol.*, 25, 1453–1461, 1991.

Tessier, C. and Blais, J. S., Determination of cadmium-metallothioneins in zebra mussels exposed to subchronic concentrations of Cd^{2+}, *Ecotoxicol. Environ. Saf.*, 33, 246–252, 1996.

Tessier, L., Vaillancourt, G., and Pazdernik, L., Comparative study of cadmium and mercury kinetics between the short-lived gastropod *Viviparus georginess* (Lea) and pelecypod *Elliptio complanata* (Light foot) under laboratory conditions, *Environ. Pollut.*, 85, 271–282, 1994.

Tessier, L., Vaillancourt, G., and Pazdernik, L., Laboratory study of Cd and Hg uptake by two freshwater molluscs in relation to concentration, age and exposure time, *Water Air Soil Pollut.*, 86, 347–357, 1996.

Thomann, R. V. and Komlos, J., Model of biota-sediment accumulation factor for polycyclic aromatic hydrocarbons, *Environ. Toxicol. Chem.*, 18, 1060–1068, 1999.

Thorsen, W. A., Bioavailability of particulate-sorbed polycyclic aromatic hydrocarbons, PhD Thesis, North Carolina State Univ., Raleigh, NC, 2003.

Thorsen, W. A., Forestier, D., Sandifer, T., Lazaro, P. R., Cope, W. G., and Shea, D., Elimination rate constants of 46 polycyclic aromatic hydrocarbons in the unionid mussel, *Elliptio complanata*, *Arch. Environ. Contam. Toxicol.*, 47, 332–340, 2004a.

Thorsen, W. A., Cope, W. G., and Shea, D., Bioavailability of PAHs: Effects of soot carbon and PAH source, *Environ. Sci. Technol.*, 38, 2029–2037, 2004b.

Uno, S., Shiraishi, H., Hatakeyama, S., and Otsuki, A., Uptake and depuration kinetics and BCFs of several pesticides in three species of shellfish (*Corbicula leana*, *Corbicula japonica*, and *Cipangopludina chinensis*): Comparison between field and laboratory experiment, *Aquat. Toxicol.*, 39, 23–43, 1997.

Van Haelst, A. G., Zhao, Q., van der Wielen, F. W. M., Govers, H. A. J., and de Voogt, P., Determination of bioconcentration factors of eight tetrachlorobenzyltoluenes in the zebra mussel *Dreissena polymorpha*, *Ecotoxicol. Environ. Saf.*, 34, 35–42, 1996a.

Van Haelst, A. G., Loonen, H., van der Wielen, F. W. M., and Govers, H. A. J., Comparison of bioconcentration factors of tetrachlorobenzyltoluenes in the guppy (*Poecilia reticulata*) and zebra mussel (*Dreissena polymorpha*), *Chemosphere*, 32, 1117–1122, 1996b.

Van Slooten, K. B. and Tarradellas, J., Accumulation, depuration and growth effects of tributyltin in the freshwater bivalve *Dreissena polymorpha* under field conditions, *Environ. Toxicol. Chem.*, 13, 755–762, 1994.

Vesk, P. A. and Byrne, M., Metal levels in tissue granules of the freshwater bivalve *Hyridella depressa* (Unionida) for biomonitoring: The importance of cryopreparation, *Sci. Total Environ.*, 225, 219–229, 1999.

Wang, W. X., Comparison of metal uptake rate and absorption efficiency in marine bivalves, *Environ. Toxicol. Chem.*, 20, 1367–1373, 2001.

Winter, S., Cadmium uptake kinetics by freshwater mollusc soft body under hard and soft water conditions, *Chemosphere*, 32, 1937–1948, 1996.

Zimmermann, S., Alt, F., Messerschmidt, J., von Bohlen, A., Taraschewski, H., and Sures, B., Biological availability of traffic-related platinum-group elements (palladium, platinum, and rhodium) and other metals to the zebra mussel (*Dreissena polymorpha*) in water containing road dust, *Environ. Toxicol. Chem.*, 21, 2713–2718, 2002.

Zuolian, C. and Jensen, A., Accumulation of organic and inorganic tine in blue mussel, *Mytilus edulis*, under natural conditions, *Mar. Pollut. Bull.*, 20, 281–286, 1989.

9 Linking Bioaccumulation and Biological Effects to Chemicals in Water and Sediment: A Conceptual Framework for Freshwater Bivalve Ecotoxicology

Michael H. Salazar and Sandra M. Salazar

INTRODUCTION

"Without observations linking levels (of pollutants) in the water or sediment with tissue concentrations and then with effects on organisms and populations and, ultimately, with the well-being of the ecosystem as a whole, an adequate assessment of pollution is impossible" (GESAMP 1980). While this conclusion was reached about 25 years ago in the context of marine studies, it can be expanded to include freshwater bivalve ecotoxicology. It is doubtful that many unionid researchers appreciated the significance of this concept that appeared in the marine literature. Even though bioaccumulation provides a mechanistic link between environment and organism and effects beginning with accumulation of chemicals at internal receptors, it is surprising how rarely tissue chemistry and effects are included as part of a unified strategy for monitoring and assessment. Nevertheless, characterizing exposure and effects is considered the cornerstone of ecological risk assessment (ERA) (USEPA 1998), a more universal and modern day paradigm that has the basic elements of the GESAMP strategy. Even when bioaccumulation or biological effects have been used, it is often with one group of organisms used to effectively characterize exposure and another group to measure effects. A unifying conceptual framework that brings these two approaches together is long overdue. There is a real need to measure bioaccumulation and biological effects in the same organism at the same time and to view freshwater bivalve ecotoxicology from this risk assessment-based vantage point.

HISTORICAL PERSPECTIVE

Because freshwater bivalve ecotoxicology is relatively new as a science compared to marine bivalve ecotoxicology, it is helpful to make some comparisons regarding the development of each. It seems likely that bivalve ecotoxicology could benefit from parallel strategies applied to both marine and freshwater systems. Overemphasis on differences between marine and freshwater bivalves and unique qualities of freshwater bivalves that preclude their use with previously developed methodologies may have contributed to limiting progress in freshwater bivalve ecotoxicology. Freshwater bivalve ecotoxicology can be enhanced by identifying the similarities between marine and freshwater species and then applying the technologies that have been previously developed and effectively utilized for marine species. It is not clear why such models as that developed for marine bivalves by Widdows and Donkin (1992) have not been integrated into similar investigations in freshwater bivalve ecotoxicology.

In the context of biomonitoring with marine bivalves, several authors have described the Mussel Watch program as a reasonable initial attempt to integrate over time, the contaminant load at a given site (Phillips 1980; Goldberg 1975) and have acknowledged the need for additional monitoring endpoints. Bayne (1976), as one of the real pioneers in marine mussel ecotoxicology, produced a succinct two-page paper that could, in itself, be the major thesis of this entire chapter. This proclamation of almost 30 years ago stressed the importance of relating the physiological response to the concentrations of contaminants in animal tissues (Bayne 1976): "In many studies, the effects of pollution are related only to environmental concentrations, and the chance is lost of providing information covering the logical sequence of environmental load, body burden, and effect on the individual." Interestingly, Granmo (1995) described measured effects endpoints in the marine bivalve *Mytilus* as a first step and claimed that they were cheaper than chemical monitoring. He described some testing methods used in Sweden, regarding sampling, toxicity tests (such as the mussel embryo bioassays), energy budget tests scope for growth (SFG), and behavior tests and suggested that biological methods are good indicators of pollution impact. The use of such methods is often quite simple and cost-effective and provides an integration of all contaminants present, known, and unknown. This biology-led monitoring strategy may therefore be used as a first step in the investigation of an area before the more costly chemically based programs are initiated. Both bioaccumulation and effects studies are potentially cost-effective. However, the most cost-effective approach is to measure both simultaneously.

Not all marine studies have included synoptic measurements of exposure and effects. However, with the long history of Mussel Watch monitoring programs and the emergence of ERA approaches, the importance of paired measurements has become more obvious. The application of synoptic measurements has become standardized for marine, estuarine, and freshwater bivalves (ASTM 2001). As suggested by Elder and Collins (1991), there has been widespread use of measuring bioaccumulation and biological effects in freshwater bivalves, but these measurements are not routinely made together. One advantage of using the ecological risk assessment paradigm is that it helps maintain a focus on the importance of concurrent characterizations of exposure and effects (USEPA 1998).

In their review of freshwater molluscs (bivalves and gastropods) as indicators of bioavailability and toxicity, Elder and Collins (1991) identified the three most commonly used biomonitoring approaches as tissue analysis, toxicity testing, and ecological surveys. Surprisingly, they do not mention the need to integrate and harmonize these approaches as a strategy or paradigm for monitoring and assessment. The longer history of biomonitoring with marine bivalves supports their use as a template. Nevertheless, many freshwater bivalve ecotoxicologists have not embraced marine studies and often choose to ignore studies on nonunionids such as *Corbicula* and *Dreissena* that have provided insight into some basic principles of bivalve ecotoxicology. Important developments in the study of non-unionid bivalves can and should be included with any discussion of freshwater bivalve ecotoxicology to provide a context for unionid work.

Existing Models

The ecotoxicological framework provided in this chapter is based on existing models developed by Mearns (1985), McCarty (1991), Widdows and Donkin (1992), and Salazar and Salazar (1998), and a consensus-based standard guide for conducting field bioassays with caged freshwater bivalves

(ASTM 2001). Mearns (1985) advocated an integrated monitoring approach that he referred to as the exposure-bioaccumulation-effects triad. He suggested that mental tools are needed for the ultimate use of aquatic toxicological research. Unionid researchers may also benefit from this suggestion. He presented a simple conceptual diagram similar to the sediment quality triad (Chapman and Long 1983; Long and Chapman 1985) with bioaccumulation as an additional critical element. Mearns suggested that the only thread connecting exposure and effects was the concentration of chemicals in tissue. McCarty (1991) suggested that the kinetics of bioconcentration to a given body or tissue level linked with an understanding of the toxicological significance of that tissue residue level are central concepts to the development of a single bioassay methodology. The nature and time course of external exposures could then be linked with related processes in the body of exposed organisms. Widdows and Donkin (1992) explicitly referred to their strategy as an ecotoxicological framework and included elements of water chemistry, tissue chemistry, and toxic effects such as physiological energetics. Salazar and Salazar (1998) discussed the use of caged bivalves as part of an exposure–dose–response (EDR) triad to support an integrated risk assessment strategy.

The ASTM standard guide for conducting in situ field bioassays (ASTM 2001) outlines specific protocols for collecting, analyzing, and interpreting exposure and effects data for marine, estuarine, and freshwater bivalves. Bivalves are integrators at several different levels: biology and chemistry, sediment chemistry and toxicity, water, sediment, and tissues. They differ from other organisms in certain characteristics that distinguish them as good monitors of both exposure *and* effects. Because bivalves possess many characteristics of indicators of exposure and effects, they are natural candidates for enhancing links between bioaccumulation and biological effects. These inherent characteristics are key in understanding tissue residue effects. It is no longer sufficient to rely on water or sediment chemistry as lone indicators of exposure because there are too many factors that complicate interpretation and establishing links between exposure and effects. Bioaccumulation is the most direct way to estimate bioavailable chemicals.

It is important to note that the ASTM standard guide was deliberately titled "Standard Guide for Conducting In situ Field Bioassays with Marine, Estuarine, and Freshwater Bivalves" (ASTM 2001) for several reasons: (1) The term bioassay fits the definition of an experiment that includes both an estimate of toxicity and an estimate of relative potency. This use of bioaccumulation as an estimate of relative potency combined with toxicity endpoints such as survival, growth, and reproduction is the essence of the proposed conceptual framework. (2) Protocols for marine, estuarine, and freshwater invertebrates were synthesized to emphasize that there are more similarities than differences in measuring bioaccumulation and biological effects among the three bivalve groups and that the taxonomic differences are due to factors other than the way they are exposed to, accumulate, and respond to chemicals in the environment. (3) Caging bivalves in the field was advocated as a means to expose test organisms to environmentally realistic conditions in a way that cannot easily be duplicated in the laboratory. In addition to accounting for the effects of receiving waters on modifying exposure conditions, the experimental control afforded by caging facilitates virtually all measurements that are routinely conducted in the laboratory after retrieving the test animals from the field.

NEED FOR A FOCUSED CONCEPTUAL FRAMEWORK

A focused conceptual framework toward an EDR strategy in freshwater mussel ecotoxicology is advantageous over many current approaches that are too isolated and are not easily linked with either water or sediment chemistry data. An EDR strategy is consistent with ecological risk assessment ERA based monitoring because it provides a means to reduce the uncertainty commonly found in more traditional assessments that emphasize either exposure- or effects-based monitoring. Several conceptual models are currently available to link chemicals in tissues with subsequent effects, but none of these models has been successfully applied because there has not been an integration of exposure and effects measurements. A focused conceptual framework can provide links between bioaccumulation and biological effects and to help characterize those processes associated with the ecotoxicology of freshwater bivalves.

The refinement, integration, and harmonization of existing models into a single, unifying approach can help focus unionid research on the most meaningful measurements. One stated purpose of the ERA paradigm is to provide a focus. To achieve this harmonization, the ERA framework is used as an "umbrella" model to develop a more holistic approach to reduce uncertainty. This is accomplished by linking a series of sub-models that involve the EDR triad, bivalves, bioaccumulation, caging, and tissue residue effects relationships. These include a tissue residue effects model to link exposure and effects, a space and time model to demonstrate temporal and spatial links, a bioaccumulation model to link other monitoring elements, a bivalve model to facilitate making these measurements using consensus-based protocols, and an overall monitoring model that serves as a reminder that ecological processes need to be included in the monitoring and assessment scheme. A key concept to these models is the concurrent assessment of chemicals in tissues, water, and sediment. A similar approach (i.e., the harmonization of water, sediment, and tissue quality measurements) has been suggested as a way to improve water quality guidelines (Reiley et al. 2003).

The purpose of this chapter is to review, synthesize, and assess risk assessment-based approaches for establishing ecotoxicological links between chemicals in tissues and associated effects in freshwater bivalves. Through this synthesis, a conceptual framework to reduce uncertainty in the ecotoxicology of freshwater bivalves is presented. The proposed framework for establishing ecotoxicological links between chemicals in tissues and effects in bivalves places equal emphasis on characterizing exposure and effects and recommends routine measurements of external chemical exposure (chemical in water and sediment) and internal dose (chemicals in tissues). Uncertainty can be reduced and ecological links established by always measuring tissue chemistry and effects when assessing chemical bioavailability, chemical toxicity, and community structure. Tissue chemistry is proposed as a "common currency" (Mearns 1985). Copper will be used as a case study. Emphasis will be placed on the use of field experiments with caged bivalves as an evolving technique in risk assessment to assess chronic exposure and toxicity. This chapter will demonstrate how the most commonly measured endpoints, that is, survival, bioaccumulation, and growth, can be used in concert to establish links between chemicals in tissues and associated effects.

BIOACCUMULATION MODEL

BIOACCUMULATION LINKS

Perhaps it would be easiest to introduce the importance of paired bioaccumulation and effects measurements using a simple example of how the addition of tissue chemistry to existing monitoring and assessment approaches could help establish links between them. Tissue chemistry data can be used as the hub for establishing links between other ecotoxicological measurements such as water and sediment chemistry as well as laboratory and field bioassays and studies of benthic community structure (Figure 9.1). As suggested previously, we have deliberately referred to these laboratory and field tests as bioassays instead of toxicity tests because the addition of tissue chemistry provides the estimate of potency required to fit the definition of a bioassay (ASTM 2001). Unionid researchers need to move toward bioassays rather than toxicity tests and utilize this unifying concept suggested previously (McCarty 1991).

Bioaccumulation is the ultimate link between the environment and the organism and represents the integration of chemical and biological measurements. It is an important component of characterizing exposure (external exposure from water and sediment is the other). This "internal exposure," or absorbed dose, may be more relevant in an ecotoxicological context than "external exposure" because in many instances, it is the most direct way to confirm that exposure has

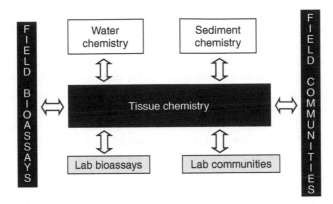

FIGURE 9.1 Diagram showing tissue chemistry as the hub of an integrated monitoring and assessment strategy to link chemical exposures from the environment with dose and response in various laboratory and field tests.

occurred. This internal or absorbed dose can be closer to the ultimate receptors of concern and can therefore help explain exposure routes better than using only measurement of chemicals in water or sediment. Bioaccumulation data can also be used for source identification using chemical finger-printing. It is commonly measured in laboratory studies and can be an essential link between bioaccumulation in other laboratory and field monitoring. Therefore, contaminant concentrations in bivalve tissues can reflect the magnitude of environmental contamination with greater accuracy. However, it would be prudent to point out that a body burden accumulated but sequestered away from the site of action poses a difficult interpretation of responses directly attributable to a quantified burden.

While we advocate measuring bioaccumulation wherever possible, it is clear that regulations have not kept pace with the use of bioaccumulation data. Developers of sediment testing, for example, recognized early the importance of measuring bioaccumulation, but regulations only include a requirement for comparison with a reference station (USEPA/US ACOE 1977). Unfortunately, this requirement has not changed substantially in the last 25 years and is not very useful. Exacerbating the problem is this unnatural dichotomy between using one group of animals for toxicity testing and another for bioaccumulation testing. This is another one of those misconceptions mentioned in the first chapter, but it is not restricted to freshwater bivalve researchers. In developing Mussel Watch monitoring programs, advocates have often emphasized the point that bivalves are "resistant" to chemical stress. This has promulgated and perpetuated the myth that bivalves are pollution tolerant. Widdows and Donkin (1992) have added an important caveat by suggesting that bivalves are "resistant but not insensitive to...." This is an extremely important caveat. Another associated problem is that the emphasis on short-term acute laboratory testing would also suggest that bivalves are relatively insensitive. However, a number of studies on freshwater bivalves suggests that they are just as sensitive or more sensitive than other species (ASTM 2001; see also Chapter 7).

The developers of the first dredge material bioassay requirements understood the potential importance of bioaccumulation, and this is why it was included as a component of the assessment (USEPA/US ACOE 1977). However, because the links were not understood, there was little connection with toxicity testing other than being conducted on the same sediments. Similarly, because marine bivalves were only required for bioaccumulation potential, no effects endpoints were required. Investigators did not know then, nor do they know now, if the test animals were in sufficiently good health to accumulate chemicals within a normal steady state environment indicative of what would be accumulated in nature. Several authors of marine studies have addressed the combination of exposure and effects endpoints but without the same ERA focus and without a focused context. The important point to be made here is that even though marine bivalve monitoring and assessment is far advanced over freshwater bivalve testing, the regulations have not kept pace with the state of the science, even for marine bivalves. This basic separation of using one group of organisms for toxicity testing and bivalves for bioaccumulation continues today, almost 30 years after the development of the first document describing an "ecological evaluation" of dredged material (USEPA/USACOE 1977). Based on the state of the science today, one would be hard pressed to justify calling these tests an "ecological evaluation."

Tissue Residue Effects

Mount (1977) recognized the number of problem chemicals in the environment that were not directly toxic but instead bioaccumulate, producing undesirable residues in the body. He also noted bioassay and toxicity test methods were far ahead of the ability to apply the results. This becomes a problem with respect to discerning tissue residue effects relationships, particularly in freshwater bivalves because of the paucity of data supporting such examination. Although tissue residues have been used more routinely to determine the potential for bioaccumulation of chemicals from sediments and dredged materials, they can also improve resolution of exposure beyond chemical measurements of water or sediment. The impact of chemical exposure is also dependent on a number of major ecological variables aside from the accumulated dose or exposure concentration that describes the hazard. With this consideration, tissue residues would seem critical to include in any integrated measure of environmental exposure.

Developing Tissue Residue Guidelines—Data Application

Critical body residue (CBR) theory (McCarty and Mackay 1993) can be combined with the effects-range paradigm (Long and Morgan 1990) to establish links with water and sediment quality guidelines and to develop tissue quality guidelines (Figure 9.2). Long and Morgan (1990) initially developed sediment quality guidelines by employing a weight-of-evidence approach, assembled from a variety of metrics (e.g., sediment chemistry, laboratory toxicity tests, and benthic community structure) using data from many geographic areas. They used a three-step evaluation approach to (1) assemble and review data where estimates of sediment concentrations were linked with adverse biological effects, (2) determine ranges in concentrations of chemicals in which effects

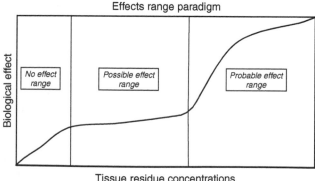

FIGURE 9.2 Using an effects range paradigm to establish tissue residue guidelines associated with no effects, possible effects, and probable effects. (Adapted from Long, E. R. and Morgan, L. G., *NOAA Technical Memorandum NOAA OMA 52*, U.S. Department of Commerce, 1990. With permission.)

were likely to occur, and (3) evaluate other data relative to these effects ranges. Based on this analysis they concluded that sediment chemistry data alone provided neither a measure of adverse biological effects or an estimate of the potential for effects.

Similar effects ranges could be established using the paradigm initially used by Long and Morgan and modifications currently being used by those authors to develop more sophisticated sediment quality guidelines (and water quality criteria) as shown in Figure 9.2. Given the ability to measure tissue residues in water and sediment exposures, it is possible to establish tissue residue guidelines based on residue–toxicity relationships. These relationships can provide a basis for criteria without the bias associated with bioavailability of chemicals from water or sediment, which is particularly true when in situ measurements provide the residue–toxicity link as with the caged bivalve approach.

COPPER AS A CASE STUDY

The best example of using the tissue residue effects approach is provided for copper because of data availability and abundance of work. It should also be pointed out, as shown in Table 9.1 and Table 9.2, that much more work has been done on marine bivalves and copper than for freshwater bivalves and copper. That concentrations associated with effects (lowest effects concentrations as LOECs) and no effects (no effects concentrations as NOECs) were similar

TABLE 9.1
Links between Tissue Copper Residues (µg/g dw) and Effects in Freshwater Bivalves

EC	NOEC	Species	Exposure	Endpoint	Citation
8.1	2.7	D. polymorpha	Lab	Regulation breakdown	Kraak et al. (1992)
6.5	2.7	D. polymorpha	Lab	Regulation breakdown	Kraak et al. (1992)
20.8	14.3	D. polymorpha	Field	Scope for growth	de Kock and Bowmer (1993)
	15	D. polymorpha	Lab	No physiological effects	Kraak et al. (1992)
41	16	D. polymorpha	Lab	Filtration rate	Kraak et al. (1992)
	46.8	C. fluminea	Artificial stream	Growth: weight, shell length	Belanger et al. (1990)
100	50	C. fluminea	Artificial stream	Growth: weight, shell length	Belanger et al. (1990)
120	65	C. fluminea	Artificial stream	Growth: weight, shell length	Belanger et al. (1990)
70		D. polymorpha	Lab	Physiological effects	Kraak et al. (1992)
93.5		Q. quadrula	Transplant	Mortality	Foster and Bates (1978)
83		D. polymorpha	Field	Tissue energy reduction	Secor et al. (1993)
64.3	26.6	Means for all bivalves (µg/g dw)			
19.1	29.6	Means for all non-unionid bivalves (µg/g dw)—zebra mussels and Asian clams			

TABLE 9.2
Links between Tissue Copper Residues and Effects in Marine Bivalves—Copper (μg/g dw)

Marine Bivalves	EC	NOEC	N
Survival and behavior	128.9	27.5	45
Growth, scope for growth, filtration, reproduction, condition, change in bioaccumulation endpoints	48.0	21.1	41
Biochemistry and histopathological endpoints	43.3	16.3	31
Means for all bivalves without "888" Calabrese value (μg/g dw)	129.0	41.0	
Means for all *Mytilus* (μg/g dw)	82.9	24.7	
Means for all bivalves without oysters and without "888" Calabrese (μg/g dw)	93.0	23.9	
Means for all Bivalves without oysters, Meretrix, and "888" Calabrese (μg/g dw)	80.3	23.9	

in both marine and freshwater species is of significant interest. Given the large number of studies conducted on marine species, mean values provided as 80 μg/g dw for effects and 24 μg/g dw for no effects provide added confidence for predicted thresholds. These data were screened to remove nine data points for oysters since oysters have been shown to be hyper-accumulators for copper (one outlier of 888 μg/g dw for *Mytilus edulis* and another apparent outlier of 1,005 μg/g dw for *Meretrix casta*). Means were provided using those values for comparative purposes. Interestingly that the mean for all *Mytilus* data (including 83 for effects and 25 for no effects) are very similar to means for remaining marine genera. It was encouraging that tissue residue effects thresholds, predicted over a decade ago and based upon *Mytilus galloprovincialis* transplants in San Diego Bay (75 and 25 μg/g dw), were very close to this *Mytilus* summary mean. These data suggest that the basic premise of predicting effects based on tissue residue effects data from controlled field experiments can be supported as a substantive monitoring and assessment tool, particularly for predicting potential effects. In this context, experimental control refers to a designated geographic location, an exposure period, and the size and number of test organisms in each replicate cage (ASTM 2001).

There were 41 copper studies on marine bivalves where tissue residues were linked to effects, but there were only 12 for freshwater bivalves. It should be noted that of these twelve freshwater studies, only two have been conducted on unionids (*Elliptio complanata* and *Quadrula quadrula*) and all the rest were on *Dreissena polymorpha* and *Corbicula fluminea*. Effects endpoints include mortality, growth, filtration rate, physiological effects, and SFG, which is the physiological evaluation of the potential for growth and not a direct measure of growth. The only two unionid studies were conducted in the late 1970s (Foster and Bates 1978) with mortality used as endpoints. All of the more sensitive effects endpoints have been measured on non-unionid freshwater bivalves (*Dreissena* and *Corbicula*), reinforcing our assertion that the paucity of effects studies on unionids is not attributable to a lack of available methods, but more a function of the researcher's range of experience and funding available to do the work. More work has been done on *Dreissena* and *Corbicula* in the last decade because there has been more money available to conduct those studies and more effort has been expended in applying methods used for marine bivalves to those two species rather than unionid species. The effects endpoints for marine bivalves include survival, growth, reproduction, condition index, physiology, SFG, and various physiological endpoints. Scope for growth has been one of the more common endpoints in the development of the tissue residue effects databases from the US Army Corps of Engineers and the USEPA (USACOE 1996,1999; Jarvinen and Ankley 1999). John Widdows and his colleagues, who developed and pioneered this method for marine bivalves (Widdows and Donkin 1992), have paired bioaccumulation and biological effects measurements for the longest period of time and have produced the most tissue residue effects data.

CBRs for Freshwater Bivalves

Although far fewer copper CBR studies have been conducted on freshwater bivalves, the overall means of 64 and 27 µg/g dw for effects and no effects, respectively, are very similar to those calculated for marine bivalves without the apparent outliers. There are not enough data to make a meaningful division by measurement endpoint as for marine bivalves, but there are some other interesting observations that might be useful to help guide future work. First, tissue accumulation values for unionids (*E. complanata* and *Q. quadrula*) are relatively higher than the overall mean for effects but within the range found for marine bivalves. Another interesting observation is that the NOEC for all non-unionid bivalves (30) is higher than the overall mean for EC (19). This appears to be a function of the number of studies comprising the data and quality represented by more sophisticated studies, suggesting that division by category would be more useful. Nevertheless, the beginning of a tissue residue effects database for freshwater bivalves is encouraging and suggests that CBRs for effects and no effects are similar in marine and freshwater bivalves.

Copper CBRs for Marine Bivalves

While the previously suggested overall copper CBR means for effects of $EC_{tissue\ (T)} = 80$ and no effects $EC_T = 24$ (Table 9.2) are virtually identical to those predicted in field studies, it may be more appropriate to group the effects by category and mode of action. These are also shown in Table 9.1. Using this approach, the EC_T for survival and behavior is 129 and the $NOEC_T$ is 28, compared to an EC_T of 48 and an $NOEC_T$ of 21. The corresponding EC_T and $NOEC_T$ are similar for biochemistry and histopathological endpoints. An important trend exists with the $NOEC_T$ for survival decreasing with sensitivity of the response. This is probably a function of the test and not a real difference in NOECs. Another important point relative to field studies and their ability to predict effects is that original predictions of effects were conservative with a wide separation between the NOEC and the EC, 25 and 75 µg/g dw. By using any elevation above the NOEC to represent potential effects, the 25 µg/g dw is still close to the 48 µg/g predicted for effects. Furthermore, it is somewhat surprising that these predictions were accurate in work done in San Diego Bay over 10 years ago when the concentrations of TBT at several stations were thought to be one of the principal factors affecting mussel growth rates. The ability to discern these copper effects, particularly when considering the effects of other factors, such as temperature and food, is encouraging. Overall means without known hyper-accumulators such as oysters and apparent outliers are also identified in Table 9.2.

Using Caged Bivalves to Establish Tissue Residue Effects Relationships

Some of the best examples of tissue residue effects theory come from studies with marine bivalves, and among those, more data are available for copper than any other metal and most other chemicals. Most if not all of the approaches used with marine bivalves could be applied to freshwater bivalves, and many of the studies were conducted using caged bivalves or other field studies. Three different effects endpoints (survival, growth, and reproductive effects) associated with CBRs for copper in marine bivalves are used as examples to demonstrate the ability of field studies to establish CBRs for various measurement endpoints (Figure 9.3). Of these CBRs, two were developed from caged bivalve studies and one from field-collected animals, stressing the utility of controlled field experiments with caged bivalves.

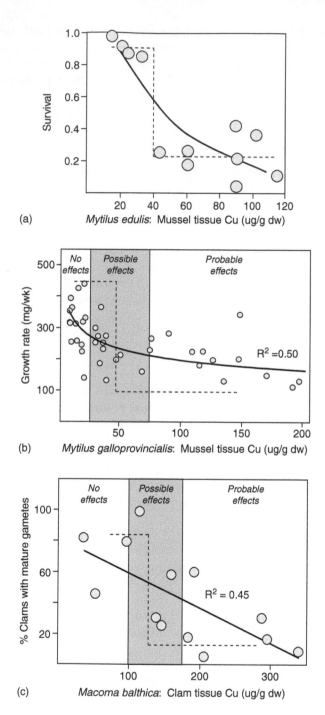

(a) *Mytilus edulis*: Mussel tissue Cu (ug/g dw)

(b) *Mytilus galloprovincialis*: Mussel tissue Cu (ug/g dw)

(c) *Macoma balthica*: Clam tissue Cu (ug/g dw)

FIGURE 9.3 Predicting effects using tissue residues. Survival (a), growth (b), and reproductive effects (c) associated with copper tissue residues in *M. edulis*, *M. galloprovincialis*, and *M. balthica*, respectively, measured in field studies.

The first example used caged mussels (*M. edulis*) to study the effects of acid mine drainage from an abandoned copper mine in Howe Sound, British Columbia, Canada (Grout and Levings 2001). These workers provide an estimated CBR for survival of 40 µg/g dw (Figure 9.3a). Their work is particularly important because they were able to link copper concentrations above a specific threshold (CBR) with declines in survival, weight growth, length growth, and condition index. Decreases in growth (weight and length) were evident at even lower copper tissue burdens near 19 μg/g dw. Perhaps even more importantly, they were among the first to compare tissue residue effects results using a threshold model rather than a typical log–linear model. They found that the threshold model produced better fits based on higher R^2 values for all relationships except weight increase. The reasons for this were not clear and require further investigation. Nevertheless, their threshold model (identified with a dashed line in Figure 9.3a) shows the drastic decline in mussel survival with increasing tissue burdens. This is consistent with CBR theory and may also help explain the poor correlations between tissue residues and effects in some other studies when the log–linear model is used.

The second example is from a series of transplant studies conducted in San Diego Bay between 1987 and 1990 (Salazar and Salazar 1995). Two CBRs were estimated from this study: an effects concentration for growth of 75 μg/g dw and a $NOEC_T$ of 25 μg/g dw (Figure 9.3b). A series of multiple regression analyses were used to determine where the relationship between copper residues and effects began to change. The R^2 value for these data (0.50) is similar to that predicted from the threshold model for tissue residue effects relationships for weight growth calculated by Grout and Levings (0.55) and somewhat lower than that calculated using the log–linear model for weight growth (0.69). We have reproduced the conceptual threshold model used by Grout and Levings (2001) with a dashed line to demonstrate a possible threshold of about 50 μg/g dw between our predicted no-effect and effective concentrations.

The third example is from a long-term study in San Francisco Bay where *Macoma balthica* were collected from a mudflat exposed to copper (Hornberger et al. 2000, Figure 9.3c). It provides an estimated CBR for reproduction ($EC_T = 100$–200 μg/g dw) and an R^2 of 0.45, which is very similar to those predicted in the other two field studies.

The three field studies are somewhat unique in that they all measured multiple effects endpoints and tissue residues. While the Grout and Levings (2001) study demonstrated the high availability of copper associated with acid mine drainage and the Salazar and Salazar (1995) study included many sites studied over several years, the Hornberger et al. (2000) study is of importance because of a single site studied over a 25-year period. Each of the field studies demonstrates the utility of a different monitoring approach with marine bivalves. Similarly, the SFG methodology has been applied to both laboratory and field exposures. These techniques should be easily applied to freshwater bivalves as well. Some methods used for assessing exposure and effects in marine bivalves have been transferred to freshwater studies, although the available literature database for freshwater mussel bivalves is not nearly as extensive as for marine species. These studies are also important because they demonstrate how the tissue residue effects approach and the EDR triad transcend media differences commonly associated with water column and sediment studies. In each case, tissue residues were used as a principal component of predicting effects, in addition to estimates of effects based on water or sediment.

Widdows and Donkin (1992) found that the tissue burdens associated with reductions in SFG in a laboratory study were similar to those measured in a caged mussel field study using absolute growth as the effects endpoint. Although data from that study were plotted on a log scale, it is still obvious that a threshold for effects can be associated with a specific water concentration (about 5.5 μg/L, top scale) and a specific tissue concentration (about 20 μg/g dw, bottom scale).

CHANGES IN THE RELATIONSHIPS AMONG EXPOSURE, DOSE, AND RESPONSE

The relationship between chemicals in water and mussels first became apparent in our work during the development of the caged bivalve bioassay in San Diego Bay. A change in the relationship between tributyltin (TBT) in seawater and in marine mussel tissues occurred above a threshold concentration of approximately 100 ng/L (Salazar and Salazar 1996). The environmental significance of bioaccumulation has been challenged by some investigators based on the following concepts: (1) bioaccumulation is not an effects endpoint and has little environmental significance; (2) many organisms, including invertebrates such as bivalves, cannot be used as effective monitoring tools because they have the ability to regulate essential metals such as copper; (3) it is more important to measure effects endpoints rather than exposure endpoints. While it is true that bioaccumulation in itself is not an effect, effects can be predicted based on the way organisms change accumulation at threshold exposure concentrations in water, sediment, and tissues, and there are a number of examples in the existing literature on marine and freshwater bivalves. The concept is depicted in Figure 9.4 and is consistent with the relationships already discussed regarding Figure 9.3. Three different bivalve species (one freshwater and two marine species) change the

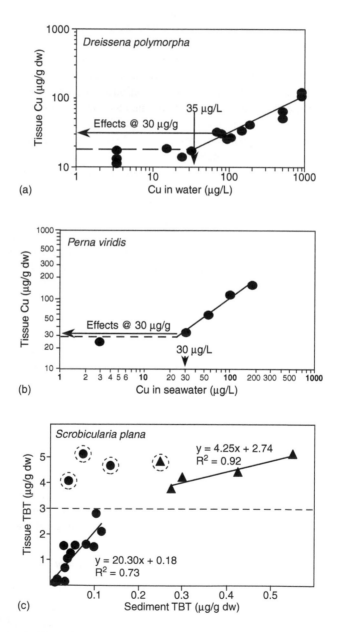

FIGURE 9.4 Changes in the relationship between exposure, dose, and response in (a) *D. polymorpha*, (b) *P. viridis*, and (c) *S. plana*.

way they accumulate copper at about 30 µg/L. Interestingly, as with most early studies that did not incorporate the tissue residue effects paradigm, these studies did not discuss the links between dose and measured response.

Figure 9.4a demonstrates a similar change in the accumulation/effects relationship for the freshwater zebra mussel, *D. polymorpha* (Kraak et al. 1992, 1993, 1994). The relationship changes near 35 µg Cu/L, and this is associated with a CBR of about 30 µg/g dry weight. In both laboratory examples, an effects endpoint was measured to demonstrate that effects were actually occurring. The effect could have been predicted from the threshold where the relationship between exposure and uptake changes. While we do not advocate this approach over measuring exposure and effects endpoints directly, it demonstrates how the process can be characterized and better understood to make these kinds of predictions in the absence of effects endpoints. There are many more data points in the *D. polymorpha* experiment than in the *Mytilus* experiment (Figure 9.4b), and the data more clearly demonstrate the change from regulation of copper by the mussel to one of increased uptake. A possible mechanistic explanation for this phenomenon involves disruption of homeostasis by increasing metal concentrations to the point where animals fail to regulate copper uptake, loss, or transformation. The data also suggest that these processes, and the CBRs, are similar in freshwater and marine mussels.

Figure 9.4b demonstrates this change in relationship for the marine mussel *Perna viridis* (Chan 1988). Although there is only one data point between 1 and 30 µg Cu/L, it appears as though the relationship changes dramatically at about 30 µg/L and above. This change in relationship has been demonstrated in many similar experiments with bivalves and other species of freshwater and marine bivalves for several different metals and is consistent with the paradigm shown in Figure 9.2.

To further illustrate the threshold concept, and the utility of field data from natural populations, Langston and Burt (1991) showed the following relationship between TBT in sediment and TBT in the tissues of the deposit-feeding marine clam *Scrobicularia plana* (Figure 9.4c). Their data demonstrated an apparent upper limit above which TBT is no longer accumulated and where mortality occurs. These results are consistent with their field surveys and the change in the relationship between 0.1 and 0.3 µg/g dw. This is also the concentration at which effects have been shown in other benthic species such as marine polychaete worms.

Although a comparable example was unavailable for copper in sediment, another interesting relationship has been reported by Bryan et al. (1987) for copper in the tissues of the estuarine alga *Fucus* and the estuarine cockle *Cerastoderma edule*. These data are from collections in natural populations and were compared to laboratory experiments and other observations to demonstrate where effects might begin to occur. In an 18-day laboratory exposure, some of the cockles had died and the survivors accumulated copper to a concentration of 242 µg/g dw. Control animals exposed to uncontaminated sediment but the same water from tanks with contaminated sediment accumulated copper concentrations up to 154 µg/g dw. These data suggest that the major exposure pathway for copper was through waterborne exposures. Bryan et al. (1987) used the waterborne exposure pathway for *Fucus vesiculosus* to show how concentrations in cockles and seaweed vary. The data also suggest that cockles are initially able to regulate copper uptake, but as the concentration increases, tissue burdens increase dramatically. They concluded that organism loss of regulatory capacity for copper was responsible for accumulation to toxic levels that precluded cockle survival, in particular, in the estuaries where they were previously found. Furthermore, these elevated concentrations in tissues are similar to those found associated with high mortalities in short-term laboratory exposures.

CAGED BIVALVE MODEL

In its simplest form, the caged bivalve approach ensures a "captive receptor" that facilitates integrative exposure and effects measurements. An added benefit compared to traditional field monitoring lies in the well-defined spatial and temporal controls on the exposure. It is somewhat surprising that in situ-based monitoring has not been utilized more in ecological risk assessments because of this ability. Alternatively, laboratory testing is based on very simple systems in which results may not be easily extrapolated to the field. Parrish et al. (1988) suggest that exposure is the real variable in hazard (and risk) assessment. The need to better characterize exposure has led to the increased use of in situ approaches to reduce the uncertainty in the exposure characterization. Focusing on effects measurements only generally precludes an accurate characterization of exposure. It could even be argued that effects assessment has less value without a robust character-ization of exposure. In situ testing provides the environmental realism necessary to make real-world comparisons. All clinical measurements can be made on the caged bivalves, and field studies with caged bivalves can be used to help establish causality. Bivalves, and the caged bivalve model, comply with the requirements of a useful biomonitoring system: versatility, practicality, integrative ability, and consistency (De Kock and Kramer 1994).

Matteson (1948) included a series of practical methods to study the life history of freshwater bivalves that are the root of several modern-day approaches. In what might be considered the first freshwater unionid transplant, Matteson (1948) held *E. complanata* in shallow water in an area marked by stakes where he returned later to collect them for life history measurements. He recog-nized the importance of making measurements on individuals and used a very rudimentary system of filing symbols on the unionid shells to track individual growth rates in adults over a period of one year. Growth measurements made with a simple, homemade caliper included length, height, width, and weight. Weights were measured with a torsion balance, and the cages for holding the bivalves were quite rudimentary and not very successful. Matteson understood the importance of holding test animals under natural conditions. To facilitate the life history studies, he attempted to hold fish with glochidia in cages but later switched to laboratory aquaria because of problems associated with wave action. He also measured growth rates on newly metamorphosed juveniles during short-term exposures in the laboratory.

The Ontario Ministry of Environment (MOE) has been using indigenous and caged *E. compla-nata* to measure water quality through bioaccumulation for almost 30 years as part of a regional monitoring program developed by the Ontario Ministry of the Environment (Hayton and Hollinger 1989; Hayton et al. 1990; Anderson et al. 1991; Richman 1992, 1997, 2003; Ontario Ministry of the Environment 1996; Ontario Ministry of the Environment and Energy, 1999). All of these studies have focused on characterizing exposure by measuring concentrations of organochlorines, such as dioxins and furans, in freshwater mussel tissues. The practicality of using the freshwater clam (*E. complanata*) as part of this program was first demonstrated by Curry (1977). Caged clams proved practical for detecting organic trace contaminants in water after a short exposure period. He discussed the advantages of using clams over fish and water samples for biomonitoring as well as the practical aspects of using caged bivalves to collect useful information and further demonstrate the utility of this organism and the caged bivalve model. Creese, Lewis, and Melkic, (1986) provided further guidance for standardized methods.

The utility of the caged bivalve model has been demonstrated in numerous studies throughout the United States and Canada. Since 1975, we have conducted over 60 caged bivalve studies, approximately 20 of these were freshwater studies using eight different bivalve species (Table 9.3). The background gained with conducting studies in the marine environment has allowed us to effectively transfer the technology to freshwater systems. Although freshwater environments possess their own suite of challenges and circumstances, the general principles and concepts developed for marine systems can be applied to freshwater. The challenges associ-ated with deploying caged bivalves at depths less than 0.3 m in Guelph, Canada are commensurate with deploying mussels at depths greater than 200 m in Port Valdez, Alaska. Caged bivalve studies have been conducted in freshwater systems as far north as Lynn Lake, Manitoba (56°N latitude), as far south as Port Arthur, Texas (30°N latitude), and on both the west coast (Bear Creek, Washington; 122°W longitude) and east coast (Augusta, Maine; 70°W

TABLE 9.3
Summary of Caged Freshwater Bivalve Studies Using ASTM Standard Protocols

Year	Location	Species	Number	Study Relevance and Importance
1994	Sudbury River, MA	*E. complanata*	900	Our first freshwater bivalve transplant study at a USEPA superfund site (sediment assessment). Freshwater mussels in compartmentalized, flexible mesh were placed directly on sediment without adversely affecting growth. Tissue burden expressed on a concentration basis alone was deceiving due to growth dilution. Different mussel metrics should be used synoptically as effects endpoints
1997	Port Arthur, TX	*C. fluminea*	2400	Bioavailability of PAHs in freshwater sediments
1997	Sault Ste. Marie, Ont., CAN	*C. fluminea*	3300	Another superfund site; results suggested that concentrations of chemicals were decreasing due to natural remediation and sedimentation
2000	Sault Ste. Marie, Ont., CAN	*C. fluminea*	3600	
1999	Red River, Winnipeg, CAN	*P. grandis*	960	First time that mussel transplant locations were directed by a dynamic model, and the results were used to help validate the model. Because of the difficulty in identifying ammonia in bivalve tissues, only effects endpoints were used here
1999	Red River, Winnipeg, CAN	*S. simile*	2100	Caged bivalves were deployed downstream of a municipal effluent. The utility of the caging methodology as a platform for conducting any clinical measurements was demonstrated (e.g., a series of biomarkers). Demonstrated the evolution of the method to new approaches. During this time, we developed a benthic cage for long-term exposures to chemicals associated in the effluent and experimentally induced sex reversal after a one-year exposure. It also showed the development of the vitellin biomarker by our colleagues to demonstrate effects on reproduction and endocrine disruption
1999	St. Lawrence R., Montreal, CAN	*E. complanata*	600	
1999	St. Lawrence R., Montreal, CAN	*D. polymorpha*	600	
2000	St. Lawrence R., Montreal, CAN	*E. complanata*	520	

(continued)

TABLE 9.3 *(Continued)*

Year	Location	Species	Number	Study Relevance and Importance
2001	St. Lawrence R., Montreal, CAN	*E. complanata*	850	
2002	St. Lawrence R., Montreal, CAN	*E. complanata*	216	
2003	St. Lawrence R., Montreal, CAN	*E. complanata*	513	
2000	Augusta, ME	*E. complanata*	1440	Demonstrated the problems associated with using the above–below experimental design and the advantages of the gradient design in field testing.
2003	Augusta, ME	*E. complanata*	1134	
2001	Guelph, Ont., CAN	*L. costata & E. complanata*	450/480	A PCB assessment marked the development of a "frameless cage" made from rigid plastic mesh that allowed us to transplant the bivalves into stream water <0.3 m in depth.
2002	Guelph, Ont., CAN	*L. costata & E. complanata*	264/180	PCB accumulation measured in two different unionid species. We were able to identify the congener distribution in the tissues and demonstrate that *Lasmigona.* accumulated concentrations of PCBs in their tissues that were considerably higher than *E. complanata*. Comparisons were also made with respect to the PCB load associated with different grain sizes
2003	Guelph, Ont., CAN	*L. costata & E. complanata*	156/255	
2004	Guelph, Ont., CAN	*L. costata & E. complanata*	286/169	
2003	Woodinville, WA	*M. falcata*	45	A very limited study to evaluate the potential impact of chemicals on the declining mussel population in an urban watershed. We found no direct evidence that any of the chemicals measured were at concentrations that should be cause for concern
2004	Lynn Lake, Manitoba, CAN	*L. radiata*	450	Conducted at a very northern latitude with a very narrow window of opportunity to work because of ice-over. It provided yet another evolution of the caging methodology by combining the "frameless cage" on top of a frame to compare bioaccumulation and growth of unionids transplanted directly on sediment in flexible mesh bags and others held approximately 30 cm above the bottom on top of a PVC, table-like frame

longitude) of North America. These studies were designed to assess the effects of metals, orga-nometals, and organic chemicals such as PCBs and PAHs. A summary of those freshwater caged bivalve studies is shown in Table 9.3, including the relevance and important advancements gained from each effort.

SPACE AND TIME, SITE-SPECIFIC CONDITIONS, NATURAL FACTORS

In situ field studies with caged bivalves can be particularly effective at characterizing both exposure and effects over space and time and under environmentally relevant, site-specific conditions. Although site-specific conditions cannot be controlled, the caged bivalve approach facilitates consistency in space, time, and animals used in the test replicates. In situ studies with caged bivalves are potentially more powerful than monitoring natural populations because the bivalves can be maintained in cages and readily transplanted to sites of interest, some of which may be sites where natural populations would not normally grow or reproduce due to various factors such as the lack of a suitable substrate. The ecotoxicological framework provided in the EDR triad serves as the foundation for using caged freshwater bivalves in controlled field experiments to establish links between chemical exposure and associated biological effects. Concurrent measurements of exposure (i.e., external exposure and dose) and biological effects are critical to the success of these field experiments.

The experimental control afforded by this approach can be used to place a large number of animals of a known size distribution in specific areas of concern to quantify exposure and effects over space and time within a clearly defined exposure period. Although a number of assessments have been conducted using bivalves to characterize exposure by measuring tissue chemistry or associated biological effects, relatively few assessments have been conducted to simultaneously characterize both exposure and biological effects (Widdows and Donkin 1992; Salazar and Salazar 1991, 1995; ASTM 2001). The procedures provided in the ASTM guide are specifically designed to help minimize the variability in tissue chemistry and response measurements by using a practical uniform size range and compartmentalized cages for multiple measurements on the same individuals and to simultaneously collect exposure and effects data. These procedures could be regarded as a guide to an exposure system to assess chemical bioavailability and toxicity under natural, site-specific conditions. It is only the site-specific conditions that are not controlled.

The ASTM Standard Guide for conducting in situ bioassays with caged bivalves contains specific guidance on various aspects of this methodology. The purpose of that guidance is to facilitate the simultaneous collection of exposure and effects data as part of an ecotoxicological framework that is consistent with the ERA paradigm of characterizing exposure and effects (USEPA 1998). Although the guide could be used for either exposure or effects measurements, the approach was designed to collect both types of data in the same organism at the same time. Our refinements to the methodology that are included in the ASTM Standard Guide include (1) the importance of minimizing the size range of test organisms to minimize variability in exposure and effects measure-ments; (2) the use of compartmentalized cages to facilitate multiple measurements on the same individuals; (3) the combined use of digital calipers, a portable analytical balance, and a laptop computer to collect digital data in the field; (4) the development of field bioassay acceptance criteria similar to those commonly used in laboratory testing; and (5) the development of an ecotoxicolo-gical framework for using the exposure and effects data (ASTM 2001).

The importance of natural factors and variability in the concentration of chemicals in water and sediment in a monitoring strategy to supplement caged bivalve biomonitoring is portrayed in Figure 9.5 (Salazar and Salazar 1996). White (1984a, 1984b) has cautioned against the arbitrary use of mussel monitoring systems without developing a model to be tested. Figure 9.5 emphasizes the importance of natural factors in modifying the environmental effects of chemical stressors and depicts the inherent cycles of natural factors, chemical stressors, and mussel biology. It is suggested

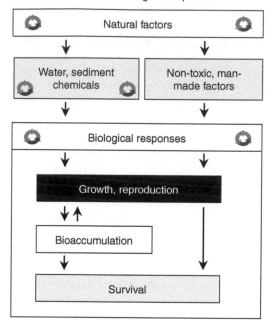

FIGURE 9.5 Identified the importance of natural factors and variability in the concentration of chemicals in water and sediment in a monitoring strategy to supplement caged bivalve biomonitoring.

that natural factors act directly on chemical stressors by altering bioavailability and directly on mussels by altering biochemistry and physiology. Other chemicals may also be involved.

GRADIENT DESIGN

One method for improving traditional monitoring approaches is the deployment of caged bivalves along suspected chemical gradients (Figure 9.6). This facilitates characterizing and understanding these gradients within the ecosystem. The gradient refers to a region of increasing chemical concentration in water or sediment. Experimental designs that account for these chemical gradients can be more appropriate for field experiments than reference sites and "above–below" comparisons (Landis 2000). As well, gradient designs increase the potential for identification of chemical stressor sources and allow for multiple comparisons and regression analyses. A series of field studies conducted by Couillard and his colleagues (Couillard, Campbell, and Tessier 1993, 1995a, 1995b) have demonstrated the utility of the gradient design using E. *complanata*, and measured a variety of exposure, dose, and response endpoints.

The classical laboratory experimental model is not easily transferable to ecological systems because ecological systems are not typically closed. Landis (2000) refers to this concept as the "Eulogy for the Reference Site." In other words, a reference site may relate more to a generalized gradient of conditions or a variety of habitat types that are deemed acceptable by stakeholders. Although reference sites have a long history in environmental science and toxicology, they are a legacy of the balance-of-nature model of ecological systems and laboratory experimental design. The balance-of-nature construct, that is, nature strives toward an ideal equilibrium state, has been falsified by field research and experimental model ecosystems. This persistent concept of the

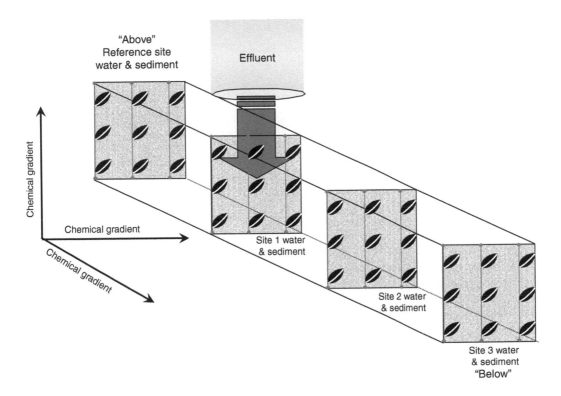

FIGURE 9.6 Caging bivalves along suspected chemical gradients, showing there is more uncertainty in "above–below" tests, because in a typical "above–below" test there is only one comparison. The gradient design allows multiple comparisons and regression analyses.

elusive reference site continues to limit effective decision-making and is perpetuated through environmental regulation and inadequate study designs. Rather, a reference condition approach attempts to define empirically the condition considered acceptable in an ecosystem, the variation of environmental attributes at spatial and temporal scales, and the deviation as a measure of the effect of stressors (Bailey, Norris, and Reynoldson 2004).

CONCEPTUAL BIVALVE MODEL

Bivalve molluscs have been proposed as both suitable response systems (Green, Singh, and Bailey 1985) and as quantitative biological indicators (Phillips 1980). The conceptual bivalve model (Figure 9.7) shows the integration of exposure, dose, and response in the context of a freshwater bivalve. Chemicals C1, C2, and C3 in water and sediment are taken up by the bivalve and are bonded to three internal receptors (R1, R2, and R3) in mussel tissues. This will be referred to as the internal chemical dose or the absorbed dose. The response associated with external exposure and internal dose is generally measured by effects endpoints such as survival, growth, and reproduction. Biomarkers are also included as part of the model because of their ability to help characterize the relationships among exposure, dose, and response.

We have routinely used growth measurements to "calibrate," characterize, and interpret bioaccumulation results. However, in a recent study with several confounding variables and significant temporal and spatial variability, the bioaccumulation data were also used to help characterize and understand the survival and growth results. This was probably the first time that we have used the data in that way. We have previously suggested that bivalve growth rates could be used as an effects

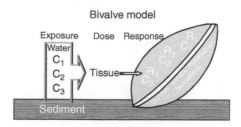

FIGURE 9.7 Bivalve model showing chemicals in water and sediment (C1, C2, and C3) taken up by a bivalve and bonding to internal receptors (R1, R2, and R3). These chemicals are integrated in bivalve tissues and effects manifested in response endpoints such as survival, growth, and reproduction, an integration of the internal concentrations at receptors of concern.

endpoint, to help "calibrate" bioaccumulation, and as a performance criterion for a successful test. We also suggest that the combination of exposure and effects endpoints can help characterize and understand processes necessary for data interpretation and for use in the ecological risk assessment framework. For example, although interval samples were not taken to estimate the rate of uptake, some inferences can be made based on background information and also can frame new questions and develop supplementary approaches to make those estimates. Further, there appears to be an anomaly between the survival and growth data that could be explained, at least in part, through the use of water, sediment, and tissue chemistry data.

BIVALVES AS INDICATORS OF EXPOSURE

Bivalves can concentrate and integrate chemicals from water, sediment, and food as they filter the water for food and provide a direct estimate of exposure (Widdows and Donkin 1992). It is important to note that sediment particles can also be a food source, as bivalves extract the organic coatings from the particles (ASTM 2001). As they do, they can also be exposed to those chemicals associated with the organic coatings. Therefore, under certain conditions, sediment particles can also be considered as food, albeit of less nutritional value. In the context of this chapter, chemical exposure can occur via the waterborne or dietary exposure pathways, and research is underway to identify those pathways and the environmental significance of those differences. Bivalves are particularly useful in that regard because of their ability to filter large quantities of water and ingest large quantities of sediment for food, and they are naturally sedentary. Collectively, measuring these various parameters can provide an estimate of chemical exposure, and in that sense, bivalves can be used as indicators of exposure.

BIVALVES AS INDICATORS OF EFFECTS

The assessment of pollution and environmental quality must ultimately be in terms of biological measurements, preferably in concert with appropriate measurements of chemical contaminants. Many of the biological responses suggested as potential techniques for monitoring the effects of marine environmental pollution (McIntyre and Pearce 1980; Bayne et al. 1985) have been applied to bivalves, particularly mussels, due to their established role as "sentinel organisms." Only those responses that are measured at the level of the whole animal are here discussed and only traditional endpoints that are most commonly measured in all chronic toxicity testing, that is, survival, growth, and reproductive effects. As discussed by Mount (1977), these chronic tests can serve as "guideposts" to stay on course. At a minimum, biological effects should be characterized by measuring survival and growth. Bivalves are particularly useful in a weight-of-evidence approach that includes the recommended multiple metrics (i.e., whole-animal wet weight, shell length, tissue weight, shell weight, percent water, and percent lipids). Collectively, these metrics yield a reasonable estimate of mussel health. Another benefit is the ability to compare beginning- and end-of-test growth metrics and pairing them with tissue chemistry results.

Survival as an Effects Endpoint. Survival can be easily quantified and often provides useful information in ecotoxicological studies. Survival is often considered an insensitive effects endpoint, but in many cases, it offers valuable insight into the relationships between exposure and effects in field studies lasting 60–90 days. Reduced survival in field monitoring can occur with the highest concentrations of metals in unionid tissues as well, in situations where laboratory toxicity tests suggest no toxicity. However, this lack of toxicity may be attributed more to the duration of the short-term laboratory toxicity test, and less to tolerance of the mussels. Test protocols that are better suited to mussel responses are needed not only for more appropriate endpoints in laboratory toxicity tests but are also critical in consideration of how water and sediment quality criteria have influenced monitoring and assessment approaches. Survival may not always be a very sensitive indicator of effects in bivalves (Salazar and Salazar 1991), but it is an important parameter to monitor. While survival may not be a sensitive endpoint for early warning of effects in the context of a short-term laboratory exposure, survival results from long-term exposures can be extremely useful in predicting potentially adverse effects, and in that sense, they can be used as an early warning indicator. Several factors can affect survival, including handling before the test begins plus physical–chemical factors at the deployment stations. Survival has been linked with tissue burdens in bivalve tissues, and examples have been provided for copper exposures.

Growth as an Effects Endpoint. Growth is another effects endpoint that has been measured for decades as part of marine bivalve aquaculture and was quickly adapted for studying toxicological effects in marine bivalves. There are obvious applications for freshwater bivalves. Growth is a sensitive indicator of sublethal effects and has been associated with population effects (Bayne et al. 1985). It is more sensitive than mortality, and reductions in growth have been related to adverse effects on bivalve populations (Bayne et al. 1985). There is a relatively large database for natural and anthropogenic factors affecting growth. Growth endpoints include, but are not limited to, whole-animal wet weight, shell length, tissue weight, and shell weight (ASTM 2001). It is recommended that growth be measured as an effects endpoint, to help calibrate and understand bioaccumulation, and as a performance criterion for a successful test. Growth has been linked with tissue burdens in bivalve tissues, and we have provided examples here for copper.

Several metrics can be measured to assess growth in bivalves, and as many endpoints as are practical should be measured. For example, it has been shown that shell growth and tissue growth are decoupled, so that measuring only one of these endpoints could provide misleading results and lead to a spurious interpretation of environmental effects on growth (Hilbish 1986; Lewis and Cerrato 1997). Whole-animal wet weights and shell lengths are non-destructive measurements and the most common growth measurements. Although measurements are normally made at the beginning and end of the test, whole-animal wet weights and shell lengths can be measured multiple times over the course of the exposure period.

Growth information for both marine and freshwater bivalves can be used in general models for freshwater bivalve ecotoxicology. Although there are obvious differences in the growth rates for the marine bivalve *M. edulis* (Rodhouse et al. 1986) and the unionid *E. complanata* (Day 1984), there are similarities in the size and time required to reach a point where more energy is allocated to reproduction and less energy is allocated to growth. It appears that the most rapid growth occurs, and is most easily measured, in a shell length up to about three years of age in both marine and freshwater bivalves. The difference is that marine bivalves average about 50 mm in length compared to only 35 mm for their unionid counterparts. This also requires a definition of the term juvenile, which many unionid researchers use to identify newly metamorphosed juveniles as opposed to juveniles not yet able to reproduce. Furthermore, measuring growth is not restricted to the shell length endpoint, such that other metrics can be used in a weight of evidence approach. For example, changes in whole-animal wet weight and tissue weights have been used as an indicator of stress in very large *Lampsilis radiata*. The key to this approach is in using the same individuals for calculating changes in whole-animal wet weight at the beginning and end of test (ASTM 2001). A number of compromises were made in the development of the ASTM Standard Guide to fit the attributes suggested earlier regarding versatility, practicality, integrative ability, and consistency (de Kock and Kramer 1994). As a practical matter, bivalves in the optimum size range for measuring growth are not always available. Given the option of providing some useful information and the alternative of doing nothing, the obvious choice is most often to utilize the available resources. It is the multiple growth metrics that provide the important information on growth that is most relevant to assessing animal health (ASTM 2001).

Reproduction as an Effects Endpoint. Like other endpoints that could be used in freshwater mussel ecotoxicology, some of the first measurements of reproduction in marine bivalves were associated with aquaculture, where it was important to understand the factors associated with changes in tissue mass, relative health, and spawning. Some of the earliest measurements of reproductive state were made in oysters (Galstoff 1930) with the thickness of gonadal layers serving as an indicator of reproductive condition. The volume and weight of the gonad was measured, and histological sections were made to determine the degree of development. Chipperfield (1953) developed a classification system based on the shape of ova for the stage of development in marine mussels that has been used and is still used in one form or another throughout the world. At about the same time, Iwata (1951) discovered that spawning in *M. edulis* could be induced by injecting potassium chloride into the visceral cavity and by electrical stimulation. These methods, combined with measurement of rapid temperature changes, were subsequently used to develop a bivalve embryo test published in 1980 (ASTM 1998). Relatively crude, low-tech observations and measurements have been used for decades to assess reproductive conditions in marine bivalves, and many of these techniques have been applied to marine mussel ecotoxicology. Surprisingly, many of these methods have not been developed for concurrent use in freshwater mussel ecotoxicology, and therefore await ongoing progress with life history research to support their broader application. Although methodology to evaluate the reproductive status of freshwater bivalves is improving, confidence has not yet been established for their inclusion in toxicological procedures. Reproduction has been linked with tissue burdens in bivalve tissues, and we have provided examples here for copper. Recently, the vitellin biomarker has been used as an indicator of effects on reproduction and endocrine disruption (Gagne et al. 2001a, 2001b, 2001c; Blaise et al. 2003) and sex reversal in *E. complanata* deployed for one year in the St. Lawrence River (Blaise et al. 2003).

BIOMARKERS AS INDICATORS OF EXPOSURE AND EFFECTS

It is often difficult to determine whether specific biomarkers are indicators of exposure, effects, or both. Rather than attempting to make this distinction, it is proposed to use biomarkers to establish links between external exposure, internal exposure, and associated biological effects. Bivalve biomarkers can be used to bridge the gap between exposure and effects and help to characterize and understand processes that are essential to each. In this way, they can be used to help establish causality in those relationships and also improve the quality of ecological risk assessments and the development of water and sediment quality criteria. Biomarkers are reasonably well-developed for freshwater bivalves, and many more could be applied from marine bivalves.

We have been working with Environment Canada scientists at the St. Lawrence Center in Montreal over the last five years to develop a suite of biomarkers for marine and freshwater bivalves that have been tested upstream and downstream of a municipal effluent and other sites. These biomarkers include an assay for immunocompetence (Blaise et al. 1999, 2002), cytochrome P450, DNA damage (Gagne et al. 2002), a vitellin assay that was linked to possible endocrine disruption and concentrations of coprostanol in caged mussel tissues (Gagne et al. 2001a, 2001b, 2001c), and experimentally-induced sex reversal in mussels caged downstream of a municipal effluent for a period of one year (Blaise et al. 2003). Mussel growth was used to help calibrate the sensitivity of the various biochemical responses. A benthic cage was designed to hold mussels in bottom sediment for a period of one year to facilitate measuring each of these responses (Blaise et al. 2003). These studies demonstrate that most effects endpoints commonly measured in fish can also be measured in caged bivalves. Similar effects on hepatic vitellin and reproductive function were demonstrated in spottail shiners at sites downstream of the same municipal effluent (Aravindakshan, Marcogliese, and Cyr 2002; Aravindakshan et al. 2004a, 2004b). Caged freshwater mussels also showed significant endocrine and reproductive effects downstream of a pulp and paper mill in Florida that were similar to those reported for largemouth bass (Kernaghan et al. 2004). The advantage of using bivalves in this instance includes the measurement of responses in controlled field experiments over space and time and under site-specific conditions. Perhaps more importantly, it is relatively easy to measure exposure and effects in a stationary organism at the same time and more easily characterize exposure and effects that are environmentally relevant.

REFINING, INTEGRATING, AND HARMONIZING THE MODELS

Widdows and Donkin (1992) proposed an ecotoxicological framework based on their work with marine bivalves that is also appropriate for freshwater mussel ecotoxicology. They combined measurements of chemicals in water, chemicals in tissues, and associated effects. We refined their ecotoxicological framework to include sediment chemistry as part of external exposure and suggested the exposure characterization should include measurement of chemicals in both external and internal compartments (Salazar and Salazar 1998). Integrated and harmonized assessments should include chemical measurements of the three major environmental compartments—water, sediment, and tissues (Reiley et al. 2003). These measurements can be used to establish links with ecological effects and as Mearns (1985) suggested, the EDR approach is direct.

ECOLOGICAL RISK ASSESSMENT AS AN UMBRELLA MODEL

The ERA model was chosen as an "umbrella" model for conceptualization (Figure 9.8) because it provides a structured process for collecting, organizing, and analyzing information to estimate the likelihood of adverse effects on individual organisms, populations, or ecosystems (Suter et al. 2000). However, it may be considered too general in its ability to link chemicals in tissues with

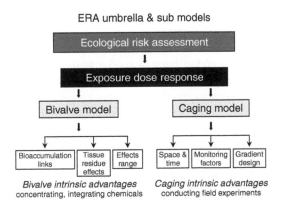

FIGURE 9.8 The relationship between the generic ERA umbrella model, the EDR conceptual model, and the subsequent sub-models.

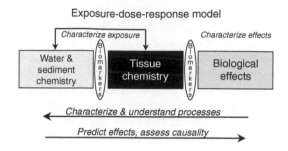

FIGURE 9.9 The relationship exposure, dose, and response emphasizing the importance of both water and sediment chemistry and bioaccumulation.

associated effects. In the context of the generic ERA model, most researchers consider exposure as chemicals in water and sediment only. The ERA was used as a basis for the EDR model, which emphasizes characterizing both external and internal exposure because they are inexorably linked, with equal emphasis on each. The EDR model specifically differentiates between internal and external exposure in contrast to the generic ERA model in providing general guidance for characterizing exposure and effects.

The relationship between the generic ERA umbrella model, the EDR conceptual model, and the subsequent sub-models is shown in Figure 9.8. The intent of the EDR model (Figure 9.9), the conceptual model proposed in this chapter, is the concurrent assessment of exposure and effects in the same test organisms. The relationship between the generic ERA umbrella model, the EDR conceptual model, and the subsequent sub-models is shown in Figure 9.8.

THE EXPOSURE–DOSE–RESPONSE MODEL

We integrated existing ecotoxicological frameworks into an EDR model to focus on the importance of bioaccumulation and biological effects (Figure 9.9). The EDR model establishes links between external exposure (water and sediment chemistry), internal exposure (tissue chemistry), and associated biological effects (e.g., survival, growth, and reproduction). This approach can be used to characterize and understand processes, and thereby predict effects and assess causality (Widdows and Donkin 1992), particularly when used in concert with biomarkers. Biomarkers help establish the links between external exposure and the absorbed dose as well as between the absorbed dose and response.

Whereas the major links in the EDR model are exposure, dose, and response, the major monitoring components of the EDR model are the bivalve model and the caging model (Figure 9.8). The bivalve model demonstrates how bivalves can integrate chemicals from multiple environmental compartments into their tissues and manifest measurable responses to these exposures. The major advantage of the bivalve model is that tissue chemistry and effects are easily measured in the same organism at the same time. Bioaccumulation links, tissue residue effects, and effects range are components of the bivalve model. The caging model demonstrates how in situ deployments of caged bivalves can be used to evaluate a wide range of exposure conditions by measuring multiple effects endpoints. The caging model facilitates logistical issues associated with collecting sufficient tissue and specimens for tissue chemistry and effects measurements. Space and time, monitoring factors, and the gradient design are components of the caging model. In other words, caged unionids can be used to characterize exposure and effects over space and time, under environmentally realistic conditions.

The approach discussed here represents the integration of the bivalve model and the caging model. While there are no "ideal" monitoring tools or species, bivalves have many of the characteristics of a good organism for monitoring, which include such factors as sedentary nature, the ability to concentrate chemicals in their tissues, and relevant sensitivity to chemical stress. Using the bivalve model facilitates use of other models such as bioaccumulation links, tissue residue effects, and effects range determinations for tissue quality guidelines. Bivalves are also easy to collect, cage, and measure, which facilitates the use of the caging model.

LINKS BETWEEN TISSUE RESIDUES AND EFFECTS

The EDR model is consistent with the tissue residue effects model. Development of this conceptual approach was influenced considerably by the work of McCarty and Mackay (1993) to link critical body residues (CBRs) with measured effects to help predict where effects begin to occur. A CBR is defined as the concentration of a chemical accumulated in tissues of an aquatic organism that corresponds to a specific adverse biological effect such as mortality, reduced growth, or reduced reproduction. Specifically, a tissue residue concentration is considered a "critical" residue if it is consistently associated with a specific toxicity endpoint, independent of aqueous exposure conditions. The body residue for a specific chemical, therefore, cannot be considered a CBR if exposure conditions or other chemical, biological, or environmental variables substantially modify the body residue concentration associated with the specific toxicity endpoint.

McCarty (1986, 1987) examined quantitative structure activity relationships (QSARs) of the acute toxicity and bioconcentration of organic chemicals and concluded that chemicals should accumulate to a CBR. He also defined the CBR as the molar tissue concentration (e.g., mmol/kg) of a toxic chemical that consistently produces a defined toxic effect such as mortality or reduced growth or impaired reproduction. According to the theory of McCarty and others (e.g., McCarty and Mackay 1993), a CBR can be defined for either individual chemicals or for classes of chemicals that share the same mode of action and have been hypothesized to be relatively constant across a wide range of aquatic species, taxonomic groups, and exposure conditions. Although the relationships derived by McCarty (1986) were theoretical, they were based on the use of QSARs to estimate the toxic internal concentrations of organic chemicals in fish. Whereas McCarty estimated critical body residues, Donkin et al. (1989) actually measured critical body residues in mussel tissues and associated them with effects on filtration. The relationships between hydrophobicity, sublethal toxicity, and critical tissue residues established in mussels were virtually identical to those offered for fish. The relationship between toxicological effects and exposure can be described as a CBR if tissue residue concentrations consistently produce the same toxicological effect. It is difficult, if not impossible, to predict effects and establish causality without first establishing ecotoxicological links, and tissue chemistry is a key element in establishing those links (Widdows and Donkin 1992). Tissue residues have been proposed to be a more appropriate indicator of adverse effects in aquatic biota than external water concentrations because tissue residues should represent a more toxicologically relevant "dose" or absorbed dose (McCarty and Mackay 1993).

The exposure–dose–response triad and critical body residue theory distinguish between external and internal chemical exposure; that is, measuring chemicals in water, sediment, and the tissues of organisms. Although this paradigm is based largely on information gained by using caged marine bivalves as a monitoring and assessment tool, the fundamentals of this model can also be directly applied to freshwater bivalves. One of the main premises behind the tissue residue effects paradigm is concurrent characterization of exposure and effects. Concurrent measurements allow for better predictions regarding the fate and effects of chemicals in aquatic ecosystems. They increase the ability to relate tissue chemistry with various acute and chronic effects measured in laboratory toxicity tests, benthic community studies, and mussel watch monitoring. Concurrent measurements are also needed to build a tissue residue effects database. As the database becomes refined and includes more data for a variety of species, it should be possible to predict effects from bioaccumulation data alone.

Asking the Right Questions

The following four questions have been utilized by Canada's Aquatic Effects Technology Evaluation Program (AETE 1997) and Uwe Borgmann (of Environment Canada) (Borgmann 2000; Borgmann and Norwood 2000) to evaluate effectiveness of key elements in monitoring and assessment programs:

1. Are contaminants entering the system?
2. Are contaminants bioavailable?
3. Is there a measurable response?
4. Are contaminants causing this response?

Traditional approaches such as the sediment quality triad, successfully address questions 1 and 2 (associated with exposure) but do not directly address questions 3 and 4 (associated with effects). A growing number of scientists agree with that assessment, which is one supportive reason for the development of the EDR triad; that is, to place more emphasis on bioaccumulation and directly address all four questions.

Borgmann et al. (2001) suggested that: (1) bioaccumulation is the most direct and reliable method of estimating bioavailable chemicals, (2) comparison with critical body residues provides a means for identifying the cause of toxicity, and (3) this approach has less uncertainty than relying on chemical concentrations in water or sediment.

The sediment quality triad is generally considered an effects-based approach because two of the three basic elements quantify effects (Figure 9.10a). It generally consists of sediment chemistry, a biological response measured in the laboratory, and field studies that generally utilize benthic community structure. By contrast, we have referred to the EDR triad as an ERA-based approach because it places equal emphasis on characterizing exposure and effects as does the ERA paradigm (Figure 9.10b). In the context of a sediment assessment, it would consist of sediment chemistry (could include water as well), tissue chemistry (dose), and effects measures such as bioassays or benthic community structure. Rather than two effects measures, it includes one element of external exposure (sediment chemistry), one element of internal exposure (tissue chemistry), and one element of response (biological effects). In the context of weight of evidence or multiple lines of evidence, it seems as though the EDR triad generally includes different kinds of information that

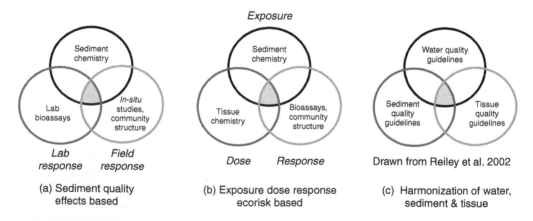

FIGURE 9.10 The relationship exposure, dose, and response in: (a) Effects-based, (b) EcoRisk-based, and (c) Harmonization of guidelines-based approaches to show the relative importance in the use of bioaccumulation.

are more consistent with ecological risk assessment. In a larger context, a recent SETAC Pellston Workshop concluded that a re-evaluation of water quality criteria should take a holistic approach and include harmonization of water, sediment, and tissue quality guidelines (Figure 9.10c) (Reiley et al. 2003).

EXAMINING AVAILABLE EXPOSURE–DOSE–RESPONSE DATA THROUGH A DIFFERENT LENS

In addition to data collected for a specific study, other information exists that can be used to help interpret study results. While we have frequently used growth to help interpret bioaccumulation results, water, sediment, and tissue chemistry results can be used to help interpret growth results. Bioaccumulation data from Mussel Watch monitoring, freshwater monitoring, and results from monitoring indigenous populations can be used to establish first-order approximations of the tissue burdens associated with living populations of freshwater bivalves. For example, in the absence of other data, or with limited data, it would be possible to develop a crude ERA for some chemicals at specific sites by using monitoring data for marine species. Fortunately, other data exist for freshwater bivalves to guide the assessment. From a mechanistic perspective, there is a need to integrate monitoring and research in order to characterize and understand processes, particularly in relating data from freshwater and marine bivalves as well as with results from non-bivalve species. In addition, life history information, tissue chemistry data from long-term monitoring studies, and growth rate data are just a few examples that might offer insight.

USING SYNOPTIC DATA FOR INTERPRETING RESULTS

We have previously suggested that bivalve growth rates could be used as an effects endpoint, to help "calibrate" bioaccumulation and as a performance criterion for a successful test (ASTM 2001). We also suggest that the combination of exposure and effects endpoints can help characterize and understand processes necessary for data interpretation and for use in the ERA framework. For example, although interval water samples were not taken to estimate the rate of uptake in a recent 67-day study with *L. radiata* (Applied Biomonitoring 2005), some inferences can be made based on background information, and new questions can be framed to develop supplementary approaches to make those estimates. Further, we were able to use those data to help explain an apparent anomaly between the survival and growth data through the use of water and tissue chemistry data at the beginning and end of the test. This highlights the importance of using the EDR triad as an ecotoxicological model. There was an obvious correlation between the elevated concentrations of aluminum, copper, and nickel in bivalves at one particular station, but the data raised four obvious questions: (1) Was this correlation real or a pseudo-correlation? (2) Why did the bivalves die? (3) When did they die? (4) Why were the growth effects not more pronounced at the station with low survival? Organisms at this site likely did not live long enough to produce growth effects. We speculated that as the metal concentrations (particularly copper) and time under stress increased, bivalves at the other stations would begin to die. This also demonstrated the importance of an extended exposure period.

The higher concentrations of metals (particularly copper) measured at the end of the test were key in explaining the observed growth differences. If those differences existed, all metal concentrations increased during the exposure period along with the metals in the bivalve tissues. However, the higher concentrations in water at the station with the lowest survival suggested bivalves acquired a significant portion of their metal burden from sediment, perhaps early in the test. Then, as the water concentrations increased, the combination of water and sediment exposure caused the observed mortalities. At the other stations, sediment concentrations may have been lower (sediment was not measured in this experiment), and it may have been later that the concentrations in bivalve tissues increased and effects on growth began. If the exposure had continued, any observed mortalities at the station with the lowest survival would have likely been attributed to different pathways of exposure. The effects at the station with the lowest survival would have been more attributable to legacy metal discharges, while those at the station closest to the surface water discharge would have been more attributable to current discharges. This also demonstrates the importance of relating exposure pathways in data interpretation and how measurements in caged bivalves could be used to assess those pathways.

USING ALL AVAILABLE DATA

Reported concentrations of copper in natural populations of freshwater bivalves (Stewart and Malley 1997; Metcalfe-Smith, Merriman, and Batchelor 1992) were useful for interpreting data for the above-mentioned example. Of the 25 data points included in Table 9.4, five exceeded or were very close to the predicted lower effects level, and they were all associated with sediment concentrations above 100 µg/g dw. Interestingly, one study found concentrations near the lower effects level associated with a sediment concentration of only 32.7 µg/g dw. Furthermore, the

TABLE 9.4
Copper Concentrations in Sediment and Bioaccumulation Factors (Concentrations in mg/kg = µg/g Dry Weight)

Sediment	Mussels	BAF	Species	Reference
4.0	14.00	3.5	P. grandis	Heit, Klusek, and Miller (1980)
4.0	9.00	2.3	E. complanata	Heit, Klusek, and Miller (1980)
4.0	18.00	4.5	L. radiata	Heit, Klusek, and Miller (1980)
7.1	15.05	2.1	E. complanata	Friant (1979)
7.1	7.45	1.1	P. cataracta	Friant (1979)
7.1	19.80	2.8	E. complanata	Friant (1979)
19.0	12.00	0.6	A. plicata	Mathis and Cummings (1973)
19.0	17.00	0.9	Q. quadrula	Mathis and Cummings 1973
19.0	17.00	0.9	F. flava	Mathis and Cummins 1973
23.7	13.80	0.6	E. complanata	Tessier et al. (1984)
26.0	20.00	0.8	L. complanata	Wren, MacCrimmon, and Loescher (1983)
26.4	13.09	0.5	L. siliquoidea	Dermott and Lum (1986)
32.7	2.82	0.1	A. marginata	Anderson (1977)
32.7	7.41	0.2	L. ventricosa	Anderson, 1977
32.7	9.41	0.3	E. complanata	Anderson (1977)
32.7	22.35	0.7	E. complanata	Anderson (1977)
35.0	16.00	0.5	E. complanata	Tessier et al. (1984)
40.7	13.60	0.3	E. complanata	Tessier et al. (1984)
59.4	5.96	0.1	E. complanata	Dermott and Lum (1986)
106.0	59.10	0.6	E. complanata	Tessier et al. (1984)
107.0	23.30	0.2	E. complanata	Tessier et al. (1984)
142.0	46.80	0.3	E. complanata	Tessier et al. (1984)
145.0	28.20	0.2	E. complanata	Tessier et al. (1984)
180.0	68.20	0.4	E. complanata	Tessier et al. (1984)
278.0	14.00	0.1	E. complanata	Dermott and Lum (1986)

Source: Adapted from Stewart, A. R. and Malley, D. F., *Technical Evaluation of Molluscs as a Biomonitoring Tool for the Canadian Mining Industry*, Ottawa, Ont., p. 248, 1997 and Metcalfe-Smith, J. L., *Environ. Toxicol. Chem.*, 13 (9), 1433–1443, 1994.

highest sediment concentration in this list (278 µg/g dw) was associated with one of the lowest bivalve tissue burdens (14 µg/g dw). Collectively, these data demonstrate two important points. First, as suggested previously, it is difficult to predict either bioavailability or toxicity based on the concentrations of metals in sediment. Second, there is a general inverse relationship in bioaccumulation factor (BAF) that is similar to bioconcentration factor (BCF) associations shown between water concentrations of metals and tissue concentrations. As the concentration of copper in sediment increases, the BAF decreases (Table 9.4). It is interesting to note that at the lower sediment concentrations found in the Lynn River, corresponding bivalve tissue burdens were comparable to those found in natural populations. However, at higher concentrations above 100 µg/g dw, the corresponding tissue burdens failed to reach the same high levels reported for other species and other environments. While this could be attributed to species and site-specific differences, it is also possible that the transplanted bivalves were under stress (as evidenced by their weight losses) and unable to accumulate metals to the same high concentrations found elsewhere. These and other hypotheses have been developed and will be tested in future studies.

We have also relied on data from a U.K. Web site (http://www.soc.staffs.ac.uk/research/groups/ cies2/taxapages/tx22info.htm) to interpret factors affecting unionid distributions. This database provides information on mussel distributions related to dissolved and total copper concentrations, calcium, cadmium, chromium, iron, magnesium, nickel, lead and zinc. Additionally, information is available on alkalinity, ammonia nitrogen and total ammonia, BOD, chlorophyll, conductivity, depth, hardness, nitrogen, oxygen, phosphorus, pH, suspended solids, sand, silt, slope, temperature, and width. These data could be useful in helping to interpret results from a variety of studies, including the current greater emphasis upon long-term observations.

ENVIRONMENTAL SIGNIFICANCE

Continuing with the copper example, the results from several laboratory, field, and monitoring studies with *M. edulis* and *M. galloprovincialis* have significant bearing on proposed lowering of the marine ambient water quality criterion for copper (USEPA 2003). The results are also potentially significant for establishing freshwater quality criteria for copper. Results suggested that much lower sensitivity of west coast mussel larvae are attributable to differences in water chemistry associated with laboratory exposures, rather than actual differences in larval sensitivity. This is also relevant to freshwater bivalve ecotoxicology in terms of advances in laboratory toxicity testing and suggests that the juvenile stages may be more sensitive than the adults. The weight of evidence strongly suggests that the west coast species *M. galloprovincialis* is much more resistant to copper than the east coast species *M. edulis*. In all of our studies with juveniles (not larvae) and adults, we never found significant effects on survival even when exposed to some of the highest concentrations of TBT and copper ever measured in the world. While it is possible that the larvae have a different sensitivity, it seems more likely that the differences are attributable to differences in water chemistry. This field validation is another reason for conducting caged bivalve studies with larval, juvenile, and adult bivalves.

It is possible to establish ecotoxicological links, develop working hypotheses, and develop ranges for effects based on water, sediment, and tissue concentrations found in laboratory and field studies, particularly the latter. Martin (1979) was one of the first to suggest this by noting that the concentrations of copper in *M. edulis* seldom if ever exceed 60 µg/g dw in natural populations. This is also evident in the summary by Widdows and Donkin (1992) and the NOAA Status and Trends Mussel Watch program. In other words, this was the first clue that tissue residues above this concentration may never be reached in field situations without the animal succumbing to the stress. Interestingly, this was a median value in our predictions (Salazar and Salazar 1991, 1995, 1996). Clearly, the data suggest that many bivalves are capable of accumulating much higher concentrations. However, an important distinction was apparent when comparing accumulation of copper in San Diego, where we routinely measured concentrations as high as 200 μg/g dw in both transplants and natural populations. The survival endpoint was significantly different in these two species in the same genus. If these data, or other conceptual data are then compared with unionid studies that employed measurements of dissolved and total copper with population distributions, an additional link can be established between water exposure and tissue residues. Finally, if these are then compared with data linking tissue residues with sediment copper concentrations, another version of the EDR triad emerges from existing data that measured water, sediment, and tissue copper, albeit non-synoptically. While we have advocated measuring chemicals in all three environmental compartments, it is possible to compare data collected at different times to help develop an understanding of exposure, dose, and response. The response in this case was survival in environmental settings over extended periods of time.

THE IMPORTANCE OF MECHANISTIC STUDIES AND RESULTS FROM DIFFERENT SPECIES

Valuable insight can be gained by broadening the scope of comparisons while interpreting freshwater bivalve data. Many researchers have realized the utility in examining mechanisms in unrelated species to help understand the processes affecting freshwater bivalves. There should be a merging of monitoring and research—monitoring for regulation purposes versus basic studies to understand processes. The following are examples of how researchers have used results and information from others to help formulate their conclusions.

DIETARY AND WATERBORNE METAL EXPOSURE IN *ELLIPTIO* AND *MYTILUS*

Our most recent caged bivalve studies provide a comparison between dietary and waterborne metal exposure in a unionid and a marine mussel. The data are being used in the development of a biotic ligand model for bivalves, and they provide important information regarding the similarities in exposure, bioaccumulation, and biological effects in the two species. We have recently compared dietary versus waterborne exposure pathways in a unionid (*E. complanata*) (Gagnon et al. 2006) and a marine mussel (*M. galloprovincialis*) (Salazar and Salazar in preparation) by using the relative uptake in the digestive gland to estimate dietary metals and the gill to estimate waterborne metals. The results were surprisingly similar. Results for both species showed consistently higher zinc concentrations in the gill, which was indicative of waterborne exposure and dissolved zinc. On the other hand, copper was more variable. In contrast to zinc, however, the copper concentration in the digestive gland and dietary exposure increased significantly at some sites and suggested that dietary exposure became more important when the particulate fraction of copper increased in the water. The freshwater study also demonstrated the importance of the Montreal municipal effluent in changing the exposure pathway. The biological availability of metals was strongly influenced by the physical and chemical conditions of the receiving waters where bivalves were exposed to both dissolved and particulate (food ingestion) forms of these metals. Results of metal bioaccumulation showed that gills and digestive gland were generally the most important target tissues, and the gill:digestive gland metal ratios demonstrated that both exposure routes must be considered to adequately assess freshwater and marine bivalve ecotoxicology. Results also showed that silver and cadmium in the dispersion plume near the effluent outfall were more associated with colloids and were generally less bioavailable with respect to the reference site in the St. Lawrence River.

COMPARISON OF EFFECTS ON *ELLIPTIO, MUSCULIUM,* AND *MYTILUS*

The comprehensive study conducted by Anderson, Sparks, and Paparo (1978) was particularly important because of the early date of the work and the comparison in effects between several freshwater bivalve species and a marine species. By using a relatively new method to monitor ciliary beating, they demonstrated that the gills of larger animals were more sensitive than smaller animals. Anderson, Sparks, and Paparo (1978) were probably the first freshwater bivalve ecotoxicologists to note that the gill filaments and cilia of marine mussels (*M. edulis*) were similar to those in fingernail clams (*Musculium transversum*). In addition, they measured the responses of the lateral cilia to water quality factors in three different bivalve species—marine mussel *M. edulis*, the freshwater fingernail clam *M. transversum*, and the unionid *E. complanata*. They also evaluated size-related differences in responses. Anderson, Sparks, and Paparo (1978) conducted a caged fingernail clam study to compare the responses of clams exposed to raw water in the field with those exposed to river water diluted with well water in the laboratory. They had the foresight to point out that there were certain deficiencies in their chronic testing system, and that these were related primarily to improvements necessary in the culture method. They also investigated the effects of temperature on ciliary response and found a maximum beating response at 20°C. Interestingly, this is the optimum temperature for growth in *M. edulis* and *M. galloprovincialis* in laboratory and field studies. They also found that ciliary beating was inhibited by light in both freshwater and marine bivalves.

From a toxicological perspective, the observation that large fingernail clams were more sensitive than smaller clams illustrates a differentiation between morphologic and metabolic processes. Fingernail clam gills were reported as being extremely sensitive to copper, lead, and zinc. They suggested that gills from large clams were more sensitive to metals than gills from small clams, and thus *M. transversum* become more sensitive to metals as they grow and mature. They acknowledged, however, that this tentative conclusion was based on tests with sensitive gill preparations and should be verified with bioassays using intact fingernail clams. It was not clear if this was ever accomplished. However, since copper inhibits sodium transport and in turn challenges organism ionoregulation, differences associated with size might be attributable to the gill surface area to organism volume ratio.

We have observed this in *M. galloprovincialis* (Salazar and Salazar in preparation), and it has been reported by others for *M. edulis* (Widdows and Donkin 1992; Luoma 1995). A mechanistic explanation has been provided based on adults with a longer period of lysosomal latency in copper exposures (Hole, Moore, and Bellamy 1992). This is one of the few cases where it has been reported for a freshwater bivalve. The prevailing opinion among many freshwater bivalve ecotoxicologists is that juveniles are more sensitive than adults. However, this could mean newly metamorphosed glochidia are not juveniles, defined as those animals incapable of reproduction. There is a large gap between metamorphosis and "adult," and more research should be directed at clarifying these potential differences in sensitivity and the reasons for the observed differences.

Comparison of Sub-Cellular Partitioning in *Pyganodon* and *Perca*

The work of Bonneris et al. (2005a), and Bonneris, Perceval, and Masson (2005b) is very relevant and important to freshwater bivalve ecotoxicology because of their mechanistic studies. They have combined the use of monitoring indigenous and transplanted unionids (*Pyganodon grandis*), measurement of biochemical changes (biomarkers), bioaccumulation, sub-cellular distribution, and comparisons with other species (fish). Most interesting of all is that none of these scientists generally consider themselves to be freshwater bivalve ecotoxicologists. Perhaps it is their focus on the scientific method, an appreciation of the effects of the receiving water on exposure, bioaccumulation, biological effects, and an emphasis on field exposures in their work that sets them apart from many others.

They worked on the sub-cellular partitioning of metals (cadmium, copper, and zinc) in the gills of *P. grandis* and evaluated the role of calcium concretions in metal sequestration using both indigenous and caged animals (Bonneris et al. 2005a; Bonneris, Perceval, and Masson 2005b). A polymetallic gradient was used to create differential exposures in the field. They concluded that, in nature, metals in *P. grandis* are bound differently in the gills and in the digestive gland and that metal detoxification in the former organ may be less effective than in the latter. They found evidence for oxidative stress in the gills but not in the digestive gland. Steady-state distribution of metals among metallothionein and other cytosolic ligands and links to cytotoxicity in bivalves living along a polymetallic gradient. Their study was designed to assess the environmental effects of metals in a field setting. They investigated exposure-bioaccumulation-effects (EDR) relationships in freshwater molluscs exposed to metals in their natural habitat.

The results of the present study using chronically exposed freshwater mussels along a metal contamination gradient differ significantly from what would have been predicted on the basis of laboratory exposures. They suggest that this demonstrates the need to study metal exposure-bioaccumulation-effects (EDR) relationships in natural populations or in transplant studies.

Finally, they have studied the influence of lake chemistry and fish age on cadmium, copper, and zinc concentrations in various organs of indigenous yellow perch (*Perca flavescens*) using the same methodology as for *P. grandis* (Campbell et al. 2005). Results suggest that cadmium concentrations in some organs increase with fish age. They interpreted this to be linked to changes in fish growth rate associated with age. They compared cadmium-handling strategies in two chronically exposed indigenous species—*P. flavescens* and *P. grandis*. Sub-cellular cadmium partitioning was determined in target tissues (bivalve gills and digestive gland; perch liver). Cadmium-handling strategies were similar in the bivalve digestive gland and perch liver, in that cadmium was mainly associated with the heat-stable protein (HSP) fraction. Furthermore, in these organs the contributions from the "mitochondria" and "lysosomes + microsomes" fractions were consistently higher than in the gill tissue. In the bivalve gill, the HSP fraction could only account for a small proportion (1.03%) of the total cadmium burden, and the metal was instead largely sequestered in calcium concretions (5.81%).

INTEGRATION OF UNIONIDS INTO A COMPREHENSIVE MONITORING STRATEGY

Humphrey, Bishop, and Brown (1990) demonstrated how both juvenile and adult unionids play an integral role in the development of a monitoring program. Short-term effects were assessed with creekside mesocosms, artificial streams, and in situ monitoring methods. The detection and assessment of any longer-term impacts, however, were based primarily on comparisons of post-release data with those of historical baselines. These baseline data were provided by studies on the structure of macroinvertebrate and fish communities and bioaccumulation by indigenous or transplanted organisms. In selecting bioindicator organisms, they identified long-lived species as desirable in bioaccumulation studies because they integrate long-term exposures to particular substances and thus provide information on equilibrium concentrations, particularly for substances with low biological turnover rates. They concluded that fish and bivalve molluscs have been considered the most suitable organisms for study but that bivalves have been used much more frequently than fish. For short-term, they recommended field bioassays and other early-detection systems. For acute studies they recommended in situ enclosures using mortality in freshwater unionid mussels as an effects endpoint. For chronic and sublethal exposures, they also recommended in situ enclosures with growth and reproduction in unionids as endpoints. Exposure endpoints included bioaccumulation in mussels and fish. Approaches used to study post-release included long-term effects and a comparison with a baseline and population studies on unionids. Long-term exposure included a comparison with a baseline and comparisons among stations using unionids and fish.

Although not providing a method for continuous observation (compared to "creekside" monitoring, i.e., artificial streams/mesocosms), in situ systems provided the most realistic information on the effects of releases in the short-term because organisms were held in cages or containers sited within the stream itself. Therefore, organisms responded to fluctuations of the creek environment. Advantages over creekside testing methods are that in situ monitoring avoids the constraints of need for elaborate equipment, it is relatively maintenance-free, and "systems" may easily be placed at strategic sites where they can most effectively monitor releases. Studies on freshwater mussels and fish migrations in Magela Creek have been underway since 1985 and have provided the most comprehensive information available on the early detection of short-term effects arising from ranger wastewater releases. The responses of juvenile and adult freshwater mussels (*Velesunio angasi*) have shown the most promise in both types of in situ containers.

During in situ trials, growth rates were dependent on the mesh sizes of container covers that presumably offered higher turnover of water and greater availability of potential food items. The only requirement for successful field rearing of one-day-old mussels was a constant turnover of creek water, and thus, this method differs from static laboratory rearing methods where additions of silt to culture waters are necessary (Hudson and Isom 1984). Not since Howard (1922) first reared newly transformed mussels in situ in floating crates in the Mississippi River (U.S.) has similar success been achieved. Howard (1922) did not report on the survival of mussels in his trials, the present report being the first to document high natural survival in containers in the field and to evaluate the potential of juvenile mussels for field monitoring of water quality.

As has been suggested previously, Humphrey, Bishop, and Brown (1990) recognized the importance of quantifying bioaccumulation as a significant element in monitoring and assessment. Because significant increases in concentrations of metals in organs of animals are regarded as harmful until proven otherwise, there is a need to identify whether this is occurring in this region and, if so, to assess the risk (e.g., by histological examination or laboratory experimentation) that any given resulting concentrations represent. Monitoring has included the concentrations of stable and radioactive metals in the soft parts of the mussel (*V. angasi*) from Magela Creek billabongs. These long-lived organisms, being sedentary, can be used to assess "overall" long-term water quality in that particular part of the creek system in which they are found but, in contrast to "wild" fishes, can also be utilized in "short-term" monitoring, the concentration in an organism of substances of short biological half-life being very much influenced by the recency of exposure. Evidence of exposure to such substances is rapidly lost—although this does not mean that such exposure might not have had some adverse metabolic or physiological impact.

Integrating Bivalve Ecotoxicology, Statistics, and Caging Studies

Roger Green was probably one of the first freshwater bivalve ecotoxicologists to propose bivalves as a response indicator and to emphasize field studies (Green, Singh, and Bailey 1985). However, he does not make a distinction between using marine and freshwater bivalves in his seminal paper. Green was able to integrate his experience and expertise in statistics, freshwater and marine bivalve ecotoxicology, and a caged bivalve methodology to make important contributions to the field. Other papers discuss the relative merits of measuring bioaccumulation and growth, although he does not specifically state that the two should be integrated. Green, Singh, and Bailey (1985) discuss the need for a "controlled experiment in the field if possible, or in the laboratory if necessary." They further suggest that experiments carried out in the field achieve the best of both worlds by combining the validity of experiment and experimental design with random assignment of true replicates to treatment levels and environmentally realistic exposure conditions.

He and his colleagues have investigated an arctic intertidal population of *M. balthica* for genotypic and phenotypic components of population structure (Green et al. 1983) and the role of a unionid clam population in the calcium budget of a small arctic lake (Green 1980). Green has provided guidelines for the design of biological monitoring programs in the marine environment (Green 1984). He and his colleagues have recommended bivalve molluscs as response systems for modeling spatial and temporal environmental patterns (Green, Singh, and Bailey 1985) and the use of freshwater mussels to monitor the nearshore environment of lakes (Green et al. 1989). They have related sets of variables in environmental studies using the sediment quality triad as a paradigm (Green et al. 1993).

POTENTIAL PITFALLS

In his comprehensive treatise on quantitative biological indicators, Phillips (1980) clearly states that the use of bivalves as an indicator of chemical exposure by measuring bioaccumulation is a potentially powerful tool. He also makes it very clear, however, that there are many pitfalls to this approach. Although not specifically stated, his statements apply to effects measurements such as survival, growth and reproduction as well. Clearly, as identified in Chapter 1, neither freshwater nor marine bivalves are ideal indicators. As suggested previously, however, bivalves probably have more of the characteristics of ideal indicators of both exposure and effects than other organisms, making them a potentially powerful monitoring and assessment tool, particularly when paired with the other models presented in this chapter. Although freshwater bivalves are good candidates for biological monitoring, some researchers have suggested problems regarding the approaches suggested here. What follows are some of the generic problems suggested and our responses.

Potential Problem #1) Lack of standardized test organisms. While there are really no standardized freshwater bivalves, the ASTM standard guide on field bioassays includes a list of several species that have been used in caging studies (ASTM 2001). There have been a substantial number of studies on several common freshwater bivalve species including *E. complanata*, *Pyganodon* (*Anodonta*) *grandis*, *L. radiata*, *C. fluminea*, and *D. polymorpha*. So much work has been done on these genera and species, that they could be considered standardized test organisms in this context.

Potential Problem #2) Lack of well-established culturing methods. While the use of cultured test organisms is preferable for most laboratory toxicity testing, it is not a requirement for in situ transplant studies. Field-raised or indigenous specimens from relatively clean environments are preferred because the animals have become acclimated to natural factors and conditions that cannot be duplicated in most culture facilities. Open water culturing techniques have been developed for several species of marine bivalves, but similar techniques are less well-developed for freshwater species, primarily because of their specialized requirement for a unique fish host. Nevertheless, the unionid *Margaritifera margaritifera* has been cultured in specialized media as part of a conservation program (Buddensiek 2004), and other methods are being developed for conservation purposes that could be used in laboratory and field bioassays.

Potential Problem #3) Lack of sensitivity information. We believe that there is an adequate amount of sensitivity information available (ASTM 2001), but even if there was not, it is always possible to use other available data as previously suggested.

Potential Problem #4) Variability in source of test organisms. This variability is an inherent part of testing with living organisms. It can be accounted for by assessing a variety of endpoints. Furthermore, this natural variability should be embraced as part of the natural physical, chemical, and biological processes that need to be characterized and understood.

Potential Problem #5) Variability in field testing. Variability is natural. Using this natural variability should be emphasized in testing because this is where the real effects occur. One reason for less variability in laboratory studies is that exposure conditions are highly controlled. Laboratory exposure conditions seldom, if ever, approach natural field conditions and as a result, growth rates of bivalves in the laboratory seldom, if ever, approach those measured in the field. Bivalves are particularly sensitive to food, flow rate, and temperature. Although exposure to natural factors may confound data interpretation, they are important in understanding how bivalves accumulate chemicals and the subsequent effects these chemicals have on bivalve biology and physiology. Field testing is largely misunderstood by many scientists who point to the variability in field studies and suggest that this reduces their utility. It is this real world variability, however, that should be embraced because this is nature and this is what needs to be characterized and understood. This is one of the real advantages of field testing. It is also interesting to note all of the important information that has been gained through field testing when standardized protocols were not available. For example, of the biomarker studies summarized in Chapter 10, 18 of the 20 examples (90%) came from field studies!

SUMMARY AND CONCLUSIONS

A conceptual framework has been proposed for freshwater bivalve ecotoxicology that includes making necessary links between bioaccumulation and biological effects as well as between chemicals in water and sediment. This EDR triad is based on a series of models previously developed, the more universal ERA paradigm, and experience and expertise gained through more than 60 caged bivalve studies. The EDR model can be used in both basic (e.g., mechanistic) and applied (e.g., impact assessment) studies, as well as routine monitoring. Just as the ERA paradigm is suggested as an overall umbrella model, the EDR model contains a bivalve model that includes the intrinsic advantages of bivalves for concentrating and integrating chemical exposure and the caging model, which includes the intrinsic advantages of conducting field experiments that control geographic position, exposure duration, and the size, age, and number of test animals in each replicate.

It is important to note that while there are no ideal indicator organisms, freshwater and marine bivalves possess many of the characteristics of an ideal bioindicator characterizing exposure with those considered ideal for characterizing effects. This is what distinguishes them from other organisms. The use of caged bivalves in the EDR triad facilitates making links between bioaccumulation and biological effects as well as between chemicals in water and sediment. As demonstrated in the example provided for copper, the critical body residues associated with the onset of effects and those associated with no effects appear to be virtually identical in freshwater and marine bivalves. We can see no apparent reason to suggest that the basic elements of exposure, uptake, and effects should be any different between the two groups. Further evidence for this similarity is provided in a number of other studies, including the synthesis in Chapter 8 that describes similar uptake-elimination kinetics among a number of marine and freshwater bivalves. Therefore, in the absence of data and considering the practical application of conducting an ecological risk assessment, data from other species (even marine or non-unionid species) could be used as a surrogate to develop a rough approximation. We suggest that the versatility, practicality, integrative ability, and consistency of the approaches suggested here provide a focus for establishing vital links in freshwater bivalve ecotoxicology.

In an attempt to bring this conceptual model back to the real world, it is further suggested that this approach will result in a more unified approach in freshwater bivalve ecotoxicology and eventually will lead to better science and ultimately better decisions (Benson 1995). Benson (1995) discusses the need for an understanding of the basic mechanisms of fundamental processes considered essential in traditional toxicology and specifically mentions links between exposure, dose, and response as a necessity for developing of meaningful environmental assessments that will stand the test of time. He also mentions a primary goal of the Society of Environmental Toxicology and Chemistry (SETAC) to promote the use of multi-disciplinary approaches to assess the effects of chemicals on the environment. As suggested in Chapter 1, SETAC meetings have been instrumental in stimulating discussions about freshwater bivalve ecotoxicology and promulgated the necessity for this book. Similarly, we view our ASTM Standard Guide on conducting field bioassays with caged marine, estuarine, and freshwater bivalves as a work in progress. Every time we conduct a caged bivalve study, we learn something new and learn by our mistakes as well as the peer review process. This is one of the advantages of working through a consensus-based standards organization such as ASTM. Similarly, discussions resulting from this book will no doubt lead to other questions and other innovations in the caged bivalve methodology. Underestimating the effect of chemicals may result in serious environmental consequences, whereas overestimating potential effects can result in inefficient use of available resources (Benson 1995). If nothing else, we suggest that assessments could be improved if those conducting laboratory or field studies added bioaccumulation to their approach to establish the links described here. However, this will necessitate changing the way they think about bioaccumulation. Benson (1995) puts such a paradigm shift in this context, "All things are ready, if our minds be so" (Shakespeare, W.).

RECOMMENDATIONS FOR FUTURE RESEARCH

Freshwater mussels can be used as effective indicators of exposure and effects, and the most useful information is provided by measuring exposure and effects endpoints in the same organism at the same time to establish links between exposure, dose, and response. There are no perfect or ideal monitoring and assessment tools, but selection should favor the most desirable characteristics, appropriate use, and careful interpretation of results. The following considerations are proposed for future unionid research:

- Include tissue chemistry measurements in freshwater bivalve studies whenever possible.
- Conduct more mechanistic studies to characterize and understand fundamental processes.
- Provide comparisons of freshwater bivalve results with results from other species.
- Integrate biomarker responses with survival, growth, and reproduction responses.
- Develop applications of the caged bivalve methodology for glochidia, juveniles, and adults.
- Investigate the relative sensitivity of glochidia, juveniles, and adults.
- Assess differences in exposure, dose, and response between the laboratory and the field.

REFERENCES

Anderson, R. V., Concentration of cadmium, copper, lead, and zinc in six species of freshwater clams, *Bull. Environ. Contam. Toxicol.*, 18, 492–496, 1977.

Anderson, K. B., Sparks, R. E., and Paparo, A. A., Rapid assessment of water quality, using the fingernail clam, *Musculium transversum*. WRC Res. Rep. No. 1, Water Resources Center, University of Illinois, Urbana, Illinois, 1978.

Anderson, J., Hayton, A., Rodrigues, A., and Richman, L., *Niagara River Biomonitoring Study*, Ontario Ministry of Environment and Energy, April 1991.

Applied Biomonitoring, *Lynn Lake 2004 Caged-Bivalve Study*, DRAFT Final Report, Prepared for TetrES Consultants, Winnipeg, MB, Canada, 4 February 2005.

Aquatic Effects Technology Evaluation (AETE), *Aquatic Effects Technology Evaluation Program, Background Document*, Canada Centre for Mineral and Energy Technology, Natural Resources Canada, 1997.

Aravindakshan, J. P., Marcogliese, D., and Cyr, D. G., Effects of Xenoestrogens From The St. Lawrence on Sperm Motility in Spot Tail Shiners, MITE 2002 Annual Symposium Abstracts, 2002.

Aravindakshan, J., Gregory, M., Dufresne, J., Marcogliese, D. J., Fournier, M., and Cyr, D. G., Consumption of xenoestrogen-contaminated fish during lactation alters adult male reproductive function, *Toxicol. Sci.*, 81, 179–189, 2004a.

Aravindakshan, J., Paquet, V., Gregory, M., Dufresne, J., Fournier, M., Marcogliese, D. J., and Cyr, D. G., Consequences of xenoestrogen exposure on male reproductive function in spottail shiners (*Notropis hudsonius*), *Toxicol. Sci.*, 78, 156–165, 2004b.

ASTM, E 724-94, Standard guide for conducting static acute toxicity tests starting with embryos of four species of saltwater bivalve molluscs, In *1998 Annual Book of ASTM Standards*. Vol. 11.05. *Biological Effects and Environmental Fate; Biotechnology, Pesticides*, American Society for Testing and Materials, Conshohocken, PA, pp. 192–209, 1998.

ASTM, E-2122, Standard guide for conducting in-situ field bioassays with marine, estuarine and freshwater bivalves, *2001 Annual Book of ASTM Standards,* Vol. 11.05, American Society for Testing and Materials (ASTM), Conshohocken, PA, pp. 1546–1575, 2001.

Bailey, R. C., Norris, R. H., and Reynoldson, T. B., Eds., *Bioassessment of Freshwater Ecosystems: Using the Reference Condition Approach*, Kluwar, Boston, MA, p. 170, 2004.

Bayne, B. L., Watch on mussels, *Mar. Pollut. Bull.*, 7(12), 217–218, 1976.

Bayne, B. L., Brown, D. A., Burns, K., Dixon, D. R., Ivanovici, A., Livingstone, D. R., Lowe, D. M., Moore, M. N., Stebbing, A. R. D., and Widdows, J., Eds., *The Effects of Stress and Pollution on Marine Animals*, Praeger Special Studies, Praeger Scientific, New York, p. 374, 1985.

Belanger, S. E., Farris, J. L., Cherry, D. S., and Cairns, J., Validation of *Corbicula fluminea* growth reductions induced by copper in artificial streams and river systems, *Can. J. Fish. Aquat. Sci.*, 47, 904–914, 1990.

Benson, W. H., Better science makes for better decisions, *Environ. Toxicol. Chem.*, 14(11), 1811–1812, 1995.

Blaise, C., Gagné, F., Pellerin, J., and Hansen, P.-D., Determination of vitellogenin-like properties in *Mya arenaria* hemolymph (Saguenay Fjord, Canada): A potential biomarker for endocrine disruption, *Environ. Toxicol.*, 14(5), 455–465, 1999.

Blaise, C., Trottier, S., Gagne, F., Lallement, C., and Hansen, P.-D., Immunocompetence of bivalve hemocytes as evaluated by a miniaturized phagocytosis assay, *Environ. Toxicol.*, 17, 160–169, 2002.

Blaise, C., Gagne, F., Salazar, M., Salazar, S., Trottier, S., and Hansen, P.-D., Experimentally-induced feminisation of freshwater mussels after long-term exposure to a municipal effluent, *Fresenius Environ. Bull.*, 12(8), 865–870, 2003.

Bonneris, E., Giguère, A., Perceval, O., Buronfosse, T., Masson, S., Hare, L., and Campbell, P. G. C., Sub-cellular partitioning of metals (Cd, Cu, Zn) in the gills of a freshwater bivalve, *Pyganodon grandis*: Role of calcium concretions in metal sequestration, *Aquat. Toxicol.*, 71, 319–334, 2005a.

Bonneris, E., Perceval, O., and Masson, S., Sub-cellular partitioning of Cd, Cu and Zn in tissues of indigenous unionid bivalves living along a metal exposure gradient and links to metal-induced effects, *Environ. Poll.*, 135, 195–208, 2005b.

Borgmann, U., Methods for assessing the toxicological significance of metals in aquatic ecosystems: Bioaccumulation–toxicity relationships, water concentrations and sediment spiking approaches, *Aquat. Ecosys. Health Manage.*, 3(3), 277–290, 2000.

Borgmann, U. and Norwood, W. P., What is wrong with the sediment quality triad?, In *27th Annual Aquatic Toxicity Workshop*, Hotel Newfoundland October 1–4, 2000, St. John's Newfoundland, Program and Abstracts, 2000.

Borgmann, U., Norwood, W. P., Reynoldson, T. B., and Rosa, F., Identifying cause in sediment assessments: Bioavailability and the Sediment Quality Triad, *Can. J. Fish. Aquat. Sci.*, 58(5), 950–960, 2001.

Bryan, G. W., Gibbs, P. E., Hummerstone, L. G., and Burt, G. R., Copper, zinc, and organotin as long-term factors governing the distribution of organisms in the Fal Estuary in southwest England, *Estuaries*, 19(3), 208–219, 1987.

Buddensiek, V., The culture of juvenile freshwater pearl mussels *Margaritifera margaritifera* L. in cages: A contribution to conservation programs and the knowledge of habitat requirements [Web Page]. Available at http://www.volkerbuddensiek.de/culture.htm, 2004.

Campbell, P. G. C., Giguere, A., Bonneris, E., and Hare, L., Cadmium-handling strategies in two chronically exposed indigenous freshwater organisms—the yellow perch (*Perca flavescens*) and the floater mollusk (*Pyganodon grandis*), *Aquat. Toxicol.*, 72, 83–97, 2005.

Chan, H. M., Accumulation and tolerance to cadmium, copper, lead and zinc by the green mussel *Perna viridis*, *Mar. Ecol. Prog. Ser.*, 48, 295–303, 1988.

Chapman, P. M. and Long, E. R., The use of bioassays as part of a comprehensive approach to marine pollution assessment, *Mar. Pollut. Bull.*, 14(3), 81–84, 1983.

Chipperfield, P. N. J., Observations on the breeding and settlement of *Mytilus edulis* (L.) in British waters, *J. Mar. Biol. Ass. U.K.*, 32, 449–476, 1953.

Couillard, Y., Campbell, P. G. C., and Tessier, A., Response of metallothionein concentrations in a freshwater bivalve (*Anodonta grandis*) along an environmental cadmium gradient, *Limnol. Oceanogr.*, 38(2), 299–313, 1993.

Couillard, Y., Campbell, P. G. C., Pellerin-Massicotte, J., and Auclair, J. C., Field transplantation of a freshwater bivalve, *Pyganodon grandis*, across a metal contamination gradient. II. Metallothionein response to Cd and Zn exposure, evidence for cytotoxicity, and links to effects at higher levels of biological organization, *Can. J. Fish. Aquat. Sci.*, 52, 703–715, 1995a.

Couillard, Y., Campbell, P. G. C., Pellerin-Massicotte, J., and Auclair, J. C., Field transplantation of a freshwater bivalve, *Pyganodon grandis*, across a metal contamination gradient. I. Temporal changes in metallothionein and metal (Cd, Cu, and Zn) concentrations in soft tissues, *Can. J. Fish. Aquat. Sci.*, 52, 690–702, 1995b.

Creese, E., Lewis, D., and Melkic, A., Toward the development of a standard clam biomonitoring methodology: Preliminary results, *Proc. Technology Transfer Conference, Part B, Water quality Res. Toronto*, Dec. 8–9, 1986, pp. 205–218, 1986.

Curry, C. A., The freshwater clam (*Elliptio complanata*), a practical tool for monitoring water quality, *Water Pollut. Res. Can.*, 13, 45–52, 1977.

Day, M. E., The shell as a recording device: Growth record and shell ultrastructure of *Lampsilis radiata radiata* (Pelecypoda: Unionidae), *Can. J. Zool.*, 62, 2495–2504, 1984.

De Kock, W. C. and Bowmer, C. T., Bioaccumulation, biological effects, and food chain transfer of contaminants in the zebra mussel (*Dreissena polymorpha*), In *Zebra Mussels. Biology, Impacts, and Control*, Nalepa, T. F. and Schloesser, D. W., Eds., Lewis Publishers, Boca Raton, FL, pp. 503–533, 1993.

De Kock, W. C. and Kramer, K. J. M., Active biomonitoring (ABM) by translocation of bivalve molluscs, In *Biomonitoring of Coastal Waters and Estuaries*, Kramer, K. J. M., Ed., CRC Press, Boca Raton, FL, pp. 51–84, Chapter 3, 1994.

Dermott, R. M. and Lum, K. R., Metal concentration in the annual shell layers of the bivalve *Elliptio complanata*, *Environ. Pollut. Ser. B*, 12, 131–143, 1986.

Donkin, P., Widdows, J., Evans, S. V., Worrall, C. M., and Carr, M., Quantitative structure–activity relationships for the effect of hydrophobic organic chemicals on rate of feeding by mussels *Mytilus edulis*, *Aquat. Toxicol.*, 14(3), 277–294, 1989.

Elder, J. F. and Collins, J. J., Freshwater molluscs as indicators of bioavailability and toxicity of metals in surface–water systems, *Rev. Environ. Contam. Toxicol.*, 122, 37–79, 1991.

Foster, R. B. and Bates, J. M., Use of freshwater mussels to monitor point source industrial discharges, *Environ. Sci. Technol.*, 12, 958–962, 1978.

Friant, S. L., Trace metal concentrations in selected biological, sediment, and water column samples in a northern New England river, *Water Air Soil Pollut.*, 11, 455–465, 1979.

Gagne, F., Blaise, C., Lachance, B., Sunahara, G. I., and Sabik, H., Evidence of coprostanol estrogenicity to the freshwater mussel *Elliptio complanata*, *Environ. Pollut.*, 15, 97–106, 2001.

Gagne, F., Blaise, C., Salazar, M., Salazar, S., and Hansen, P. D., Evaluation of estrogenic effects of municipal effluents to the freshwater mussel *Elliptio complanata*, *Comp. Biochem. Physiol. C*, 128, 213–225, 2001.

Gagne, F., Marcogliese, D. J., Blaise, C., and Gendron, A. D., Occurrence of compounds estrogenic to freshwater mussels in surface waters in an urban area, *Environ. Toxicol.*, 16(3), 260–268, 2001.

Gagne, F., Blaise, C., Aoyama, I., Luo, R., Gagnon, C., Couillard, Y., and Salazar, M., Biomarker study of a municipal effluent dispersion plume in two species of freshwater mussels, *Environ. Toxicol.*, 17, 149–159, 2002.

Gagnon, C., Gagne, F., Turcotte, P., Saulnier, I., Blaise, C., Salazar, M. H., and Salazar, S. M., Exposure of caged mussels to metals in a primary-treated municipal wastewater plume, *Chemosphere*, 62 (6), 998–1010, 2006.

Galstoff, P. S., The fecundity of the oyster, *Science*, 72(1856), 97–98, 1930.

GESAMP (Joint Group of Experts on the Scientific Aspects of Marine Pollution), Monitoring biological variables related to marine pollution. Rept. Stud. No. 12, UNESCO-IOC, Paris, p. 22, 1980.

Goldberg, E. D., The mussel watch—a first step in global marine monitoring, *Mar. Pollut. Bull.*, 6(7), 111, 1975.

Granmo, A., Mussels as a tool in impact assessment, *Phuket Mar. Biol. Cent. Spec. Publ. No.*, 15, 215–220, 1995.

Green, R. H., Role of a unionid clam population in the calcium budget of a small arctic lake, *Can. J. Fish. Aquat. Sci.*, 37, 219–224, 1980.

Green, R. H., Some guidelines for the design of biological monitoring programs in the marine environment, In *Concepts in Marine Pollution Measurements*, White, H. H., Ed., Maryland Sea Grant Publication, University of Maryland, College Park, pp. 647–655, 1984.

Green, R. H., Singh, S. M., Hicks, B., and McCauig, J. M., An arctic intertidal population of *Macoma balthica* (Mollusca, Pelecypoda): Genotypic and phenotypic components of population structure, *Can. J. Fish. Aquat. Sci.*, 40, 1360–1371, 1983.

Green, R. H., Singh, S. M., and Bailey, R. C., Bivalve molluscs as response systems for modelling spatial and temporal environmental patterns, *Sci. Total Environ.*, 46, 147–169, 1985.

Green, R. H., Bailey, R. C., Hinch, S. G., Metcalfe, J. L., and Young, V. H., Use of freshwater Mussels (Bivalvia: Unionidae) to monitor the nearshore environment of lakes, *J. Great Lakes Res.*, 15(4), 635–644, 1989.

Green, R. H., Boyd, J. M., and Macdonald, J. S., Relating sets of variables in environmental studies: The sediment quality triad as a paradigm, *Environ. Metrics*, 4(4), 439–458, 1993.

Grout, J. A. and Levings, C. D., Effects of acid mine drainage from an abandoned copper mine Britannia Mines, Howe, Sound, B.C., on transplanted blue mussels (*Mytilus edulis*), *Mar. Environ. Res.*, 51(3), 265–288, 2001.

Hayton, A. and Hollinger, D., *In-situ Clam Exposure in the Rainy River to Determine the Sources of Organochlorine contaminants*, Ontario Ministry of the Environment, 1989.

Hayton, A., Hollinger, D., Tashiro, C., and Reiner, E., Biological monitoring of chlorinated dibenzo–dioxins in the Rainy River using introduced mussels (*Elliptio complanata*), *Chemosphere*, 20, 1687–1693, 1990.

Heit, M., Klusek, C. S., and Miller, K. M., Trace element, radionuclide, and polynuclear aromatic hydrocarbon concentrations in unionidae mussels from northern Lake George, *Environ. Sci. Technol.*, 14(4), 465–468, 1980.

Hilbish, T. J., Growth trajectories of shell and soft tissue in bivalves: Seasonal variation in *Mytilus edulis* L, *J. Exp. Mar. Biol. Ecol.*, 96, 103–113, 1986.

Hole, L. M., Moore, M. N., and Bellamy, D., Age-related differences in the recovery of lysosomes from stress-induced pathological reactions in marine mussels, *Mar. Environ. Res.*, 34, 75–80, 1992.

Hornberger, M. I., Luoma, S. N., Cain, D. J., Parchaseo, F., Brown, C. L., Bouse, R. M., Wellise, C., and Thompson, J. K., Linkage of bioaccumulation and biological effects to changes in pollutant loads in South San Francisco Bay, *Environ. Sci. Technol.*, 34(12), 2401–2409, 2000.

Howard, A. D., Experiments in the culture of fresh-water mussels, *US Fish Wildl. Serv. Bull.*, 37, 63–90, 1922.

Hudson, R. G. and Isom, B. G., Rearing juveniles of the freshwater mussels (Unionidae) in a laboratory setting, *Nautilus*, 98(4), 129–137, 1984.

Humphrey, C. L., Bishop, K. A., and Brown, V. M., Use of biological monitoring in the assessment of mining wastes on aquatic ecosystems of the Alligator Rivers Region, tropical northern Australia, *Environ. Monitor. Assess.*, 14, 139–181, 1990.

Iwata, K. S., Spawning of *Mytilus edulis*. Discharge by KCL injection, *Bull. Jpn. Soc. Sci. Fish.*, 16, 393–394, 1951.

Jarvinen, A. W. and Ankley, G. T., Linkage of effects to tissue residues: Development of a comprehensive database for aquatic organisms exposed to inorganic and organic chemicals, SETAC, Pensacola, FL, 1999.

Kernaghan, N. J., Ruessler, D. S., Holm, S. E., and Gross, T. S., An evaluation of the potential effects of paper mill effluents on freshwater mussels in Rice Creek, Florida, In *Fate and Effects of Pulp and Paper Mill Effluents. Proceedings of the 5th International Conference*, Seattle, Washington, June 1–4, 2003, Borton, D., Fisher, R., Hall, T., and Thomas, J., Eds., DEStech Publications, Inc., Lancaster, PA, pp. 455–463, 2004.

Kraak, M. H. S., Lavy, D., Peeters, W. H. M., and Davids, C., Chronic ecotoxicity of copper and cadmium to the zebra mussel *Dreissena polymorpha*, *Arch. Environ. Contam. Toxicol.*, 23, 363–369, 1992.

Kraak, M. H. S., Lavy, D., Toussaint, M., Schoon, H., and Peeters, W. H. M., Toxicity of heavy metals to the zebra mussel (*Dreissena polymorpha*), In *Zebra Mussels. Biology, Impacts, and Control*, Nalepa, T. F. and Schloesser, D. W., Eds., Lewis Publishers, Boca Raton, FL, pp. 491–502, 1993.

Kraak, M. H. S., Lavy, D., Schoon, H., Toussaint, M., Peeters, W. H. M., and Van Straalen, N. M., Ecotoxicity of mixtures of metals to the zebra mussel *Dreissena polymorpha*, *Environ. Toxicol. Chem.*, 13, 109–114, 1994.

Landis, W. G., Eulogy for the reference site, In *SETAC 21st Annual Meeting. Environmental Sciences in the 21st Century: Paradigms, Opportunities, and Challenges*, SETAC, Nashville, TN, Pensacola, Fl, pp. 144–145, 2000.

Langston, W. J. and Burt, G. R., Bioavailability and effects of sediment-bound TBT in deposit-feeding clams, *Scrobicularia plana*, *Mar. Environ. Res.*, 32, 61–77, 1991.

Lewis, D. E. and Cerrato, R. M., Growth uncoupling and the relationship between shell growth and metabolism in the soft shell clam *Mya arenaria*, *Mar. Ecol. Prog. Ser.*, 158, 177–189, 1997.

Long, E. R. and Chapman, P. M., A sediment quality triad: Measures of sediment contamination, toxicity and infaunal community composition in Puget Sound, *Mar. Pollut. Bull.*, 16(10), 405–415, 1985.

Long, E. R. and Morgan, L. G., The potential for biological effects of sediment-sorbed contaminants tested in the national status and trends program, NOAA Technical Memorandum NOAA OMA 52, U.S. Department of Commerce, 1990.

Luoma, S. N., Prediction of metal toxicity in nature from bioassays: Limitations and research needs, In *Metal Speciation and Bioavailability*, Tessier, A. and Turner, D., Eds., Wiley, Chichester, pp. 609–659, 1995.

Martin, J. L. M., Schema of lethal action of copper on mussels, *Bull. Environ. Contam. Toxicol.*, 21, 808–814, 1979.

Mathis, B. J. and Cummings, T. F., Selected metals in sediments, water, and biota of the Illinois River, *J. Water Poll. Cont. Fed.*, 45, 1573–1583, 1973.

Matteson, M. R., Life history of Elliptio complanata (Dillwyn 1817), *Am. Midland Nat.*, 40(3), 690–723, 1948.

McCarty, L. S., The relationship between aquatic toxicity QSARs and bioconcentration for some organic chemicals, *Environ. Toxicol. Chem.*, 5, 1071–1080, 1986.

McCarty, L. S., Relationship between toxicity and bioconcentration for some organic chemicals—II. Application of the relationship, In *QSAR in Environmental Toxicology—II*, Kaiser, K. L. E., Ed., Reidel Publishing Company, Dordrecht, Holland, pp. 207–220, 1987.

McCarty, L. S., Toxicant body residues: Implications for aquatic bioassays with some organic chemicals, In *Aquatic Toxicology and Risk Assessment*, Mayes, M. A. and Barron, M. G., Eds., Vol. 14, American Society for Testing and Materials, ASTM STP 1124, Philadelphia, PA, pp. 183–192, 1991.

McCarty, L. S. and Mackay, D., Enhancing ecotoxicological modeling and assessment, *Environ. Sci. Technol.*, 27(9), 1719–1728, 1993.

McIntyre, A. D., and Pearce, J. B., Biological effects of marine pollution and the problems of monitoring. Rapports & Proces-Verbaux Reunions Conseil Int. Explor. Mer., Copenhagen, Denmark, 1980.

Mearns, A. J., Biological implications of the management of waste materials: The importance of integrating measures of exposure, uptake, and effects, In *Aquatic Toxicology and Hazard Assessment: Seventh Symposium*, Cardwell, R. D., Purdy, R., and Bahner, R. C., Eds., Vol. ASTM STP 854, American Society for Testing and Materials, Philadelphia, PA, pp. 335–343, 1985.

Metcalfe-Smith, J. L., Influence of species and sex on metal residues in freshwater mussels (family Unionidae) from the St. Lawrence River, with implications for biomonitoring programs, *Environ. Toxicol. Chem.*, 13(9), 1433–1443, 1994.

Metcalfe-Smith, J. L., Merriman, J. C., and Batchelor, S. P., Relationships between concentration of metals in sediment and two species of freshwater mussels in the Ottawa River, *Water Pollut. Res. J. Canada*, 27(4), 845–869, 1992.

Mount, D. I., Present approaches to toxicity testing—a perspective, In *Aquatic Toxicology and Hazard Evaluation. ASTM STP 634. Proceedings of the First Annual Symposium on Aquatic Toxicology*, Mayer, F. L. and Hamelink, J. L., Eds., American Society for Testing and Materials, Philadelphia, PA, pp. 5–16, 1977.

Ontario Ministry of the Environment, *Niagara River Mussel Biomonitoring Program, 1997.* Queen's Printer for Ontario, Ontario, Report No.: ISBN 0-7778-9097-6, October 1999.

Ontario Ministry of the Environment and Energy, Canagagigue Creek mussel and leech biomonitoring study. A Report Submitted to the Uniroyal Public Advisory Committee, January 1996.

Parrish, P. R., Dickson, K. L., Hamelink, J. L., Kimerle, R. A., Macek, K. J., Mayer Jr. F. L., Mount, D. I., *In Aquatic toxicology: Ten years in review and a look at the future, In Aquatic Toxicology and Hazard Assessment: 10th Volume*, ASTM STP 971, Adams, W. J., Chapman, G. A., and Landis, W. G., American Society for Testing and Materials, Philadelphia, PA, pp. 7–25, 1988.

Phillips, D. J. H., *Quantitative Aquatic Biological Indicators—Their Use to Monitor Trace Metal and Organochlorine Pollution. 1980*, Applied Science Publishers Ltd, London, 1980.

Reiley, M. C., Stubblefield, W. A., Adams, W. J., Di Toro, D. M., Hodson, P. V., Erickson, R. J., and Keating, F. J., *Reevaluation of the State of the Science for Water-Quality Criteria Development 2003*, SETAC press, p. 197, 2003.

Richman, L., The Niagra River mussel and leech biomonitoring study, Ontario: Queen's Printer for Ontario, October 1992, Report No.: ISBN 0-7778-0228-7, 1992.

Richman, L. A., Niagara River Mussel Biomonitoring Program, 1995, Ontario: Queen's Printer for Ontario, November 1997, Report No.: ISBN 0-7778-9097-6.

Richman, L. A., Niagara River Mussel Biomonitoring Program, 2000, 2003.

Rodhouse, P. G., McDonald, J. H., Newell, R. I. E., and Koehn, R. K., Gamete production, somatic growth and multiple-locus enzyme heterozygosity in *Mytilus edulis*, *Mar. Biol.*, 90, 209–214, 1986.

Salazar, M. H. and Salazar, S. M., Assessing site-specific effects of TBT contamination with mussel growth rates, *Mar. Environ. Res.*, 32(1–4), 131–150, 1991.

Salazar, M. H. and Salazar, S. M., An evaluation of dietary and waterborne copper exposures and effects on growth in caged *Mytilus galloprovincialis* in San Diego Bay, in preparation.

Salazar, M. H. and Salazar, S. M., In situ bioassays using transplanted mussels: I. Estimating chemical exposure and bioeffects with bioaccumulation and growth, In *Environmental Toxicology and Risk Assessment—Third Volume*, Hughes, J. S., Biddinger, G. R., and Mones, E., Eds., American Society for Testing and Materials STP 1218, Philadelphia, PA, pp. 216–241, 1995.

Salazar, M. H. and Salazar, S. M., Mussels as bioindicators: Effects of TBT on survival, bioaccumulation and growth under natural conditions, In *Tributyltin: Environmental Fate and Effects*, Champ, M. A. and Seligman, P. F., Eds., Chapman and Hall, London, pp. 305–330, 1996.

Salazar, M. H. and Salazar, S. M., Using caged bivalves as part of an exposure–dose–response triad to support an integrated risk assessment strategy, In *Proceedings, Ecological Risk Assessment: A meeting of Policy and Science*, SETAC Special Publication, de Peyster, A. and Day, K., Eds., SETAC Press, pp. 167–192, 1998.

Secor, C. L., Mills, E. L., Harshbarger, J., Kunts, H. T., Gutenmann, W. H., and Lisk, D. J., Bioaccumulation of toxicants, element and nutrient composition, and soft tissue histology of zebra mussels (Dreissena polymorpha) from New York State waters, *Chemosphere*, 26(8), 1559–1575, 1993.

Stewart, A. R. and Malley, D. F., *Technical Evaluation of Molluscs as a Biomonitoring Tool for the Canadian Mining Industry*, Ottawa, Ontario. p. 248, 1997.

Suter, G. W., Efroymson, R. A., Sample, B. E., and Jones, D. S., *Ecological Risk Assessment for Contaminated Sites*, CRC Press, Boca Raton, FL, p. 438, 2000.

Tessier, A., Campbell, P. G. C., Auclair, J. C., and Bisson, M., Relationships between the partitioning of trace metals in sediments and their accumulation in the tissues of the freshwater mollusc Elliptio complanata in a mining area, *Can. J. Fish. Aquat. Sci.*, 41, 1463–1472, 1984.

US ACOE, Proposed New Guidance for Interpreting the Consequences of Bioaccumulation from Dredged Material. Environmental Effects of Dredging Technical Notes. EEDP-01-41. August 1996.

US ACOE (Army Corps of Engineers), Environmental Residue Effects Database Home Page [Web Page], Available at http://www.wes.army.mil/el/ered/index.html#misc, 1999.

USEPA, Guidelines for Ecological Risk Assessment, EPA/630/R-95/002F, Final. Risk Assessment Forum, US Environmental Protection Agency, Washington, DC, April 1998.

USEPA, 2003 Draft update of ambient water quality criteria for copper, U.S. Env. Protect. Agency, Office Wat., Office Sci. Technol., Washington, DC, EPA 822-R-03-026, p. 86, 2003.

USEPA and US ACOE, Ecological evaluation of proposed discharge of dredged material into ocean waters. Environmental Effects Laboratory, US Environmental Protection Agency and US Army Corps of Engineers, Vicksburg, Mississippi, July 1977.

White, H. H., Ed, *Concepts in Marine Pollution Measurements*, Maryland Sea Grant Publication, University of Maryland, College Park, p. 743, 1984.

White, H. H., Mussel madness: Use and misuse of biological monitors of marine pollution, In *Concepts in Marine Pollution Measurements*, White, H. H., Ed., Maryland Sea Grant Publication, University of Maryland, College Park, Maryland, pp. 325–337, 1984b.

Widdows, J. and Donkin, P., Mussels and environmental contaminants: Bioaccumulation and physiological aspects, In *The Mussel Mytilus: Ecology, Physiology, Genetics and Culture*, Gosling, E., Ed., Elsevier Science Publishers, Amsterdam, pp. 383–424, 1992.

Wren, C. D., MacCrimmon, H. R., and Loescher, B. R., Examination of bioaccumulation and biomagnification of metals in a precambrian shield lake, *Water Air Soil Pollut.*, 19, 277–291, 1983.

10 Biomarker Responses of Unionid Mussels to Environmental Contaminants

Teresa J. Newton and W. Gregory Cope

INTRODUCTION

Unionid mussels are ecologically and economically important in aquatic ecosystems. The biomass of unionids can exceed the biomass of all other benthic organisms by an order of magnitude (Negus 1966; Layzer, Gordon, and Anderson 1993), and production (range, 1–20 g dry mass/m^2/yr) can equal that by all other macrobenthos in many streams (Strayer et al. 1994). Thus, unionids may play important roles in particle processing, nutrient release, and sediment mixing (Vaughn and Hakenkamp 2001; Strayer et al. 2004). Mussels also serve as food for aquatic mammals, including raccoons, muskrats, and otters (Van der Schalie and Van der Schalie 1950). Historically, unionids provided a supplemental food source to indigenous peoples prior to European settlement, but large-scale commercial interest in unionids did not develop until the mid-1800s (Claassen 1994). Soon thereafter, unionids were extensively harvested for the production of pearl buttons, and presently, unionid shells are used in the multimillion dollar Asian cultured-pearl industry (Anthony and Downing 2001).

Overharvesting, widespread habitat destruction, pollution, land-use change, and invasive species introductions have caused many unionid populations to decline or disappear. In North America, most species are now extinct or imperiled, and unionids are widely recognized as one of the most imperiled plants or animals on the continent (Master et al. 2000). Although not sufficiently documented, exposure to toxic contaminants may also be contributing to these declines. There are few instances where chemical spills and other point sources of contaminants have caused localized mortality (Sheehan, Neves, and Kitchel 1989; Fleming, Augspurger, and Alderman 1995); however, widespread decreases in density and diversity are more likely to result from the subtle, pervasive effects of chronic, low-level contamination (Naimo 1995).

There is convincing evidence that unionids, and glochidial and juvenile life stages in particular, are sensitive to many contaminants relative to other aquatic species (e.g., Newton et al. 2003, Chapter 5 and Chapter 7). However, for a given chemical, toxicity can vary by an order of magnitude among life stage and species (Cherry et al. 2002; Augspurger et al. 2003). This is not surprising given the diversity of life history adaptations present in this faunal group. For example, differences in longevity (30–130 years), habitat requirements (silt to gravel), feeding strategies (filter-, deposit-, and pedal-feeding), and reproduction (hermaphrodites, dioecious) all contribute to this diversity. Although there are about 1,000 species of unionids worldwide, and about 300 species in North America, only about 11 species have been reported in the peer-reviewed literature to assess the effects of contaminants on biomarker responses in this imperiled faunal group. Clearly,

257

ecotoxicological research on unionids needs to expand to encompass the breadth of life histories found in this group.

The development of physiological and biochemical tests or "biomarkers" of sublethal exposure are critical in assessing the condition of unionids. Several textbooks and review articles have been written on the use of biomarkers in a wide range of aquatic organisms (e.g., McCarthy and Shugart 1990; Huggett et al. 1992; Van der Oost, Beyer, and Vermeulen 2003)—many of these have application in unionids.

Freshwater bivalve ecotoxicology gained momentum in the early 1970s with potassium and copper bioassays with unionids (Imlay 1971). Once initial studies revealed that unionids were indeed sensitive to a variety of contaminants relative to other invertebrates and fishes, there has been growing interest in using unionids to evaluate the toxicity of chemicals. Most studies in freshwater bivalve ecotoxicology have either been short-term laboratory tests with single chemicals (Chapter 5 and Chapter 7) or monitoring of biomarker responses in the field (Chapter 6 and Chapter 9). Both of these approaches are necessary to understand basic biological responses. However, methods had to be, and are continuing to be, developed to culture and maintain adults and juveniles in suitable physiological condition before and during testing and to evaluate suitable sublethal response endpoints (ASTM 2006). The field has made substantial progress in many of these areas in the past two decades, but it could further benefit from lessons learned in other ecotoxicological areas of study, such as with marine bivalves (ASTM 2002).

Although there are substantial differences in reproductive strategies between unionids and marine bivalves that may confound certain direct comparisons, the field of unionid ecotoxicology could benefit from some of the approaches commonly used in marine studies. For example, a framework used in marine bivalve ecotoxicology recognized that this field needed to move beyond simple acute tests to multidisciplinary studies that gather knowledge at several levels of biological organization, conduct persistent and systematic field studies, and provide iteration between experiment and field observations (Luoma 1996). Further, this framework stressed that standardized "simple" ecotoxicologic approaches lack power in explaining implications of contaminants in complicated circumstances. Additional lessons from marine bivalves that should be incorporated into a future framework for unionid ecotoxicology include the following (from Luoma 1996):

1. Viewing contaminants as just one of the several influential physical, chemical, or biological variables in many aquatic systems.
2. Recognizing that contaminants are distributed among solution, suspended particles, sediments, pore waters, and food resources and that each species or life-history stage may "sample" differently from this complex matrix.
3. Toxicity databases derived from water-only exposures were developed to support regulatory criteria, but these probably underestimate the exposures of bivalves in many circumstances.
4. Biological responses to contaminants in nature can be much more complex than the responses observed in the laboratory.
5. When conducted alone, simplistic toxicity tests, a single biomarker, whole organism analysis, or studies that exclude variables other than contaminants will probably be insensitive to all but the most extreme influences of contaminants.

Contaminants can influence unionids at many different levels of biological organization (Figure 10.1). Usually, the most robust approach is the use of a number of different indices that span the levels of biological complexity, such as biochemical, cytological, physiological, and autecological. In theory, a contaminant first exerts its effects at the molecular level, and the changes at this level lead to changes in organelles and cellular structures, and so forth. However,

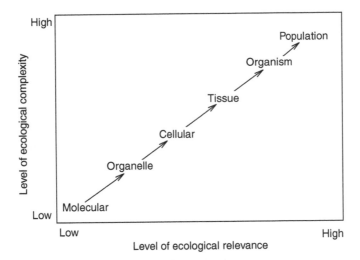

FIGURE 10.1 A generalized approach of biological organization that illustrates how a given contaminant may exert its influence on unionids (Modified from Stebbing A. R. D., *The Effects of Stress and Pollution on Marine Animals*, Praeger Scientific, New York, 1985. With permission.)

in nature, things are rarely this simple because at each level of biological organization, there probably exists homeostatic mechanisms to counteract these disruptive effects. Also, because organisms are rarely exposed to a single contaminant at a time, certain combinations of contaminants may act in an additive or synergistic manner to amplify the effects of one another. Lastly, this approach implies a linear response to contaminants that may not include a detectable threshold effect. Certain contaminants, like essential metals, may be required for basic biological functions at low concentrations, whereas above a threshold level, they may become toxic. This concept should not be overlooked in studies of the effects of contaminants on unionids.

This review of biomarker responses in unionids exposed to environmental contaminants focuses on studies that (1) reported measured contaminant concentrations; (2) had robust experimental designs, including the replication of control and contaminant treatments; and (3) were published in the peer-reviewed literature. These criteria effectively removed about 50% of the papers published in this area, primarily because many authors reported only nominal concentrations. However, we believe these criteria are critical for the objective evaluation of the effects of contaminants on unionids. For example, if only nominal concentrations are reported, the actual concentration of a given contaminant that is available for uptake is basically unknown. In some studies that only report nominal concentrations, the exposure concentrations are so high that they exceed the known solubility of the contaminant. In these instances, the actual amount of the contaminant that is available for uptake may be a small fraction of the nominal concentration, which could seriously underestimate toxicity. Also, many contaminants may adhere to the walls of exposure chambers or can be lost to volatilization—both of which may effectively reduce the actual exposure concentration. Similarly, treatments need to be replicated to get an estimate of the variation associated with the exposure. Future studies should strive to report measured contaminant concentrations (whenever possible) and replicate contaminant and control treatments to ensure a robust design and analysis.

BIOMARKER CONCEPT

A biomarker is a change in a biological response (at the molecular, cellular, biochemical, physiological, or behavioral level) that can be related to exposure to, or toxic effects of, environmental chemicals (Peakall 1994). As defined, biomarkers can span several levels of biological organization (Figure 10.1), but biomarkers that have been investigated most extensively have been enzymes involved in the detoxification of xenobiotics and their metabolites (biotransformation and antioxidant enzymes). Biomarkers are generally classified into those that indicate exposure, effects, or susceptibility. Although connections must be established between exposure to contaminants and effects on biota, biomarkers show promise as indicators demonstrating that contaminants have entered organisms, have been distributed among tissues, and are eliciting toxic effects at critical targets (McCarthy and Shugart 1990). In a recent review of biomarkers in the fisheries literature, Van der Oost, Beyer, and Vermeulen (2003) proposed six criteria that should be established for each candidate biomarker:

1. The assay should be reliable, relatively inexpensive, and easy to perform.
2. The response should be sensitive to pollutant exposure and/or effects in order to serve as an early warning parameter.
3. Baseline data of the biomarker should be well defined in order to distinguish between natural variability and contaminant-induced stress.
4. The impacts of the confounding factors should be well established.
5. The underlying mechanism of the relations between the response and pollutant exposure should be established.
6. The relation between the biomarker response and its long-term impact to the organism should be established.

Although there are numerous classification systems for biomarkers, we will follow the one recently used by Van der Oost, Beyer, and Vermeulen (2003) that groups biomarkers into 1 of 10 categories (Table 10.1). Many of these biomarker types have already been applied to unionids, but additional studies with multiple species are needed. The available data suggest that this is a promising avenue for future research.

BIOTRANSFORMATION ENZYMES

To our knowledge, there have only been two studies that have examined the effects of contaminants on enzymes associated with Phase I of biotransformation. This first phase of metabolism involves the exposing or adding of reactive functional groups, through oxidation, reduction, or hydrolysis. The activity of 7-ethoxyresorufin O-deethylase (EROD) has been used as a biomarker in fish, and these data suggest that EROD activity may not only indicate chemical exposure (primarily to organic contaminants such as polycyclic aromatic hydrocarbons [PAHs], polychlorinated biphenyls [PCBs], polychlorinated dibenzo-*p*-dioxins [PCDDs] and polychlorinated dibenzofurans), but may also precede effects at various levels of biological organization (Whyte et al. 2000; Van der Oost, Beyer, and Vermeulen 2003). In unionids, a 1.5-fold increase in EROD activity was observed in the digestive gland of *Elliptio complanata* deployed for 62 days downstream of a municipal waste effluent, relative to mussels deployed upstream (Gagné et al. 2002). These findings are consistent with elevated levels of PAHs—a common constituent in sewage effluent and urban runoff. These initial data suggest that EROD activity may be a valuable indicator of exposure to organic contaminants in unionids (Table 10.1). More recently, exposure of *Unio tumidus* to diethylhexylphthalate induced the expression of CYP4 (a cytochrome P450 enzyme), however, no CYP1A sequence was amplified in Aroclor-treated mussels (Chaty, Rodius, and Vasseur 2004).

The bulk of the published data on biotransformation enzymes in unionids involves those enzymes associated with the second phase of biotransformation. This phase involves the conjugation (the addition of large and often polar chemical groups) of the xenobiotic parent compound or its metabolite with an endogenous ligand. The addition of more polar groups

TABLE 10.1
Categories of Biomarkers with Known or Potential Application to Unionids

Category of Biomarker	Examples	Reference in Unionids
Biotransformation enzymes		
A. Phase I	Cytochrome P450	Chaty, Rodius, and Vasseur (2004)
	Ethoxyresorufin O-deethylase (EROD)	Gagné et al. (2002)
	Aryl hydrocarbon hydroxylase (AHH)	NA[a]
B. Phase II	Reduced (GSH) and oxidized (GSSG) glutathione	Cossu et al. (1997, 2000), Doyotte et al. (1997)
	Glutathione S-transferases (GST)	Mäkelä, Lindström-Seppä, and Oikari (1992)
	UDP-glucuronyl transferases	NA[a]
Oxidative stress	Superoxide dismutase (SOD)	Cossu et al. (1997), Doyotte et al. (1997)
	Catalase (CAT)	Cossu et al. (1997), Doyotte et al. (1997)
	Glutathione peroxidase (GPOX)	Cossu et al. (1997, 2000), Doyotte et al. (1997)
	Glutathione reductase (GRED)	Cossu et al. (1997, 2000), Doyotte et al. (1997)
	Lipid peroxidation (LPOX)	Cossu et al. (1997, 2000), Doyotte et al. (1997)
Biotransformation products	Polyaromatic hydrocarbon metabolites in bile	NA[a]
Amino acids and proteins	Amino acids	Gardner, Miller, and Imlay (1981), Day, Metcalfe, and Batchelor (1990)
	Stress proteins	NA[a]
	Metallothioneins (MT)	Holwerda (1991), Couillard et al. (1993, 1995a, 1995b), Malley et al. (1993), Wang et al. (1999), Gagné et al. (2002), Perceval et al. (2002)
Hematological	Serum transaminases	NA[a]
	Alterations in the heme pathway	Chamberland et al. (1995)
Immunological	Cell- and humoral-mediated immunity	NA[a]
	Phagocytosis	Blaise et al. (2002)
	Lysosomal activity	NA[a]
Reproductive and endocrine	Imposex	NA[a]
	Vitellogenin	Gagné et al. (2001a, 2001b, 2001c), Riffeser and Hock (2002), Blaise et al. (2003)
	Sexual competence	NA[a]
Neuromuscular	Cholinesterases	Doran et al. (2001)
Genotoxic	DNA damage	Black et al. (1996), Gagné et al. (2002), Rodius, Hammer, and Vasseur (2002)
	Irreversible genotoxic events	NA[a]
Physiological and morphological	Histopathology	Lasee (1991)
	Osmotic and ion regulation	Malley, Huebner, and Donkersloot (1988), Hemelraad et al. (1990)
	Digestive processes	Mäkelä, Lindström-Seppä, and Oikari (1992), Naimo, Atchison, and Holland-Bartels (1992), Milam and Farris (1998)

(continued)

TABLE 10.1 *(Continued)*

Category of Biomarker	Examples	Reference in Unionids
	Condition indices	Mäkelä, Lindström-Seppä, and Oikari (1992), Naimo, Atchison, and Holland-Bartels (1992), Couillard et al. (1995a, 1995b), Hickey, Roper, and Buckland (1995), Hyötyläinen, Karels, and Oikari (2002), Blaise et al. (2003)
	Energetics	Mäkelä, Lindström-Seppä, and Oikari (1992), Naimo, Atchison, and Holland-Bartels (1992), Hickey, Roper, and Buckland (1995), Gagné et al. (2002, 2001b), Hyötyläinen, Karels, and Oikari (2002)
	Valve activity	Balogh and Salanki (1984); Huebner and Pynnönen (1992), Englund and Heino (1996), Kádár et al. (2001), Markich, (2003)
	Growth	Manly and George (1977); Foster and Bates (1978), Muncaster, Hebert, and Lazar (1990), Lasee (1991); Couillard et al. (1995a), Beckvar et al. (2000), Gagné et al. (2001b), Bartsch et al. (2003), Newton et al. (2003)

[a] Not available.

Source: Modified from Van der Oost, R., Beyer, J., and Vermeulen, N. P. E., *Environ Toxicol Pharmacol*, 13, 57–149, 2003. With permission.

generally facilitates the excretion of these chemicals by biota. Many biotransformation enzymes can be induced or inhibited upon exposure to contaminants. Enzyme induction is an increase in the amount or activity of these enzymes (or both), while inhibition refers to the blocking of enzymatic activity, usually because of binding or complex formation with the inhibitor (Van der Oost, Beyer, and Vermeulen 2003). The primary Phase II enzymes that have been examined in unionids include reduced (GSH) and oxidized (GSSG) glutathione and glutathione S-transferase (GST, Table 10.1).

Reduced glutathione is a tripeptide whose major functions are to conjugate electrophilic intermediates and to serve as an antioxidant. This enzyme ensures the reduction of oxidants, the quenching of free radicals, the neutralization of organic peroxides, and the elimination of hydrocarbons by conjugation (Cossu et al. 1997). Most of the research on the utility of Phase II enzymes as potential biomarkers in unionids comes from research on *U. tumidus* in which reduced levels of GSH in cytosolic and particulate fractions of the gills and digestive gland were reported after 15- and 30-day deployment at sites near and downstream from the outfall of a cokery (primarily contaminated with PCBs and PAHs; Cossu et al. 1997; Doyotte et al. 1997). For example, GSH concentrations were reduced in the cytosolic fraction by 79% in the gills and by 59% in the digestive gland at the most polluted site (Cossu et al. 1997). These decreases paralleled lipid peroxidation in the gills (see Oxidative Stress section), which reflected cell injury and toxic effects in this tissue. Similar results were found when *U. tumidus* were transplanted to other areas contaminated by effluents from a laundry and a foundry (Cossu et al. 2000).

One of the detoxification enzymes that has been assayed in unionids is GST. This is a family of enzymes that catalyze the initial step of mercapturic acid synthesis—the conjugation of GSH with

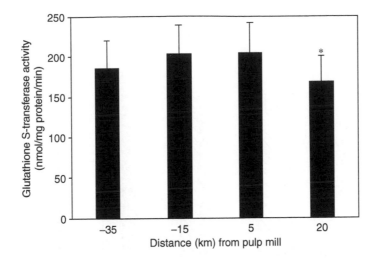

FIGURE 10.2 Glutathione S-transferase activity in the digestive gland of *A. anatina* after a 4-month deployment to sites upstream (-35 and -15 km) and downstream (5 and 20 km) from a bleached kraft pulp and paper mill. Asterisk indicates a significant difference from the upstream reference sites (Adapted from Mäkelä, T. P., Lindström-Seppä, P., and Oikari A. O. J., *Aqua. Fenn.*, 22, 49–55, 1992.)

xenobiotics and their metabolites. Their primary roles are defense against oxidative damage and peroxidative products of DNA and lipids. GST activity was significantly reduced (~20%) in the digestive gland of *Anodonta anatina* deployed for 4 months at a site 20 km downstream of a bleached kraft pulp and paper mill, but not at a site 5 km downstream, compared to control sites (Figure 10.2; Mäkelä, Lindström-Seppä, and Oikari 1992). Given that the spatial extent of this effect was limited, the authors concluded that organically bound chlorine (a major constituent of the effluent from bleached kraft pulp and paper mills) did not consistently induce GST activity in unionids.

OXIDATIVE STRESS

Many environmental contaminants have been shown to exert toxic effects through oxidative stress. Antioxidant defenses include antioxidant enzymes (superoxide dismutase [SOD], catalase [CAT], glutathione peroxidase [GPOX], and glutathione reductase [GRED]) and free radical scavengers (vitamins C and E, carotenoids, and glutathione), whose function is to remove reactive oxygen species thus protecting organisms from oxidative stress (Doyotte et al. 1997). SOD catalyzes the conversion of reactive superoxide anions into hydrogen peroxide, which in turn is detoxified by CAT. Hydrogen peroxide and hydroperoxides are destroyed by GPOXs in the presence of GSH. Glutathione is regenerated by GRED. When antioxidant systems are impaired, oxidative stress may also produce lipid peroxidation (LPOX), or the oxidation of polyunsaturated fatty acids.

Catalase is often one of the earliest antioxidant enzymes to be induced and has been shown to be induced in *Mytilus* sp. exposed to organic pollution (Porte et al. 1991). Catalase and SOD activity were measured in *U. tumidus* after in situ deployment at sites upstream and downstream of effluent from a cokery. In one study, CAT and SOD (Figure 10.3) were significantly reduced at the most polluted site, relative to mussels deployed upstream (Cossu et al. 1997). In the second study, SOD and CAT were markedly unchanged upon exposure to the cokery effluent, suggesting that these enzymes were not sensitive to short-term exposure to the chemicals contained therein (Doyotte et al. 1997). Interestingly, the experimental design for these two studies were nearly identical, with the exception that mussels in the latter study were deployed for 7 days, whereas mussels in the Cossu et al.

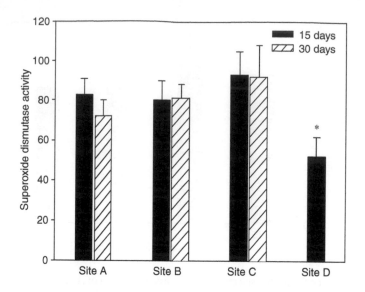

FIGURE 10.3 Superoxide dismutase activity in the cytosolic fraction of the gills of *Unio tumidus* deployed for 15 or 30 days to four sites along the Fensch River, France. Site A was a reference, Site B was upstream of a complex cokery effluent, Site C was downstream near the outfall, and Site D was about 2.5 km downstream. Asterisk indicates a significant difference from control (Adapted from Cossu, C., Doyotte, A., Jacquine, M. C., Babut, M., Exinger, A., and Vasseur, P., *Ecotoxicol. Environ. Saf.*, 38, 122–131, 1997.)

(1997) study were deployed for 15 and 30 days. These data suggest that the temporal variation in these oxidative enzymes may make it difficult to detect treatment effects at shorter exposure durations.

Selenium-dependent GPOX, GRED, and GSH levels appear to be early biomarkers of exposure to pollutants in unionids. For example, deployment of *U. tumidus* to sites upstream and downstream of effluent from a cokery resulted in significant decreases in Se-dependent GPOX and GRED activities (Cossu et al. 1997; Doyotte et al. 1997). Similar exposure of *U. tumidus* to effluents from other sources (a laundry and a foundry) also resulted in significant reductions in Se-dependent GPOX and GRED activities (by 70 and 80%, respectively) (Cossu et al. 2000). However, in the latter study, a relation among the antioxidant response and the degree and type of contamination in sediments was not consistently observed, suggesting that these effects could result from unidentified contaminants and/or issues associated with contaminant bioavailability.

Lipid peroxidation is an important outcome of oxidative stress because it can demonstrate the ability of a single radical species to propagate a number of adverse biochemical reactions (Van der Oost, Beyer, and Vermeulen 2003). Studies suggest that LPOX has considerable potential as a biomarker for environmental risk assessment (Stegeman et al. 1992), although it can result as a consequence of cellular damage because of a variety of stressors other than exposure to contaminants. Numerous studies in the fisheries literature have demonstrated an enhancement in LPOX as a function of contaminant exposure (see references in Van der Oost, Beyer, and Vermeulen 2003), and the preliminary data on unionids are promising. In the first of three studies on antioxidant enzymes in *U. tumidus*, LPOX (as expressed by malonaldehyde content, MDA) did not differ between reference unionids and those exposed to a complex industrial effluent (Doyotte et al. 1997). It is possible that the lack of an effect in this study was a function of the 7-day deployment period; in subsequent studies, unionids were deployed for at least 15 days. In a follow-up study, decreases in Se-dependent GPOX and GRED activities and GSH levels were associated with a three-fold increase in MDA content in the gills and with a high level of contamination of sediments by PAHs and PCBs (Cossu et al. 1997). Similarly,

Cossu et al. (2000) found elevated levels of MDA in unionids deployed in four rivers in France with various pollution sources. In particular, in one river primarily contaminated by metals, deficiencies in antioxidant defenses resulted in dramatic lipid peroxidation—MDA concentrations ranged from 29 to 85 ng/mg protein downstream of the source, while the controls did not exceed 5 ng/mg protein. The elevated levels of MDA, coupled with decreases in Se-dependent GPOX and GRED activities and GSH levels, suggest that these antioxidant parameters may be useful biomarkers of exposure.

AMINO ACIDS AND PROTEINS

Changes in the concentrations of free amino acids in the gill, mantle, and adductor muscle in marine bivalves have been used as biomarkers of exposure to contaminated environments (Livingstone 1985 and references therein). For example, the ratio of taurine to glycine has been used in marine bivalves to indicate biochemical responses to hydrocarbons (Widdows et al. 1982). To our knowledge, two studies have measured free amino acid concentrations in unionids and neither measured this ratio; thus, the utility of this ratio in unionids is unclear but deserves future study. Both unionid studies suggest that increases or decreases in total-free amino acids in some tissues (especially the mantle and adductor muscle) may be indicative of generalized stress induced by a variety of environmental factors (e.g., starvation or increased temperature) and may be useful as an in situ biochemical index of toxicity (Gardner, Miller, and Imlay 1981; Day, Metcalfe, and Batchelor 1990). As with many aspects of unionid biochemistry and physiology, the concentrations of most free amino acids vary seasonally. Thus, the baseline seasonal variation needs to be characterized in a given population prior to attributing changes in amino acid concentrations to contaminant-induced effects (biomarker criterion 3 in Biomarker Concept).

The molecules that probably offer the greatest potential for monitoring biological effects and have attracted the most attention are enzymes and other functional proteins such as metallothioneins (MT) (Livingstone 1985). Freshwater and marine molluscs are known to accumulate metals from their environment. The tolerance of the resulting body burden has been attributed to the existence of an effective detoxification system (Viarengo 1989). Metallothioneins are low molecular weight, cysteine-rich metal binding proteins that function as a detoxification mechanism by sequestering divalent metals through specific ligands present in the cytosol. They are thought to provide one of two protective functions (1) interception and binding of free metal ions that are initially taken up by the cell and (2) removal of metals from non-thioncin ligands that include cellular targets of toxicity (Van der Oost, Beyer, and Vermeulen 2003).

One of the first studies to examine the role of metal-binding proteins in unionids (*A. cygnea*) found that Cd and Cu were bound to different fractions upon laboratory exposure. For example, Cu was generally bound to the high molecular weight fraction, whereas Cd was mainly bound to a specific metal-binding, carbohydrate-containing protein fraction of M_r ~11,000 (Holwerda 1991). The Cd-binding protein was similar to the metal-binding proteins observed in *Mytilus* (Roesijadi and Hall 1981) and *Crassostrea* (Ridlington and Fowler 1979). In one of the first reports of MT-induction in unionids, Malley et al. (1993) showed that a 22-day exposure of *A. grandis grandis* to waterborne Cd induced MT in the gills. Although MT was measured in the mantle, gills, foot, kidney, visceral mass, and whole body, only MT in the gills increased significantly with increasing exposure to Cd. More recently, the time course for MT induction has been examined. Before induction of MT (~14 d) in *Pyganodon grandis*, Cd was primarily bound to the high molecular weight fraction, and after induction (~90 d) all the Cd had apparently shifted to fractions of moderate molecular size (15 to 3 kDa)—which corresponds to the expected MT-fraction (Couillard et al. 1995a).

Field studies support experimental data that MTs sequester heavy metals and that MT levels correlate with tissue levels of heavy metals. Induction of MT in unionids has been observed in field locations where mussels were exposed to Cd (Malley et al. 1993; Couillard et al. 1995b) and urban effluents (Gagné et al. 2002). Concentrations of MT generally correlate with tissue Cd concentrations, but not with tissue concentrations of Cu or Zn (Figure 10.4; Couillard, Campbell, and Tessier 1993; Wang et al. 1999). Variation in MT concentrations among field locations is strongly correlated with free Cd^{2+} (Couillard, Campbell, and Tessier 1993; Wang et al. 1999). Most recently, Perceval et al. (2002) examined the relative influence of limnological and geochemical confounding factors on MT synthesis in natural populations of *P. grandis* in lakes along a Cd concentration gradient (biomarker criterion 4 in Biomarker Concept). Predictive models found that dissolved Ca ($-$) and free Cd^{2+}($+$) explained 62% of the variation in MT. These data can be used in monitoring programs to select field sites to reduce the relative influence of factors that confound MT concentrations (Perceval et al. 2004).

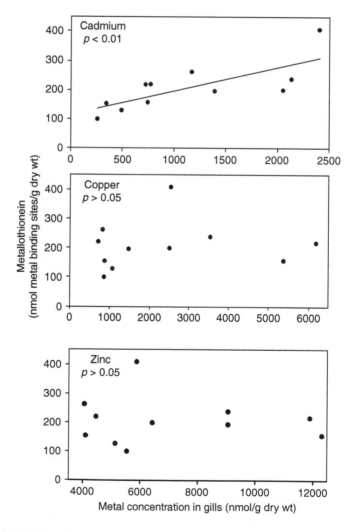

FIGURE 10.4 Concentration of metallothionein as a function of Cd, Cu, and Zn in the gills of *A. grandis* from 11 lacustrine sites along a geochemical gradient of pH, Cd, Cu, and Zn in an area influenced by mining and smelting (Adapted from Couillard, Y., Campbell, P. G. C., and Tessier, A., *Limnol. Oceanogr.*, 38, 299–313, 1993.)

HEMATOLOGICAL

The heme biosynthesis pathway is sensitive to organic and inorganic contaminants, and porphyrin intermediates of this pathway have been used as biomarkers of chemical stress in mammals. Chamberland et al. (1995) identified and quantified seven porphyrin profiles in *A. g. grandis* in two reference lakes and in a lake in which Cd (57–177 ng/L) had been added for 5 years. Abnormal porphyrin profiles (especially hepatocarboxyporphyrin) were apparent in mussels from the Cd-treated lake, relative to mussels in the reference lakes, and these changes were consistent with a Cd-induced biochemical response. These data suggest that a mussel's porphyrin profile has potential as an ecotoxicological biomarker for Cd and possibly other metals.

IMMUNOLOGICAL

The highly regulated nature of the immune system renders it vulnerable to exposure to environmental contaminants. In marine bivalves, hemocytes circulating in the hemolymph are the main components of the immune system—these cells are capable of several immune functions, including phagocytosis and cytotoxicity through the generation of reactive oxygen intermediates (Cheng 1977). Contaminants that adversely affect the immune system can ultimately make unionids more susceptible to infectious diseases. Blaise et al. (2002) developed a miniaturized immunocompetence assay that examines the effects of contaminants on three critical targets of the immune system: capacity to ingest bacteria (phagocytosis activity), cell viability, and hemolymph bacteria levels. In the laboratory portion of that study, exposure of *E. complanata* to a municipal effluent led to immunosuppression (phagocytosis activity and cell viability were reduced). In the field, *E. complanata* deployed upstream and downstream of the effluent for 62 days resulted in both immunosuppressive and immunostimulative effects; low concentrations of heavy metals and microorganisms presumably caused immunostimulation, whereas immunosuppressive effects appeared after exposure to higher levels of contaminants and microorganisms. This study suggests that contaminants may compromise the immune system of unionids. Future studies should determine the contaminant concentrations that contribute to irreversible damage to the immune system and how these effects may be translated into population-level effects (biomarker criterion 6 in Biomarker Concept).

REPRODUCTIVE AND ENDOCRINE

Since the early 1990s, considerable effort has been directed towards understanding the effects of endocrine-disrupting chemicals on aquatic biota. For example, high levels of serum vitellogenin (an estrogen-inducible egg protein precursor) were found in male carp in a river contaminated by a municipal effluent (Folmar et al. 1996). Much of this research has focused on surface waters downstream from municipal effluents and agricultural fields because the concentrations of potential xenoestrogens in these environments may be sufficient to produce biological effects, such as alterations in sexual differentiation, gonad development, and oocyte growth. In certain invertebrates, the presence of estrogens trigger the synthesis of vitellogenin. In marine gastropods, low levels of organotin compounds (e.g., 0.02 µg/L tin) can induce imposex—the manifestation of male morphological sex characters in females (Bryan et al. 1986). Comparatively less is known about the effects of endocrine-disrupting chemicals on unionids, although advances in this area are ongoing and seem promising.

Most of our knowledge on the effects of contaminants on reproductive-related processes in unionids comes from several studies evaluating the effects of a municipal effluent on *E. complanata*. In one study, extracts from surface waters upstream and downstream of a municipal effluent were injected into mussels and the effluent emitted relatively high levels of estradiol competitors that were able to compete with estradiol-binding sites in gonad cytosols and induced vitellogenin (Gagné et al. 2001a). Moreover, the estrogen-competing potential of the extracts was significantly correlated with total and fecal coliform bacteria in the effluent and with levels of vitellogenin in the hemolymph. In a combined laboratory and field study, Gagné et al. (2001b) found that injections of 17β-estradiol and p-nonylphenol and exposure to the effluent increased vitellogenin-like proteins in the hemolymph of male and female E. complanata. Interestingly, females responded more readily to effluent treatment than males, which suggests that continuous exposure of mussel populations to urban effluents not only has the potential to elicit estrogen-mediated effects that include vitellogenin synthesis, but also may cause a shift in the male-to-female sex ratio. Recently, contaminants in the effluent plume were reported to lead to feminization of E. complanata populations (Blaise et al. 2003). Adults deployed for 1 year at sites 8 and 10 km downstream from the effluent plume had an elevated gonado-somatic index, increased condition factor, and significantly elevated levels of vitellogenin-related proteins in gonads, relative to mussels deployed upstream. Also, the proportion of females increased from 41% at the upstream site to 66% at the furthest downstream site. Environmental chemicals with the potential to interfere with the endocrine system can mimic estrogen or inhibit androgen. Most of the current research is on estrogen; future research should evaluate both estrogenic and androgenic effects on unionids.

One compound of concern in municipal effluents is coprostanol (5β [H]-cholestan-3β ol)—a reduced metabolite of cholesterol that is produced by microorganisms in the intestinal tract of mammals. Coprostanol was found in significantly elevated concentrations in adult E. complanata after a 62-day deployment downstream of a municipal effluent (Gagné et al. 2001c). In the laboratory, coprostanol was shown to have a relatively weak binding potential to estradiol-binding sites, although its injection into the adductor muscle led to increased levels of vitellogenin-like proteins in mussel gonads and hemolymph. This implies that coprostanol may be biologically transformed into a more potent estrogenic metabolite and may be metabolized in the gonads of unionids. These studies suggest that continuous exposure to the numerous contaminants and microorganisms in municipal effluents may have deleterious effects on reproductive processes in unionids that may ultimately be manifested in population-level effects.

In contrast to these findings, no changes in protein patterns were observed in A. cygnea in response to exposure to 17β-estradiol or effluents from a wastewater treatment plant (Riffeser and Hock 2002). These authors suggested that hemolymph may not be the carrier of major egg yolk proteins in bivalves, rather the presence of vitellogenin-like proteins in hemolymph may originate from break-down products of gonadal tissue. Future mechanistic-based research with unionids should help clarify this issue.

Female unionids are generally classified into one of two reproductive categories (1) those that carry glochidia over winter and release them in the spring or early summer (long-term brooders) and (2) those that carry glochidia for a short duration in the spring and release them in summer (short-term brooders; Chapter 3). Thus, it is often possible to choose a species for study that is gravid during the time frame of interest. Because females of certain species use either the inner or outer gills, or both, to hold the developing glochidia, a gill index (mass of outer gill divided by the mass of the inner gill) has been proposed as an index of reproductive potential in those species who hold glochidia in their outer gills (Mäkelä, Lindström-Seppä, and Oikari 1992). However, in situ exposure of A. anatina (a long-term brooder) did not adversely affect the gill index among individuals deployed upstream and downstream of a pulp mill effluent, nor were any gross histological changes in gonad development observed (Mäkelä, Lindström-Seppä, and Oikari 1992). This index requires additional study in other unionid species.

NEUROMUSCULAR

Cholinesterases are a group of enzymes whose function is to transmit nerve impulses. Inhibition of cholinesterase prevents the metabolic breakdown of choline neurotransmitters and prolongs signal

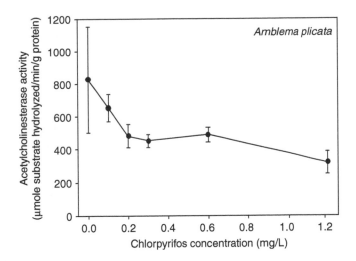

FIGURE 10.5 Acetylcholinesterase activity in *Amblema plicata* after 96-hour exposure to chloropyrifos, an organophosphorus insecticide (Adapted from Doran, W. J., Cope, W. G., Rada, R. G., and Sandheinrich M. B., *Ecotoxicol. Environ. Saf.*, 49, 91–98, 2001.)

transmissions through the synapse, which effectively causes paralysis through over-stimulation of the nervous system. In the field, organophosphate and carbamate insecticides have been implicated in the mortality of ~1000 *Elliptio* spp. in North Carolina, presumably by inhibiting the mussels' cholinesterases (Fleming, Augspurger, and Alderman 1995). In the laboratory, exposure of *Amblema plicata* to chlorpyrifos, an organophosphorus insecticide, resulted in a significant reduction in acetylcholinesterase activity in the adductor muscle after 96 hours (Figure 10.5), but the exposure did not follow a concentration-response relation (Doran et al. 2001). The authors noted that acetylcholinesterase activity was highly variable among individuals and suggested that either sample sizes need to be increased or different tissues should be assayed. This is a promising biomarker of exposure, and additional research with cholinesterase inhibition in unionids is needed.

GENOTOXIC

In marine bivalves, the Comet assay (alkaline single-cell gel electrophoresis) has been a frequently used biomarker to assess the effects of contaminants on DNA strand breakage (Mitchelmore et al. 1998; Shaw et al. 2000). Although substantially less research has been done with freshwater molluscs, recent genotoxicity studies show that contaminants can indeed adversely effect genomic DNA. For example, exposure of *A. grandis* to 50µg Pb/L in the laboratory showed significant DNA strand breakage and a reduction in strand length after 28 days (Black et al. 1996). However, no damage was observed upon exposure to higher Pb concentrations (500 and 5,000 µg/L), suggesting a threshold effect for DNA damage and repair. In contrast, Gagné et al. (2002) observed that levels of DNA damage (measured as DNA strand breaks) in the digestive gland of *E. complanata* were significantly reduced in individuals exposed for 62 days to a municipal effluent plume compared to mussels deployed upstream. Recently, a technique termed RNA arbitrarily primed polymerase chain reaction was developed to detect DNA damage and variations in gene expression in *U. tumidus* in response to exposure to toxic contaminants (Rodius, Hammer, and Vasseur 2002). These authors report variations corresponding to the loss of polymerase chain reaction products in some individuals deployed in situ for 14 days in a river polluted by PAHs, PCBs, and metals, relative to controls. Although sample sizes were small, this method shows promise as a new technique to explore the genotoxic effects of contaminants on unionids. More methods development, concentration-response studies with model chemicals, and documentation that genotoxic effects on individuals ultimately influence populations are still required (biomarker criterion 6 in Biomarker Concept).

PHYSIOLOGICAL AND MORPHOLOGICAL

From the mid-1970s, aquatic ecotoxicology has increasingly used the tools of physiologists. Although this was partly done to understand why an organism is physiologically stressed, it was also because of the realization that many sublethal effects were apparent before death of the individual (biomarker criterion 2 in Biomarker Concept). Widdows (1985) suggested that physiological responses have three important attributes in providing an assessment of an individual's condition: (1) they represent an integration of the many cellular and biochemical processes that can alter in response to changes in the environment, (2) they represent non-specific responses to the sum of environmental stimuli that are complementary to more specific responses at the biochemical levels, and (3) they are capable of reflecting deterioration in the environment before effects manifest themselves in the population or the community. In this section, we review the responses of unionids to contaminants on histopathology, osmotic and ion regulation, digestive processes, condition indices, energetics, valve activity, and growth.

HISTOPATHOLOGY

We are aware of only one study that documented the histopathological effects of contaminants on unionids. In that study, Lasee (1991) evaluated the effects of Cd on the histopathology of 0-day-old juvenile *Lampsilis ventricosa* in 7-day laboratory tests. Exposures as low as 30 µg Cd/L resulted in the dissolution and loss of the crystalline style; vacuolization, necrosis, and tissue separations in mantle, ganglia, and digestive glands; altered lipid catabolism; hypersecretion of mucus; and reduced feeding. While this study suggested that Cd may adversely affect several tissue and organ features, it is unclear how, or if, these changes translate into individual- or population-level effects. Substantial research needs to be directed at closing the gap between histological events, such as the loss of the crystalline style, and its influence on feeding and ultimately growth.

OSMOTIC AND ION REGULATION

Because unionids inhabit freshwater systems with low ionic strength, they must accumulate and retain solutes (i.e., Na^+ and Cl^-). To preserve their water balance, they need to excrete a volume of water equivalent to that taken up osmotically in order to maintain ionic homeostasis. Under normal conditions, unionids are able to obtain ions from food and through epithelial ion transport systems, thereby compensating for ion losses (Murphy and Dietz 1976). Exposure of *A. cygnea* to 50 µg Cd/L for 12 weeks resulted in substantial changes in the composition of 27 elements (Hemelraad et al. 1990). In particular, Na^+ concentrations in hemolymph declined 55% from 20 to 9 mM. The disruption of Na^+ transport may have additional consequences because the transport of salts (including Na^+) has been intimately linked to acid–base balance in unionids (Dietz 1985). The 3-day addition of aluminum sulfate (alum) to an experimental lake resulted in a substantial increase in Ca^{2+} (from 8 to 15 meq/L) but decreases in Na^+ and Cl^- in the hemolymph of *A. g. grandis*; most values returned to near normal after 21 days (Malley, Huebner, and Donkersloot 1988). The increase in blood Ca^{2+} was probably a result of the dissolution of Ca^{2+} stores (as a result of the lower pH associated with the alum addition) in the shell or mantle. These data suggest that contaminants can disrupt the osmotic balance in unionids, at least for some period of time. Additionally, because ion balance in unionids is probably coupled to acid–base balance, contaminants whose mode of action includes disruption of the transepithelial transport of ions (e.g., Cd) may also adversely affect the acid–base balance.

DIGESTIVE PROCESSES

Bivalves have a relatively simple digestive system that is comprised of the stomach, crystalline style and sac, gastric shield, digestive diverticula, and midgut, hindgut, and rectum (Morton 1983). Relatively little research has been conducted on the effects of contaminants on the digestive system, perhaps because of the difficulty in isolating specific components for study. In unionids, the entire digestive tract is embedded in the main visceral mass. However, the effects of contaminants on three aspects of unionid digestion have been explored: development of a digestive gland index, changes in assimilation efficiencies, and the measure of cellulase activity. The digestive gland somatic index has been proposed as an indicator of nutritive status in marine bivalves (Widdows 1985). A significant increase in this index [(dry weight of digestive gland/total soft tissue dry weight)×100] was found in A. anatina after a 4-month deployment to a site 20 km downstream of a bleached kraft pulp mill (Mäkelä, Lindström-Seppä, and Oikari 1992). Application of this index may require additional research because the authors state that it was difficult to dissect out this diffuse organ. The efficiency of food adsorption by the digestive gland in L. ventricosa was not adversely affected by 28-day exposure to Cd (Naimo, Atchison, and Holland-Bartels 1992). Although the assimilation efficiency dropped from 32% in the controls to 18% after exposure to 305 µg Cd/L, the high individual variability masked statistical significance. Additional research needs to be focused on this endpoint because it is one of the components of the "scope for growth (SFG)" equation that has been successfully used in marine bivalves (see ENERGETICS).

Cellulase is an enzyme that hydrolyzes cellulose. Cellulase activity has been measured in non-unionid mollusks (Corbicula and Musculium) as an indirect measure of feeding because it measures the rate of breakdown of complex sugars into simple molecules (Farris et al. 1988; Naimo et al. 2000). Controlled laboratory and field exposures have provided evidence that reductions in enzyme activity are related to the eventual survival of the animal and to more subtle changes that occur in filtration and growth rates (biomarker criterion 6 in Biomarker Concept; Farris et al. 1994; Milam and Farris 1998). The use of cellulase activity as a biomarker of contaminant exposure has only recently been examined in unionids. For example, cellulolytic activity was measured in Quadrula quadrula at five sites (three upstream, two downstream) after the controlled release of a partly treated mine-water discharge dominated by iron in the Ohio River (Milam and Farris 1998). Although cellulase activity varied significantly among sites, differentiation between the upstream and downstream sites was not apparent. These data suggest that although cellulase activity shows promise in delineating contaminant-related effects before the onset of mortality or declines in physiological measures in some molluscs, its utility as a biomarker in unionids needs further assessment.

CONDITION INDICES

Condition indices have been used extensively as an indicator of nutritive status in marine bivalves, especially commercially important shellfish. These indices are often measured as the proportion of internal shell volume that is comprised of tissue, although numerous wet- and dry-weight-based variations have been used. Generally, the dry-weight-based indices are more reliable because they avoid the difficulties of removing and measuring wet tissue. Further, wet tissues can contain variable amounts of water that can increase the variability in estimated condition indices. Recently, these condition indices have been applied in unionids as a biomarker of exposure to contaminants (Table 10.2). These studies have had variable results. For example, Naimo, Atchison, and Holland-Bartels (1992) reported that condition indices in L. ventricosa

TABLE 10.2
Summary of the Effects of Environmental Contaminants on Condition Indices in Unionids

Species	Contaminant(s)	Condition Index Calculation	Condition Index	References
E. complanata	Effluent plume	$\dfrac{\text{tissue wet weight (g)}}{\text{shell length (mm)}}$	Upstream: 0.49 ± 0.01 Downstream: 0.50–0.53	Blaise et al. (2003)
P. grandis	Cd, Zn	$\dfrac{\text{tissue dry weight (g)}}{\text{shell weight (g)}}$	0–30 days after relocation: 0.12–0.16, 90–400 days after relocation 0.01–0.11	Couillard et al. (1995a, 1995b)
H. menziesi	As, Hg	$\dfrac{\text{tissue dry weight (mg)}}{\text{shell weight (g)}}$	90–160	Hickey, Roper, and Buckland (1995)
A. anatina	Creosote-contaminated lake	$\dfrac{\text{tissue dry weight (g)}}{\text{shell length (mm)}} \times 100$	Upstream: 0.11–0.13 Downstream: 0.07–0.10	Hyötyläinen, Karels, and Oikari (2002)
A. anatina	Pulp mill effluent	$\dfrac{\text{tissue wet weight (g)}}{\text{shell length (cm)}^3} \times 100$	Upstream: 4.61 ± 0.19[a] Downstream: 4.34–4.68[a]	Mäkelä, Lindström-Seppä, and Oikari (1992)
L. ventricosa	Cd	$\dfrac{\text{tissue dry weight (g)}}{\text{shell dry weight (g)}} \times 100$	Reference: 7.91 ± 0.37 Cd-exposed: 5.94–6.78	Naimo, Atchison, and Holland-Bartels (1992)

[a] Measured in males only.

were similar (range, 5.9–6.7) after a 28-day exposure to 0–300 μg Cd/L in a laboratory study and did not exhibit a concentration-response relation. However, the condition of mussels used in the toxicity test was significantly less than the mean condition of 7.9 from a subsample of mussels taken from the field prior to the test. Similarly, condition indices in *P. grandis* decreased over time in both control mussels and those relocated into a lake contaminated by Cd and Zn (Couillard et al. 1995a, 1995b). These two studies suggest that condition indices declined over time in response to handling or seasonal changes in condition, or some other cumulative stressor, and may not be related to Cd or Zn exposure.

Condition indices of unionids were unchanged after a 4- or 10-month deployment downstream from a pulp mill effluent and a creosote plant (Mäkelä, Lindström-Seppä, and Oikari 1992; Hyötyläinen, Karels, and Oikari 2002, respectively). In contrast, condition indices were significantly elevated in *E. complanata* caged downstream of a municipal effluent (Blaise et al. 2003) and in resident *Hyridella menziesi* over an 80-km reach of the Waikato River, New Zealand (Hickey, Roper, and Buckland 1995). In the latter study, the study area encompassed potential effects from pulp and paper mills and geothermal, sewage, and industrial discharges, but the increase in condition indices was consistent with higher levels of food, as evident by a three-fold increase in chlorophyll concentrations across the study area. Hickey, Roper, and Buckland (1995) hypothesized that the increase in food availability was a response to the discharge of sewage rather than a contaminant-related effect. The available data for condition indices in unionids suggest that because they are a nonspecific indicator, changes in condition may or may not be directly related to contaminant exposure. Further evidence for this is found in studies that have simultaneously compared condition indices and glycogen concentrations. Data from oysters (Gabbott and Walker 1971) and unionids (Monroe and Newton 2001) indicate that because measures of a simple condition index correlate well with measures of glycogen content, for routine monitoring purposes condition indices may suffice as an estimate of nutritional status. The utility of this concept for unionids requires additional research and validation.

ENERGETICS

In the past 10 years, there has been a substantial focus on measures of energy reserves (glycogen, lipids, and proteins) in unionids after exposure to numerous stressors, including contaminants (see citations below), relocation (Patterson, Parker, and Neves 1997; Naimo and Monroe 1999; Newton et al. 2001), and introduction of the exotic zebra mussel (Haag et al. 1993; Hallac and Marsden 2000). The results of the effects of contaminants on energetic resources in unionids is summarized in Table 10.3. Lipid content increased after deployment of unionids downstream of a municipal effluent plume (Gagné et al. 2001b, 2002) and a pulp mill effluent (Mäkelä, Lindström-Seppä, and Oikari 1992). In the former studies, lipid levels were sex-dependent, with males having more lipids than females. Protein concentrations were similar in *H. menziesi* over an 80-km reach of the Waikato River, New Zealand, which was contaminated by As, Hg, and municipal, industrial, and geothermal discharges (Hickey, Roper, and Buckland 1995) and in *A. anatina* after a 4-month deployment at sites up to 40 km downstream from a pulp mill effluent (Mäkelä, Lindström-Seppä, and Oikari 1992). In contrast, protein concentrations were significantly reduced in *A. anatina* at a site downstream from a creosote plant that had the highest concentrations of PAHs in water and settled particulate matter (Hyötyläinen, Karels, and Oikari 2002).

Comparison of carbohydrate concentrations among studies are complicated by variations in analytical methods, units of measurement, and the tissues analyzed. For example, studies have measured glucose concentrations, glucose concentrations per unit protein, glucose concentrations converted to glycogen, and total glycogen concentrations in the mantle, foot, and whole body. For comparative purposes, the use of similar methods, or at least the expression of endpoints in a similar manner (e.g., Naimo and Monroe 1999; Greseth et al. 2003), would be beneficial for measuring

TABLE 10.3
Carbohydrate, Protein, and Lipid Concentrations in Unionids after Exposure to Contaminants

Species	Contaminant(s)	Carbohydrate	Protein	Lipid	Reference
A. anatina	PAHs[a]	Decrease (adductor muscle)	Decrease (adductor muscle)	—[b]	Hyötyläinen, Karels, and Oikari (2002)
A. anatina	Pulp mill effluent	No change (digestive gland and adductor muscle)	No change (adductor muscle)	Increase (digestive gland)	Mäkelä, Lindström-Seppä, and Oikari (1992)
E. complanata	Effluent plume	Increase (gonads)	—[b]	Increase (gonads)	Gagné et al. (2002, 2001b)
H. menziesi	As, Hg	Increase (whole animal)	No change (whole animal)	—[b]	Hickey, Roper, and Buckland (1995)

Note: Tissues sampled shown in parentheses.

[a] Polycyclic aromatic hydrocarbons.
[b] Not measured.

carbohydrate concentrations in unionids. Additional research is warranted in this area before specific recommendations can be made; however, the method for glycogen determination in Naimo et al. (1998) underwent a rigorous quality assurance and quality control assessment. Concentrations of carbohydrates have increased, decreased, and remained similar in response to contaminant exposure (Table 10.3). In locations where municipal effluent was the suspected contaminant, carbohydrate concentrations generally increased (Hickey, Roper, and Buckland 1995; Gagné et al. 2001b, 2002). Glycogen concentrations in *A. anatina* did not differ among four sites above and below a pulp mill effluent (Mäkelä, Lindström-Seppä, and Oikari 1992). In contrast, exposure of *A. anatina* to the effluent from a creosote plant resulted in a decrease in glycogen concentrations that may have resulted from chemical stress caused by PAH-contaminated sediments, although numerous other chemicals from the plant may have also exerted a physiological effect on unionids (Hyötyläinen, Karels, and Oikari 2002).

Few studies have examined the effects of contaminants on physiological energetics such as excretion rates, respiration rates, and filtration rates. Hickey, Roper, and Buckland (1995) found that filtration and respiration rates in *H. menziesi* were highly variable and did not differ among three sites that spanned a gradient in sedimentary As and Hg. Naimo, Atchison, and Holland-Bartels (1992) measured five physiological variables in *L. ventricosa* after 0, 14, and 28-day exposures to Cd concentrations ranging from 0 to 300 µg/L. Respiration rates, the least variable measure, were significantly reduced in the presence of Cd and showed a concentration-response relation (Figure 10.6). In contrast, excretion rate, clearance rate, assimilation efficiency, and the O:N ratio did not differ among treatments. Both studies suggest that physiological measures were highly variable among individual mussels and masked detection of treatment differences at acceptable statistical levels. For example, un-ionized ammonia (NH_3) excretion rates varied 82% from 0.004 to 0.022 mg NH_3–N/h/g dry weight and the O:N ratio varied from 24.7 to 125.5 (Naimo, Atchison, and Holland-Bartels 1992). A common way to reduce variation is to increase sample size. In the Naimo, Atchison, and Holland-Bartels (1992) study, physiological measures were made on 20 individuals per treatment—a sample size that was unable to reduce variation to acceptable levels. Alternatively, the sensitivity of methods used to make these physiological measures should be improved.

The "(SFG)" is an integration of several physiological responses that has been used extensively to assess the effects of contaminants on marine molluscs (e.g., Widdows et al. 1981, 1990, 1995,

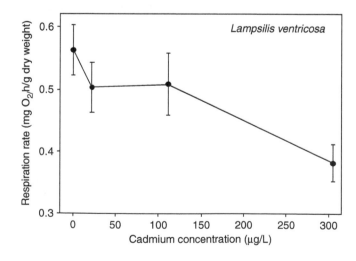

FIGURE 10.6 Respiration rate in *L. ventricosa* as a function of cadmium exposure during a 28-day laboratory study (Adapted from Naimo, T. J., Atchison, G. J., and Holland-Bartels, L. E., *Environ. Toxicol. Chem.*, 11, 1013–1021, 1992.)

1997). The SFG is not measured directly but is derived by subtraction of the energy respired and excreted from the energy absorbed from food, as listed below:

$$P = A - (R + U)$$

where P is the energy incorporated into somatic growth and gamete production, A is the energy absorbed from food, R is the energy respired, and U is the energy excreted (Widdows 1985). The direct measurement of growth and production is difficult in bivalves because a large proportion of the total production can be lost in the form of gametes and because the measurement of tissue somatic growth or weight change is impractical because of the shell (Widdows 1985).

Even though SFG has been used extensively in marine bivalves, it has only recently been attempted in unionids. The physiological measures made by Naimo, Atchison, and Holland-Bartels (1992) were chosen because they were input parameters in the SFG bioenergetics model for *L. ventricosa*. However, the authors found that the physiological measures proved highly variable in unionids. SFG shows promise as an integrator of physiological responses in unionids after contaminant exposure; however, additional research to adapt methods for measuring physiological processes in freshwater mussels are needed.

VALVE ACTIVITY

In the field, adult unionids open and close their valves in a daily rhythm as a normal behavioral function. The duration of activity in the open and closed positions has been quantified and used as an endpoint after exposure to contaminants. Because it is generally assumed that filtration rate is an increasing function of valve gaping, if a contaminant reduces the duration of open periods, then filtration rate may be reduced, mussels may lose weight, and contaminant uptake may be affected. Exposure to 10 µg Hg/L or 16 µg Cd/L for 35 days substantially altered the daily rhythm of valve activity in *A. cygnea* (Balogh and Salanki 1984). In reference mussels, the average duration of the open periods varied between 10 and 25 hours; after exposure to either metal, the average duration of the open periods decreased from 5 to 10 hours for several days but returned to normal within 14 days. Similarly, the average duration of the closed periods was 3–7 hours in reference mussels, but exposure to either metal resulted in a three- to seven-fold increase in the duration of the closed periods such that by the end of the experiment, mussels were in open and closed positions for about the same portion of the time. Kádár et al. (2001) also used shell gape as an endpoint in *A. cygnea* after exposure to environmentally relevant concentrations of Al (250 and 500 µg/L). In that study, a 15-day exposure to 250 µg/L produced no detectable change in valve activity, but exposure to 500 µg/L reduced the mean duration of shell opening by 50%. This effect was irreversible over a 15-day recovery period. Interestingly, tissue levels of Al were an order of magnitude higher in the lower exposure concentrations than in the higher concentration, which is consistent with the inhibition of filtering activity due to valve closure at the higher concentration. These data suggest that *A. cygnea* is able to detect and respond to increased Al concentrations by closing its shell and reducing uptake. However, after prolonged closure, which is facilitated by the ability to respire anaerobically, the build up of lactic acid eventually necessitates that filtration is resumed, and contaminants can re-enter the body. In contrast, the rhythm of valve movements in *A. anatina* was correlated with the photoperiod but not with exposure to 2,4,6-trichlorophenol (Englund and Heino 1996). In a recent study, the valve movement responses of *Velesunio angasi* to U were independent of gender but were size-and age-dependent; smaller and younger individuals were 22% more sensitive than larger and older individuals (Markich 2003). In summary, valve activity has promise as a biological response to contaminants because it is relatively easy and inexpensive to monitor (biomarker criterion 1 in Biomarker Concept), shows responses at environmentally realistic concentrations and may be proportional to filtration activity. This endpoint should be evaluated in other species and with other contaminants. Moreover, the relations among valve activity, filtration rate, and contaminant uptake rate should be investigated further.

Another type of valve activity is the closing of the glochidial valve as they attach to the appropriate fish host. The short, but critical period between when glochidia are released from the adult but before they attach to a fish host is a susceptible period for contaminant exposure (Chapter 5). If exposure to contaminants either prohibits closure or causes premature and irreversible closure of glochidial valves, then these larvae would be unable to attach to a host fish and unable to complete development. Thus, glochidial viability, as assessed through valve closure, has been used as an endpoint in studies with metals. For example, Huebner and Pynnönen (1992) ranked the toxicities of several metals based on the concentrations needed to reduce the closure response by 50% (effective concentration, EC50) in *A. cygnea*. After a 72-h exposure, the EC50s for Cu, Cd, and Zn were 2.1, 4.8, and 45.8 µg/L, respectively. These data suggest that exposure to environmentally realistic metal concentrations during this short, but critical, period in the life cycle of unionids has the potential to diminish reproductive success.

GROWTH

Early studies showed conflicting data on whether the growth of adult unionids was related to the body burden of several inorganic and organic contaminants (Table 10.4). For example, Manly and George (1977) correlated concentrations of six metals in *A. anatina* from seven locations along the

TABLE 10.4
Summary of the Effects of Environmental Contaminants on Growth of Unionids

Species	Life Stage	Contaminant(s)	Effect	Reference
L. ventricosa	Juvenile	Cd	Reduction in shell growth at 10 µg/L	Lasee (1991)
L. cardium	Juvenile	NH$_3$	Reduction in shell height at 30 µg/L	Newton et al. (2003)
L. cardium	Juvenile	NH$_3$	No relation between growth and NH$_3$	Bartsch et al. (2003)
A. anatina	Adult	Cd, Cu, Hg, Ni, Pb, Zn	Positive correlation between Cd and Pb concentrations and dry tissue weight, negative correlation for Cu and Hg, no correlation for Ni and Zn	Manly and George (1977)
Q. quadrula	Adult	Cu	Accumulation was inversely related to wet tissue weight	Foster and Bates (1978)
L. radiata	Adult	Hexachlorobenzene, PCB congeners 110, 118[a]	Negative correlation with shell length	Muncaster, Hebert, and Lazar (1990)
P. grandis	Adult	Cd, Zn	Shell length was reduced after a 400-day deployment	Couillard et al. (1995a)
E. complanata	Adult	Hg	Growth was negatively correlated with tissue concentrations of total Hg	Beckvar et al. (2000)
E. complanata	Adult	Municipal effluent	Reduction in shell length in mussels deployed downstream of effluent	Gagné et al. (2001b)

[a] Polychlorinated biphenyls.

Thames River (England) with dry tissue weight and found significant correlations in 12 of 42 possible situations. Further, correlations between concentrations of Pb and Cd and dry weight were positive, whereas correlations for Cu and Hg were negative—suggesting metal-specific relations between contaminants and growth. In contrast, Cu accumulation in *Q. quadrula* was inversely related to wet tissue weight (Foster and Bates 1978) and accumulation of hexachloro-benzene, and PCB congeners 110 and 118 were negatively correlated with shell length in *L. radiata* (Muncaster, Hebert, and Lazar 1990).

Recent studies transplanting unionids across a contamination gradient have shown some success with using the growth of adults as an endpoint, especially in studies with test durations in excess of 60 days (Table 10.4). Beckvar et al. (2000) assessed the effects of Hg contamination on the growth of adult *E. complanata* transplanted in situ for 12 weeks at various distances from a Superfund Site. Mussel growth (calculated from changes in whole-animal wet weight) was nega-tively correlated with tissue concentrations of total Hg (Figure 10.7) but not with total Hg in sediments—a finding that demonstrates that tissue concentrations of contaminants may be more predictive of growth effects than contaminant concentrations in sediments or water. Gagné et al. (2001b) observed that changes in shell length, whole animal growth rate, and soft tissue weight were significantly different in adult *E. complanata* transplanted for 62 days upstream and down-stream of a municipal effluent plume. Interestingly, mussels transplanted downstream of the municipal effluent plume grew less in shell length, but added tissue weight, relative to those deployed upstream. This suggests that the mussels deployed downstream put available energy into soft tissue weight at the expense of shell growth. Lastly, growth of adult *P. grandis* was significantly reduced in mussels transplanted from a relatively unpolluted lake to one contaminated by Cd and Zn for 400 days, relative to mussels in a reference lake (mean change in shell length 1.7 and 3.9 mm, respectively) (Couillard et al. 1995a).

The use of growth as an endpoint of contaminant exposure has been demonstrated in both freshwater and marine bivalves. In all growth-related studies with adults previously cited, the magnitude of the changes in shell length ($\pm$ 2.5 mm) and tissue weight ($\pm$ 3 g) after contaminant exposure were minimal and were similar to those that could be attributed to measurement error (Downing and Downing 1993). Therefore, a suite of growth measures (e.g., shell metrics and bio-chemical and physiological responses such as RNA/DNA ratios) would be beneficial in demonstrating growth effects above and beyond measurement error. Even though these studies suggest that growth of adults can be a sensitive endpoint after contaminant exposure, few

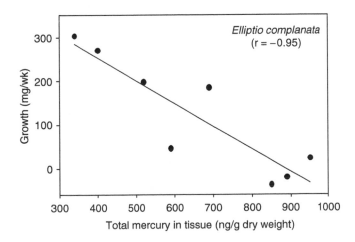

FIGURE 10.7 Growth rate of adult *E. complanata* deployed for 3 months in a river contaminated by mercury as a function of total mercury in tissue (Adapted from Beckvar, N., Salazar, S., Salazar, M., and Finkelstein K., *Can. J. Fish Aquat. Sci.*, 57, 1103–1112, 2000.)

uncertainties still exist. First, because unionids can hold varying levels of water within their mantle cavity, whole animal wet weights may inconsistently overestimate body mass. Second, interpretation of growth data can be problematic because some individuals put on mass at the expense of shell length and vice versa. Experiments with fast-growing species and young individuals (as recommended in ASTM 2002) may provide good estimates of growth. Alternatively, traditional measures of growth in short-term studies with older and slow-growing adults may be inconclusive (Newton et al. 2001).

Recent advances in computer-aided optical imaging systems allows the growth of juvenile unionids to be measured with a high degree of precision. For example, growth of juvenile *L. cardium* was a sensitive indicator of exposure to NH_3 in laboratory sediment toxicity tests (Newton et al. 2003). In that study, the growth rate of juveniles in the highest NH_3 treatment was reduced by 65 and 83% relative to the controls after 96 hours and by 57 and 100% relative to the controls after 10 days (two replicate tests per exposure duration). Similarly, Lasee (1991) observed that exposure to 10 ±g Cd/L resulted in significantly reduced anterior shell growth in juvenile *L. ventricosa* after 7 days in the laboratory. Recently, growth rate was used as an endpoint in a study that deployed juvenile *L. cardium* in situ across an ammonia gradient in the St. Croix River (Bartsch et al. 2003). In that study, NH_3 concentrations in sediment pore water were positively related to growth at 4 and 28 days but unrelated at 10 days. It was unclear why NH_3 concentrations were positively correlated with growth, but at certain concentrations, NH_3 may function more as a nutrient and less as a contaminant. Because juveniles are fast-growing, it seems intuitive that growth may be a sensitive, and easily measured, endpoint. Further, advances in optical imaging systems allow precise measures in juveniles that may help avoid human-induced measurement error. However, as for adults, this endpoint needs to be further evaluated in other species and with additional contaminants. Lastly, because reductions in growth may be the direct result of contaminant exposure or a secondary effect produced by starvation or altered metabolism, the mechanisms by which contaminants affect growth (e.g., impaired biomineralization processes) needs to be determined.

SUMMARY AND RECOMMENDATIONS

We found a total of 43 papers, published between 1977 and 2003, in the peer-reviewed literature that examined the biomarker responses of unionid mussels to environmental contaminants (Figure 10.8). Only studies that reported measured contaminant concentrations and had robust

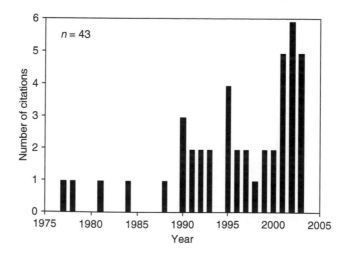

FIGURE 10.8 Number of citations examining the biological responses of environmental contaminants to unionid mussels (1977 through April 2003). Only publications that measured contaminant concentrations, replicated treatments, and were published in the peer-reviewed literature were included.

experimental designs, including replication of control and contaminant treatments, were included. Our assessment has shown that a combination of field and laboratory studies has substantially advanced the field of unionid ecotoxicology. Several general conclusions emerged as a result of this review. First, the overwhelming majority of the published data on the biomarker responses of unionids to environmental contaminants has been on metals—comparatively less is known about the effects of organic chemicals on unionids. Second, although there are roughly 300 species of unionids in North America, only 11 species (3%) have been used to evaluate the effects of environmental contaminants. Further, almost 50% of the research on these 11 species was done with *Anodonta* sp. and another 20% was conducted with *E. complanata*. Thus, most of the reported data were only examined in one species and usually with only one contaminant. Third, over 88% of this research has been published since 1990 (Figure 10.8). Encouragingly, 40% of the research has been done since 2000, which parallels the recent interest in unionids after they were listed as the most imperiled fauna in North America (Master et al. 2000). This interest will probably continue as unionids are recognized to be among the most sensitive genera tested to specific contaminants (e.g., Augspurger et al. 2003).

Lastly, most of the research on biological responses in unionids after exposure to environmental contaminants has focused on physiological processes and amino acids and proteins (the latter is largely driven by the recent research on MTs; Table 10.1). While some of these measures show promise as biomarkers in unionids (Table 10.5), many only meet one or two of the criteria of a successful biomarker proposed by Van der Oost, Beyer, and Vermeulen (2003, Biomarker Concept). As with many other groups of organisms, linkages between a given biological response and the effects on populations or communities have been difficult to establish. The current research effort on unionids is largely focused in a few areas (physiology and proteins) and could benefit from expanding to biomarkers at several levels of biological organization (Figure 10.1). In this regard, unionid ecotoxicologists should look to the marine bivalve literature for biomarkers that have been successfully applied in the marine environment. For example, biomarkers such as EROD, DNA damage, and oxidative stress enzymes have potential for use in unionids based on their preliminary

TABLE 10.5
Biomarker Responses That Have Been Used Successfully to Assess the Effects of Contaminants on Unionids and Those That Show Promise Based on Their Preliminary Use in Unionids or Those That Have Been Extensively Used in Marine Bivalves

Biomarker Category	Successful Responses in Unionids	Potential Responses in Unionids
Biotransformation enzymes		Ethoxyresorufin O-deethylase (EROD), glutathione
Oxidative stress		Catalase, superoxide dismutase, lipid peroxidation
Biotransformation products		
Amino acids and proteins	Metallothioneins	Ratio of taurine:glycine
Hematological		Porphyrin profile
Immunological		Phagocytosis, lysosomal destabilization
Reproductive and endocrine		Vitellogenin
Neuromuscular		Cholinesterases
Genotoxic		DNA damage (comet assay), micronucleus assay
Physiological and morphological	Growth (especially in juveniles), valve activity	Histopathology, osmotic and ion regulation, scope for growth

use in unionids or their use in marine bivalves (Table 10.5). In addition, a framework for ecotoxicological research on unionids, similar to the one proposed for marine bivalves (Luoma 1996) would help to advance this area of study by identifying data gaps, needed research, and developing a comprehensive, but realistic, plan for accomplishing these tasks.

REFERENCES

[ASTM] American Society for Testing and Materials, Standard guide for conducting in-situ field bioassays with marine, estuarine, and freshwater bivalves. E2122-01, In *Annual Book of ASTM Standards*, *Volume. 11.05*, West Conshohocken, PA, 2002.

[ASTM] American Society for Testing and Materials, Standard guide for conducting laboratory toxicity tests with freshwater mussels. E2455-06, In *Annual Book of ASTM Standards*, *Volume 11.05*, West Conshohocken, PA, 2006.

Anthony, J. L. and Downing, J. A., Exploitation trajectory of a declining fauna: A century of freshwater mussel fisheries in North America, *Can. J. Fish Aquat. Sci.*, 58, 2071–2090, 2001.

Augspurger, T., Keller, A. E., Black, M. C., Cope, W. G., and Dwyer, F. J., Water quality guidance for protection of freshwater mussels (Unionidae) from ammonia exposure, *Environ. Toxicol. Chem.*, 22, 2569–2575, 2003.

Balogh, K. V. and Salanki, J., The dynamics of mercury and cadmium uptake into different organs of *Anodonta cygnea* L., *Water Res.*, 18, 1381–1387, 1984.

Bartsch, M. R., Newton, T. J., Allran, J. W., O'Donnell, J. A., and Richardson, W. B., Effects of pore water ammonia on in situ survival and growth of juvenile mussels (*Lampsilis cardium*) in the St. Croix Riverway, Wisconsin, USA, *Environ. Toxicol. Chem.*, 22, 2561–2568, 2003.

Beckvar, N., Salazar, S., Salazar, M., and Finkelstein, K., An in situ assessment of mercury contamination in the Sudbury River, Massachusetts, using transplanted freshwater mussels (*Elliptio complanata*), *Can. J. Fish Aquat. Sci.*, 57, 1103–1112, 2000.

Black, M. C., Ferrell, J. R., Horning, R. C., and Martin, L. K., DNA strand breakage in freshwater mussels (*Anodonta grandis*) exposed to lead in the laboratory and field, *Environ. Toxicol. Chem.*, 15, 802–808, 1996.

Blaise, C., Trottier, S., Gagné, F., Lallement, C., and Hansen, P. D., Immunocompetence of bivalve hemocytes as evaluated by a miniaturized phagocytosis assay, *Environ. Toxicol.*, 17, 160–169, 2002.

Blaise, C., Gagné, F., Salazar, M., Salazar, S., Trottier, S., and Hansen, P. D., Experimentally-induced feminisation of freshwater mussels after long-term exposure to a municipal effluent, *Fresenius Environ. Bull.*, 12, 865–870, 2003.

Bryan, G. W., Gibbs, P. E., Hummerstone, L. G., and Burt, G. R., The decline of the gastropod *Nucella lapillus* around south-west England: evidence for the effect of tributlytin from antifouling paints, *J. Mar. Biol. Assoc. UK*, 66, 611–640, 1986.

Chamberland, G., Bélanger, D., Lavière, N., Vermette, L., Klaverkamp, J. F., and Blais, J. S., Abnormal porphyrin profile in mussels exposed to low concentrations of cadmium in an experimental Precambrian Shield lake, *Can. J. Fish. Aquat. Sci.*, 52, 1286–1293, 1995.

Chaty, S., Rodius, F., and Vasseur, P., A comparative study of the expression of *CYP1A* and *CYP4* genes in aquatic invertebrate (freshwater mussel, *Unio tumidus*) and vertebrate (rainbow trout, *Oncorhynchus mykiss*), *Aquat. Toxicol.*, 69, 81–93, 2004.

Cheng, T. C., Biochemical and ultrastructural evidence for the double role of phagocytosis in molluscs: Defense and nutrition, *Comp. Pathobiol.*, 3, 21–30, 1977.

Cherry, D. S., Van Hassel, J. H., Farris, J. L., Soucek, D. J., and Neves, R. J., Site-specific derivation of the acute copper criteria for the Clinch River, Virginia, *Hum. Ecol. Risk Assess.*, 8, 591–601, 2002.

Claassen, C., Washboards, pigtoes, and muckets: historic musseling in the Mississippi watershed, *Hist. Archaeol.*, 28, 1–145, 1994.

Cossu, C., Doyotte, A., Jacquine, M. C., Babut, M., Exinger, A., and Vasseur, P., Glutathione reductase, selenium-dependent glutathione peroxidase, glutathione levels, and lipid peroxidation in freshwater bivalves, *Unio tumidus*, as biomarkers of aquatic contamination in field studies, *Ecotoxicol. Environ. Saf.*, 38, 122–131, 1997.

Cossu, C., Doyotte, A., Babut, M., Exinger, A., and Vasseur, P., Antioxidant biomarkers in freshwater bivalves, *Unio tumidus*, in response to different contamination profiles of aquatic sediments, *Ecotoxicol. Environ. Saf.*, 45, 106–121, 2000.

Couillard, Y., Campbell, P. G. C., and Tessier, A., Response of metallothionein concentrations in a freshwater bivalve (*Anodonta grandis*) along an environmental cadmium gradient, *Limnol. Oceanogr.*, 38, 299–313, 1993.

Couillard, Y., Campbell, P. G. C., Pellerin-Massicotte, J., and Auclair, J. C., Field transplantation of a freshwater bivalve, *Pyganodon grandis*, across a metal contamination gradient. II. Metallothionein response to Cd and Zn exposure, evidence for cytotoxicity, and links to effects at higher levels of biological organization, *Can. J. Fish Aquat. Sci.*, 52, 703–715, 1995a.

Couillard, Y., Campbell, P. G. C., Tessier, A., Pellerin-Massicotte, J., and Auclair, J. C., Field transplantation of a freshwater bivalve, *Pyganodon grandis*, across a metal contamination gradient. I. Temporal changes in metallothionein and metal (Cd, Cu, and Zn) concentrations in soft tissues, *Can. J. Fish Aquat. Sci.*, 52, 690–702, 1995b.

Day, K. E., Metcalfe, J. L., and Batchelor, S. P., Changes in intracellular free amino acids in tissues of the caged mussel, *Elliptio complanata*, exposed to contaminated environments, *Arch. Environ. Contam. Toxicol.*, 19, 816–827, 1990.

Dietz, T. H., Ionic regulation in freshwater mussels: a brief review, *Am. Malacol. Bull.*, 3, 233–242, 1985.

Doran, W. J., Cope, W. G., Rada, R. G., and Sandheinrich, M. B., Acetylcholinesterase inhibition in the threeridge mussel (*Amblema plicata*) by chlorpyrifos: Implications for biomonitoring, *Ecotoxicol. Environ. Saf.*, 49, 91–98, 2001.

Downing, W. L. and Downing, J. A., Molluscan shell growth and loss, *Nature*, 362, 506, 1993.

Doyotte, A., Cossu, C., Jacquin, M. C., Babut, M., and Vasseur, P., Antioxidant enzymes, glutathione and lipid peroxidation as relevant biomarkers of experimental or field exposure in the gills and the digestive gland of the freshwater bivalve *Unio tumidus*, *Aquat. Toxicol.*, 39, 93–110, 1997.

Englund, V. P. M. and Heino, M. P., The freshwater mussel (*Anodonta anatina*) in monitoring of 2,4,6-trichlorophenol: behaviour and environmental variation considered, *Chemosphere*, 32, 391–403, 1996.

Farris, J. L., Van Hassel, J. H., Belanger, S. E., Cherry, D. S., and Cairns, J., Application of celluloytic activity of Asiatic clams (*Corbicula* sp.) to in-stream monitoring of power plant effluents, *Environ. Toxicol. Chem.*, 7, 701–713, 1988.

Farris, J. L., Grudzien, J. L., Belanger, S. E., Cherry, D. S., and Cairns, J., Molluscan celluloytic activity responses to zinc exposure in laboratory and field stream comparisons, *Hydrobiologia*, 287, 161–178, 1994.

Fleming, W. J., Augspurger, T. P., and Alderman, J. A., Freshwater mussel die-off attributed to anticholinesterase poisoning, *Environ. Toxicol. Chem.*, 14, 877–879, 1995.

Folmar, L. C., Denslow, N. D., Rao, V., Chow, M., Crain, D. A., Enblom, J., Marcino, J., and Guilette, L. J., Vitellogenin induction and reduced serum testosterone concentrations in feral male carp (*Cyprinus carpio*) captured near a major metropolitan sewage treatment plant, *Environ. Health Perspect*, 104, 1096–1101, 1996.

Foster, R. B. and Bates, J. M., Use of freshwater mussels to monitor point source industrial discharges, *Environ. Sci. Technol.*, 12, 958–962, 1978.

Gabbott, P. A. and Walker, A. J. M., Changes in the condition index and biochemical content of adult oysters (*Ostrea edulis* L.) maintained under hatchery conditions, *J. Cons. Int. Explor. Mer.*, 34, 99–106, 1971.

Gagné, F., Marcogliese, D. J., Blaise, C., and Gendron, A. D., Occurrence of compounds estrogenic to freshwater mussels in surface waters in an urban area, *Environ. Toxicol.*, 16, 260–268, 2001aa.

Gagné, F., Blaise, C., Salazar, M., Salazar, S., and Hansen, P. D., Evaluation of estrogenic effects of municipal effluents to the freshwater mussel *Elliptio complanata*, *Comp. Biochem. Physiol.*, 128C, 213–225, 2001bb.

Gagné, F., Blaise, C., Lachance, B., Sunahara, G. I., and Sabik, H., Evidence of coprostanol estrogenicity to the freshwater mussel *Elliptio complanata*, *Environ. Pollut.*, 115, 97–106, 2001cc.

Gagné, F., Blaise, C., Aoyama, I., Luo, R., Gagnon, C., Couillard, Y., Campbell, P., and Salazar, M., Biomarker study of a municipal effluent dispersion plume in two species of freshwater mussels, *Environ. Toxicol.*, 17, 149–159, 2002.

Gardner, W. S., Miller, W. H., and Imlay, M. J., Free amino acids in mantle tissue of *Amblema plicata*: Possible relation to environmental stress, *Bull. Environ. Toxicol. Chem.*, 26, 157–162, 1981.

Greseth, S. L., Cope, W. G., Rada, R. G., Waller, D. L., and Bartsch, M. R., Biochemical composition of three species of unionid mussels after emersion, *J. Molluscan Stud.*, 69, 101–106, 2003.

Haag, W. R., Berg, D. J., Garton, D. W., and Farris, J. L., Reduced survival and fitness in native bivalves in response to fouling by the introduced zebra mussel (*Dreissena polymorpha*) in western Lake Erie, *Can. J. Fish Aquat. Sci.*, 50, 13–19, 1993.

Hallac, D. E. and Marsden, J. E., Differences in tolerance to and recovery from zebra mussel (*Dreissena polymorpha*) fouling by *Elliptio complanata* and *Lampsilis radiata*, *Can. J. Zoolog.*, 78, 161–166, 2000.

Hemelraad, J., Holwerda, D. A., Wijnne, H. J. A., and Zandee, D. I., Effects of cadmium in freshwater clams. I. Interaction with essential elements in *Anodonta cygnea*, *Arch. Environ. Contam. Toxicol.*, 19, 686–690, 1990.

Hickey, C. W., Roper, D. S., and Buckland, S. J., Metal concentrations of resident and transplanted freshwater mussels *Hyridella menziesi* (Unionacea: Hyriidae) and sediments in the Waikito River, New Zealand, *Sci. Total Environ.*, 175, 163–177, 1995.

Holwerda, D. A., Cadmium kinetics in freshwater clams. V. Cadmium-copper interaction in metal accumulation by *Anodonta cygnea* and characterization of the metal-binding protein, *Arch. Environ. Contam. Toxicol.*, 21, 432–437, 1991.

Huebner, J. D. and Pynnönen, K. S., Viability of glochidia of two species of *Anodonta* exposed to low pH and selected metals, *Can. J. Zoolog.*, 70, 2348–2355, 1992.

Huggett, R. J., Kimerle, R. A., Mehrle, P. M., and Bergman, H. L., *Biomarkers: Biochemical, Physiological, and Histological Markers of Anthropogenic Stress*, Lewis, Boca Raton, FL, 1992.

Hyötyläinen, T., Karels, A., and Oikari, A., Assessment of bioavailability and effects of chemicals due to remediation actions with caging mussels (*Anodonta anatina*) at a creosote-contaminanted lake sediment site, *Water Res.*, 36, 4497–4504, 2002.

Imlay, M., Bioassay tests with naiads, In *Rare and Endangered Mollusks (Naiads) of the U.S.*, Jorgenson, S. E. and Sharp, R. W., Eds., U.S. Fish and Wildlife Service, Bureau of Sport Fisheries and Wildlife, Twin Cities, MN, pp. 38–41, 1971.

Kádár, E., Salánki, J., Jugdaohsingh, R., Powell, J. J., McCrohan, C. R., and White, K. N., Avoidance responses to aluminum in the freshwater bivalve *Anodonta cygnea*, *Aquat. Toxicol.*, 55, 137–148, 2001.

Lasee B. A., Histological and ultrastructural studies of larval and juvenile *Lampsilis* (Bivalvia) from the upper Mississippi River [PhD thesis], Iowa State Univ., Ames, IA, 1991.

Layzer, J. B., Gordon, M. E., and Anderson, R. M., Mussels: The forgotten fauna of regulated rivers. A case study of the Caney Fork River, *Regul. Rivers Res. Manag.*, 8, 63–71, 1993.

Livingstone, D. R., Responses of the detoxication toxication enzyme-systems of mollusks to organic pollutants and xenobiotics, *Mar. Pollut. Bull.*, 16, 158–164, 1985.

Luoma, S. N., The developing framework of marine ecotoxicology: Pollutants as a variable in marine ecosystems, *J. Exp. Mar. Biol. Ecol.*, 200, 29–55, 1996.

Mäkelä, T. P., Lindström-Seppä, P., and Oikari, A. O. J., Organochlorine residues and physiological condition of the freshwater mussel *Anodonta anatina* caged in River Pielinen, eastern Finland, receiving pulp mill effluent, *Aqua. Fenn.*, 22, 49–58, 1992.

Malley, D. F., Huebner, J. D., and Donkersloot, K., Effects on ionic composition of blood and tissues of *Anodonta grandis grandis* (Bivalvia) of an addition of aluminum acid to a lake, *Arch. Environ. Contam. Toxicol.*, 17, 479–491, 1988.

Malley, D. F., Klaverkamp, J. F., Brown, S. B., and Chang, P. S. S., Increase in metallothionein in freshwater mussels *Anodonta grandis grandis* exposed to cadmium in the laboratory and the field, *Water Pollut. Res. J. Can.*, 28, 253–273, 1993.

Manly, R. and George, W. O., The occurrence of some heavy metals in populations of the freshwater mussel *Anodonta anatina* (L.) from the River Thames, *Environ. Pollut.*, 14, 139–154, 1977.

Markich, S. J., Influence of body size and gender on valve movement responses of a freshwater bivalve to uranium, *Environ. Toxicol.*, 18, 126–136, 2003.

Master, L. L., Stein, B. A., Kutner, L. S., and Hammerson, G. A., Vanishing assets: Conservation status of U.S. species, In *Precious Heritage: The Status of Biodiversity in the United States*, Stein, B. A., Kutner, L. S., and Adams, J. S., Eds., Oxford Univ. Press, Oxford, pp. 93–118, 2000.

McCarthy, J. F. and Shugart, L. R., Biological markers of environmental contamination, In *Biomarkers of Environmental Contamination*, McCarthy, J. F. and Shugart, L. R., Eds., Lewis, Boca Raton, FL, pp. 3–16, 1990.

Milam, C. D. and Farris, J. L., Risk identification associated with iron-dominated mine discharges and their effect upon freshwater bivalves, *Environ. Toxicol. Chem.*, 17, 1611–1619, 1998.

Mitchelmore, C. L., Birmelin, C., Livingstone, D. R., and Chipman, J. K., Detection of DNA strand breaks in isolated mussel (*Mytilus edulis* L.) digestive gland cells using the "comet" assay, *Ecotoxicol. Environ. Saf.*, 41, 51–58, 1998.

Monroe, E. M. and Newton, T. J., Seasonal variation in physiological condition of *Amblema plicata* in the Upper Mississippi River, *J. Shellfish Res.*, 20, 1167–1171, 2001.

Morton, B., Feeding and digestion in Bivalvia, In *The Mollusca. Physiology. Part 2*, Saleuddin, A. S. M. and Wilbur, K. M., Eds. Vol. 5, Academic Press, New York, NY, pp. 65–147, 1983.

Muncaster, B. W., Hebert, P. D. N., and Lazar, R., Biological and physical factors affecting the body burden of organic contaminants in freshwater mussels, *Arch. Environ. Contam. Toxicol.*, 19, 25–34, 1990.

Murphy, W. A. and Dietz, T. H., The effects of salt depletion on blood and tissue ion concentrations in the freshwater mussel *Ligumia subrostrata* (Say), *J. Comp. Physiol.*, 108, 233–242, 1976.

Naimo, T. J., A review of the effects of heavy metals on freshwater mussels, *Ecotoxicology*, 4, 341–362, 1995.

Naimo, T. J. and Monroe, E. M., Variation in glycogen concentrations within mantle and foot tissue in *Amblema plicata*: Implications for tissue biopsy sampling, *Am. Malacol. Bull.*, 15, 51–56, 1999.

Naimo, T. J., Atchison, G. J., and Holland-Bartels, L. E., Sublethal effects of cadmium on physiological responses in the pocketbook mussel, *Lampsilis ventricosa*, *Environ. Toxicol. Chem.*, 11, 1013–1021, 1992.

Naimo, T. J., Damschen, E. D., Rada, R. G., and Monroe, E. M., Nonlethal evaluation of the physiological health of unionid mussels: methods for biopsy and glycogen analysis, *J. North Am. Benthol. Soc.*, 17, 121–128, 1998.

Naimo, T. J., Cope, W. G., Monroe, E. M., Farris, J. L., and Milam, C. D., Influence of diet on survival, growth, and physiological condition of fingernail clams *Musculium transversum*, *J. Shellfish Res.*, 19, 23–28, 2000.

Negus, C. L., A quantitative study of growth and reproduction of unionid mussels in the River Thames at Reading, *J. Anim. Ecol.*, 35, 513–532, 1966.

Newton, T. J., Monroe, E. M., Kenyon, R., Gutreuter, S., Welke, K. I., and Thiel, P. A., Evaluation of relocation of unionid mussels into artificial ponds, *J. North Am. Benthol. Soc.*, 20, 468–485, 2001.

Newton, T. J., Allran, J. W., O'Donnell, J. A., Bartsch, M. R., and Richardson, W. B., Effects of ammonia on juvenile unionid mussels (*Lampsilis cardium*) in laboratory sediment toxicity tests, *Environ. Toxicol. Chem.*, 22, 2554–2560, 2003.

Patterson, M. A., Parker, B. C., and Neves, R. J., Effects of quarantine times on glycogen levels of native freshwater mussels (Bivalvia: Unionidae) previously infested with zebra mussels, *Am. Malacol. Bull.*, 14, 75–79, 1997.

Peakall, D. W., Biomarkers: The way forward in environmental assessment, *Toxicol. Ecotoxicol. News*, 1, 55–60, 1994.

Perceval, O., Pinel-Alloul, B., Méthot, G., Couillard, Y., Giguère, A., Campbell, P. G. C., and Hare, L., Cadmium accumulation and metallothionein synthesis in freshwater bivalves (*Pyganodon grandis*): Relative influence of the metal exposure gradient versus limnological variability, *Environ. Pollut.*, 118, 5–17, 2002.

Perceval, O., Couillard, Y., Pinel-Alloul, B., Giguère, A., and Campbell, P. G. C., Metal-induced stress in bivalves living along a gradient of Cd contamination: Relating sub-cellular metal distribution to population-level responses, *Aquat. Toxicol.*, 69, 327–345, 2004.

Porte, C., Solé, M., Albaiges, J., and Livingstone, D. R., Responses of mixed-function oxygenase and anti-oxidase enzyme system of *Mytilus* sp. to organic pollution, *Comp. Biochem. Physiol.*, 100C, 183–186, 1991.

Ridlington, J. W. and Fowler, B. A., Isolation and partial characterization of a cadmium-binding protein from the American oyster (*Crassostrea virginica*), *Chem-Biol. Interact.*, 25, 127–138, 1979.

Riffeser, M. and Hock, B., Vitellogenin levels in mussel hemolymph-a suitable biomarker for the exposure to estrogens?, *Comp. Biochem. Physiol.*, 132C, 75–84, 2002.

Rodius, F., Hammer, C., and Vasseur, P., Use of RNA arbitrarily primed PCR to identify genomic alterations in the digestive gland of the freshwater bivalve *Unio tumidus* at a contaminated site, *Environ. Toxicol.*, 17, 538–546, 2002.

Roesijadi, G. and Hall, R. E., Characterization of mercury-binding proteins from the gills of marine mussels exposed to mercury, *Comp. Biochem. Physiol.*, 70C, 59–64, 1981.

Shaw, J. P., Large, A. T., Chipman, J. K., Livingstone, D. R., and Peters, L. D., Seasonal variation in mussel *Mytilus edulis* digestive gland cytochrome P4501A- and 2E-immunoidentified protein levels and DNA strand breaks (Comet assay), *Mar. Environ. Res.*, 50, 405–409, 2000.

Sheehan, R. J., Neves, R. J., and Kitchel, H. E., Fate of freshwater mussels transplanted to formerly polluted reaches of the Clinch and North Fork Holston Rivers, Virginia, *J. Freshw. Biol.*, 5, 139–149, 1989.

Stebbing, A. R. D., A possible synthesis, In *The Effects of Stress and Pollution on Marine Animals*, Bayne, B. L., Brown, D. A., Burns, K., Dixon, D. R., Ivanovici, A., Livingstone, D. R., Lowe, D. M., Moore, M. N., Stebbing, A. R. D., and Widdows, J., Eds., Praeger Scientific, New York, NY, pp. 301–314, 1985.

Stegeman, J. J., Brouwer, M., Richard, T. D. G., Förlin, L., Fowler, B. A., Sanders, B. M., and van Veld, P. A., Molecular responses to environmental contamination: Enzyme and protein systems as indicators of chemical exposure and effect, In *Biomarkers: Biochemical, Physiological, and Histological Markers of Anthropogenic Stress*, Huggett, R. J., Kimerly, R. A., Mehrle, P. M., and Bergman, H. L., Eds., Lewis, Chelsea, MI, pp. 235–335, 1992.

Strayer, D. L., Hunter, D. C., Smith, L. C., and Borg, C., Distribution, abundance, and role of freshwater clams (Bivalvia:Unionidae) in the freshwater tidal Hudson River, *Freshw. Biol.*, 31, 239–248, 1994.

Strayer, D. L., Downing, J. A., Haag, W. R., King, T. L., Layzer, J. B., Newton, T. J., and Nichols, S. J., Changing perspectives on pearly mussels, North America's most imperiled animals, *BioScience*, 54, 429–439, 2004.

Van der Oost, R., Beyer, J., and Vermeulen, N. P. E., Fish bioaccumulation and biomarkers in environmental risk assessment: A review, *Environ. Toxicol. Pharmacol.*, 13, 57–149, 2003.

Van der Schalie, H. and Van der Schalie, A., The mussels of the Mississippi River, *Am. Midl. Nat.*, 44, 448–466, 1950.

Vaughn, C. C. and Hakenkamp, C. C., The functional role of burrowing bivalves in freshwater ecosystems, *Freshw. Biol.*, 46, 1431–1446, 2001.

Viarengo, A., Heavy metals in marine invertebrates: mechanisms of regulation and toxicity at the cellular level, *CRC Crit. Rev. Aquat. Sci.*, 1, 295–317, 1989.

Wang, D., Couillard, Y., Campbell, P. G. C., and Jolicoeur, P., Changes in subcellular metal partitioning in the gills of freshwater bivalves (*Pyganodon grandis*) living along an environmental cadmium gradient, *Can. J. Fish Aquat. Sci.*, 56, 774–784, 1999.

Whyte, J. J., Jung, R. E., Schmitt, C. J., and Tillitt, D. E., Ethoxyresorufin-O-deethylase (EROD) activity in fish as a biomarker of chemical exposure, *Crit. Rev. Toxicol.*, 30, 347–570, 2000.

Widdows, J., Physiological measurements, In *The Effects of Stress and Pollution on Marine Animals*, Bayne, B. L., Brown, D. A., Burns, K., Dixon, D. R., Ivanovici, A., Livingstone, D. R., Lowe, D. M., Moore, M. N., Stebbing, A. R. D., and Widdows, J., Eds., Praeger Scientific, New York, pp. 3–45, 1985.

Widdows, J., Phelps, D. K., and Galloway, W., Measurement of physiological condition of mussels transplanted along a pollution gradient in Narragansett Bay, *Mar. Environ. Res.*, 4, 181–194, 1981.

Widdows, J., Bakke, T., Bayne, B. L., Donkin, P., Livingstone, D. R., Lowe, D. M., Moore, M. N., Evans, S. V., and Moore, S. L., Responses of *Mytilus edulis* L. on exposure to the water accomodated fraction of North Sea oil, *Mar. Biol.*, 67, 15–31, 1982.

Widdows, J., Burns, K. A., Menon, N. R., Page, D. S., and Soria, S., Measurement of physiological energetics (scope for growth) and chemical contaminants in mussels (*Arca zebra*) transplanted along a contamination gradient in Bermuda, *J. Exp. Mar. Biol. Ecol.*, 138, 99–117, 1990.

Widdows, J., Donkin, P., Brinsley, M. D., Evans, S. V., Salkeld, P. N., Franklin, A., Law, R. J., and Waldock, M. J., Scope for growth and contaminant levels in North Sea mussels *Mytilus edulis*, *Mar. Ecol. Prog. Ser.*, 127, 131–148, 1995.

Widdows, J., Nasci, C., and Fossato, V. U., Effects of pollution on the scope for growth of mussels (*Mytilus galloprovincialis*) from the Venice Lagoon, Italy, *Mar. Environ. Res.*, 43, 69–79, 1997.

11 Case Study: Comparison of Asian Clam (*Corbicula fluminea*) in Situ Testing to Several Nontarget Test Organism Responses to Biocidal Dosing at a Nuclear Power Plant

Donald S. Cherry and David J. Soucek

INTRODUCTION

Bivalve mollusks have been widely used as indicators of environmental stress. They dominate some aquatic environments, often making up a large portion of the biomass, and can control how an ecosystem functions because of the processes they perform (Vaughn and Hakenkamp 2001). They are useful as biomonitors because they are sedentary and accumulate contaminants, but are not efficient in metabolizing chemicals compared to some species ([ASTM] American Society for Testing and Materials 2001). In addition, they usually provide sufficient tissue for chemical analysis, and exhibit measurable sublethal effects, such as reduced growth, tissue condition and glycogen levels, reduced shell length, DNA strand breakage, reduced cellulolytic activity, and valve movement (Belanger et al. 1986, 1990; Doherty and Cherry 1988; Farris et al. 1988; Doherty 1990; Allen et al. 1996; Black 1997; Naimo et al. 1998; [ASTM] American Society for Testing and Materials 2001; Newton et al. 2001). Furthermore, their immature stages have been found to be among the most sensitive of test species in acute laboratory bioassays (Jacobson et al. 1993a, 1993b, 1997; Cherry et al. 2002).

In situ field bioassays have gained popularity as replacements for, or supplements to, standard bioassessment tools such as laboratory toxicity testing and benthic macroinvertebrate sampling (Burton 1999). They are thought to provide more environmental realism than laboratory tests by incorporating continuous exposure over an extended period of time to the multiple stressors that regulate indigenous communities, and are less time and labor intensive than macroinvertebrate community sampling (Cherry 1996). Numerous researchers have used caged bivalves in field bioassays in both marine and freshwater systems (e.g., Widdows, Phelps, and Galloway 1981; Couillard et al. 1995; Salazar and Salazar 1995; Black and Belin 1998; Smith and Beauchamp 2000), and a standard guide for test methods has been developed ([ASTM] American Society for Testing and Materials 2001).

Caged Asian clams (*Corbicula fluminea* [Müller]) have been used extensively in field bioassays (e.g., Farris et al. 1988; Allen et al. 1996; Black 1997; Soucek, Schmidt, and Cherry

2001; Hull et al. 2002), and they are particularly useful in areas they have colonized because of their availability, sensitivity, and relatively short life span of 1–3 years (Doherty and Cherry 1988; Doherty 1990). The survivorship and growth responses of transplanted bivalves to relatively concentrated pollutants have been strongly correlated with measures of benthic macroinvertebrate community structure (Smith and Beauchamp 2000; Soucek, Schmidt, and Cherry 2001), whereas in a larger system with dilute contamination, these responses were associated more with impairment of naturally-occurring bivalves than with other benthic macroinvertebrate indices (Hull et al. 2002; Hull, Cherry, and Merricks in press).

Because of its high reproductive capacity, the Asian clam is a prominent pest of industries that use raw water intakes (Cherry et al. 1980), and various biocides have been employed to control it (Cherry et al. 1991; Cherry et al. 1992; Cherry et al. 1993; Bidwell, Farris, and Cherry). We conducted a multi-year study at the Beaver Valley Power Station (BVPS, Shippingport, Pennsylvania) to evaluate potential in-stream impacts of a chemical additive for Asian clam control using a three-tiered approach: (1) formal laboratory toxicity testing, (2) on-site toxicity testing in experimental streams, and (3) in-river benthic macroinvertebrate community sampling, plus in situ bioassays using field-caged Asian clams. The BVPS used a chemical additive (Clam-Trol or CT-1) and a detoxification agent (bentonite clay or DT-1) from Betz Laboratories, Inc., Trevose, Pennsylvania, as a molluscicide or biocide.

While ecotoxicological data collected at different levels of biological organization can be useful in quantifying the effects of pollutants on aquatic ecosystems (e.g., Adams et al. 1992; Clements 2000), it has been frequently demonstrated that such data can vary considerably from one level to the next (e.g., Cherry et al. 2001; Cherry et al. 2002; Hull et al. 2002; Kennedy, Cherry, and Currie 2003; Kennedy, Cherry, and Currie in press). Our objectives were (1) to determine the efficacy of the biocidal addition/detoxification process (i.e., addition of CT-1 followed by DT-1) utilized to control biofouling at BVPS with a three-tiered bioassessment approach, and (2) to provide a review of practical uses of the Asian clam as an in situ monitoring test organism. Our findings underscore the increasing importance of integrating in situ bioassays using field-caged bivalves with traditional measures of ecological integrity.

MATERIALS AND METHODS

SAMPLING SITES

The study site was located near Shippingport, Pennsylvania, USA, along the Ohio River at the BVPS of Duquesne Light Company. Four sampling stations for in situ studies were selected, one above the plant outfall near the power plant intake area (reference site), and three below the outfall at ~350 (Site P5), 700 (Site 2B) and 1,050 m (Site P10) downstream, respectively (Figure 11.1).

BIOCIDE DESCRIPTION

The biocide used (CT-1) was a nonoxidizing, surfactant-based product containing two active ingredients: dodecyl guanidine hydrochloride (DGH) and n-alkyl dimethyl ammonium chloride (Quaaternary ammonium compound or Quat). After use in the plant, the biocide was bound for detoxification purposes with bentonite clay at the beginning of the outfall runway that ran several hundred meters before reaching the river. The CT-1 was dosed at ~15 ppm during each 24-hour dosing and ~45 ppm bentonite clay was added for detoxification purposes. Various plant mega-dosings occurred during the study period, all of which resulted in the biocide/bentonite clay being released into the same plant outfall area.

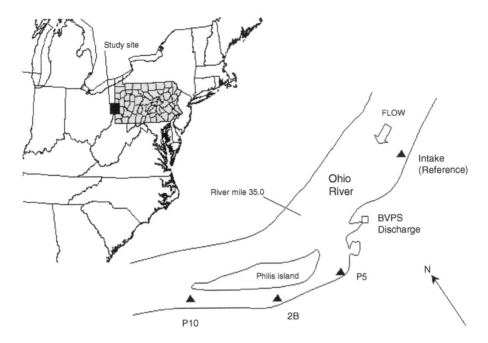

FIGURE 11.1 Map of the study area: Beaver valley power station, Duquesne Light Company, near Shipping-port, Pennsylvania.

LABORATORY TOXICITY TESTING WITH STANDARD TEST ORGANISMS

Effluent toxicity testing. Effluent toxicity tests were conducted with *Daphnia magna, Ceriodaphnia dubia*, and *Pimephales promelas*, according to USEPA (1985, 1989) methods. Test organisms were obtained from laboratory-reared cultures at Virginia Tech. Effluent concentrations were 0, 2.5, 5, 10, 20, 40, and 100% effluent, and filtered Ohio River water (1.5 μm filters) was used as the dilution water.

Chironomus sediment testing. The 10-day survival and growth impairment test for *Chironomus riparius* was used to evaluate the toxicity of river sediments prior to CT-1 dosing in the plant, immediately after dosing, and 35 days later. Testing procedures were based upon methods described by Giesy, Leversee, and Williams (1988) and Nelson, Ingersoll, and Dwyer (1988). Second instar *C. riparius* larvae were exposed to river sediments collected in bioboxes (described under In Situ Toxicity Testing) for 10 days. After the test ended, the number of surviving organisms was recorded and organisms were dried and weighed to the nearest 0.001 mg. Pre-dose (6/19) and post-dose (7/30) tests were conducted in 1990. In 1991, sediments were collected for three tests in association with two separate dosings: pre-dose (8/21, 11/5), during-dose (8/23, 12/11), and post-dose (9/26/91, 1/13/92).

EXPERIMENTAL STREAM EXPERIMENTS

Part of the three-tiered testing program was to develop a laboratory experimental stream system on site. The experimental stream testing was performed during the summer and fall biocide dosings for clam control. A series of 12 oval, paddle-driven streams was designed and housed inside a trailer located adjacent to the effluent outfall into the river. Each stream capacity was 68 l.

River water from a submersible pump in the Ohio River entered via a set of three head-boxes. Gravity provided constant pressure through a head-box drainpipe that led to each of the streams.

Drain pipes, constructed with 19 mm schedule 40 PVC pipe, were fitted with 19 mm straight valves that allowed regulation of water flow to each stream. Inflow rates were checked daily.

Effluent from the plume was pumped into one head-box and then was delivered into three sets of three streams at 5% effluent-95% river water, 50% effluent-50% river water, and a third set received 100% effluent. We also included a fourth set of streams that received 100% river water. River water was pumped from the bottom of a traveling screen well at the cooling-water intake through a fire hose line to the on-site laboratory. Separate head-boxes above each set of three streams received the river water from which pipelines entered each stream.

Stream-water depth was regulated by 19 mm diameter PVC standpipes mounted in bulkhead and male PVC adapters. Current was provided by a series of plexiglass paddle wheels attached to a steel rod, powered by a 1/4 hp continuous-use motor.

Midges (*C. riparius*) were exposed to the three effluent concentrations by placing second instar larvae into 250 mL square Nalgene® bottles, modified such that the sides were replaced with Nytex® mesh to allow water to flow through. Bottles were suspended in the artificial streams for 10- and 13-day exposures in November 1990, after which, midges were evaluated for survival and dry weight.

Snails (*Physa* in June 1990 and *Goniobasis* in November 1990 and August 1991) were introduced into the experimental streams just prior to dosing and were monitored for survival over 35 days. Snails were collected from a small stream (Sinking Creek) in Montgomery County, Virginia, and delivered to the on-site laboratory in less than 24 hours. Mayflies (*Isonychia bicolor*) were also obtained from Sinking Creek and incorporated into the experimental streams after acclimation in fiberglass mesh containers. Survivorship and emergence were evaluated over 40 days.

In the summer of 1991, bluegill sunfish (*Lepomis macrochirus*) were placed in each of the 12 experimental streams, 10 fish/stream, after arrival at the laboratory four days earlier. In the December 1991 experiment, bluegill were held in the laboratory for ~four weeks prior to testing. Fish were shipped overnight from Kurtz Fish Hatchery, Elverton, Pennsylvania. The fish were housed in $300 \times 150 \times 170$ mm^3 deep containers and fed a Tetra-min and ground trout chow mixture twice daily. A sample of 20 fish was weighed (dry weight) and measured (fork length) prior to testing. At the end of the test (40 days) all fish were weighed and measured. Fish lengths were measured immediately after sacrificing the organisms in ethanol. Fish weights were obtained after drying the organisms at 60°C for 24 hours.

Asian clams also were exposed to the various effluent concentrations in the experimental streams for 40 days in two separate experiments in August and November 1991. Clams were evaluated for survival and growth. For the growth measurements, clams were individually marked with a file and measured both before and after the experiments to the nearest 0.001 mm from the umbo, or beak, to the ventral margin. Therefore, a growth record was available for each individual clam used in the study.

IN SITU TOXICITY TESTING

The bioboxes used in the 1990–1991 in situ toxicity studies were developed from milk crates by Robert L. Shema and coworkers at Aquatic Systems Corp. (ASC) located in Pittsburgh, Pennsylvania. Crates were lined with fiberglass mesh and filled with several centimeters of glass marbles that served as ballast to keep the bioboxes in place on the river bottom. The open top was also covered with coarse fiberglass mesh screen, which permitted the infiltration of particulate matter from the water column. Rope attached to the four corners of the bioboxes was secured at the river bank as the bioboxes were lowered to the river floor.

In 1992, the Pennsylvania Department of Environmental Resources (PA DER) granted the use of the biocide and bentonite clay detoxifying agent to be used again at the BVPS for mega-plant dosing of Asian clam control. However, rather than repeat all the testing done in 1991–1992, only one test was deemed appropriate, based upon the previous results. The PA DER required a long-term in situ Asian clam study, using the same sampling sites previously identified, over a period of several months bracketing the plant biocide megadoses. Survivorship and growth were measured seven times over the 162-day study period. The same sampling sites in the river used in 1991–1992 were also used in the expanded clam growth study.

In the 162-day in situ study, a total of 16 bioboxes were used. Each site contained four bioboxes labeled as A, B, C, and D, respectively. Two bioboxes (A, B) were removed from each of the four sites three days prior to biocide dosing and placed upstream above the reference area at a barge slip deemed as a refugium or "safe area." On the day prior to biocide dosing, bioboxes A and B were returned to each site and kept in place during biocide dosing. Bioboxes A and B were designated as "dosed" clams. Bioboxes C and D from each site were removed and placed in the refugium during the days of each biocide dosing and returned after dosing was over and were referred to as "nondosed" clams. The strategy of moving bioboxes before/during biocide dosings was used to address potential experimental bias of handling the clams, and to segregate potential effects of biocide dosing from that of the routine plant effluent operations alone.

Each biobox contained 20 clams, which averaged ~14 mm from umbo to ventral margin at the onset of the study. Clams were obtained from the Unit 2 cooling tower of the BVPS, and individually marked with a file. Clams were initially marked on June 6, 1992, and held in the refugium for 16 days for acclimation to the Ohio River and handling process.

Clams were measured for growth from the onset of the initial acclimation on June 6, 1992, and were again measured after 16 days to determine their condition prior to the initial biocide dosing. The plant was dosed with biocide on July 23, 1992, and the clams were assessed for survivorship and growth after 30, 58, 104, 122 and 162 days. The final clam measurements were made 57 days after the second biocide treatment, and 35 days after the third.

BENTHIC MACROINVERTEBRATE MONITORING

Sediment samples in the river were collected with a ponar dredge (0.25×0.25 m^2), dropped by block and tackle from a boat. Collections were taken above BVPS at river mile 34.5 (1), within the effluent back channel at river mile 35.2 (P5) and 35.4 (2B), which was 0.2 and 0.4 miles below the effluent release to the river at river mile 35.0 (Figure 11.1). A fourth station was located in the main river channel (2A) at river mile 35.4, adjacent to Phillis Island. Three sample replicates were taken in the same area for river mile station 34.5 (1) and 35.4 (2A), while in the effluent back channel, the samples were taken at left, middle, and right locations across each transect. The benthic sampling sites were the same in the 1991–1992 studies with those in the in situ study.

The substrate at each sample was characterized at the time of collection, washed within a US Standard No. 30 sieve, preserved with 10% formalin, and returned to the laboratory of ASC. Macroinvertebrates were sorted from each sample, identified to the lowest possible taxon and counted. Subsampling was used, when appropriate, according to USEPA (1973) methodologies. Mean densities (numbers/m^2) for each taxon were calculated for each station. Three taxon diversity indices were calculated: Shannon–Weiner, evenness, and richness (number of taxa). Invertebrate samples were collected on six dates in the summer and winter of both 1990 and 1991. Collections were made prior to, shortly after, and approximately one month after each dosing period in the summer and winter of each year.

STATISTICAL ANALYSIS

The mean clam width of 20 clams per biobox and that of the 10 clams closest in size were initially measured as "trimmed data". The latter approach was used to narrow the initial variability in clam sizes across treatments at the start of a test, and to follow the growth of these clams throughout the 162-day test. Also, the growth increment of each clam group was determined between each measuring interval. Overall, the accumulative growth increment was tabulated as a running score of clam growth development over time from various biocide dosings. The same statistical treatment was applied for the laboratory chronic toxicity tests for *Ceriodaphnia, D. magna*, etc.

The Shapiro–Wilks statistic was used to test whether the data were normally distributed (Sokal and Rohlf 1981). Because the majority of the data were not normally distributed, non-parametric procedures were used (Hollander and Wolfe 1979). A Wilcoxon's Rank–Sum test was used to evaluate differences in shell growth and size of non-dosed and dosed clam groups. The Kruskal–Wallis test was used to perform a non-parametric, one-way analysis of variance between sites. Duncan's Multiple Range test was then performed on the rank-transformed data to determine significant differences between groups ($\alpha = 0.05$).

RESULTS

EFFLUENT CHRONIC TOXICITY

During the first two-year period of study, the plant megadosing occurred five times, each with a 24-hour duration (Table 11.1). Three megadosings occurred in 1990 during June, September, and October, and two in 1991 during August and December. *Ceriodaphnia* were much more sensitive to both the dosed and undosed effluent than fathead minnows. The NOAEC (No Observable Adverse Effects Concentration) for the fish was 100% effluent for the nine tests conducted. Dosing the plant effluent with the combined biocide/detoxifier caused significant reductions in *Ceriodaphnia* reproduction, compared to when the effluent was not influenced by biocide dosing during routine plant operations. The average LOAEC (Lowest Observable Adverse Effects Concentration) during baseline conditions was 100% effluent, whereas under dosed conditions, the average LOAEC was 48%. The greatest toxicity occurred on the last biocide dosing (December 10–11, 1991), where the NOAEC and LOAEC were 5 and 10%, respectively. The Instream Waste Concentration (IWC) was 5%, which is equal to the NOAEC, so no impairment in the Ohio River was anticipated. The LOAEC of 10% effluent obtained in the December 1991 test was the closest where the effluent could have fallen below NPDES permit regulations.

On three occasions, dosed effluent was held or aged for 7–35 days, and tested again for toxicity. The NOAEC for *Ceriodaphnia* (reproduction) exposed to these effluents was always 100%, indicating that the toxicity of the biocide declined with age. Initially, range-finding acute toxicity tests were conducted with *C. dubia* and *D. magna* on Ohio River water and plant effluent, but no 48-hour LC50's were ascertained. Thereafter, *C. dubia* chronic, seven-day testing was implemented.

CHIRONOMUS SEDIMENT TOXICITY

Midge chronic toxicity to Ohio River sediments was evaluated in the laboratory twice during 1990 and six times in 1991 (Table 11.2). In preliminary studies during 1990, sediments collected downstream of the effluent were not toxic relative to those collected at the intake site, as mean midge weights were nominally higher in the downstream sediment treatments after dosing.

Significant effects to midges were observed in 1991 (Table 11.2). One day prior to biocide dosing on August 21, 1991, midge weights at all sampling sites were not significantly different from each other. Sediments collected after the 24-hour plant-dosing period had ended on August 23, 1991, resulted in midge weight impairment at the first two sites (P5 and 2B) below the effluent. Approximately one month after biocide dosing, sediment was tested again on September 26, 1991, which resulted in significantly impaired midges at all three sites below the outfall. Sediments tested again ~six weeks after the August dosing had no significant effects on midge weights relative to sediments collected upstream of the effluent. After the second biocide dosing ended on December 11, 1991, and on January 13, 1992 (postdose), sediments were tested again and had no significant impairment to midge growth.

TABLE 11.1
No Observable Adverse Effects Concentration (NOAEC, % Effluent) and Lowest Observable Adverse Effects Concentration (LOAEC, % Effluent) of BVPS Effluent for *C. dubia* and *P. promelas* Prior to (Baseline) and during (Dosed) Molluscacide/Bentonite Clay Exposures

Date/Effluent Condition	*Ceriodaphnia* Reproduction		*Pimephales* Survival/Growth
	NOAEC (%)	LOAEC (%)	NOAEC (%)
5/5–12/90 baseline	100	>100	100
5/18–25/90 baseline	40	100	100
7/14–21/90 baseline	40	100	100
12/14–21/90 baseline	40	100	100
Mean (std. dev.) of baseline tests	55(30)	100	100
6/22–29/90 dosed	20	40	100
9/12–19/90 dosed	<50	50	—
11/21–28/90 dosed	40	100	100
8/22–29/91 dosed	20	40	100
12/10–17/91 dosed	5[a]	10	100
Mean (std. dev.) of dosed tests	27(15)	48(33)[b]	100

[a] In-stream Waste Concentration (IWC) = 5% for effluent into Ohio River.
[b] Mean is significantly different ($p = 0.0165$) from that for baseline tests.

CHIRONOMUS TESTING IN EXPERIMENTAL STREAMS

In fall 1990, chronic tests were conducted with midges contained in bioboxes housed in the experimental streams that received discharge from the effluent-influenced river water, which was continuously pumped into the streams (Table 11.3). In the November 20–30, 1990 test, midge

TABLE 11.2
Mean Weights of the Midge, *C. riparius*, After 10-day Chronic Laboratory Tests with Sediments Collected from In Situ Bioboxes at Various Sampling Stations before (Pre-Dose), during (First and Second Doses), and after (Post-Dose) Dosing with Molluscacide/Bentonite Clay

River Station	1990—Midge Weights (mg)	
	Pre-Dose	Post-Dose
Intake	0.991	0.847
P5	1.016	0.905
2B	0.920	0.930

River Station	1991—Midge Weights (mg)					
	Pre-Dose 8/21/91	1st Dose 8/23/91	Post-Dose 9/26/91	Pre-Dose 11/5/91	2nd Dose 12/11/91	Post-Dose 1/13/92
Intake	0.414	0.579	0.712	0.872	1.445	1.222
P5	0.452	0.278[a]	0.522[a]	0.889	1.572	1.330
2B	0.382	0.322[a]	0.571[a]	0.883	1.519	1.427
P10	0.429	0.418	0.406[a]	0.827	1.451	1.317

[a] Significantly different from control at $\alpha = 0.05$.

TABLE 11.3
Mortality and Growth of the Midge (*C. riparius*) after a 10- and 13-day Exposure in Experimental Stream Sediments at the BVPS Laboratory

Stream	% Mortality	N	Mean Dry Weight	Temperature (°C)
		November 20–30, 1990		
Control	25	2	0.1515 (0.0146)	7.5–11.5
5%	30	3	0.1377 (0.0226)	7.9–11.5
50%	0	3	0.2694 (0.0508)	11.1–17.0
100%	0	2	0.3365 (0.0625)	15.0–22.0
		November 30–December 13, 1990		
Control	67	3	0.1026 (0.0216)	5.0–12.0
5%	57	3	0.0900 (0.0113)	5.0–12.0
50%	53	3	0.2639 (0.1302)	9.0–12.0
100%	33	3	0.3069 (0.0184)[a]	12.0–14.0

Tests were run November 20–30, 1990 and November 30 to December 13, 1990.

[a] Significantly greater than control at $\alpha = 0.05$.

weights increased as stream effluent increased, and were highest in the 100% effluent stream. Although there was 25% mortality in the control, no mortality was recorded in either of the two highest effluent streams. Experimental stream water temperatures were lowest (5.0–12.0°C) in the control stream and increased accordingly to 13.5–22.0°C in the 100% effluent stream. The lower, falling temperatures in the fall likely impaired the survivorship capability of the test organisms in the streams receiving no to minimal heated effluent.

In the second test conducted just after the first midge test ended, midge weights increased as the effluent concentration increased and were significantly higher in the 100% effluent stream than in the others (Table 11.3). However, mortality was extremely high in the control (67%), but was markedly lower in the other treatments, as values ranged from 57% in the lowest effluent stream, to 33% in highest effluent streams. Again, temperature likely affected the survivorship and growth of midges.

Laboratory Experimental Stream Toxicity with Snails, Mayflies, Fish, and Clams

Three test species were evaluated on site in laboratory experimental streams for 40 days, during three 24-hour biocide dosings on June 22, 1990, November 21, 1990, and August 21, 1991 (Table 11.4). In the June 22–23, 1990 biocide dosing, the snail, *Physa* sp., was not sensitive to the effluent, with only 4.0–7.3% mortality in the two treatments, including the control. For the mayfly (*I. bicolor*), emergence was minimal for all treatments, but control mortality was higher than in the June test. After seven days of exposure, control mortality was 20%, but stabilized throughout the rest of the exposure duration and only increased to 30%. Mayfly emergence was not sensitive to the effluent, as mean values ranged from 60 to 66.7% in the treatments.

In the fall 1990 tests, another snail (*Goniobasis* sp.) had extremely low mortality (0–3.3%) throughout the test in all experimental stream conditions (Table 11.4). For the mayfly, *I. bicolor* emergence was low in all treatments; however, control mortality was substantially higher than in the June test. From days 2 to 15 of the test, mayfly control mortality ranged from 0 to 10% in the control stream to 0–40% in the 100% effluent streams. After that, mayfly mortality increased in all streams. At the end of the test, mayfly mortality had a dose independent pattern, as 53.3% died in the control and 23.3–56.7% were dead in the 50–100% effluent streams. In August 1991,

TABLE 11.4
Percent Mortality of Three Test Species in BVPS Laboratory Experimental Streams after Exposure to June and November Plant Megadosing of a Biocide in 1990 and 1991—*Isonychia* Data Are Presented as Percent Mortality and Emergence

Plant Dosed on 6/22/90 Treatment (% Effluent)	*Isonychia*		*Physa* % Mortality
	% Mortality	% Emergence	
Control	30.0	60.0	4.0
5	33.3	66.7	7.3
50	36.7	60.0	4.7

Plant Dosed on 11/21/90 Treatment (% Effluent)	*Isonychia*		*Goniobasis* % Mortality
	% Mortality	% Emergence	
Control	53.3	0	0
5	53.3	3.3	0
50	23.3	0	0
100	56.7	0	3.3

Plant Dosed on 8/21/91 Treatment (% Effluent)			*Goniobasis* % Mortality
Control	—	—	3.3
5	—	—	0
50	—	—	6.7
100	—	—	0

Goniobasis sp. was tested again for 40 days. Results were similar to those in November 1990, in that mortality was extremely low (0–6.7%) throughout the test at all stream effluent concentrations.

Bluegill sunfish mortality in August 1991 was rather low in experimental streams from days 1 to 30, but by day 40, control mortality remained low (3.3%), while in the two highest effluent concentrations, mortality increased 36.7–40.0% (Table 11.5). Length and dry weight gain was also nominally lower in these upper concentrations, relative to controls, but not significantly so. In the December 1991 experimental stream tests, the same species had minimal mortality (0–3.3%) for the duration of the test. No significant differences were observed in fish length for all effluent concentrations, but fish weight was significantly enhanced in the 100% effluent streams.

In 1991, two laboratory 40-day experimental stream tests were conducted with the Asian clam, resulting in minimal (0–3.3%) mortality (Table 11.6). In the summer study, clam growth was highest in the control and 5% effluent streams, significantly lower in the 50% effluent stream, and low (but not significantly) in the 100% stream. Water temperature ranged from 18.3 to 26.0°C in all streams. In the fall study, however, clam growth was significantly lowest in the control and 5% effluent streams, and then was significantly enhanced in the 50% effluent streams, and more so in the 100% streams. It was presumed that increased effluent temperature enhanced clam growth as river water ranged from a low of 5.8 to 11.1°C in the control and 5% effluent streams, to much higher in the 50% (11.0–17.8°C) and 100% (15.3–23.2°C) treated streams.

CORBICULA GROWTH IN SITU

In 1991, clam growth studies were carried out twice in situ, once in the summer during the first biocide dosing and again in latter fall during the second dosing (Table 11.6). In the summer study,

TABLE 11.5
Mortality and Growth of Bluegill Sunfish (*L. macrochirus*) in BVPS Experimental Streams for 40 Days after an Initial Exposure to Effluent from the Biocide Dosing in August and December, 1991—Fish Were Measured for Increase in Growth by Fork Length and Dry Weight (SE in Parentheses)

Effluent Concentration in Streams (%)	n	August 1991 % Mortality	Length Gain (mm)	Dry Weight (mg)
Control	39	3.3	30.37 (1.55)	0.111 (0.023)
5	30	0.0	30.21 (1.36)	0.108 (0.019)
50	18	40.0[a]	29.50 (4.13)	0.095 (0.024)
100	19	36.7[a]	28.31 (3.75)	0.083 (0.018)

Effluent Concentration in Streams (%)	n	December 1991 % Mortality	Length Gain (mm)	Dry Weight (mg)
Control	29	3.3	30.68 (0.43)	0.077 (0.004)
5	30	0.0	31.27 (0.31)	0.078 (0.003)
50	30	0.0	30.78 (0.30)	0.077 (0.003)
100	29	0.0	31.53 (0.41)	0.091 (0.004)[a]

[a] Significantly different ($p < 0.05$) from control.

clam growth was highest at the reference (intake) site, and was significantly lower in the first two sites (P5 and 2B) below the outfall but not at the farthest downstream site (P10). Clam mortality was minimal (0–3.3%) during both summer and fall studies.

In the fall dosing effort, growth was highest at the reference site, significantly lowest at the first outfall site (P5), and was still significantly reduced at the next two downstream sites (Table 11.6). In the river, the thermal influence from outfall 001 skimmed along the upper water column and had no vertical mixing influence 8 m below the surface at the three downriver sites, where clam bioboxes resided on the river sediment. The thermal enrichment consequence observed in the laboratory experimental stream tests was not a factor in the in situ bioboxes. Consequently, the laboratory experimental stream clam tests provided a false-positive result, while the in situ tests were more reliable about the actual conditions existing in the riverine receiving system.

11.3.6 BENTHIC MACROINVERTEBRATE MONITORING

Benthic macroinvertebrate samples were collected before the first biocide dosing on June 19, 1990, and then twice after the dosing was completed (June 25 and July 31) and then three times again on November 19, 1990–January 14, 1991 (Figure 11.2). The mean and total abundance of taxa, taxon diversity, and taxon richness indicated minimal differences found between the reference and outfall influenced sampling sites. The mean number of taxa was 13.3–13.7 on June 19, 1990, at all three sampling sites and declined to 9.7–7.0 at sites P5 and 2B on June 21, 1990. After that, taxon numbers increased to 12.0 and 10.0 at P5 and 2B, and higher by November 25, 1990, where values ranged from 16.0, 16.7, to 15.3 at Sites 1, P5, and 2B. Taxon diversity and richness had the same trends as number of taxa.

In 1991, benthic macroinvertebrate river samplings were conducted three times around the August megadosing effort and three times again at the December dosing (Figure 11.3). In the summer samplings, taxa numbers varied from 19 to 24 at Site 1 for samplings on 8/19, 8/26 and 9/30/91. The number of taxa increased at Site P5 (21–45) and remained about the same (15–25) at

TABLE 11.6
Growth of *Corbicula* Held in Bioboxes for 40 Days on the Bottom of the Ohio River and in On-Site Laboratory Experimental Streams during the Summer (8/15-9/30/91) and Fall (11/28/91-1/14/92) Studies ($n = 30$)

Parameter	% Dead	Mean Growth (mm ± 56)	Duncan's Multiple Range Test	Temperature Range (°C)
% Effluent		Laboratory Experimental Streams—Summer Study		
0	0	1.81 (0.21)	A	18.3–26.0
5	0	1.89 (0.16)	A	18.3–26.1
50	0	1.37 (0.21)	B	18.4–27.2
100	0	1.56 (0.24)	A	16.5–28.9
Station		In Ohio River—Summer Study		
Intake	0	0.87 (0.52)	A	18.3–26.0
P5	3.3	0.27 (0.22)	B	18.3–26.0
2B	3.3	0.16 (0.14)	B	18.3–26.0
P10	0	0.75 (0.42)	A	18.3–26.0
% Effluent		Laboratory Experimental Streams—Fall Study		
0	0	0.632 (0.028)	A	5.8–11.1
5	0	0.685 (0.020)	A	5.9–10.6
50	0	0.948 (0.025)	B	11.0–17.8
100	3.3	1.320 (0.032)	C	15.3–23.2
Station		In Ohio River—Fall Study		
Intake	0	0.340 (0.026)	A	5.8–11.1 at all sites because the outfall thermal influence stayed along the surface, not ~8 m below at the river bottom
P5	0	0.184 (0.021)	C	
2B	0	0.219 (0.021)	B C	
P10	0	0.276 (0.039)	B	

Site 2B. Diversity index values were actually higher at Sites P5 and 2B than at Site 1. In the fall sampling effort of 1991, the number of taxa was substantially enhanced at downstream Sites P5 and 2B, over Site 1. The number of taxa either tripled or quadrupled at Sites P5 and 2B, relative to Site 1. The Shannon–Weiner Index also substantiated these trends.

EFFICACY OF BIOCIDE

The biocide dosing lasted over a 24-hour period and clams were held in bioboxes in the cooling tower during the dosing and monitored for mortality several days thereafter. In the fall 1990 dosing effort, the plant heat exchanger system was dosed to a level up to 32.5 mg/L CT-1 and all clams died within seven days thereafter. Biocide dosing was also conducted in the cooling tower basin where the maximum concentration reached 13.0 mg/L. Juvenile clams took eight days to reach 100% mortality, while 95% of the adults died after the 18th day.

In summer (August) 1991, the Cooling Tower No. 2 basin was dosed to a maximum concentration of 13.0 mg/L and 100% of the juvenile and adult clams died four and five days later. In the fall (December) 1991 dosing, the maximum biocide concentration was 12.3 mg/L. Efficacy of the biocide was less efficient, as 87% of the juveniles died by day nine and only 43% of the adults were killed. Additional surveillance continued for 25 days after dosing but no additional juvenile or adult clams died.

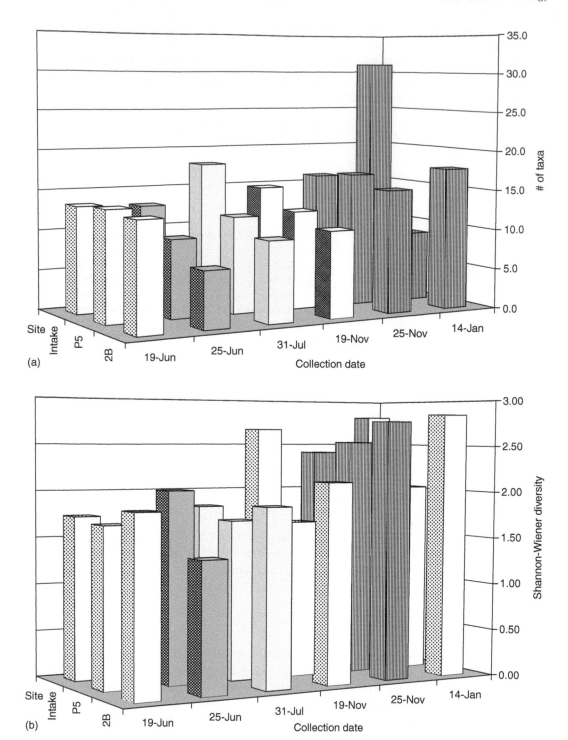

FIGURE 11.2 (a) Taxonomic richness and (b) Shannon-Wiener diversity indices for three sampling sites on six sampling dates in 1990.

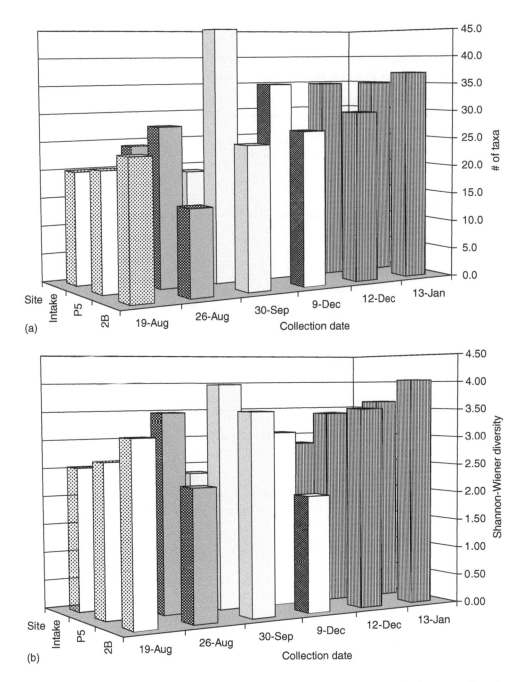

FIGURE 11.3 (a) Taxonomic richness and (b) Shannon-Wiener diversity indices for three sampling sites on six sampling dates in 1991.

IN SITU 162-DAY ASIAN CLAM TEST IN 1992

At the start of the 162-day in situ study, mean clam size was smallest at the intake site (14.13 mm) followed by those at P5 (14.50 mm) while clams at 2B and P10 were significantly larger (Table 11.7). The growth increment 16 days after acclimation in the river, prior to dosing, resulted in growth being lowest at 2B, where the clams initially were the largest. Thirty days after the first

TABLE 11.7
Mean Width of *Corbic*ula Shells Held in the "Dosed" Group (Clams Located at River Station throughout Each 24-Hour Biocide Dosing with CT-1:DT-1)

Station	Mean Clam Size (mm)	Significant Differences	Growth Incre- ment (mm)	Significant Differences[a]	Accumulative Growth Increments (mm)[a]	
			Day 0 (6/6/92)			
Int	14.13	B	—		—	
P5	14.50	B	—		—	
2B	15.88	A	—		—	
P10	15.35	A	—		—	
		16 Days after River Acclimation Prior to Dosing (6/22/92)				
Int	15.11	B	0.98	A	0.98	A
P5	15.51	B	1.00	A	1.00	A
2B	16.73	A	0.85	B	0.85	B
P10	16.32	A	0.96	A	0.98	A
		30 Days after First Dosing (7/23/92)				
Int	17.05	C	1.93	A	2.93	A
P5	17.43	B C	1.93	A	2.93	A
2B	18.40	A	1.67	B	2.52	B
P10	17.85	B	1.54	B	2.53	B
		58 Days after First Dosing (8/20/92)				
Int	18.09	B	1.04	A	3.97	A
P5	18.26	B	0.83	B	3.77	A
2B	19.17	A	0.77	B	3.29	B
P10	18.75	A	0.89	B	3.43	B
		104 Days after First Dosing (10/5/92)				
Int	19.65	B	1.74	A	5.70	A
P5	19.73	B	1.55	A B	5.12	B C
2B	20.53	A	1.37	B	4.66	C
P10	20.42	A	1.68	A	5.12	B C
		122 Days after First Dosing, 17 Days after Second Dosing (10/23/92)				
Int	19.83	B	0.17	A	5.88	A
P5	19.86	B	0.13	A B	5.45	A B
2B	20.64	A	0.11	B	4.77	C
P10	20.56	A	0.14	A B	5.25	B C
		162 Days after First Dosing, 57 Days after Second Dosing, 35 Days after Third Dosing, (12/2/92)				
Int	19.91	B	0.08	C	5.96	A
P5	19.98	B	0.12	A	5.56	A B
2B	20.71	A	0.07	C	4.84	C
P10	20.66	A	0.10	A B	5.36	B C

[a] Data with the same letters are not significantly different from each other.

dosing, clam growth was the highest at the intake and first site (P5) below the plant. After 58 days from the first biocide dosing, clam growth was significantly highest at the intake. After 104 days since the first dosing, the accumulative growth increment (5.70 mm) was significantly highest at the intake. Following 122 days after the first dosing and 17 days from the second one, the clam accumulative growth was highest at the intake, but not significantly higher than those at P5. At the end of the study, 162 days since the first dosing, to 57 and 35 days after the second and third dosings, clam accumulative growth was significantly higher at the intake and P5 sites than at the two sites further downriver.

TABLE 11.8
Mean Width of *Corbi*cula Shells Held in the "Nondosed" Group (Clams Located at a Refugium during Each 24-Hour Biocide Dosing with CT-1:DT-1)

Station	Mean Clam Size (mm)	Significant Differences	Growth Increment (mm)	Significant Differences[a]	Accumulative Growth Increments (mm)[a]	
			Day 0 (6/6/92)			
Int	14.37	B	—		—	
P5	15.07	A	—		—	
2B	14.99	A	—		—	
P10	14.83	A	—		—	
		16 Days after River Acclimation Prior to Dosing (6/22/92)				
Int	15.61	B	1.24	A	1.24	A
P5	16.23	A	1.24	A	1.24	A
2B	16.21	A	1.23	B	1.23	A
P10	15.99	A B	1.16	A	1.16	A
		30 Days after First Dosing (7/23/92)				
Int	17.32	B	1.72	A	2.99	A
P5	17.77	A	1.54	B	2.71	B
2B	17.88	A	1.67	A B	2.90	A B
P10	17.60	A B	1.61	A B	2.79	B
		58 Days after First Dosing (8/20/92)				
Int	18.24	B	0.92	A	3.92	A
P5	18.61	A	0.84	A	3.55	B
2B	18.83	A	0.94	A	3.84	A B
P10	18.50	A B	0.90	A	3.65	B
		104 Days after First Dosing (10/5/92)				
Int	20.33	A	2.12	A	6.04	A
P5	20.20	A	1.59	B	5.17	B
2B	20.33	A	1.50	B	5.34	B
P10	20.17	A	1.68	B	5.33	B
		122 Days after First Dosing, 17 Days after Second Dosing (10/23/92)				
Int	20.49	A	0.16	A	6.20	A
P5	20.36	A	0.16	A	5.33	B
2B	20.47	A	0.14	A	5.48	B
P10	20.32	A	0.15	A	5.48	B
		162 Days after First Dosing, 57 Days after Second Dosing, 35 Days after Third Dosing (12/2/92)				
Int	20.56	A	0.07	A	6.27	A
P5	20.49	A	0.06	A	5.40	B
2B	20.54	A	0.07	A	5.55	B
P10	20.38	A	0.06	A	5.54	B

[a] Data with the same letters are not significantly different from each other.

Upon analyzing "non-dosed" clam groups to evaluate effluent influence without biocidal treatment, clams were significantly smaller at the intake during Day 0 (Table 11.8). After 16 days of acclimation in the river, accumulative clam growth was not significantly different between sites. It was not until 104 days into the study where clam growth increment (2.12 mm) and accumulative growth (6.04 mm) were significantly highest at the intake. This pattern continued through the duration of the study where accumulative clam growth (6.27 mm) was significantly highest at the intake. Clam growth declined substantially at the last (December 2, 1992) and next to last (October 23, 1992) measurements, as growth ranged from 0.06 to 0.07 mm and from 0.14 to 0.16 mm, respectively. The clam growth increment was substantially higher (1.50–2.12 mm) during the October 5, 1992 measurement, before river temperatures declined rapidly into the fall season. In essence, it only required 30 days after the first plant dosing for clam growth to become significantly lower at the first downstream site (P5) relative to the reference.

SAMPLING/TOXICITY TESTING EFFICIENCY

The overall results of 17 testing efforts from acute to chronic exposure in the laboratory, to within on-site laboratory experimental streams, and in-river (in situ) testing are presented in Figure 11.4. The most sensitive testing parameters appeared to be laboratory *Ceriodaphnia*, seven-day chronic reproduction toxicity testing, and in situ chronic (30–40 days) clam growth studies, followed by chronic *Chironomus* growth studies from river biobox sediments tested in the laboratory, and bluegill mortality/growth in experimental streams during the summer. The most insensitive tests were in situ benthic macroinvertebrate sampling and several on-site laboratory experimental stream tests (i.e., *Physa, Goniobasis* survival, and bluegill sunfish survival/growth in the fall), followed by the fathead minnow effluent chronic and acute (*C. dubia, D. magna*) toxicity tests conducted in the laboratory. Those that were questionable due to high control mortality were the *I. bicolor* and *C. riparius* tests carried out in the experimental streams, although midge growth was highly robust for the survivors in 100% effluent. Bluegill survival/growth in the fall experimental stream study indicated a robust fish response to increasing effluent concentrations, compared to the reverse in the summer. Clam survivorship in experimental streams and in situ bioboxes also were insensitive parameters, while clam growth in the fall stream study suggested a false-positive response due to elevated effluent temperature influences in the 50% and 100% effluent-treated streams.

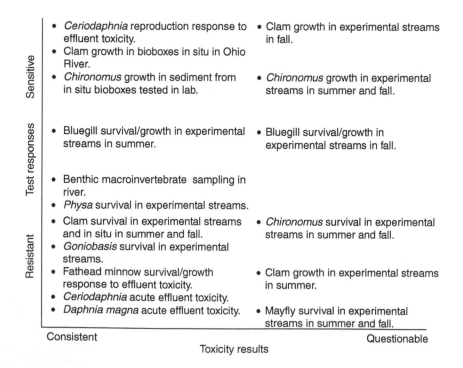

FIGURE 11.4 Summary of testing thresholds and their reliability during this research project. Formal laboratory tests were acute and chronic effluent toxicity carried out at Virginia Tech while experimental stream testing was done at the laboratory using flow-through effluent from the Power Plant.

DISCUSSION

The USEPA test organisms had mixed responses in laboratory tests with BVPS effluents containing the biocide and detoxifying agent. As expected, *Ceriodaphnia* was more sensitive to the plant effluent than fathead minnows were, with substantially lower LOAECs after biocide dosing, and on one occasion, the LOAEC of 10% nearly reached the critical IWC of 5%. Fathead minnows had NOAECs of 100% after all tests, indicating their lack of sensitivity to this particular effluent. Similarly, when *Ceriodaphnia* and fathead minnows were tested for sensitivity to copper, along with 17 and 15 other genera in acute and chronic exposures, respectively, the fathead minnow was relatively tolerant, ranking 14th in genus mean acute/chronic values in both cases (Farris, Cherry, and Neves 1991; Cherry et al. 2002). In fact, the genus mean acute value for *P. promelas* was 4-fold higher than that for *C. dubia*, and nearly 10-fold higher than the most sensitive unionid mussel species (*Lampsilis fasciola*) tested (Cherry et al. 2002).

Midge responses to the BVPS effluent were highly variable among the different dosing applications. In laboratory tests, no significant effects were observed in 1990, but sediments collected both directly after dosing in August 1991, and approximately one month later, indicated significant impairment up to ~1,000 m downstream of the outfall. The same trend, however, was not observed later in the same year. When *C. riparius* sediment tests were carried out in the laboratory experimental streams, control survivorship was a problem, reaching 25–67% in two tests run in the fall. However, river water temperature played a key role in these experimental stream tests, as temperature increased in the 100% effluent to 12.0–14.0°C in the second set of tests, when the effluent temperature controller malfunctioned. In the control and lower effluent streams, temperature remained much lower at 5.0–12.0°C. Apparently, thermal enrichment encouraged midge growth in the 100% effluent streams. While it is possible that warmer temperatures in the higher effluent concentrations of the experimental streams negated any toxic effects from the sediment, the laboratory tests conducted in 1990 also indicated an absence of effluent toxicity.

Isonychia responded poorly in two experimental stream tests housed in the laboratory located at the plant. Control mortality was high (30–53.3%) in both tests and mayfly emergence was high in the spring test; therefore, the results of toxicity testing with this organism are inconclusive.

Other bioassessment studies have been conducted using *I. bicolor*. Peters, Cherry, and Cairns (1985) evaluated responses of this mayfly to alkaline pH exposures in field and laboratory experimental streams, and found 96-hour LC50s ranged from 9.54 to 10.37, respectively. They concluded that the differences in LC50 values were due to variations in water quality and temperature in the field-located experimental streams; however, control survivorship was above 90% in both tests. In 1989–1990, we exposed *I. bicolor* to copper in rigid baskets in outdoor experimental streams at the Clinch River Plant, Carbo, Virginia. We observed minimal control mortality over 96 hours, and the mayflies ranked fourth in sensitivity out of the 17 species tested (Cherry et al. 2002). *Isonychia* also were exposed to copper in laboratory experimental streams for 20 days, and had high control survivorship (≥80%) and ranked third in sensitivity of 15 species tested (Farris, Cherry, and Neves 1991). In recent laboratory chronic tests with simulated coal processing effluent, Kennedy et al (in press) had >90% control survivorship over seven days and the mayfly survival LOAEC first occurred at 1,562 µS/cm, which was much lower than that for *C. dubia* (3,730 µS/cm). Despite their apparent sensitivity to various toxicants, the poor control survival in the present study may have been due to rigors of transport from Virginia to Pennsylvania and not enough time for acclimation to testing conditions in the Ohio River.

The two snail species tested were insensitive to the "detoxified" biocide in experimental stream exposures, with a maximum of only ~7% mortality. In a previous study comparing responses of various organisms to copper, two snails (both prosobranchs) ranked 9th and 13th in sensitivity out of 17 (Cherry et al. 2002). The rankings placed them behind *Ceriodaphnia, Isonychia,* and several freshwater mussel genera, but ahead of fathead minnows and bluegill sunfish, relative to sensitivity.

In the present study, we used one pulmonate and one prosobranch snail, but there did not appear to be a difference in sensitivity based upon their subclass.

The bluegill sunfish varied in sensitivity to the dosed effluent in experimental stream exposures. During the August 1991 dosing, 36–40% mortality was observed in the 50 and 100% effluent treatments, but in the early December test, fish survivorship was high throughout all test concentrations. Growth of the fish, in terms of both fork length and dry weight, was not influenced by exposure to the dosed effluent, except that in the December exposure, dry weight was significantly higher in the 100% effluent treatment than in the others. This may have been due to the slightly elevated temperatures in the 100% stream, when the laboratory cooling system could not negate the thermal enrichment of the effluent. Bluegills are rather tolerant test organisms, in general, to environmental stressors. For example, they were extremely resistant to acute copper exposure in experimental streams with a genus mean acute value (GMAV) = 4,030 ug/L, which was more than two orders of magnitude above the most sensitive mussel species' GMAVs, and one order of magnitude more tolerant than *Corbicula* (Cherry et al. 2002). In contrast, the high bluegill mortality in the August 1991 study was more sensitive to the dosed effluent in 50–100% effluent streams than that of *Corbicula*, which had 0% mortality. However, clam growth was most adversely influenced by the effluent in the in situ biobox studies, where effluent temperature was not a factor.

The benthic macroinvertebrate assemblages in this large, deep river system were comprised largely of collector-gatherers, as would be predicted by the River Continuum Concept (Vannote et al. 1980). The two most abundant taxonomic groups were the oligochaetes and chironomids. Chironomid midges often comprise a large percentage of benthic macroinvertebrate communities in this type of system, and the many genera of this family have a wide range of sensitivities to environmental contaminants, especially metals (Clements 1991; Swansburg et al. 2002). Therefore, in addition to overall diversity and taxonomic richness, chironomid richness may serve as an indicator of environmental contamination. In this study, total taxonomic richness appeared to decrease downstream of the dosed effluent in June of 1990, but beyond that no clear trend was observed. Shannon–Wiener diversity indices did not indicate in-stream impairment in either 1990 or 1991, and no clear trend was observed for chironomid richness (data not presented). A number of oil spills have occurred over the years, and in sampling the sediment by ponar dredge, oil deposits were common. Furthermore, sampling by ponar dredge made it difficult to observe impairment due to spatial heterogeneity of habitat within each site. However, these confounding variables can be eliminated and in-stream impacts of the dosed effluent can be assessed by use of in situ toxicity testing with caged organisms, such as bivalves (Cherry 1996; Hull 2002; Soucek, Schmidt, and Cherry 2001; Hull, Cherry, and Merricks 2004; Hull, Cherry, and Neves 2006).

BIVALVES AS INDICATOR SPECIES

Mollusks have long been used as biological indicators, especially for trace elements. Studies began by Merlini (1966) and Merlini, Cadano, and Oregioni (1978) in the 1960s and expanded thereafter into the 1970s. Bivalves are good bioindicators because they accumulate contaminants from three available sources: (1) from solution, (2) from sorbed to fine particulate organic matter, and (3) from ingestion of inorganic materials through pedal feeding (Moore 1971). Furthermore, they efficiently metabolize contaminants so they often are a better indicator of the presence of pollution than other organisms. Phillips (1977) developed a list of 10 criteria necessary for an organism to be an effective monitor of metal contamination and specifically worked with mussels in marine and estuarine environments (Phillips 1976a, 1976b; 1979). These criteria included: (1) the organism accumulate pollutants without suffering mortality, (2) the organism must be sedentary, (3) the organism's life span must be sufficient to allow for sampling of more than one year class, (4) the organism must be abundant in the study region,(5) the size of the organism must be adequate to allow tissue samples for contaminant analysis, (6) the organism must be easy to collect and hardy enough to survive in the laboratory, (7) the organism must tolerate brackish water, (8) high metal concentration factor should be exhibited by the organism, (9) correlation should exist between metal content of organisms and average metal concentration in the surrounding water, and (10) all organisms in a survey from all locations studied should exhibit the same correlation between the metal contents and those of the surrounding water under all conditions. Not all of these criteria are relevant to the present study, but Asian clams certainly have been found to fit each of them by other researchers (see Chapter 2). An important additional consideration is the fact that Asian clams have a number of sublethal responses to environmental contaminants. These will be described below.

The Asian clam was first discovered in the US in 1935 in the Columbia River, Washington (Gregg 1947). It became well established in many western states by the late 1940s (Ingram 1959) and invaded the Mississippi River and Ohio River drainages in the 1950s (Bickel 1966). This rapidly reproducing bivalve was initially studied as a biofouling pest, first in agricultural irrigation ditches (Prokopovitch and Herbert 1965), and then in condensor and service water systems of power plants and other industrial facilities (Sinclair and Isom 1963; Sinclair 1971; McMahon 1977; Eng II 1979; Mattice 1979; Morton 1979; Cherry et al. 1980). Because of their high reproductive capacity and relative resistance to some biocidal measures, clams can clog condenser tubes and industrial water intake systems, even to the point of reducing power plant efficiency (Goss and Cain 1977). By 1976, *Corbicula* had invaded the New River at Glen Lyn, Virginia, in outdoor experimental stream studies, so our research with these clams began then (Rodgers et al. 1977a, 1977b). By 1981, *Corbicula* had invaded an industrial facility several river miles upstream and found river sediment to be highly compatible for clam proliferation (Cherry et al. 1986).

Later, researchers used *Corbicula* as a biological indicator of environmental stress, initially in experimental stream bioassays (Rodgers et al. 1980; Graney et al. 1980; Graney, Cherry, and Cairns 1983; Graney, Cherry, and Cairns 1984). In addition to mortality, growth, and tissue contaminant concentrations, numerous endpoints have been used to monitor environmental contamination with *Corbicula*, including valve closure (Doherty, Cherry, and Cairns 1987; Allen et al. 1996), metallothionein induction (Doherty, Failla, and Cherry 1988), tissue condition and glycogen levels (Naimo et al. 1998; Newton et al. 2001), DNA strand breakage (Black 1997), and reduced cellulolytic activity (Farris et al. 1988, 1989). For a comprehensive review of the responses of Asian clams to various contaminants see Doherty and Cherry (1988). A recent review of methods for conducting toxicity tests with *Corbicula* as a surrogate species to native unionids has been developed by Ingersoll et al. (Chapter 5). Although primarily reviewing the unionid literature, they provided advantages and disadvantages of using toxicity tests with *Corbicula* versus unionids. One recent study indicated that young adult (10–12 mm) *Corbicula* were more sensitive to the Clinch River Plant effluent than juvenile (two months old) *Villosa iris* in both in situ biobox and laboratory experimental stream studies (Hull 2002). In the studies and review by Salazar and Salazar (see Chapter 9), they recommend a need to better characterize and understand the influence of chemical input with biological processes by using several directives in improving freshwater mussel ecotoxicology, especially the use of critical body residues of bivalves from toxicants.

In the Cherry Lab at Virginia Tech, experimental stream studies continued with evaluation of *Corbicula* responses to sediment-bound contaminants (Belanger et al. 1985), followed by responses to halogen control measures (Doherty et al. 1986), and then with benthic macroinvertebrates (Clements, Cherry, and Cairns 1990). Then both laboratory and field experimental stream studies were conducted evaluating clam growth responses to zinc (Belanger et al. 1986; Belanger et al. 1990) and cellulolytic activity reductions caused by power plant effluents in the Clinch River (Farris et al. 1988, 1989; Farris, Cherry, and Cairns 1994). Eventually, we used *Corbicula* as an in situ test organism in different types of bioboxes that were placed below various industrial effluent releases in aquatic receiving systems. Beginning in 1973 at the Savannah River Project near Aiken, South Carolina we (Guthrie and Cherry) used stainless steel mesh bioboxes to study the effects of fly ash effluent released into a stream/swamp receiving system with an array of test organisms (e.g., grass shrimp, freshwater clams, salamander, crayfish, insects, and fish) (Guthrie, Cherry, and Rodgers 1974; Guthrie and Cherry 1976).

Our first in situ enclosures with *Corbicula* were plastic mesh bags, each containing several individually marked clams that were secured to the stream substrate using rebar. In the present study, clams were placed in plastic milk bottle cases loaded with glass beads as ballast (Cherry et al. 1991, 1992, 1993). Hull et al. (2002) recently compared growth of *Corbicula* enclosed in mesh bags with that of clams placed in more structurally rigid bioboxes. Both enclosure types were deployed in the Clinch River power plant effluent dissipation zone, and clams grew significantly larger and had a lower within-experimental unit coefficient of variation in the rigid bioboxes. Other recent in situ studies with Asian clams by our laboratory include evaluating acute/chronic toxicity of high levels of dissolved solids from an active coal mining effluent (Kennedy, Cherry, and Currie 2003), and assessment of various acid mine drainage impacted sites in sub-watersheds of the Powell River, Virginia (Soucek, Schmidt, and Cherry 2001; Cherry et al. 2002; Schmidt, Soucek, and Cherry 2002).

In the present study, we investigated the in-stream impacts of a detoxified biocide used to control an exotic bivalve. Using caged Asian clams, we found that the bentonite clay needed for detoxification of the biocide was not always effective, particularly under low flow conditions. Variable flow conditions are the likely explanation for the difference in clam responses from year to year, because in August of 1991, Ohio River levels were extremely low, resulting in reduced dilution of the detoxified biocide. In 1992, flow conditions were higher throughout the year, and accordingly, no clear substantial effects were observed on in situ clam growth after 162 days. Despite the fact that the effluent studied contained a toxicant designed to control Asian clams, the combination of the biocide and bentonite clay had effects on other nontarget organisms including daphnids, midges, and bluegill sunfish. All three of these organisms were significantly impacted in formal laboratory bioassays or experimental stream exposures during the August 1991 dosing period, coincident with the observation of reduced growth of clams in situ. The incorporation of in situ exposures enabled us to determine the presence of occasional in-stream impacts, while the observation of effects in the laboratory exposures strengthened our confidence in the fact that the treated effluent was the cause of observed in situ Asian clam impacts.

The exotic Asian clam has a number of adverse impacts on native aquatic environments. Because of its tendency to pedal-feed, *Corbicula* has been associated with a decrease in abundance of benthic bacteria and flagellates that may serve as food to other organisms (Hakenkamp et al. 1999, 2001). Pedal feeding also allows *Corbicula* to grow at a faster rate than it would by filter feeding alone, and it has been shown to contribute significantly to total community respiration (carbon dioxide production) in some streams. This exotic species is also thought to have had an adverse impact on native freshwater mussel populations in the US, either by reducing levels of phytoplankton available and thereby starving mussels, by depleting benthic food resources available to juvenile mussels, or by producing high levels of toxic ammonia during mass die-offs (Scheller 1997; Strayer 1999). In addition, adult Asian clams (15–25 mm) have been found to readily ingest mussel glochidea in 10-day flow-through laboratory sediment tests and concluded that high density Asian clams can adversely affect survival and growth of newly metamorphosed juvenile mussels (Yeager, Neves, and Cherry 1999). These clam impacts on native fauna, along with its potential to decrease efficiency of numerous industries that use raw water intakes by biofouling (Cherry et al. 1980), make control of this organism necessary. However, we have shown that some control measures may have adverse impacts on nontarget organisms, and the pros and cons of controlling the Asian clam must be weighed carefully.

ACKNOWLEDGMENTS

These studies were funded by Duquesne Light Company, Shippingport, Pennsylvania, 15077, through the coordinated efforts of Mr. J. Wayne McIntire, Director of Safety and Environmental Services. Special thanks are forwarded to Mr. Robert L. Shema, former president of Aquatic Systems Corporation (ASC) in Pittsburgh, Pennsylvania. Through his efforts and those of ASC personnel at that time, (William R. Cody, Gary J. Kenders, Michael F. Davison, Michael R. Noel, and Gregory M. Stybonski), they assisted in maintaining the on-site experimental stream laboratory and developing the database. They also were responsible for making the clam bioboxes and conducting the in situ clam and benthic macroinvertebrate studies. In the third year of study, they basically generated the 162-day clam in situ database, whereas the data were statistically analyzed by Dr. Michael G. Dobbs, formerly of Virginia Tech. Other former graduate students from Virginia Tech assisted in this work. Dr. Jerry Farris was instrumental in developing the experimental stream laboratory on site. Dr. Joseph Bidwell and Dr. Mindy Yeager contributed to various aspects in laboratory toxicity tests and experimental stream monitoring. Dr. Donald S. Cherry coordinated the three-tiered ecotoxicological study, trained all participants in generating the database, and wrote all progress reports with input from Wayne McIntire.

REFERENCES

Adams, S. M., Crumby, W. D., Greeley, M. S., Ryon, M. G., and Schilling, E. M., Relationships between physiological and fish population responses in a contaminated stream, *Environ. Toxicol. Chem.*, 11, 1549–1557, 1992.

Allen, H. J., Waller, W. T., Acevedo, M. F., Morgan, E. L., Dickson, K. L., and Kennedy, J. H., A minimally invasive technique to monitor valve-movement behavior in bivalves, *Environ. Tech.*, 17, 501–507, 1996.

[ASTM] American Society for Testing and Materials, *Standard Guide for Conducting In-Situ Field Bioassays with Marine, Estuarine, and Freshwater Bivalves*, ASTM E 2122-01, West Conshohocken, PA, p. 1546–1575, 2001.

Belanger, S. E., Farris, J. L., Cherry, D. S., and Cairns, J., Sediment preference of the Asiatic freshwater clam, *Corbicula fluminea*, *Nautilus*, 99, 66–73, 1985.

Belanger, S. E., Farris, J. L., Cherry, D. S., and Cairns, J. Jr., Growth of Asiatic clams (*Corbicula* sp.) during and after long-term zinc exposure in field-located and laboratory artificial streams, *Arch. Environ. Contam. Toxicol.*, 15, 427–434, 1986.

Belanger, S. E., Farris, J. L., Cherry, D. S., and Cairns, J. Jr., Validation of *Corbicula* growth as a stress response to copper in artificial and natural streams, *Can. J. Fish. Aquat. Sci.*, 47, 904–914, 1990.

Bickel, D., Ecology of *Corbicula manilensis* in the Ohio River at Louisville, KY, *Sterkiana*, 23, 19–24, 1966.

Bidwell, J. R., Farris, J. L., and Cherry, D. S., Comparative responses of the zebra mussel and Asiatic clam to DEH/QUAT, a nonoxidizing molluscicide, *Aquat. Toxicol.*, 33, 183–200, 1995.

Black, M. C., Biomarker assessment of environmental contamination with freshwater mussels, *J. Shellfish. Res.*, 16, 323, 1997.

Black, M. C. and Belin, J. I., *Environmental Toxicology and Risk Assessment*, Vol. 7, ASTM STP 1333, Fredericksburg, VA, pp. 76–91, 1998.

Burton, G. A., Jr., Realistic assessments of ecotoxicity using traditional and novel approaches, *Aquat. Ecosyst. Health Manage.*, 2, 1–8, 1999.

Cherry, D. S., State of the art in situ testing (transplant experiments) in hazard evaluation, *SETAC News*, 16, 24–25, 1996.

Cherry, D. S., Rodgers J. H. Jr., Graney, R. L., and Cairns J. Jr., *Dynamics and Control of the Asiatic Clam in the New River, Virginia*, Virginia Water Resources Research Center, Virginia Polytechnic Institute & State University, Blacksburg, VA, Bulletin 123, p. 72, 1980.

Cherry, D. S., Roy, R. L., Lechleitner, R. A., Dunhardt, P. A., Peters, G. T., and Cairns J. Jr., *Corbicula* fouling and control measures at the Celco Plant, Virginia, In *Proceedings of the Second International Corbicula Symposium*, Prexant, R. S., Ed., *Am. Malacol. Bull.*, Spec. ed. No. 2, pp. 69–81, 1986.

Cherry, D. S., Farris, J. L., Bidwell, J. R., Mikailoff, A., Shema, R. L., and McIntire, J. W., *Corbicula* control program environmental fate and effects studies—baseline, spring and fall dosing studies, Duquesne Light Co., Beaver Valley Power Station, Annual Report 15077, Shippingport, PA, p. 356, 1991.

Cherry, D. S., et al., *Corbicula* control program environmental fate and effects studies, Duquesne Light Company, Beaver Valley Power Station, Annual Report 15077; Shippingport, PA, p. 330, 1992.

Cherry, D. S., Dobbs, M. G., Shema, R. L., Cody, W. R., Kenderes, G. J., Davison, M. F., Noel, M. R., Styborski, G. M., and McIntire, J. W., Duquesne Light Company Beaver Valley Power Station Ohio River In-Situ *Corbicula* growth study, Final Report 15077, Shippingport, PA, p. 55, 1993.

Cherry, D. S., Currie, R. J., Soucek, D. J., Latimer, H. A., and Trent, G. C., An integrative assessment of a watershed impacted by abandoned mined land discharges, *Environ. Pollut.*, 11, 377–388, 2001.

Cherry, D. S., Van Hassel, J. H., Farris, J. L., Soucek, D. J., and Neves, R. J., Site-specific derivation of the acute copper criteria for the Clinch River, Virginia, *Hum. Ecol. Risk Assess.*, 8, 591–601, 2002.

Clements, W. H., Community responses of stream organisms to heavy metals; a review of observational and experimental approaches, In *Metal Ecotoxicology: Concepts and Applications*, Newman, M. C. and McIntosh, S. W., Eds., Lewis Publications, Boca Raton, FL, 1991.

Clements, W. H., Integrating effects of contaminants across levels of biological organization: An overview, *J. Aquat. Ecosyst. Stress Recov.*, 7, 113–116, 2000.

Clements, W. H., Cherry, D. S., and Cairns, J., Jr., Macroinvertebrate community responses to copper in laboratory and field experimental streams, *Environ. Contam. Toxicol.*, 19, 361–365, 1990.

Couillard, Y., Campbell, P. G. C., Pellerin-Massicotte, J., and Auclair, J. C., Field transplantation of a freshwater bivalve, *Pyganodon grandis*, across a metal contamination gradient. I. Temporal changes in metallothionein and metal (Cd, Cu, and Zn) concentrations in soft tissues, *Can. J. Fish. Aquat. Sci.*, 52, 690–702, 1995.

Doherty, F. G., The Asiatic clam, *Corbicula* spp., as a biological monitor in freshwater environments, *Environ. Monit Assess.*, 15, 143–181, 1990.

Doherty, F. G. and Cherry, D. S., Tolerances of the Asiatic clam *Corbicula* spp. to lethal levels of toxic stressors—a review, *Environ. Pollut.*, 51, 269–313, 1988.

Doherty, F. G., Farris, J. L., Cherry, D. S., and Cairns, J. Jr., Control of the freshwater fouling bivalve, *Corbicula fluminea*, by halogenation, *Arch. Environ. Contam. Toxicol.*, 153, 535–542, 1986.

Doherty, F. G., Cherry, D. S., and Cairns, J. Jr., Valve closure responses of the Asiatic clam, *Corbicula fluminea*, exposed to cadmium and zinc, *Hydrobiologia*, 153, 159–167, 1987.

Doherty, F. G., Failla, M. L., and Cherry, D. S., Metallothionein-like heavy metal binding protein levels in Asiatic clam are dependent on the duration and mode of exposure to cadmium, *Water Res.*, 22, 927–932, 1988.

Eng L. L., Population dynamics of the Asiatic clam, *Corbicula fluminea* (Muller), in the concrete lined Delta-Mendota canal of central California, In *Proceedings, First International Corbicula Symposium*, Britton, J. C., Ed., Texas Christian University Research Foundation, Fort Worth, TX, pp. 34–68, 1979.

Farris, J. L., Belanger, S. E., Cherry, D. S., and Cairns, J. Jr., Application of cellulolytic activity of *Corbicula* to in-stream monitoring of power plant effluents, *Environ. Toxic. Chem.*, 7, 701–715, 1988.

Farris, J. L., Belanger, S. E., Cherry, D. S., and Cairns, J. Jr., Cellulolytic activity as a novel approach to assess long-term zinc toxicity to *Corbicula*, *Water Res.*, 23, 1275–1283, 1989.

Farris, J. L., Cherry, D. S., and Neves, R. J., Validation of copper concentrations in laboratory testing for site-specific copper criteria in the Clinch River. Report prepared for American Electric Power Company. Virginia Polytech. Inst. State Univ., Biol. Dept., Univ. Center Environ. Haz. Mat. Stud., Blacksburg, VA, p. 54, 1991.

Farris, J. L., Cherry, D. S., and Cairns, J. Jr., Mollusc can cellulolytic activity responses to zinc exposure in laboratory and field stream comparisons, *Hydrobiologia*, 287, 161–178, 1994.

Giesy, J. B., Leversee, G. J., and Williams, D. R., Effects of naturally occurring aquatic organic fractions on cadmium toxicity to *Simocephalus serrulatus* (Daphonidae) and *Gambusia affinis* (Poeciliidae), *Water Res.*, 11, 1013–1020, 1988.

Goss, L. B. and Cain, C., Jr., Power plant condenser and service water system fouling by *Corbicula*, the Asiatic clam, In *Biofouling Control Procedures*, Jenson, L. D., Ed., Marcel Dekker, Inc., New York, pp. 11–17, 1977.

Graney, R. L. Jr., Cherry, D. S., Rodgers, J. H. Jr., and Cairns, J. Jr., The influence of thermal discharge and substrate comparisons on the population structure and distribution of the Asiatic clam, *Corbicula fluminea*, in the New River, Virginia, *Nautilus*, 94, 130–134, 1980.

Graney, R. L. Jr., Cherry, D. S., and Cairns, J. Jr., Heavy metal indicator potential of the Asiatic clam (*Corbicula fluminea*) in artificial stream systems, *Hydrobiologia*, 102, 81–88, 1983.

Graney, R. L. Jr., Cherry, D. S., and Cairns, J. Jr., The influence of substrate, pH, diet and temperature upon cadmium accumulation in the Asiatic clam (*Corbicula fluminea*) in laboratory artificial streams, *Water Res.*, 18, 833–842, 1984.

Gregg, W. O., Conchological Club of Southern California, *Minutes, May*, 69, 3–4, 1947.

Guthrie, R. K. and Cherry, D. S., Pollutant removal from coal ash basin effluent, *Water Res. Bull.*, 12, 889–902, 1976.

Guthrie, R. K., Cherry, D. S., and Rodgers, J. H., *The Impact of Ash Basin Effluent on Biota in the Drainage System*, Proceedings of 7th Mid-Atlantic Indus. Waste Conf., Drexel Univ, Philadelphia, PA, 1974.

Hakenkamp, C. C. and Palmer, M. A., Introduced bivalves in freshwater ecosystems: The impact of *Corbicula* on organic matter dynamics in a sandy stream, *Oecologia*, 119, 445–451, 1999.

Hakenkamp, C. C., Ribblett, S. G., Palmer, M. A., Swan, C. M., Reid, J. W., and Goodison, M. R., The impact of an introduced bivalve (*Corbicula fluminea*) on the benthos of a sandy stream, *Freshwat. Biol.*, 46, 491–501, 2001.

Hollander, M. and Wolfe, D. A., *Nonparametric Statistical Methods*, Wiley, New York, p. 503, 1979.

Hull, M. S., An ecotoxicological recovery assessment of the Clinch River following coal industry-related disturbances in Carbo, Virginia (USA): 1967–2002, [M.Sc. Thesis]. Biology Dept., Virginia Tech, Blacksburg, VA, p. 176, 2002.

Hull, M. S., Cherry, D. S., Soucek, D. S., Currie, R. J., and Neves, R. J., Comparison of Asian clam field bioassays and benthic community surveys in quantifying effects of a coal-fired power plant effluent on Clinch River biota, *J. Aquat. Ecos. Stress Recov.*, 9, 271–283, 2002.

Hull, M. S., Cherry, D. S., and Merricks, T. C., Effect of cage design on growth of transplanted Asian clams: Implications for assessing bivalve responses in streams, *J. Environ. Monit. Assess.*, 96, 1–14, 2004.

Hull, M. S., Cherry, D. S., and Neves, R. J., Use of bivalve metrics to quantify influence of coal-related activities in the Clinch River Watershed, Virginia, *Hydrobiologia*, 556, 341–355, 2006.

Ingram, W. W., Asiatic clams as potential pests in California water supplier, *J. Am. Water Works Assoc.*, 51, 363–367, 1959.

Jacobson, P. J., Farris, J. L., Neves, R. J., and Cherry, D. S., Juvenile freshwater mussel (Bivalia: Unionidae) responses to acute toxicity testing with copper, *Environ. Toxic. Chem.*, 12, 879–883, 1993a.

Jacobson, P. J., Farris, J. L., Neves, R. J., and Cherry, D. S., Use of neutral red to assess survival of juvenile freshwater mussels (Bivalvia: Unionidae) in bioassays, *Trans. Am. Microsc. Soc.*, 112, 72–80, 1993b.

Jacobson, P. J., Farris, J. L., Neves, R. J., and Cherry, D. S., Sensitivity of glochidial stages of freshwater mussels to copper, *Environ. Toxic. Chem.*, 16, 2384–2392, 1997.

Kennedy, A. J., Cherry, D. S., and Currie, R. J., Field and laboratory assessment of a coal processing effluent in the Leading Creek watershed, Meigs County, Ohio, *Arch. Environ. Contam. Toxicol.*, 44, 324–331, 2003.

Kennedy, A. J., Cherry, D. S., and Currie, R. J., Evaluation of ecologically relevant bioassays for a lotic system impacted by a coal mining effluent using *Isonychia bicolor*, *Environ. Monit. Assess.*, in press.

Mattice, J. S., Interactions of *Corbicula* with power plants, In *Proceedings, First International Corbicula Symposium*, Britton, J. C., Ed., Texas Christian University Research Foundation, Texas, TX, pp. 119–138, 1979.

McMahon, R. F., Shell size-frequency distributions of *Corbicula manilensis* Philippi from a clam-fouled stream condenser, *Nautilus*, 91, 54–59, 1977.

Merlini, M., The freshwater clam as a biological indicator of radiomanganese, In *Radioecological Concentration Processes, Proceedings of International Symposium*, Aberg, B. and Hungate, F. P., Eds., pp. 977–982, 1966.

Merlini, M., Cadano, G., and Oregioni, B., The unionid mussel as a biogeochemical indicator of metal pollution, In *Environmental Biogeochemistry and Geomicrobiology*, Krumbein, W. E., Ed., Vol. 3, pp. 955–965, 1978.

Moore, H. J., The structure of the lateral frontal cirri on the gills of certain lamellibranch molluscs and their role in suspension feeding, *Mar. Biol.*, 11, 23–27, 1971.

Morton, H., Freshwater fouling bivalves. In *Proceedings First International Corbicula Symposium*, Britton, J. C., Ed., Texas Christian University Research Foundation, Texas, TX, pp. 11–15, 1979.

Naimo, T. J., Damschen, E. D., Rada, R. G., and Monroe, E. M., Nonlethal evaluation of the physiological health of unionid mussels: Methods for biopsy and glycogen analysis, *J. North Am. Benthol. Soc.*, 17, 121–128, 1998.

Nelson, M. K., Ingersoll, C. G., and Dwyer, F. J., Proposed guide for conducting solid-phase sediment toxicity tests with freshwater invertebrates. Draft No. 2. ASTM Comm. E-47, 1988.

Newton, T. J., Monroe, E. M., Kenyon, R., Gutreuter, S., Welke, K. I., and Thiel, P. A., Evaluation of relocation of unionid mussels into artificial ponds, *J. North Am. Benthol. Soc.*, 20, 468–485, 2001.

Peters, G. T., Cherry, D. S., and Cairns, J. Jr., The responses of *Isonychia bicolor* to alkaline pH: An evaluation of survival, oxygen consumption and chloride cell ultrastructure, *Can. J. Fish. Aquat. Sci.*, 42, 1088–1095, 1985.

Phillips, D. J. H., The common mussel *Mytilus edulis* as an indicator of pollution by zinc, cadmium, lead, and copper I. Effects of environmental variables on uptake of metals, *Mar. Biol.*, 38, 59–60, 1976a.

Phillips, D. J. H., The common mussel *Mytilus edulis* as an indicator of pollution by zinc, cadmium, lead, and copper. I. Relationship of metals in the mussels to those discharged by industry, *Mar. Biol.*, 38, 71–80, 1976b.

Phillips, D. J. H., The use of biological indicator organisms to monitor trace metal pollution in marine and estuarine environments—a review, *Environ. Pollut.*, 13, 281–317, 1977.

Phillips, D. J. H., Trace metals in the common mussel *Mytilus edulis* (L.), and in alga *Fucus vesiculosus* (L.) from the region of the Sound (Oresund), *Environ. Pollut.*, 18, 31–43, 1979.

Prokopovitch, N. P. and Herbert, D. J., Sedimentation in the Delta-Mendota Canal, *J. Am. Water Works Assoc.*, 57, 375–382, 1965.

Rodgers, J. H., Jr., Cherry, D. S., Clark, J. R., Dickson, K. L., and Cairns, J., Jr., The invasion of Asiatic Clam, *Corbicula manilensis*, in the New River, Virginia, *Nautilus*, 91, 43–45, 1977a.

Rodgers J. H. Jr., Cherry, D. S., Dickson, K. L., and Cairns J. Jr., Invasion, population dynamics and elemental accumulation of *Corbicula fluminea* in the New River at Glen Lyn, Virginia, *Proceedings of First International Corbicula Symposium*, Texas Christian University, Fort Worth TX, pp. 99–110, 1977b.

Rodgers, J. H. Jr., Cherry, D. S., Graney, R. L., Dickson, K. L., and Cairns, J. Jr., Comparison of heavy metal interactions in acute and artificial stream bioassay techniques for the Asiatic clam (*Corbicula fluminea*), In *Aquatic Toxicology, ASTM STP 707*, Eaton, J. B., Parrish, P. R., and Hendricks, A. C., Eds., ASTM, Philadelphia, PA, pp. 266–280, 1980.

Salazar, M. H. and Salazar, S. M., In situ bioassays using transplanted mussels. I. Estimating chemical exposure and bioeffects with bioaccumulation and growth, In *Environmental Toxicology and Risk Assessment*, Hughes, J. S., Biddinger, G. R., and Mones, E., Eds., American Society for Testing and Materials, Philadelphia, PA, pp. 216–241, 1995.

Scheller, J. L., The impact of dieoffs of the Asian clam (*Corbicula fluminea*) on native freshwater unionids [M.Sc. Thesis], Dept. of Biology, VA Tech., Blacksburg, VA, p. 116, 1997.

Schmidt, T. S., Soucek, D. J., and Cherry, D. S., Integrative bioassessment of small acid mine drainage impacted watersheds in the Powell River Watershed, *Environ. Toxic. Chem.*, 21, 2233–2246, 2002.

Sinclair, R. J., Annotated bibliography on the exotic bivalve *Corbicula* in North America, *Sterkiana*, 43, 11–18, 1971.

Sinclair, R. M. and Isom, B. G., Further studies on the introduced Asiatic clam (*Corbicula*) in Tennessee. Tennessee Stream Pollution Control Board, Tennessee Department of Public Health, 1963.

Smith, J. G. and Beauchamp, J. J., Evaluation of caging designs and a fingernail clam for use in an in situ bioassay, *Environ. Monit. Assess.*, 62, 205–230, 2000.

Sokal, R. R. and Rohlf, F. J., *Biometry*, W.H. Freeman Co., San Francisco, CA, p. 776, 1981.

Soucek, D. J., Schmidt, T. S., and Cherry, D. S., In situ studies with Asian clams (*Corbicula fluminea*) detect acid mine drainage and nutrient inputs in low order streams, *Can. J. Fish. Aquat. Sci.*, 58, 602–608, 2001.

Strayer, D. L., Effects of alien species on freshwater mollusks in North America, *J. North Am. Benthol. Soc.*, 18, 74–98, 1999.

Swansburg, E. O., Fairchild, W. L., Fryer, B. J., and Ciborowski, J. J. H., Mouthpart deformities and community composition of Chironomidae (Diptera) larvae downstream of metal mines in New Brunswick, Canada, *Environ. Toxicol. Chem.*, 21, 2675–2684, 2002.

USEPA, In *Biological, Field and Laboratory Methods for Measuring the Quality of Surface Waters and Effluents*, Weber, C. I., Ed., USEPA, Cincinnati, OH, p. 216, 1973.

USEPA, In *Short-Term Methods for Estimating the Chronic Toxicity of Effluents and Receiving Waters to Freshwater Organisms*, Horning, W. B. and Weber, C. I., Eds., USEPA, Cincinnati, OH, p. 162, 1985.

USEPA, In *Short-Term Methods for Estimating the Chronic Toxicity of Effluents and Receiving Waters to Freshwater Organisms*, Weber, C. I., Ed., USEPA, Cincinnati, OH, p. 249, 1989.

Vannote, R. L., Minshall, G. W., Cummins, K. W., Sedell, J. R., and Cushing, C. E., The river continuum concept, *Can. J. Fish. Aquat. Sci.*, 37, 130–137, 1980.

Vaughn, C. C. and Hakenkamp, C. C., The functional role of burrowing bivalves in freshwater ecosystems, *Freshwat. Biol.*, 46, 1431–1446, 2001.

Widdows, J., Phelps, D. K., and Galloway, W., Measurement of physiological condition of mussels transplanted along a pollution gradient in Narragansett Bay, *Mar. Environ. Res.*, 4, 181–194, 1981.

Yeager, M. M., Neves, R. J., and Cherry, D. S., Competitive interactions between early life stages of *Villosa iris* (Bivalvia:Unionidae) and adult Asian clams (*Corbicula fluminea*), *Proceedings of First Freshwater Conservation Society Symposium*, Ohio Biol. Soc., pp. 253–259, 1999.

12 Case Study: Discrimination of Factors Affecting Unionid Mussel Distribution in the Clinch River, Virginia, U.S.A.

John H. Van Hassel

INTRODUCTION

The unionid mussel fauna of the upper Clinch River has been well documented as one of the best remaining epicenters of the highly diverse and endemic Cumberlandian fauna. Surveys of 1912–1913 by Ortmann (1918) established the presence of 49 mussel species in the upper Clinch River. Stansbery (1973) added an additional six species to this number based on collections between 1963 and 1971, for a historical total of 55 species. Subsequent surveys were performed during 1972–1975 by Bates and Dennis (1978), during 1978–1983 by Ahlstedt (1984), and in 1988 and 1994 by Ahlstedt and Tuberville (1997). According to the latter authors, the recent mussel fauna of the upper Clinch River consists of 39 species, to which one species, the tan riffleshell (*Epioblasma florentina walkeri*), was subsequently added. Of these 40 species, 10 (25%) are currently federally listed as endangered.

These surveys indicate that as many as 15 (27%) of the 55 previously known species may be extirpated from the upper Clinch River. In addition, some of the remaining species appear to have undergone recent declines in abundance (Dennis 1987; Bruenderman and Neves 1993; Ahlstedt and Tuberville 1997). This situation parallels the global trend of mussel imperilment described in Chapter 1 of this volume. Specific causes for the faunal decline in the Clinch River have not been identified, but there are a number of likely candidates. Active logging occurred throughout the watershed from 1870 to 1920 (Masnik 1974), with concomitant sedimentation impacts. Extensive habitat modification and pollutant loading from coal mining activities has occurred within the region (Helfrich et al. 1986), particularly from poorly regulated activities prior to enactment of the Surface Mining Control and Reclamation Act of 1977. Additional potential sources of impact were associated with development of industry and agriculture during the past century (Ahlstedt 1983; Neves 1984); however, the proportion of forested land in the watershed has actually increased, from 42% in 1941; to 57% in 1960; and to 67% in 2000 (Masnik 1974; Hampson et al. 2000), with a corresponding decrease in land devoted to agriculture. Additional important considerations were the impacts of predators such as muskrats (Neves and Odom 1989) and the loss or decline of fish host species for the parasitic glochidial mussel life stage (Ahlstedt 1983; Neves 1984). Thus, the extirpation of mussel species in the Clinch River has multiple suspected causes, but quantitative elucidation of specific causes is relatively rare.

Studies aimed at establishing cause and effect of impacts on freshwater mussels are faced with several difficulties. The characteristic patchy distribution of mussels necessitates careful consideration of sampling design and level of effort to enable quantitative differentiation of important measured characteristics (i.e., mussel individuals, populations, or assemblages; Strayer and Smith 2003). If these differences are not quantified, then the relative influence of various environmental factors can only be speculative.

Another limitation is that few structural and functional measures of mussel condition have been sufficiently developed to enable dependable impact assessments (see Chapter 2). Population density and size demographics are most often used. Growth is commonly measured in studies on fish and a variety of other aquatic organisms but is problematic for use with freshwater mussels. The slow growth rate of most species requires impractically long study periods in order to obtain quantifiable increases in growth. Physiological measures provide a means of real-time documentation of mussel condition; however, few have been widely used on mussels (see Chapter 10 for further discussion). Cellulolytic enzyme activity in mussels and Asian clams (*Corbicula fluminea*) was successfully applied in some previous site-specific Clinch River studies (Farris et al. 1988, 1991), but there is a need for additional physiological measures or assays.

A third difficulty encountered when conducting impact assessments on freshwater mussels is that many types of desired measures (e.g., physiological indices, bioaccumulation, and toxicity tests) require that the collected organisms be sacrificed (Naimo and Monroe 1999). This is often undesirable when dealing with rare and imperiled species. If it can be scientifically justified, the use of surrogate species (e.g., common unionids, other bivalves, or other invertebrates) should be considered. Previous Clinch River impact studies by Cherry et al. (1991) found that Asian clams were useful surrogates for toxicity testing of unionids.

The objective of this study was to quantify factors affecting freshwater mussel distribution and abundance within Virginia waters of the Clinch River. To achieve this objective, differences in mussel species occurrence, abundance, age distribution, physiological condition, and contaminant body burden were measured at selected impact and reference locations. Sampling sites were carefully selected for good mussel habitat based on previous surveys in order to minimize the influence of natural habitat characteristics in defining sources of impact. The biological information was evaluated in conjunction with concurrent, site-specific measurements of water and sediment quality, habitat quality, riparian land use, point and nonpoint discharge occurrence, and fish host availability. The environmental measures ranged in scale from microhabitat to watershed level to provide a comprehensive assessment of potential factors influencing Clinch River mussel distribution.

METHODS

SAMPLING LOCATIONS

The Clinch River arises in extreme southwest Virginia, flowing into Tennessee and joining with the Powell River before entering the Tennessee River. The Virginia waters of the Clinch River lie within the steep-sloped Ridge and Valley and Cumberland Plateau physiographic provinces of the central Appalachian Mountains. The average gradient of the upper, free-flowing portion of the river covering 188 miles (302 km) from its source near Tazewell, Virginia, to Norris Reservoir in Tennessee is 9.3 ft./mi (1.8 m/km) (Masnik 1974). The river is characterized by extensive pool-riffle development, including several islands and braided-channel segments. There are also large beds of water willow (*Justicia americana*) along the shorelines and shallows throughout the upper Clinch River. The geology of the region is dominated by exposed limestone and dolomite formations, which produce a carbonate-rich system with pH in the range of 7.5–8.5.

Sampling stations were established at 12 mainstem Clinch River locations (Figure 12.1), including two sites at Station 6 (left channel and right channel). Stations were selected at

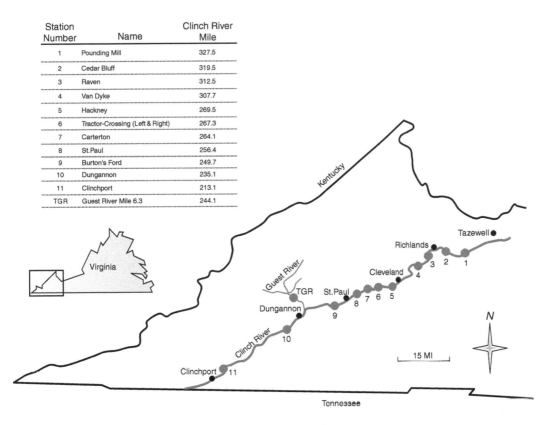

Station Number	Name	Clinch River Mile
1	Pounding Mill	327.5
2	Cedar Bluff	319.5
3	Raven	312.5
4	Van Dyke	307.7
5	Hackney	269.5
6	Tractor-Crossing (Left & Right)	267.3
7	Carterton	264.1
8	St. Paul	256.4
9	Burton's Ford	249.7
10	Dungannon	235.1
11	Clinchport	213.1
TGR	Guest River Mile 6.3	244.1

FIGURE 12.1 Map of the Clinch River, Virginia, showing site locations for sampling of freshwater mussels and associated environmental measures.

representative riffle-run habitats encompassing the length of the Virginia portion of the river, which extends approximately 146 river miles (235 km) from the Tennessee line to the confluence of the North Fork and South Fork Clinch River headwaters near Tazewell, Virginia. Sampling stations were selected to provide a mix of known productive mussel sites with sites that would provide a gradient of potential impacts from large communities and point source dischargers. An additional sampling station was established on the Guest River at Coeburn, Virginia, 6.3 river miles (10.1 km) upstream of its confluence with the Clinch River at Clinch River Mile (CRM) 244.1. The Guest River, as the largest Virginia tributary to the Clinch River, represented a major potential influence on water quality. Two sampling stations were established at CRM 267.3 in order to evaluate possible effects from American Electric Power's Clinch River electric-generating plant, which is the largest industrial facility in the study area. Station 6L was located within the plant effluent mixing zone, while Station 6R was located on the opposite bank outside of effluent influence.

Each of the 13 sampling stations was characterized by quantitative measurements of flow direction, maximum hours of direct summer sunlight, bank height, bank slope, bank stability (percentage of bare soil), riparian vegetation height, riffle length, upstream pool length, downstream pool length, width-to-depth ratio (average width divided by average thalweg depth of runs and pools), riffle-to-riffle ratio (average distance between riffles divided by average width), and bend-to-bend ratio (average distance between bends divided by average width).

Substrate composition was determined from four 1-m^2 quadrat samples for mussels. Bedrock (solid slabs) and boulders (greater than 256 mm) were measured by hand. The volume of rubble (64–256 mm) in the quadrat was measured by displacement. Gravel (2–64 mm), sand (0.0625–2 mm), silt (0.0039–0.0625 mm), and clay (less than 0.0039 mm) were excavated using a portable sand dredge, with gravel and sand volumes measured by displacement in the field, and silt and clay samples returned to the laboratory for volume determination by centrifugation at 750 rpm for three minutes. Substrate penetrability was determined using three randomly spaced measurements within each quadrat using a sliding hammer on a half-inch (1.27 cm) steel rod. Depth (cm) of penetration of the rod into the substrate was measured following three standardized hammer strikes. Surface and bottom current velocity (cm/sec) was measured within each quadrat using a Mead Instruments flow meter.

Riparian land use influencing each site was determined by a review of USGS 7.5-minute topographic maps and aerial photographs of the Clinch River watershed, from which the drainage area, percentage of forest, percentage of agricultural, percentage of urban, and percentage of mine lands were calculated for 10-mile (16 km) segments upstream of each sampling location. Discharge permit holders were reviewed to determine the number of major NPDES discharges and sewage treatment plants within the 10-mile (16 km) segment, including tributaries, upstream of each site.

WATER CHEMISTRY

Water samples were collected at each sampling station on eight occasions between October 1991 and 1994. Field water column measurements included pH, temperature, conductivity, alkalinity, hardness, and turbidity. Samples were preserved for laboratory analysis of ammonia, total organic carbon, phosphate, and nitrate–nitrite.

Sediment interstitial water samples were obtained by pushing a 45-cm length of 5-cm ID PVC pipe with a PVC cone attached to the bottom end into the substrate. A series of 0.3-cm holes around the circumference of the pipe near the bottom end were positioned at a depth of 5–8 cm below the substrate surface. Interstitial water that flowed into the pipe was sampled using a polypropylene hand pump that was used to initially flush the pipe and then collect two 500-mL samples. One of these samples was immediately preserved using 1:1 HNO_3 for total recoverable metals analysis. The second sample was placed on ice, filtered (0.45 μm) within eight hours, and preserved using 1:1 HNO_3. Field measurements on the interstitial water included pH and conductivity.

Laboratory analysis of ammonia, total organic carbon, phosphate, nitrate–nitrite, and dissolved and total recoverable concentrations of Al, As, Cu, Mn, Ni, Pb, and Zn were performed according to USEPA (1979). Metal concentrations were determined using an AES–ICP analyzer. This list of parameters was selected based on their significance in previous Clinch River sampling efforts.

SEDIMENT CHEMISTRY

Sediment samples were collected twice at each sampling station—in July and October of 1992. Samples were 500-mL composites collected from depositional sediments using a stainless steel scoop and were preserved intact on ice for laboratory analysis according to USEPA (1986) methods. In the laboratory, samples were dried for 48 hours at 50–60°C, sieved through 12 mesh (1.4 mm) to remove debris and large particles, then milled and sieved through 100 mesh (0.15 mm).

The portion of each sample to be analyzed for aluminum (as Al_2O_3) was digested in a closed PFA vessel using hydrofluoric, hydrochloric, nitric, and boric acids in a microwave digestion system. Another portion of each sample to be analyzed for As, Cu, Mn, Ni, Pb, and Zn was similarly digested using hydrochloric and nitric acids. The vessel pressures were not allowed to exceed 95 psi during the digestion. Metal concentrations were measured using an AES–ICP analyzer.

FAUNAL SURVEYS

Unionid mussel assemblages at each sampling station were surveyed quantitatively between July 1992 and October 1994 using a stratified (by habitat) random sampling approach. At each site, four 1-m^2 quadrats were placed randomly in riffle/run habitat and excavated by hand and by a portable sand dredge to a depth of 30 cm or bedrock. Gravel and sand substrates were rinsed through two nested sieves (19 and 5 mm). All unionids were removed, identified to species, measured for length (mm) with digital calipers, and returned to the river. The percentage of each species that was age five or less was determined from age-length keys developed for the Clinch River by Scott (1994). Other bivalves (*C. fluminea* and fingernail clams) were counted and returned to the river.

Qualitative searches were also conducted at each sampling station by snorkeling and the collection of fresh shells from middens. These surveys were performed two to three times at each site between October 1991 and 1994, with a total level of effort of approximately 4.5 man-hours per site. Unionids were identified to species, counted, and returned to the river.

Fish were surveyed at each sampling station on one occasion between July 1992 and October 1994 by backpack electrofishing. All species collected in 45 minutes of sampling time were identified to species, counted, examined for deformities (such as eroded fins, lesions, external tumors, and glochidia infestation), and then returned to the river. Index of Biotic Integrity (IBI) scores were calculated for each sampling station using the 12 metrics adapted to the Clinch River by Angermeier and Smogor (1993).

TISSUE MEASUREMENTS

Tissue metal residues were measured in *C. fluminea* collected from each of the 13 sampling stations on three occasions per site between October 1991 and July 1992. Each sample consisted of 10 clams of approximately 10 mm in length, which were placed on ice and returned to the laboratory for digestion and metals analysis according to ASTM Method D4638-86. In the laboratory, clam soft tissues were excised from the shells, oven-dried at 60°C for 24 hours, and individually weighed and digested in 1:1 HNO$_3$. Concentrations of Al, As, Cu, Mn, Ni, Pb, and Zn were determined using an AES–ICP analyzer.

Cellulolytic enzyme activity (Farris et al. 1989) was measured on *C. fluminea* collected from each of the 13 sampling stations on a minimum of six occasions per site between October 1991 and September 1995. Each sample consisted of 10 clams of approximately 10 mm in length. Soft tissues were excised in the field, placed in individual vials, and put on dry ice for transport to the laboratory. One-gram samples of each clam were homogenized in a phosphate buffer (0.15 M at pH 6.1) at a wet mass to buffer ratio of 0.02 g/mL. Samples were centrifuged for 15 minutes at 15,000 rpm, supernatants were decanted for endo/exocellulase analysis, and the pellets were recovered for dry mass measurements.

Corbicula fluminea collected from each site were also measured for tissue levels of cholinesterase activity on two occasions per site in 1995. Ten clams per site were collected, frozen for transport to the laboratory, and analyzed as described by Fleming et al. (1995).

TOXICITY TESTING

In situ toxicity tests were performed at each of the 13 sites on two occasions in 1995. On each occasion, 130 juvenile *Villosa iris* ranging from 9 to 15 mm length obtained from Station 1 were measured (mm) using digital calipers and randomly placed in 13 mesh bags (10 per bag) with approximately 1.5-mm openings. The bags were attached to rocks in riffle/run areas at each site for approximately six weeks. Endpoints were survival and growth (shell length).

The toxicity of accumulated silt (0.0039–0.0625 mm particle size) at each site was tested in the laboratory using a 24-hour acute toxicity test using the freshwater rotifer *Brachionus calyciflorus*

(Snell and Persoone 1989). Silt was collected from each site by placing $14 \times 14 \times 2.5$ cm^3 slotted plastic boxes in depositional areas of each site for approximately six weeks on two occasions in 1995. Collected silt was transferred to plastic bags, placed on ice, and returned to the laboratory for toxicity testing within seven days. Testing consisted of the placement of 50 cysts of *B. calyciflorus* in 30-mL polyethylene vials containing 5 mL of silt and 20 mL of moderately hard reconstituted water (USEPA 1990). Replicate vials were prepared for each silt sample. The vials were placed in a rack and rotated at a rate of two times per hour at a temperature of 25°C and a 16:8, light:dark photoperiod. Hatching success of each replicate was determined after 24 hours.

DATA ANALYSIS

Statistical analysis of study results for water and sediment chemistry, habitat and land use measurements, faunal surveys, and toxicity tests was performed using Spearman correlation analysis and the distribution-free Wilcoxon test for one-way layouts on SAS (SAS Institute 1985). A multivariate analysis of the relationship between mussel structural indices and measured environmental variables was performed using canonical correspondence analysis (Ter Braak 1986).

RESULTS

SAMPLING LOCATIONS

The 13 sampling locations exhibited distinct differences in a number of instream habitat and riparian land use characteristics. The most obvious difference was an increase in mean annual stream flow from 187 cfs (5.3 m^3/sec) at Station 1–1637 cfs (46.3 m^3/sec) at Station 11. The mean annual flow of the Guest River at Station TGR was 144 cfs (4.1 m^3/sec). Mean riffle/run current velocities during summer sampling events were typically in the range of 28–37 cm/sec, with a low of 8 cm/sec at Station 4 and a high of 65 cm/sec at Station 5. The Clinch River flows in general from northeast to southwest; however, the direction of flow at individual sampling sites varied greatly. As a result, with the additional influence of riparian vegetation height, the hours of summer sunlight received by individual sites varied from as little as 2–3 hours at sites with north–south orientations (e.g., Stations 5 and 9) to 8–10 hours at sites with east–west orientations (e.g., Stations 2 and 6). The height of streamside vegetation varied greatly between sites, and from bank to bank at individual sites, ranging from grasses to mature hardwoods.

The bank height at most sites during normal summer flows ranged from 1 to 3 m before leveling out onto the flood plain, with the exception of sharp relief present on one side each of Stations 6, 7, 8, and 10. Banks at all sites were fairly steep (30–40°) with the exception of Station 2, which had a very gradual left-bank slope of about 10°. Bank erosion was minimal at most sites, with a limited amount of exposed soil observed at Stations 2, 5, 6, 9, 11, and TGR.

The riffle-run areas that represented the primary sampling locations at each site ranged from 14 m in length at Station 4 to an extensive 180-m segment at Station 11. The corresponding upstream and downstream pools tended to increase in length from upriver sites to downriver sites, ranging from 40 m at Station 1 to 275 m at Station 10. Width-to-depth ratios were fairly uniform, ranging from 26.3 m/m at Station 4 to 43.4 m/m at the Guest River site. Riffle-to-riffle ratios varied by a factor of five between sites (from 1.6 m/m at Station 11 to 5.6 m/m at Station 4). Bend-to-bend ratios were all in the range of 14–20 m/m except Stations 4 (29.6 m/m) and 11 (37.2 m/m).

Substrate composition from quadrat samples of the riffle/run areas of each site was dominated by the sand/gravel/rubble fractions (Table 12.1). Silt was greater than 5% at Stations 4 and 11, while the boulder fraction represented greater than 5% at Stations 4, 6L, and TGR. The rubble fraction was exceptionally high at Station 9. Substrate penetration, measured by standard hits to a steel rod, was typically 9–14 cm. The exceptions were Station TGR, where the sediment was most

TABLE 12.1
Clinch River Quadrat Substrate Composition (Mean Percent by Volume)

Station	Boulder (>256 mm)	Rubble (64–256 mm)	Gravel (2–64 mm)	Sand (0.0625–2 mm)	Silt (<0.0625 mm)
1	1.0	28.3	26.6	42.1	2.0
2	0	37.0	29.2	30.9	2.9
3	0	31.4	27.5	36.6	4.5
4	7.8	20.8	23.2	40.7	7.5
5	2.4	45.5	19.7	30.5	1.9
6R	0	38.1	19.2	38.7	4.0
6L	12.0	32.0	24.6	30.6	0.8
7	0	43.3	23.1	33.3	0.3
8	0	35.4	25.4	35.1	4.1
9	0	59.5	12.5	22.8	5.2
10	2.0	39.7	23.3	30.5	4.5
11	0	30.8	21.2	41.0	7.0
Guest River	5.5	33.4	32.4	26.7	2.0

compacted (mean penetration of 5.3 cm), and Station 1, where the sediment was most unconsolidated (mean penetration of 16.5 cm).

Riparian land use influencing each site was arbitrarily defined as the drainage area to the stream segment 10 miles (16 km) upstream of each sampling site. This drainage area ranged from 8740 to 15,500 ha for all sites except Station 1, for which the area was 3550 ha. Within these areas, forestland made up the greatest percentage, ranging from 64 to 77%; the exceptions were Stations 1 (46%), 5 (52%), and 9 (57%). Most of the remaining area (9.7–54.9%) was agricultural, particularly grazing land. The percentage of land apportioned to urban development was typically low, ranging from 1 to 3% except at Stations 3 (8%), 4 (6%), and TGR (7%). Likewise, mined land comprised fairly low percentages of the land use at each site, typically less than 1% except at Stations 3 (3%), 4 (3%), 6 (2%), 7 (2%), and TGR (2%).

WATER CHEMISTRY

Clinch River water chemistry did not vary significantly between sampling sites for most measured parameters, either in the water column or sediment interstitial water (Table 12.2). Noteworthy results included a mean phosphate concentration of 1.4 mg/L (interstitial water) at Station 1, which was substantially higher than measurements at the other sites, and significantly ($\alpha = 0.05$) higher than concentrations at Stations 3, 8, 9, 10, and 11. The Guest River site had significantly higher concentrations of dissolved manganese (mean = 1240 µg/L) than all other sampling stations. Other parameters showing statistically significant differences between sites included conductivity, hardness, and total recoverable zinc, all of which were higher at the Guest River site than most or all other sites.

SEDIMENT CHEMISTRY

Two rounds of sediment sampling revealed few notable differences between sampling stations (Table 12.2). The Guest River site exhibited higher concentrations of manganese, nickel, lead, and zinc than the Clinch River mainstem sites. Stations 4 and 11 also had fairly high levels of zinc, while Station 4 also exhibited a higher concentration of sediment aluminum than the other sites. Comparison of these concentrations to screening values for determining possible areas of sediment

TABLE 12.2
Mean Water and Sediment Chemistry Measurements at 13 Clinch River Sites, 1992–1994

Parameter	Units	Sampling Station												
		1	2	3	4	5	6L	6R	7	8	9	10	11	TGR
Water Column (n = 8)														
Temperature	°C	18.3	19.5	18.4	19.6	17.8	19.4	17.4	18.8	19.5	19.0	17.5	17.4	17.3
pH	s.u.	8.4	8.5	8.3	8.4	8.3	8.3	8.4	8.3	8.5	8.4	8.3	8.2	8.2
Conductivity	µmhos/cm	294	297	300	299	300	315	320	316	313	301	297	303	373
Alkalinity	mg/L	141	146	148	174	153	142	150	132	147	137	130	123	116
Hardness	mg/L	152	161	165	177	160	159	181	158	168	178	200	186	250
Turbidity	NTU	5.3	5.5	6.0	5.3	5.0	6.7	4.8	5.2	5.5	5.9	3.4	7.3	4.7
NH_3-N	mg/L	0.08	<0.05	<0.05	<0.05	<0.05	<0.05	<0.05	<0.05	<0.05	<0.05	<0.05	<0.05	<0.05
TOC	mg/L	5	3	4	5	2	4	4	1	1	1	2	2	3
PO_4-P	mg/L	0.09	0.07	0.09	0.02	<0.01	0.01	<0.01	<0.01	<0.01	<0.01	<0.01	<0.01	0.12
NO_3-NO_2	mg/L	0.99	0.78	0.66	0.56	0.74	0.66	0.70	0.75	0.77	0.74	0.68	0.50	0.41
Interstitial Water (n = 8)														
pH	s.u.	7.8	7.9	8.0	7.8	7.9	8.0	7.9	7.9	8.0	8.0	7.9	7.9	7.9
Conductivity	µmhos/cm	271	266	312	304	303	327	303	284	315	320	329	298	465
NH_3-N	mg/L	0.05	<0.05	<0.05	0.06	<0.05	<0.05	<0.05	<0.05	<0.05	<0.05	<0.05	<0.05	<0.05
TOC	mg/L	6	3	4	4	4	5	6	5	8	7	3	3	4
PO_4-P	mg/L	1.40	0.45	0.14	0.97	0.25	0.30	0.34	0.32	0.14	0.12	0.06	0.15	0.75
NO_3-NO_2	mg/L	0.71	0.50	0.23	0.46	0.44	0.53	0.46	0.52	0.39	0.44	0.96	0.60	0.59
Al-diss.	µg/L	<50	<50	<50	<50	<50	<50	<50	<50	<50	<50	<50	<50	<50
Total	µg/L	4570	10,425	3500	22,250	12,725	16,550	5550	11,950	8050	6305	17,825	21,600	7780
As-diss.	µg/L	<4	<4	<4	<4	<4	<4	<4	<4	<4	<4	<4	<4	<4
Total	µg/L	<4	<4	<4	5	<4	<4	<4	<4	<4	<4	<4	<4	<4
Cu-diss.	µg/L	5	1	1	1	1	3	1	2	1	1	1	<1	1
Total	µg/L	15	14	7	25	14	48	23	23	15	15	14	22	29
Mn-diss.	µg/L	<10	25	<10	10	<10	<10	10	<10	10	<10	12	12	1240
Total	µg/L	1280	452	310	1120	768	727	760	773	515	842	2530	1500	3205
Ni-diss.	µg/L	<3	<3	<3	<3	<3	<3	<3	<3	<3	<3	<3	<3	3
Total	µg/L	11	10	4	25	13	17	12	14	8	10	14	25	57
Pb-diss.	µg/L	<2	<2	<2	<2	<2	<2	<2	<2	<2	<2	<2	<2	<2

Total	µg/L	43	31	8	50	24	16	19	19	12	22	20	49	55
Zn-diss.	µg/L	6	6	6	7	11	7	6	7	5	5	6	6	7
Total	µg/L	79	66	28	155	74	73	77	72	62	61	72	118	201
Sediment (n = 2)														
Al	%	1.16	2.66	0.68	5.67	2.47	2.67	3.69	3.73	2.05	1.22	2.88	3.49	2.43
As	mg/kg	<4	<4	<4	<4	<4	<4	<4	<4	<4	<4	<4	<4	<4
Cu	mg/kg	17	15	7	26	18	23	37	27	16	19	18	36	33
Mn	mg/kg	890	310	220	780	280	210	220	280	360	310	730	430	1160
Ni	mg/kg	14	12	5	31	12	26	19	19	19	9	21	40	77
Pb	mg/kg	18	13	5	21	9	7	17	28	5	8	9	44	81
Zn	mg/kg	48	40	37	94	41	45	73	53	38	34	44	112	148

contamination indicated that nickel concentrations at the Guest River site (mean = 77 mg/kg) exceeded the probable effect level of 48.6 mg/kg (MacDonald et al. 2000).

FAUNAL SURVEYS

A combined total of 29 species of unionid mussels were collected from 1992 to 1994 at the 13 sampling sites (Table 12.3). Results at individual sites varied dramatically, with no live specimens collected from Stations 3, 6L, and TGR, while 23 species were collected at Station 5. Species dominance among sites conformed to an upstream–downstream trend. Dominant species at Stations 1–4 (CRM 307.7–327.5) were the wavyrayed lampmussel (*Lampsilis fasciola*), Cumberland moccasinshell (*Medionidus conradicus*), Tennessee clubshell (*Pleurobema oviforme*), kidneyshell (*Ptychobranchus fasciolaris*), and rainbow (*Villosa iris*). Downstream, from Stations 5 to 11 (CRM 213.1–269.5), the dominant species were the mucket (*Actinonaias ligamentina*), pheasantshell (*Actinonaias pectorosa*), threeridge (*Amblema plicata*), spike (*Elliptio dilatata*), and Cumberland moccasinshell. Four federally listed endangered species were collected: tan riffleshell (*E. florentina walkeri*), shiny pigtoe (*Fusconaia cor*), finerayed pigtoe (*Fusconaia cuneolus*), and rough rabbitsfoot (*Quadrula cylindrica strigillata*), all of which were locally common at one or more sites.

Mussel densities in quadrat samples also varied greatly between sites, with the highest densities at Stations 1 and 5 in the upper half of the study area. Examining the occurrence of young mussels aged five years or less at the 13 sites provided another perspective on mussel populations along the river. The percentage of young specimens at three upstream sites (Stations 1, 2, and 5) was much higher than downstream percentages, while two sites with fairly good mussel assemblages in terms of species richness or density (Stations 4 and 11) produced no young specimens.

Asian clam (*C. fluminea*) density varied significantly ($P < .01$) between sites, ranging from a mean density of 78 clams/m^2 at Station 9 to 715 clams/m^2 at Station TGR. Fingernail clams (Sphaeriidae) also exhibited significant differences in density between sites, ranging from zero at several sites to 10.4 clams/m^2 at Station 4. The fingernail clams were much more prevalent at the five upstream sites than farther downstream.

Electrofishing results indicated that good to excellent fish assemblages existed at all sampling sites, with the exception of the Guest River location. IBI scores ranged from 48 to 58 (maximum possible score = 60) at the Clinch River sites, while Station TGR scored a 32, for a poor rating. Species composition among the Clinch River mainstem sites was very similar.

TISSUE MEASUREMENTS

Tissue metal concentrations in Asian clams exhibited little spatial variation in three samples from 1991 to 1992. Mean tissue concentration ranges (µg/g dry weight) among the 13 sites were: As, 2–5; Al, 1236–3070; Cu, 19–41; Pb, 2–8; Mn, 92–520; Ni, 2–9; and Zn, 234–380. The only sites significantly different from one or more other sites were Stations 2 and 10 for aluminum (respective means of 3030 and 3070 µg/g compared to 1236–2295 µg/g at remaining sites) and Station TGR for manganese (mean = 520 µg/g compared to 92–253 µg/g at remaining sites).

Measurement of Asian clam cellulolytic activity as an indicator of sublethal stress resulted in extremely variable results, both spatially between sampling sites and temporally among the six to nine samples collected at each site from 1991 to 1995. Station 9 produced the best results consistently, with clam cellulolytic activities averaging 97.8% of the activity of control clams. Sites with results that were significantly different from Station 9 included Stations 1 (mean = 49.8%), 4 (17.7%), 6L (45.2%), 8 (42.8%), 10 (39.4%), 11 (27.2%), and TGR (45.6%). For the two samples measuring cholinesterase activity, no significant difference was found for any sampling station compared to control activity.

TABLE 12.3
Density (#/m²) of Unionid Mussels at 13 Clinch River, Virginia, Sites, 1992–1994

Species	Sampling Site												
	1	2	3	4	5	6R	6L	7	8	9	10	11	Guest River
River Mile:	327.5	319.5	312.5	307.7	269.5	267.3	267.3	265.0	256.4	249.7	235.1	213.1	6.3
A. ligamentina					Q				Q	0.25	0.5	0.5	
A. pectorosa		Q[a]			2.0				Q	0.5	0.5	0.25	
Alasmidonta marginata				Q						Q			
A. plicata					Q			0.5	Q	0.5	0.5	0.25	
Cumberlandia monodonta[b]					Q							Q	
Cyclonaias tuberculata										0.25	0.25	Q	
E. dilatata					2.0			0.25			Q		
Epioblasma capsaeformis												Q	
E. florentina walkeri		Q								Q			
Fusconaia barnesiana		Q											
F. cor		Q			Q			Q		Q		Q	
F. cuneolus					0.5					0.25	Q		
F. subrotunda					Q			Q	0.5	Q	Q		
L. fasciola	0.5	Q		Q	0.25					Q			
L. ovata					0.25					Q	Q	0.25	
Lasmigona costata					0.25	Q				Q	0.25		
Leptodea fragilis					0.25								
Lexingtonia dolabelloides					Q					Q			
M. conradicus	0.25	0.5			3.25	1.0		0.25				0.25	
Plethobasus cyphyus					0.75								
P. oviforme	1.25	1.5		1.0	Q					0.5			
Potamilus alatus					0.25					0.25	0.25	0.25	
P. fasciolaris	0.25	Q		0.5	Q	Q				Q	0.25	0.25	
P. subtentum		Q			0.25						Q	Q	
Q. cylindrica strigillata					Q								
Q. quadrula													
Strophitus undulates					Q						Q	Q	
V. iris	8.25	6.0		3.5	0.25	Q			0.5	Q	Q	Q	
V. v. vanuxemensis					Q					Q			
Total Density	10.5	8.0	0	5.0	10.25	1.0	0	1.0	1.0	2.5	2.5	2.0	0
Total No. Species	5	10	0	5	23	4	0	5	6	18	14	14	0
% Young (≤5 Years)	23.5	19.4	0	0	25.0	0	0	0	10.0	5.6	2.0	0	0

[a] Collected in qualitative surveys.

[b] Family Margaritiferidae.

TOXICITY TESTING

In situ toxicity testing of the rainbow (*V. iris*) on two occasions in 1995 resulted in no significant effects on either survival or growth over six-week exposures. In both tests, survival was 80–100% at all sites, while mean growth (shell length) at each site ranged from 0.8 to 1.8 mm. Laboratory toxicity testing of silt samples collected from each site on two occasions in 1995 resulted in significant effects on hatching success of the rotifer, *B. calyciflorus*, for three sites in at least one of the tests. In the first test, Stations 1 (53.1% of controls), 4 (56.6%), and 11 (67.4%) exhibited hatching rates significantly below the control rate. In the second test, only Station 11 (62.8% of controls) hatching rates were significantly below the control rate.

DISCUSSION

Significant associations of environmental variables with three measures of mussel community structure (density, species richness, percentage of young) using Spearman's rank correlation procedure were dominated by habitat variables; however, land use characteristics and water/sediment quality measures were also important (Table 12.4). The strongest associations were found for greater substrate penetrability and higher fingernail clam (*Musculium* and *Pisidium* spp.) density (positive correlations); proximity to sewage treatment plant discharges, however, was a significant negative association.

The relationship between environmental variables that exhibited significant differences or correlations between sampling sites and the mussel community structure at those sites was further explored using canonical correspondence analysis (CCA), which examines groups of variables simultaneously, rather than single-pair correlation analysis (Ter Braak 1986). Environmental variables were dropped from the analysis when co-linearity with other variables was detected, retaining the variable with the strongest statistical relationship to mussel community structure.

The resulting relationship of environmental variables to mussel community structure is depicted in Figure 12.2. Eigen values of the first two canonical correspondence analysis axes were 0.248 and 0.049. These two axes explained 83.4% of the cumulative variance of the three mussel community measures and 83.2% of the mussel–environment relationship. The 13 sampling

TABLE 12.4
Spearman Rank Correlation Coefficients (ρ) of Statistically Significant Environmental Variables Associated with Three Mussel Community Measures

Environmental Variable	Mussel Density	Species Richness	% Young
Mussel tissue Zn	−0.546[a]	−0.286	−0.634[a]
Fingernail clam density	0.676[b]	0.573[a]	0.500
Direct sunlight	−0.456	−0.510[a]	−0.423
Substrate penetrability	0.770[b]	0.633[a]	0.458
Riffle length	−0.455	−0.513[a]	−0.571[a]
Upper pool length	−0.611[a]	−0.288	−0.584[a]
Phosphates	0.602[a]	0.389	0.662[b]
% Urban	−0.605[a]	−0.505	−0.501
% Mined land	−0.481	−0.496	−0.634[a]
Bend-to-bend ratio	−0.633[a]	−0.481	−0.496
STPs	−0.663[b]	−0.484	−0.499

[a] Significant at $P < .05$.
[b] Significant at $P < .01$.

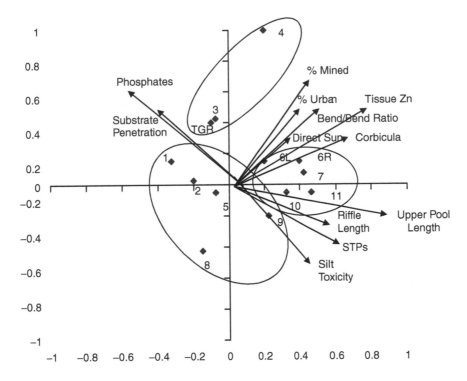

FIGURE 12.2 CCA ordination diagram relating mussel community structure at 13 Clinch River sites to 12 environmental variables.

sites fell into three distinct groups, as shown in Figure 12.2, corresponding to the combined mussel–environmental data. The group including Stations 1, 2, 5, 8, and 9 reflected the best mussel assemblages and were the sites least influenced by strongly negative environmental factors such as the percentage of mined land, percentage of urban area, and clam tissue zinc concentration. Conversely, the group containing Stations 3, 4, and TGR represented the sites with extremely poor or absent mussel fauna and were strongly influenced by proximity to mined lands, urban areas, and low substrate penetrability. The remaining five sampling sites were intermediate in the three mussel measures, with a large variety of habitat and land use influences.

The three site groups and important associated environmental variables identified by CCA are further contrasted in Table 12.5, which indicates mean values of measured variables. Clearly, the percentage of young was the community statistic that most strongly defined the separation of Stations 1, 2, 5, 8, and 9 from the other sites. The remaining two groups exhibited similar percentages of young and mussel density but were sharply separated by species richness.

The negative influence of mined land on Clinch River mussels is evident when the variables percentage of mined land, clam tissue zinc concentrations, dissolved manganese concentrations in the interstitial water, and substrate penetrability are compared between site groupings (Table 12.5). Although the tissue zinc concentrations and dissolved manganese levels were probably not high enough to cause toxicity to mussels, the elevated levels compared to the other sites are indicative of chronic exposure to contaminated runoff. Likewise, exceedance at the Guest River site of a sediment nickel consensus-based probable effect concentration of 48.6 mg/kg (MacDonald et al. 2000) and the significantly higher manganese in clam tissues from this site provide further evidence of likely mining-related effects. Elevated concentrations of manganese, nickel, and zinc, although individually known from a variety of sources, together constitute a signature of contamination from coal mine sources (see Chapter 13). The impact is further illustrated in Figure 12.3,

TABLE 12.5
Comparison of Environmental Characteristics between Groups of Sites Identified by Canonical Correspondence Analysis

Environmental Variable	Mean: Sites 1, 2, 5, 8, & 9	Mean: Sites 3, 4, & TGR	Mean: Sites 6L, 6R, 7, 10, & 11
Mussel % young	16.70	0.00	0.40
Mussel density (#/m^2)	6.50	1.70	1.30
Mussel species richness	12.40	1.70	7.40
Corbicula dens. (#/m^2)	167.00	322.00	363.00
Silt toxicity (% ref.)	85.70	85.00	90.60
Mussel tissue Zn (μg/g)	266.00	336.00	283.00
Phosphate (as P, mg/L)	0.32	0.17	0.11
Direct sunlight (hours)	5.50	7.90	8.00
Substrate pen. (cm)	11.40	6.60	10.20
Riffle length (m)	46.00	71.00	94.00
Upper pool length (m)	84.00	165.00	161.00
Bend/bend ratio (m/m)	17.00	25.00	29.00
% Urban	1.40	6.80	0.70
% Mined land	0.20	2.60	1.20
STPs	2.00	2.30	1.60
Other parameters of interest			
Fingernail clams (#/m^2)	1.9	3.5	0.0
Cellulolytic act. (% ref.)	59.1	39.4	45.8
Dissolved Mn (μg/L)	<10	43.0	10.0
% Boulder	0.7	4.4	2.8
% Silt	3.2	5.0	3.3
IBI	54.0	44.0	55.0

which shows substantial drops in mussel density, species richness, and percentage of young corresponding to increased site influences from mined land.

Past mining activities may also have been responsible for effects on habitat quality. The riffle/run substrate at Station TGR is measurably compacted, exhibiting the lowest penetrability among all sampling sites (mean = 5.3 cm). Substrate penetrability among sampling sites was significantly correlated to both mussel density and species richness (Table 12.4; Figure 12.4). Gray (1996) described the characteristic precipitation of solids and sedimentation effects associated with mining impacts. The negative effects of sedimentation on unionid mussels were reviewed by Brim Box and Mossa (1999), who found that both bed load and suspended sediments can impact mussels. The sediment compaction observed at Station TGR may limit colonization and habitation of this area by inhibiting the burrowing activity of mussels. A controlled laboratory study by Yeager et al. (1994) found that 98.5% of juvenile *V. iris* burrowed into sediment.

A study of fish populations in the Clinch River drainage by Angermeier and Smogor (1993) found that, similar to the mussel data presented here, the Guest River fish fauna was in poor condition, primarily due to impacts from nonpoint mining and agricultural runoff. A watershed-scale analysis of the Clinch River by Diamond and Serveiss (2001) determined a statistically significant negative relationship between the number of mussel species and, in decreasing order of significance, the percentage of urban land use, the proximity to mining, and the percentage of cropland. A significant association was not demonstrated in the present study between mussel community structure at the 13 sampling sites and the percentage of agricultural land use. This

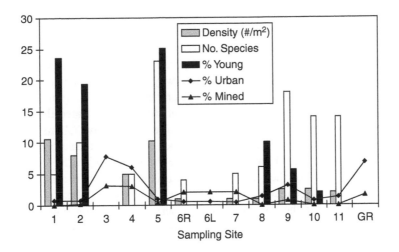

FIGURE 12.3 Relationship of urban and mine land uses with three mussel community measures at 13 Clinch River sites.

lack of influence was supported by the clam tissue measurements of cholinesterase activity, indicative of agricultural pesticide exposure, which did not vary significantly from control activities in two sampling events at each site.

As depicted in Figure 12.2 and Figure 12.3, a very strong influence of urban development on mussel populations was demonstrated in this study. Related to the strong correlation of mussel density to urban land use was a similar correlation to the number of near-upstream sewage treatment plant (STP) discharges (Table 12.4). An assessment by Neves and Angermeier (1990) concluded that, although Clinch River water quality was generally good, there was local degradation associated with STP discharges. A study of two specific STP discharges in the upper Clinch River documented adverse effects on mussels for up to 3.7 km downstream (Goudreau et al. 1993).

In addition to sewage treatment plants, urban areas represented an elevated risk to the Clinch River due to associated human activity, as manifested by storm sewer runoff and an increased incidence of transportation-related and other types of pollutant spills. These types of ephemeral

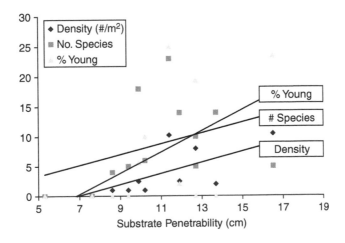

FIGURE 12.4 Relationship of substrate penetrability with three mussel community measures at 13 Clinch River sites.

events represented the primary type of impact affecting Clinch River mussels. In fact, a global perspective on biological monitoring by Karr and Chu (2000) indicated that urbanization may be the current most important influence on river ecology. These authors cited studies that correlated degraded fish and benthic communities to elevated percentages of impervious land surfaces, which cause high levels of stormwater runoff to reach waterways via ditches and culverts and lead to far more ecological damage than agriculture, logging, or other land uses. Extensive monitoring of Asian clam cellulolytic enzyme activity in the present study demonstrated the episodic nature of stressful environmental conditions, both spatially and temporally. Clams at virtually every sampling site exhibited significantly depressed enzyme activity on at least one occasion. Overall, clams at Stations 1, 4, 6L, 8, 10, 11, and TGR were the most frequently affected. These seven sites encompass the entire length of the study area. Of particular interest is a significant correlation ($\rho = 0.506$, $P < .05$) between concurrent enzyme measurements and silt toxicity (Figure 12.5). This finding provides support for the contention that observed impacts on mussels may have been associated with ephemeral, often runoff-related events. In situ toxicity testing of *V. iris* juveniles by Cherry et al. (1996) measured effects on growth and survival at several Clinch River sites when a runoff-producing storm event occurred during the test; however, no effects were measured at the same sites during an identical subsequent test when no storm event occurred.

A significant association between interstitial water phosphorus concentrations and higher mussel density and the percentage of young was found in this study (Table 12.4). Elevated phosphorus may be attributable to runoff from either urban or agricultural sources and was primarily confined to the upstream sites (Table 12.2). The mussels at these sites may have responded positively to this slight enrichment or, at least equally likely, the association may have been spurious.

Additional study findings reflect the influence of habitat, and associated fish and non-unionid bivalve distributions, on mussel distribution. Other than the influence of substrate penetrability on mussel density and species richness discussed above, substrate composition had no measurable effect on mussel community measures. An important caveat to this result is that the 13 sampling sites in this study were all riffle-run areas that were selected on the basis of previous knowledge of mussel distribution. Mussel populations are characteristically highly aggregated according to depth and current (Strayer and Ralley 1993); if sites having a less favorable habitat had been included in the study, more significant habitat relationships might have been apparent. An extensive longitudinal mussel survey of the Clinch River determined that mussels were concentrated in riffles or shoals (and the pool areas immediately upstream) but that, in general, most species exhibited no apparent

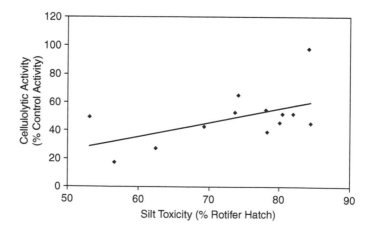

FIGURE 12.5 Relationship between Asian clam cellulolytic enzyme activity and toxicity of concurrent silt samples from 13 Clinch River sites.

microhabitat preference within these areas (Ahlstedt 1984). In this study, it was decided to reduce the influence of natural habitat factors in order to increase the likelihood of detecting anthropogenic effects.

Channel morphology, as measured by riffle length, upper pool length, and bend-to-bend ratio, was significantly correlated to all three measures of mussel community structure (Table 12.4). These results reflected a tendency for stronger mussel assemblages in areas of shorter riffle and pool lengths and smaller bend-to-bend ratios. These sites offered a wider range of habitat choices for both mussels and associated fish host species (e.g., Stations 2 and 5), as opposed to sites characterized by long, homogeneous stretches of habitat (e.g., Stations 4 and 11). Sites with shorter riffle-run reaches, particularly shoal areas, tended to concentrate mussels into smaller areas of desirable habitat.

An unexpected association of the most productive mussel sites with fewer hours of direct sunlight was found (Table 12.4 and Table 12.5). This does not appear to represent a phototactic response, but rather, it most likely reflected the observed tendency for mussels to concentrate near the bases of large riparian trees. Fish sampling demonstrated that a large majority of the collected centrarchids were caught among the rootwads of these trees. Since centrarchids are host species for most of the mussel species listed in Table 12.3 for which hosts have been identified, large numbers of juvenile mussels would be expected to drop into these areas following the parasitic stage. Consequently, areas near trees that have acceptable habitats should contain higher mussel densities than surrounding areas, subject to variations in water depth and current that could influence juvenile colonization and downstream transport, such that the relationship to sunlight was most likely a covarying factor.

The importance of the fish host in determining mussel distribution was inferred by the fish sampling results, as summarized by the IBI (Table 12.5). For the majority of sites, fish assemblages were sufficiently diverse and abundant that host availability was most likely not a limiting factor. However, the sites with the lowest IBI scores (Stations 3, 4, and TGR) were also among the least productive mussel sites. Although it is impossible from this analysis to separate the influence of factors affecting mussels directly versus indirectly via host fish effects, both pathways were probable influences on the current distribution of Clinch River mussels. Possible support for this view was provided in an analysis of watershed-scale trends in the Clinch River by USEPA (2000), which found that fish IBI scores were better predictors of mussel species richness than macroinvertebrate Ephemeroptera–Plecoptera–Trichoptera (EPT) scores. One might expect that macroinvertebrate community metrics should intuitively reflect mussel data because both groups are sediment-associated.

Less clear is the relationship between the distribution of fingernail clams and the Asian clam on unionid community structure. Fingernail clam (*Musculium* and *Pisidium* spp.) density was strongly correlated to both unionid density and species richness (Table 12.4). Although individual species within the Sphaeriidae exhibit a wide range of tolerances to various environmental stresses (Fuller 1974), the species collected in this study were possibly influenced by some of the same environmental factors associated with unionid distribution. However, statistical correlations between fingernail clam density and measured environmental variables were not significant.

Asian clam density varied significantly between sampling sites and was a prominent factor in the ordination analysis (Figure 12.2; Table 12.5). Low numbers of young mussels (aged 5 years or less) were observed at sites of both high and low Asian clam densities; however, high numbers of young mussels did not occur at sites that exhibited high Asian clam densities (Figure 12.6). Thus, the possibility of a negative influence of Asian clams on juvenile unionids cannot be dismissed. The comprehensive survey of the river by Ahlstedt (1984) produced no evidence that Asian clams were actually displacing native mussels; effects on juveniles, though, may not be reflected in the results of adult surveys for many years because of the known longevity of these species. In fact, in a follow-up survey, Ahlstedt and Tuberville (1997) noted a lack of recruitment at many sites. The nature of a possible negative influence of Asian clams on juvenile unionids is unknown.

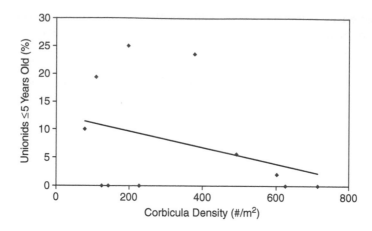

FIGURE 12.6 Relationship between Asian clam density and the percentage of unionid mussels aged 5 years or less at 13 Clinch River sites.

Vaughn and Hakenkamp (2001), in their review of the functional role of burrowing bivalves, discussed a number of ways in which Asian clams might impact native mussels, ranging from direct predation on juveniles to competition for food or space. These authors concluded, however, that there was no hard evidence that such impacts had occurred. Studies by Cherry et al. (1996) demonstrated that elevated concentrations of ammonia caused by population crashes of densely aggregated Asian clams, a frequent event in the Clinch River, could create conditions that are toxic to resident unionids. A laboratory study by Yeager et al. (2000) demonstrated density-dependent adult Asian clam impacts on juvenile *V. iris* survival, growth, and positioning. To date, there is only circumstantial evidence of negative influences of Asian clams on native mussels, and further study is required before a definitive judgment can be made.

The lack of juvenile recruitment documented in recent surveys combined with the identification of the proportion of young mussels as a key measure of environmental impact together point to juvenile recruitment as the critical element with respect to the status of Clinch River mussels. Seven of the thirteen sampling sites produced no mussels aged 5 years or less in combined quantitative and qualitative sampling. Ahlstedt and Tuberville (1997) made a qualitative observation regarding a pervasive lack of recruitment based on extensive sampling throughout the river. A mussel population study of six species at four Clinch River sites by Scott (1994) documented good recruitment of juveniles at two upstream sites (identical to Stations 1 and 5 in this study) and a third site located in Tennessee but almost no recruitment at CRM 223.5 (located between Stations 10 and 11 of this study). The particular sensitivity of mussel early life stages to environmental perturbations inferred by the data presented here agrees with the findings of other studies. Laboratory toxicity testing of several mussel life stages by Jacobson et al. (1997) demonstrated that early juveniles were the most sensitive to copper. Dunn et al. (Chapter 13) reported that the only detectable impact of a release of partially treated mine water to an Ohio River mussel bed was on young mussels aged 5 years or less.

Basin-wide studies to quantify how contaminants affect freshwater mussels have not been published. A desktop risk assessment of available Clinch River mussel survey data, regressed against land use map data and other selected geographic characteristics, indicated statistical associations between mussel species richness and urban, mining, and agricultural land uses (Diamond and Serveiss 2001). Similarly, few published site-specific mussel impact assessments, other than that of Dunn et al. (Chapter 13) summarized above, were found. Fleming et al. (1995) used tissue enzyme analysis to attribute a mussel die-off in Swift Creek, North Carolina, to agricultural pesticide use. Goudreau et al. (1993) performed a field and laboratory impact assessment of two sewage treatment plants on Clinch River mussels and attributed observed effects to chloramines in the discharges. Farris et al. (1988) measured depressed cellulolytic enzyme activity in Asian clams transplanted downstream of a power plant discharge in an area of the Clinch River devoid of native mussels.

A prominent historical event on the Clinch River was the 1967 spill from a coal fly ash treatment pond at Appalachian Power Company's (APCo's) steam-electric generating plant located at Carbo (CRM 267). This spill caused a documented fish kill covering 66 miles (106 km) of the Clinch River downstream of the facility (Crossman 1973). Subsequent to this event, authors of Clinch River mussel assessments often mention this event as also impacting mussels for a distance of approximately 18 miles (29 km), usually citing either Crossman (1973) or Cairns et al. (1971); however, no mussel kill assessment was actually performed, and the only post-spill recovery sampling by these investigators of mussels was the incidental occurrence of a few specimens collected using kicknets. The results of the recovery sampling combined mussels and snails into "mollusks," which were mostly snails. However, the recovery work and later surveys, particularly that of Ahlstedt (1984), did document a virtual absence of unionids downstream of the plant for a distance of approximately 12 miles (19 km). This finding precipitated NPDES permit-mandated studies by APCo that determined that the cause of the impact was elevated copper concentrations in a plant effluent—a situation that had most likely existed before the spill to when the facility was constructed in the 1950s. A state-of-the-art treatment facility completed in 1993 resulted in significantly reduced loadings of copper and, combined with earlier measures to reduce copper concentrations, has led to the recolonization of some of the previously impacted downstream areas, as shown for Stations 7 and 8 (Table 12.3).

The reader should consider that the results of this study represent only the conditions that existed at the time of sampling and may not reflect past influences that contributed to the present-day distribution of Clinch River mussels. From a historical perspective, species richness documented in this study was fairly similar to that measured by Ahlstedt (1984) from 1978 to 1983 at five sites in common, but both showed a substantial decline at four of the five sites when compared to the 1912–1913 survey by Ortmann (1918) (Table 12.6). Reasons for this decline postulated by a number of researchers include coal mining, pre-Clean Water Act industrial and municipal discharges and spills, and nonpoint runoff from agricultural and urban sources (Neves et al. 1980; Ahlstedt 1983; Hampson et al. 2000). In general, these are the same factors linked to mussel distribution by this study, with the main difference being the declining impact of point source discharges as they have been brought into compliance with technology-based and water quality-based standards.

In summary, evaluation of biological and environmental data from 13 sampling sites, encompassing most of the 146 miles (235 km) of the Virginia portion of the Clinch River, has

---

TABLE 12.6
Species Richness at Five Clinch River Sites Sampled in Three Surveys, 1912–1994

Sampling Station	Ortmann, 1912–1913	Ahlstedt, 1978–1983	This Study, 1992–1994
2	17	10	10
3	16	NS[a]	0
5	24	12	23
8	25	10	6
11	32	14	14

[a] Not sampled.

demonstrated strong statistical relationships between mussel community structure and the following independent variables: habitat, riparian land use, and water and sediment quality. Specifically, metals and sediment compaction associated with coal mining areas, and sewage treatment plants and ephemeral toxic stormwater runoff events from urban areas were identified as negative influences on mussels. Additional important factors were fish host distributions, habitat diversity, and possibly Asian clam density. Recruitment of young mussels was particularly sensitive to these perturbations.

Recommendations for future large-scale mussel impact assessments can be made based on the experience gained during this study. Site selection is crucial, both in terms of defining a gradient of effects and minimizing the tendency of habitat differences between sites to obscure the effects of other factors affecting mussels. A sampling guide (e.g., Strayer and Smith 2003) should be consulted to match the sampling approach and level of effort to study objectives. Basinwide studies must consider large-scale influences on the watershed in addition to measurements of site-specific environmental characteristics. To investigate the influence of toxicants, physiological indicators should be included, since mussel community indices and water quality sampling alone may miss significant influences. Effects on mussel early life stages must be investigated. As seen in this study, documenting effects related to stormwater events may be critical, as well. The use of Asian clams for biochemical exposure indicators was based on previous Clinch River research by Cherry et al. (1991), demonstrating that Asian clam cellulolytic enzyme activity responded similarly to that of unionid mussels in side-by-side in situ and artificial stream toxicity tests. Continued use of the Asian clam and other potential surrogates, such as the rotifer species used for silt toxicity testing, is encouraged for future mussel assessments requiring the sacrifice of large numbers of animals. Asian clam and rotifer results in this study were mutually supported by the instream measurements on native mussels and provided reasonable and logical interpretations of observed conditions.

The findings of this study can help better define recommendations for improvement of conditions for mussels in the Clinch River. Human activity in the watershed will continue to increase, so any recommendations must take a realistic view of available watershed management options. Unfortunately, the results of this study have shown that the entire length of the Clinch River in Virginia is at risk from ephemeral insults that can impact the mussel fauna. This means that proposed improvement options should consider the entire watershed; however, site-specific improvement activities are important, and several of them collectively can result in measurable improvement to the watershed environment. Specifically needed are active management of riparian land use practices, including the formulation of best management practice plans to address toxicants reaching the river via stormwater runoff, and a continuation of efforts already underway to correct existing situations, such as reclamation of abandoned minelands. Through such efforts, a favorable outlook for the Clinch River mussel fauna is yet possible. The documented faunal decline is not irreversible, particularly if nonpoint influences on the watershed can be brought under control.

ACKNOWLEDGMENTS

I greatly appreciate the assistance of Ken Wood in carrying out the considerable amount of field work and laboratory toxicity testing involved in this project. Don Cherry and a number of his former graduate assistants at Virginia Tech, particularly Mindy Yeager and Jennifer Scott, lent invaluable assistance in the quadrat sampling and electrofishing. Thanks to Jerry Farris of Arkansas State University for completing all of the cellulase assays. American Electric Power's John E. Dolan Laboratory provided analytical services. Malinda Hafley assisted with graphics, and Rob Reash and Tim Lohner provided insightful reviews of the manuscript.

REFERENCES

Ahlstedt, S., Recovery plan for the birdwing pearly mussel *Conradilla caelata* (Conrad, 1834), Report prepared for U.S. Fish Wildl. Serv., Southeast Region, Atlanta, GA, 1983.

Ahlstedt, S. A., Twentieth century changes in the freshwater mussel fauna of the Clinch River (Tennessee and Virginia) [MSc Thesis], Univ. Tennessee, Knoxville, TN, 1984.

Ahlstedt, S. A. and Tuberville, J. D., Quantitative reassessment of the freshwater mussel fauna in the Clinch and Powell Rivers, Tennessee and Virginia, In *Conservation and Management of Freshwater Mussels II. Initiatives for the Future*, Cummings, K. S., Buchanan, A. C., Mayer, C. A., and Naimo, T. J., Eds., UMRCC Symposium Proceedings, 1995 October 16–18, St. Louis MO, pp. 72–97, 1997.

Angermeier, P. L. and Smogor, R. A., Assessment of biological integrity as a tool in the recovery of rare aquatic species. Report prepared for Virginia Dept. Game Inland. Fish., Richmond VA., Virginia Tech., Blacksburg, VA, 1993.

Bates, J. M. and Dennis, S. D., The mussel fauna of the Clinch River Tennessee and Virginia, *Sterkiana*, 69–70, 3–23, 1978.

Brim Box, J. and Mossa, J., Sediment, land use, and freshwater mussels: prospects and problems, *J. N. Am. Benthol. Soc.*, 18, 99–117, 1999.

Bruenderman, S. A. and Neves, R. J., Life history of the endangered fine-rayed pigtoe *Fusconaia cuneolus* (Bivalvia: Unionidae) in the Clinch River, Virginia, *Am. Malacol. Bull.*, 10, 83–91, 1993.

Cairns, J., Crossman, J. S., Dickson, K. L., and Herricks, E. E., The recovery of damaged streams, *Assoc. Southeast. Biol. Bull.*, 18, 79–106, 1971.

Cherry, D. S., Farris, J. L., and Neves, R. J., Laboratory and field ecotoxicological studies at the Clinch River Plant, Virginia. Report prepared for American Electric Power Company. Virginia Polytech. Inst. State Univ., Biol. Dept., Univ. Center Environ. Stud., Blacksburg, VA, p. 228, 1991.

Cherry, D. S., Yeager, M. M., Balfour, D. L., Church, G. W., Scott, J. F., Scheller, J. L., and Neves, R. J., Sources of pollutants influencing sediment toxicity and the mussel fauna in the Clinch River drainage system—an on-site investigation. Final report. Virginia Polytech. Inst. State Univ., Biol. Dept., Univ. Center Environ. Stud., Blacksburg, VA, 1996.

Crossman, J. S., The biological recovery of the Clinch River following pollutional stress [Ph.D. Dissertation]. Virginia Polytech. Inst. State Univ., Blacksburg, VA, 1973.

Dennis, S. D., An unexpected decline in populations of the freshwater mussel, *Dysnomia* (= *Epioblasma*) *capsaeformis*, in the Clinch River of Virginia and Tennessee, *Va. J. Sci.*, 38, 281–288, 1987.

Diamond, J. M. and Serveiss, V. B., Identifying sources of stress to native aquatic fauna using a watershed ecological risk assessment framework, *Environ. Sci. Technol.*, 35, 4711–4718, 2001.

Farris, J. L., Van Hassel, J. H., Belanger, S. E., Cherry, D. S., and Cairns, Jr., J., Application of cellulolytic activity of asiatic clams (*Corbicula* sp.) to in-stream monitoring of power plant effluents, *Environ. Toxicol. Chem.*, 7, 701–713, 1988.

Farris, J. L., Belanger, S. E., Cherry, D. S., and Cairns, Jr., J., Cellulolytic activity as a novel approach to assess long-term zinc stress to *Corbicula*, *Water Res.*, 23, 1275–1283, 1989.

Farris, J. L., Cherry, D. S., and Neves, R. J., Validation of copper concentrations in laboratory testing for site-specific copper criteria in the Clinch River. Report prepared for American Electric Power Company. Virginia Polytech. Inst. State Univ., Biol. Dept., Univ. Center Environ. Haz. Mat. Stud., Blacksburg, VA, p. 54, 1991.

Fleming, W. J., Augspurger, T. P., and Alderman, J. A., Freshwater mussel die-off attributed to anticholinesterase poisoning, *Environ. Toxicol. Chem.*, 14, 877–879, 1995.

Fuller, S. L. H., Clams and mussels (Mollusca: Bivalvia), In *Pollution Ecology of Freshwater Invertebrates*, Hart, C. W. and Fuller, S. L. H., Eds., Academic Press, New York, pp. 215–273, 1974.

Goudreau, S. E., Neves, R. J., and Sheehan, R. J., Effects of wastewater treatment plant effluents on freshwater mollusks in the upper Clinch River, Virginia, USA, *Hydrobiologia*, 252, 211–230, 1993.

Gray, N. F., A substrate classification index for the visual assessment of the impact of acid mine drainage in lotic systems, *Water Resour.*, 30, 1551–1554, 1996.

Hampson, P. S., Treece, Jr., M. W., Johnson, G. C., Ahlstedt, S. A., and Connell, J. F., *Water Quality in the Upper Tennessee River Basin, Tennessee, North Carolina, Virginia, and Georgia 1994–1998*, U.S. Geol. Surv., Water Resour. Div., Circ. 1205, Reston, VA, 2000.

Helfrich, L. A., Weigmann, D. L., Neves, R. J., and Bromley, P. T., *The Clinch, Powell, and Holston Rivers of Virginia and Tennessee: Wildlife and Water Quality*, Virginia Polytech. Inst. State Univ., Dept. Fish. Wildl. Sci., Virginia Coop. Ext. Serv. Publ. 460-110, Blacksburg, VA, 1986.

Jacobson, P. J., Neves, R. J., Cherry, D. S., and Farris, J. L., Sensitivity of glochidial stages of freshwater mussels (Bivalvia: Unionidae) to copper, *Environ. Toxicol. Chem.*, 16, 2384–2392, 1997.

Karr, J. R. and Chu, E. W., Sustaining living rivers, *Hydrobiologia*, 422–423, 1–14, 2000.

MacDonald, D. D., Ingersoll, C. G., and Berger, T. A., Development and evaluation of consensus-based sediment quality guidelines for freshwater ecosystems, *Arch. Environ. Contam. Toxicol.*, 39, 20–31, 2000.

Masnik, M. T., Composition, longitudinal distribution, and zoogeography of the fish fauna of the upper Clinch system in Tennessee and Virginia [Ph.D. Dissertation]. Virginia Polytech. Inst. State Univ., Blacksburg, VA, 1974.

Naimo, T. J. and Monroe, E. M., Variation in glycogen concentrations within mantle and foot tissue in *Amblema plicata plicata*: Implications for tissue biopsy sampling, *Am. Malacol. Bull.*, 15, 51–56, 1999.

Neves, R. J., Recovery plan, fine-rayed pigtoe pearly mussel (*Fusconaia cuneolus*). Report prepared for U.S. Fish Wildl. Serv., Atlanta, GA, 1984.

Neves, R. J. and Angermeier, P. L., Habitat alteration and its effects on native fishes in the upper Tennessee River system, east-central U.S.A., *J. Fish Biol.*, 37(Suppl A), 45–52, 1990.

Neves, R. J. and Odom, M. C., Muskrat predation on endangered freshwater mussels in Virginia, *J. Wildl. Manage.*, 53, 934–941, 1989.

Neves, R. J., Pardue, G. B., Benfield, E. F., and Dennis, S. D., An evaluation of endangered mollusks in Virginia. Report prepared for Virginia Comm. Game Inland Fish., Proj. No. E-F-1., Virginia Polytech. Inst. State Univ., Blacksburg, VA, 1980.

Ortmann, A. E., The nayades (freshwater mussels) of the upper Tennessee drainage, with notes on synonymy and distribution, *Proc. Amer. Philos. Soc.*, 57, 521–626, 1918.

SAS Institute, *SAS® User's Guide*, SAS Institute, Cary, NC, 1985.

Scott, J. C., Population demographics of six freshwater mussel species (Bivalvia: Unionidae) in the upper Clinch River, Virginia and Tennessee [M.Sc. Thesis]. Virginia Polytech. Inst. State Univ., Blacksburg, VA, 1994.

Snell, T. W. and Persoone, G., Acute toxicity bioassays using rotifers. II. A freshwater test with *Brachionus rubens*, *Aquat. Toxicol.*, 14, 81–92, 1989.

Stansbery, D. H., A preliminary report on the naiad fauna of the Clinch River in the southern Appalachian Mountains of Virginia and Tennessee (Mollusca: Bivalvia: Unionoida), *Bull. Amer. Malacol. Union*, March, 20–22, 1973.

Strayer, D. L. and Ralley, J., Microhabitat use by an assemblage of stream-dwelling unionaceans (Bivalvia), including two rare species of *Alasmidonta*, *J. N. Am. Benthol. Soc.*, 12, 247–258, 1993.

Strayer, D. L. and Smith, D. R., A guide to sampling freshwater mussel populations, *Am. Fish. Soc. Monogr.*, 8, p 103, 2003. Bethesda, MD.

Ter Braak, C. J. F., Canonical correspondence analysis: A new eigenvector technique for multivariate direct gradient analysis, *Ecology*, 67, 1167–1179, 1986.

[USEPA] U.S. Environmental Protection Agency, *Methods for Chemical Analysis of Water and Wastes*, U.S. Environmental Protection Agency, Cincinnati, OH, EPA 600/4-79-020, 1979.

[USEPA] U.S. Environmental Protection Agency, *Test Methods for Evaluating Solid Waste. Volume 1A: Laboratory Manual Physical/Chemical Methods*, 3rd. ed., U.S. Environmental Protection Agency, Office of Solid Waste Emergency Response, SW846, Washington DC, 1986.

[USEPA] U.S. Environmental Protection Agency, *Methods for Measuring the Acute Toxicity of Effluents and Receiving Waters to Freshwater and Marine Organisms*, U.S. Environmental Protection Agency, Cincinnati, OH, EPA 600/4-90/027F, 1990.

[USEPA] U.S. Environmental Protection Agency, *Workshop Report on Characterizing Ecological Risk at the Watershed Scale*, U.S. Environmental Protection Agency, Washington, DC, EPA 600/R-99/111, 2000.

Vaughn, C. C. and Hakenkamp, C. C., The functional role of burrowing bivalves in freshwater ecosystems, *Freshw. Biol.*, 46, 1431–1446, 2001.

Yeager, M. M., Cherry, D. S., and Neves, R. J., Feeding and burrowing behaviors of juvenile rainbow mussels, *Villosa iris* (Bivalvia: Unionidae), *J. N. Am. Benthol. Soc.*, 13, 217–222, 1994.

Yeager, M. M., Neves, R. J., and Cherry, D. S., Competitive interactions between early life stages of *Villosa iris* (Bivalvia: Unionidae) and adult Asian clams (*Corbicula fluminea*), In *Freshwater Mollusk Symposium Proceedings*, Tankersley, R. A., Warmolts, D. I., Watters, G. T., Armitage, B. J., Johnson, P. D., and Butler, R. S., Eds., Ohio Biol. Surv., Columbus, OH, pp. 253–259, 2000.

13 Case Study: Impact of Partially Treated Mine Water on an Ohio River (U.S.A.) Mussel Bed—Use of Multiple Lines of Evidence in Impact Analysis

Heidi L. Dunn, Jerry L. Farris, and John H. Van Hassel

INTRODUCTION

In July 1993, an underground coal mine in southern Ohio underwent an emergency dewatering operation due to flooding from an adjacent abandoned mine. Because of the tremendous volume of water involved (~1 billion gallons) and the urgency of the situation in terms of preserving the mining operation, regulatory action allowed the removed mine water to bypass normal treatment. The mine water, therefore, only received partial treatment with a caustic to raise the pH. A substantial volume of mine water was released into Parker Run, a tributary of Leading Creek in Meigs County, Ohio, over a 28-day period. Aquatic life was virtually extirpated from approximately 29 km of Parker Run and Leading Creek, from the mine-water discharge point to the Leading Creek confluence with the Ohio River, due to acidic pH averaging about 4.5 and elevated concentrations of iron, copper, manganese, nickel, and zinc.

Extensive chemical and biological monitoring was undertaken to document the magnitude of impact caused by the mine-water release and to monitor stream recovery. As part of this program, potential impacts on the Ohio River were assessed. In particular, freshwater unionid mussel resources were considered at risk because of the known pollution sensitivity of these organisms and the presence of the federally endangered pink mucket, *Lampsilis abrupta*, from both upstream and downstream Ohio River pools (Zeto et al. 1987).

Mine-water chemistry and its effects on aquatic life have been extensively studied. Mine water is typically characterized by extremely acidic pH < 3 and elevated concentrations of several metals, particularly iron, which can reach concentrations in the thousands of parts per million,—three orders of magnitude above background concentrations in stream water (Parsons 1968; Short et al. 1990; Grippo and Dunson 1996). The ferrous (Fe^{2+}) form of iron predominates at acidic pH and low redox conditions and is gradually replaced by the ferric (Fe^{3+}) form as pH and oxidation potential increase. Both the direct toxic effects of the pH/metal interaction and the indirect effects of ferric iron precipitation on stream bottoms can be severe (Parsons 1968; Grippo and Dunson 1996). Parsons (1968) observed that mussels were the most sensitive group of organisms to acid mine drainage in a Missouri stream.

Measuring the potential mine-water release impact on Ohio River mussels was the logical choice because of their known sensitivity to a variety of environmental stressors compared to other groups of organisms (Simmons and Reed 1973; Green et al. 1989; Cherry et al. 2002). Mussels have provided specific evidence of impact in studies of acid mine drainage (Simmons and Reed 1973; Gardner et al. 1981). Density, species richness, size and age range, recruitment, mortality, and growth are useful metrics for assessing impacts to mussel communities. Cellulolytic enzyme activity and glycogen content of mussels are physiological measures that appear sensitive to environmental stress (Haag et al. 1993).

In this study, a combination of water and sediment chemistry, mussel community monitoring, and toxicological investigations was used to define the immediate and long-term impact of the 1993 mine-water release on Ohio River unionid mussels.

METHODS

SAMPLING LOCATIONS

Mussels (in the years 1993, 1994, 1995, and 1997), water (in 1993), and bottom sediments (in 1994 and 1995) were collected from the Ohio River at established locations upstream and downstream of the confluence of Leading Creek (Figure 13.1). These sites were selected to provide an upstream–downstream assessment of impact from mine water introduced into Leading Creek and were located in the Ohio-West Virginia portion of the upper Ohio River near Middleport, Ohio.

To assess conditions upstream of Leading Creek, a known mussel bed at Ohio River Mile (ORM) 252.6, approximately 2.4 km upstream of the Leading Creek confluence on the West Virginia side of the river, was selected in 1993 as an upstream reference site (Site 1). The site was located on a sharp outside bend, and depth was 6.1–11.0 m. Due to scouring discharges in winter 1993/1994, substrate at this site changed from predominately gravel/sand in 1993 to cobble/gravel/boulder in 1994. Mussel communities were also reduced due to scour, and diving conditions became unsafe, prompting a change of mussel sampling to ORM 254.0 for 1995 and 1997. This new upstream site (Site 1A) was 160 m upstream of the Leading Creek confluence on the Ohio side of the river and 4.5–8.7 m deep, with substrate of mostly sand and gravel. Water quality (from Aug. 2 to Sept. 2, 1993), sediment quality (in 1994 and 1995), and unionid mussels (in 1993 and 1994) were sampled at Site 1, and unionid mussels (in 1995 and 1997) were sampled at Site 1A.

The first known mussel bed downstream of Leading Creek occurred at ORM 255.5, approximately 2.25 km downstream of Leading Creek (Site 2). Depth within the bed ranged from 6.4 to 13.5 m, and substrate was gravel/cobble. Site 2 was between the water quality transects located 1.3 and 2.6 km downstream from the confluence of Leading Creek. Water quality (from Aug. 2 to Sept. 2, 1993), sediment quality (in 1994 and 1995), and unionid mussels (in the years 1993, 1994, 1995, and 1997) were sampled at Site 2. A few unionids had also been previously found at ORM 257.6 on the right descending bank of Eightmile Island, which is on the West Virginia side of the river approximately 5.6 km downstream of the Leading Creek confluence (Site 3). Water depth at Site 3 ranged from 2.7 to 4.4 m, and substrate was predominately sand and gravel. Site 3 was downstream of the downstream-most water quality transect (4.7 km downstream of Leading Creek). Unionids (in the years 1993, 1994, 1995, and 1997) and sediment (in 1994 and 1995) were sampled at Site 3.

To determine the extent of the impact area, water quality was also sampled in Leading Creek 150 m upstream of the Ohio River confluence and in the Ohio River downstream of the Leading Creek confluence at distances of 122 m, 183 m, 244 m, 366 m, 1.3 km, 2.6 km, and 4.7 km. Sediment was also sampled on the Ohio side of the river 160 m downstream of the Leading Creek confluence.

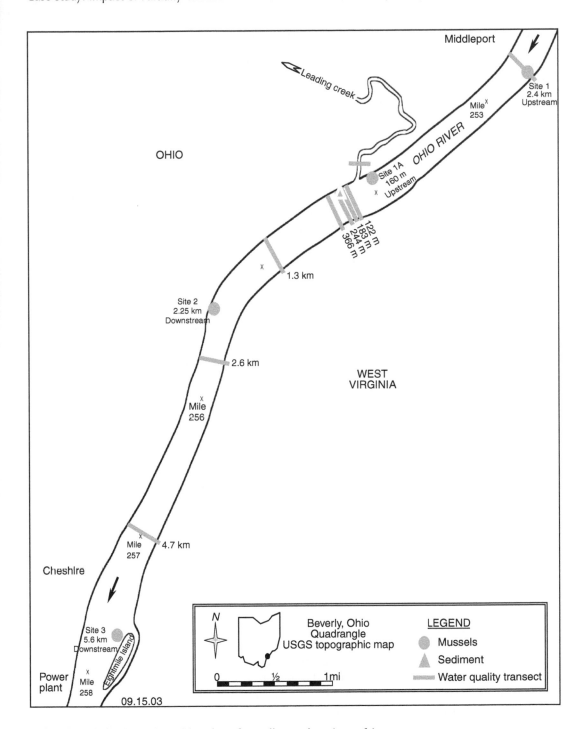

FIGURE 13.1 Study locale and location of sampling stations (mussels).

WATER CHEMISTRY

Water samples were collected daily beginning on August 2, 1993, when the partially-treated mine water first reached the Ohio River, through September 2, 1993, when sampling indicated that mine water discharging from Leading Creek was fully treated. For the first few days, each of the

Ohio River transects was sampled for temperature and conductivity from surface to bottom at approximately 0.3-m depth intervals and at several points across the width of the river. Mine water discharged from Leading Creek maintained a discrete plume at the bottom of the Ohio River due to its temperature (18–19°C) being cooler than that of the Ohio River. Sampling at each transect consisted of plume tracking using conductivity measurements. Temperature and conductivity were then sampled at approximately 0.3-km depth intervals at a point centered on the plume, and samples for further analysis were collected from the depth displaying the highest mine-water density (conductivity). Field measurement of pH and dissolved oxygen and sample preservation for laboratory analysis were performed according to USEPA (1979) methods.

In the laboratory, water hardness, conductivity, acidity, alkalinity, total suspended solids, sulfates, and the metals Ag, Al, As, Ba, Cd, Cu, Fe (total and ferrous), Hg, Mg, Mn, Ni, Pb, Se, and Zn were measured. All analyses except ferrous iron were performed according to USEPA (1979) methods, with total recoverable metal concentrations determined by an AES–ICP analyzer and FAA and GFAA spectrometers. Ferrous iron concentrations were determined by the phenanthroline method (American Public Health Association 1992).

SEDIMENT CHEMISTRY

Sediment samples were collected 2.4 km upstream and 160 m, 2.25 km, and 4.7 km downstream of Leading Creek in July, August, and September 1994, and July and August, 1995. At each site, samples were collected from depositional sediments using a ponar dredge and were preserved intact for laboratory analysis. In the laboratory, samples were dried for 48 hours at 50–60°C, sieved through 12 mesh (1.4 mm) to remove debris and large particles, then milled and sieved through 100 mesh (0.15 mm).

The portion of each sample to be analyzed for aluminum (as Al_2O_3) and iron (as Fe_2O_3) was digested in a closed PFA vessel using hydrofluoric, hydrochloric, nitric, and boric acids in a microwave digestion system. Another portion of each sample to be analyzed for Ag, As, Ba, Be, Cd, Cr, Cu, Hg, Mn, Ni, Pb, Sb, Se, and Zn was similarly digested using hydrochloric and nitric acids. The vessel pressures were not allowed to exceed 95 psi during the digestion. Metal concentrations were measured using an AES–ICP analyzer and FAA and GFAA spectrometers.

MUSSEL SURVEYS

Mussel beds located at 2.4 km (Site 1) and 160 m (Site 1A) upstream and 2.25 km (Site 2) and 4.7 km (Site 3) downstream of Leading Creek were sampled in August–September of 1993, 1994, and 1995, and June 1997, approximately two weeks, one, two, and four years following mine-water release. Sampling was attempted but not completed in 1996 because of extended high-water conditions on the Ohio River. Sampling at each site consisted of visual observation and quantitative and qualitative collecting. Before collecting, each site was visually assessed for substrate characteristics, relative mussel density, presence of dead shells, and mussel behavior (e.g., buried, siphoning, and gaping). At each site, four 100-m transect lines were established perpendicular to the riverbank at 100-m intervals. Twenty (20) quantitative samples were collected from 20 randomly selected points along each transect line in 1993. This sample size was only sufficient to detect a 50% change in unionid density; therefore, sample size was increased to 40 in subsequent years. However, the additional samples yielded only a slight decrease in sample variability. Each sample was collected by excavating a 0.25 m^2 area of substrate to a depth of 15 cm, brought to the surface in a bucket, then rinsed through a series of nested sieves with decreasing mesh size (12, 6, and 3 mm). All unionids were removed, and substrate composition was visually estimated and scored according to the Wentworth Scale.

To increase the probability of collecting rare species, each transect and areas of high unionid density were also qualitatively sampled by hand picking for two to three hours or until 100 unionids were recovered. All collected unionids, including relic shells, were identified to species, counted, measured for length (mm) and weight (g), aged (external annuli count), and returned to the river near the collection location. Age estimates were obtained by counting external annuli, which are prone to underestimate true age but is an acceptable method for younger age classes (Neves and Moyer 1988), which were of primary interest for this study.

Live mussel density (number/m^2), percent mortality (number of freshly dead shells compared to the number of live animals and freshly dead shells, with "freshly dead" defined as with or without soft parts, nacre lustrous, periostracum intact, and dead less than one year), percent *less than or equal to* three years of age, percent *less than or equal to five* years of age, and average age were calculated from quantitative samples. Species richness and species relative abundance were determined from the total of all unionids collected in quantitative and qualitative samples.

TOXICITY TESTING

Samples of the mine water that had been discharged to Leading Creek were shipped to Arkansas State University's Ecotoxicology Research Facility for acute and chronic toxicity testing during 1994–1995. Forty-eight-hour acute toxicity tests were performed on the fathead minnow (*Pimephales promelas*) and two cladocerans (*Daphnia magna* and *Daphnia pulex*) according to USEPA (1993) protocols. Two tests each were performed on the fathead minnow and *D. magna*, and one on *D. pulex*. Test treatments consisted of dilutions of the mine water, ranging from 1.25 to 100% mine water, using dechlorinated, carbon-filtered tap water for dilution and controls. Further description of test conditions is provided in Milam and Farris (1998).

In addition, two acute tests were performed on the fragile papershell mussel (*Leptodea fragilis*), measuring glochidial viability according to the methods of Jacobson et al. (1997). Gravid *L. fragilis* females were obtained from the field, and glochidia were excised from the marsupia for transfer into test chambers. Following 24-hour exposure to a series of mine-water dilutions, glochidia viability was determined by recording the number of animals with open versus closed valves before and after exposure to a saturated salt solution.

Chronic toxicity tests of the mine water were performed using the Asian clam *Corbicula fluminea* and the cladoceran *Ceriodaphnia dubia*. Two 30-day exposures of the Asian clam were conducted using an artificial stream system consisting of a series of 60-liter oval fiberglass streams (Farris et al. 1989). A current averaging 0.05 m/sec was maintained in each stream unit using fiberglass paddles powered by an electric motor. Mine-water treatments (three replicate streams per treatment) ranged from 1.25 to 20%, using dechlorinated, charcoal-filtered tap water for dilution and controls. Mine water was renewed in each treatment on a daily basis from a mine-water stock that was in turn renewed by weekly shipments from the mine.

Adult Asian clams obtained from an unimpacted area of the Saline River, Arkansas, were used in testing. Clams were held in baskets with a 1-cm^2 mesh size and were placed in each stream. During daily mine-water renewals, the baskets were removed and placed for approximately one hour in 2-liter polycarbonate vessels containing concentrated algae. A tri-algal mix (*Chlamydomonas*, *Ankistrodesmus*, and *Chlorella*) cultured in Bold's basic medium was dispensed daily to each container at a rate of 5 mL/L, which represented approximately 200 cells/mL as a suggested provisional feeding rate (USEPA 1989).

The chronic endpoints measured were cellulolytic enzyme activity for both tests and clam growth in length for Test 1. Twelve clams from each replicate stream were randomly chosen following the 30-day exposure and were dissected for cellulolytic enzyme analysis (Farris et al. 1989). The 1-g samples were homogenized in a phosphate buffer (0.15 M at pH 6.1) at a wet mass to buffer ratio of 0.02 g/mL. Samples were centrifuged for 15 minutes at 15,000 rpm, supernatants were decanted for endo/exocellulase analysis, and the pellets were recovered for dry mass measurements. Length in millimeters of ten individually marked clams in each replicate stream in Test 1 was measured at the start and end of the 30-day exposure using calipers.

TABLE 13.1
Toxicity Test Conditions (Mean $\pm 95\%$ C.I.; $n = 30$ Except for Hardness, Alkalinity, and 50–100% Exposures: $n = 4$)

Mine Water (%)	pH (s.u.)	Conductivity (µmhos/cm)	Dissolved Oxygen (mg/L)	Turbidity (NTU)	Hardness (mg/L)	Alka-linity (mg/L)	Total Fe (mg/L)	Ferrous Fe (mg/L)
0 (Control)	8.02±0.01	187±1	8.40±0.43	0.57±0.02	95±7	73±1	<0.1	<0.1
1.25	8.03±0.01	272±2	7.85±0.03	6.12±0.29	93±10	64±2	0.80±0.06	0.16±0.03
2.50	8.01±0.01	354±4	7.84±0.03	15.7±0.80	108±10	60±2	1.71±0.25	0.22±0.04
5.00	7.89±0.02	516±4	8.00±0.03	37.8±1.80	120±11	47±3	3.17±0.18	0.36±0.07
10.00	7.73±0.04	948±7	7.74±0.04	60.9±4.30	212±47	22±5	5.79±0.54	0.83±0.14
20.00	7.38±0.07	1696±15	7.83±0.03	184.0±8.20	324±41	22±6	12.40±0.87	2.54±0.39
50.00	7.27±1.49	4485±495	8.08±0.24	ND	590±55	24±8	ND	ND
75.00	7.00±2.63	6380±838	7.95±0.21	ND	930±77	18±5	ND	ND
100.00	6.83±1.65	7888±909	8.00±0.29	ND	1115±445	13±13	ND	ND

Three short-term chronic tests were performed on the mine water using *C. dubia* according to USEPA (1989) protocols. The three tests were performed consecutively during the first 30-day Asian clam test using the same stock of mine water. Treatment concentrations were the same as those used for the Asian clam exposures.

Water quality measurements for the toxicity tests are shown in Table 13.1 for the combined tests. All testing was performed at a temperature of 25 ± 1°C.

DATA ANALYSIS

Statistical analysis of study results for water and sediment chemistry (site comparisons by parameter), mussel surveys (site, year, and site times year comparisons for density and age), and the 30-day Asian clam toxicity tests (treatment vs. control) was performed using ANOVA (SAS Institute 1985). Results of the remaining acute and chronic toxicity tests were evaluated using Dunnett's test and the trimmed Spearman-Karber method (Hamilton et al. 1977). The implied significance level was $\alpha = 0.05$.

RESULTS

WATER CHEMISTRY

Ohio River water chemistry did not vary significantly between sampling sites prior to August 6, 1993, or subsequent to August 26, 1993. However, a statistically measurable influence of mine water discharging from Leading Creek was observed between these dates ($p < 0.05$). The mine-water plume at the bottom of the river was apparent as conductivity and temperature differed between the upper and bottom 0.5 m of the water column (Table 13.2). At the 183 m transect, the plume was approximately 3.7 m deep, measured vertically from the bottom and one-half of the river wide, compared to a total depth of 8.2–9.1 m and a total width of 366–457 m. The plume gradually expanded and became less concentrated with increasing distance from Leading Creek as it mixed with the ambient water, so that at the 2.6-km transect, the plume occupied approximately one-half of the water column and close to two-thirds of the river width. At the 4.7-km transect, the plume was no longer discrete, and water chemistry measurements were only slightly elevated compared to results upstream of Leading Creek. Of the suite of metals analyzed, only total and ferrous iron measurements within the mine-water plume were significantly elevated to levels expected to pose a risk to resident

TABLE 13.2
Ambient Water Chemistry for Leading Creek and the Ohio River, August 6–26, 1993 ($n = 21$; Mean $\pm 95\%$ C.I.)

Location[a]	Conductivity (Surface-µmhos/cm)	Conductivity (Bottom-µmhos/cm)	Temperature (Surface-°C)	Temperature (Bottom-°C)	Total Iron (mg/L)	Ferrous Iron (mg/L)
Ohio River Upstream	585±8	582±7	27.6±0.2	27.3±0.4	0.27±0.07	<0.1
Leading Creek	6635±956	6748±567	19.1±2.1	19.4±1.5	1362±594	554±190
Ohio River-122 m	592±11	1870±311	27.6±0.1	26.3±0.6	25.6±11.2	11.6±4.9
Ohio River-183 m	588±7	1623±309	27.7±0.3	26.5±0.4	23.0±6.40	10.9±5.5
Ohio River-244 m	588±7	1723±483	27.7±0.1	26.4±0.7	22.9±6.00	11.0±3.9
Ohio River-366 m	590±11	1450±631	27.7±0.3	26.6±0.8	ND	8.5±6.5
Ohio River-1.3 km	596±16	1185±400	27.7±0.4	26.8±0.5	9.60±6.50	5.0±2.2
Ohio River-2.6 km	628±23	835±347	27.5±0.3	27.2±0.4	3.80±3.00	0.1±0.1
Ohio River-4.7 km	666±16	667±27	27.4±0.2	27.1±0.1	1.34±0.38	<0.1

[a] Distance from confluence of the Ohio River and Leading Creek.

organisms. Of the remaining parameters, small but statistically significant increases were observed only for manganese, nickel, and zinc. Concentrations for these three metals during August 6–26 averaged 61, less than 3, and 5 µg/L at 2.4 km upstream, compared to 154, 10, and 15 µg/L, respectively, 2.25 km downstream of the Leading Creek confluence.

SEDIMENT CHEMISTRY

Sediment samples from the Ohio River upstream of Leading Creek and at various distances downstream were collected on three occasions in 1994, and an additional two samples were collected in 1995 to address concerns regarding potential residual effects from the 1993 discharge of mine water. No evidence of any residual metals was found in the sediments. Metal concentrations within sediments for the six metals that were most elevated in the mine water were not significantly different among sampled sites (Table 13.3). The only observable trend in the data was slightly higher aluminum and iron at 2.25 km downstream of Leading Creek, attributable to higher concentrations in the first two samples collected in 1994.

MUSSEL SURVEYS

Twenty-one unionid mussel species were collected from 1993 to 1997 at the four sampling sites combined (Table 13.4). Species richness and composition were very similar among the

TABLE 13.3
Selected Sediment Metals Data at Four Ohio River Sites ($n = 5$; Mean $\pm 95\%$ C.I.)

Metal	2.4 km Upstream	160 m Downstream	2.25 km Downstream	4.7 km Downstream
Al_2O_3 (%)	8.73±2.28	10.8±2.54	11.1±2.75	6.18±2.85
Fe_2O_3 (%)	5.28±0.68	4.91±0.48	7.98±2.85	5.33±1.11
Cu (ppm)	35.0±17.0	30.0±6.00	46.0±21.0	21.0±11.0
Mn (ppm)	1008±282	712±247	844±310	730±181
Ni (ppm)	44.0±14.0	32.0±9.60	44.0±20.0	30.0±5.50
Zn (ppm)	194±96.0	110±55.0	221±188	115±35.0

TABLE 13.4
Unionid Species Total Relative Abundance (1993, 1994, 1995, 1997) at Three Ohio River Sites

Species	Site 1/1A		Site 2		Site 3		Total	
	No.	%	No.	%	No.	%	No.	%
Actinonaias ligamentina	1	0.1			1	0.4	2	0.1
Amblema plicata	404	51.0	195	26.9	41	17.1	640	36.4
Ellipsaria lineolata	3	0.4			1	0.4	4	0.2
Elliptio crassidens	7	0.9	10	1.4			17	1.0
Fusconaia ebena	1	0.1	1	0.1	1	0.4	3	0.2
Fusconaia flava	6	0.8	22	3.0	3	1.3	31	1.8
Lampsilis cardium	6	0.8	3	0.4	23	9.6	32	1.8
Lampsilis siliquoidea	1	0.1			2	0.8	3	0.2
Lasmigona c. complanata	31	3.9	23	3.2	2	0.8	56	3.2
Leptodea fragilis	4	0.5	14	1.9	3	1.3	21	1.2
Ligumia recta	5	0.6	2	0.3	8	3.3	15	0.9
Megalonaias nervosa	8	1.0	7	1.0	1	0.4	16	0.9
Obliguaria reflexa	116	14.6	209	28.8	36	15.0	361	20.5
Potamilus alatus	79	10.0	80	11.0	42	17.5	201	11.4
Pyganodon grandis	1	0.1	2	0.3			3	0.2
Quadrula metanevra	3	0.4			19	7.9	22	1.3
Quadrula p. pustulosa	9	1.1	6	0.8	4	1.7	19	1.1
Quadrula quadrula	79	10.0	124	17.1	26	10.8	229	13.0
Tritogonia verrucosa			1	0.1			1	<0.1
Truncilla truncata	27	3.4	22	3.0	27	11.3	76	4.3
Utterbackia imbecillis	1	0.1	4	0.6			5	0.3
Total	792		725		240		1757	
No. Species	20		17		17		21	

three sites. Of the 21 total species collected, 20 appeared in collections at the upstream sites, and 17 each were collected at the two downstream sites. Additionally, 12 species occurred at all three sites, 8 occurred at two of the three sites, and only the pistol grip (*Tritogonia verrucosa*) was limited to a single individual at only one site (Site 2). Dominant species were also similar among sites. The four most abundant species overall and at each site were the threeridge (*Amblema plicata*), threehorn wartyback (*Obliquaria reflexa*), pink heelsplitter (*Potamilus alatus*), and mapleleaf (*Quadrula quadrula*). No federally listed endangered or threatened species were collected at any of the sites during sampling; however, West Virginia DNR recovered one pink pearly mucket (*Lampsilis abrupta*) during a reconnaissance dive at Site 2.

Species richness and mortality also varied little among years (Table 13.5). Total species richness (quantitative and qualitative samples combined) varied from 10 to 13, 12 to 14, and 9 to 12 at Sites 1/1A, 2, and 3, respectively, with no trend toward increase or decrease with time at any of the sites. Mortality was less than 5% at all sites during all study years. No freshly dead shells were recovered in quantitative samples in 1993 or 1994 at any site. Only a few freshly dead shells (3.2%) were found at Site 2 in 1995.

Changes in density over time were difficult to detect due to high variability in density estimates at all sites; however, some differences were significant. Statistical comparison of mussel densities in the quantitative samples indicated that both site and year significantly affected unionid density, but the interaction of these variables was not significant. Density was significantly higher at Site 2 than at Sites 1/1A and 3 in three of the four years. Annual variation in density was not significant at

TABLE 13.5
Comparison of Unionid Species Richness, Mortality, and Density Among Years and Sites

Metric	Year	Site 1/1A	Site 2	Site 3
No. Species	1993	12.0	12.0	12.0
	1994	10.0	14.0	9.0
	1995	13.0	14.0	12.0
	1997	13.0	14.0	10.0
Mortality (%)	1993	0	0	0
	1994	0	0	
	1995	0	3.2	0
	1997	0	0	0
Density (No./m^2)	1993	$5.2^{A,1}$	$6.2^{A,1,2}$	$2.6^{A,1,2}$
	1994	$2.2^{A,1}$	$4.7^{B,1}$	$0.5^{A,1}$
	1995	$3.0^{A,1}$	$6.0^{B,1,2}$	$2.9^{A,2}$
	1997	$4.3^{A,1}$	$9.6^{B,2}$	$1.9^{A,1,2}$

Different letters within a row indicate a significant difference ($p<0.05$); different numbers within a column indicate a significant difference ($p<0.05$).

Sites 1/1A. At Site 2, density estimates fluctuated among years, with 1997 density significantly greater than 1994, but neither 1994 nor 1997 density significantly differed from density in 1993 or 1995. At Site 3, density estimates also fluctuated, with 1995 density significantly greater than 1994, but neither 1994 nor 1995 differed significantly from 1993 or 1997.

Mean age of the collected mussels was very uniform across sampling sites and years, with significant variation occurring only at Site 2 in 1995 (Figure 13.2). The mean age of 10.0 years at this site in 1995 was significantly greater than the mean ages at this site observed for the other three years, and it was also significantly greater than the 1995 mean age at Site 3 (5.2 years), but not at Site 1A (7.3 years).

A change in average age became apparent when age classes were examined separately (Figure 13.3). The percentage of *less than or equal to* three-year-olds, four- and five-year-olds, and greater than five-year-olds at Site 1/1A remained fairly consistent during all four years. In 1993,

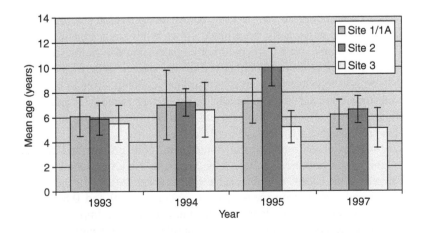

FIGURE 13.2 Estimated annular age (years ±95% C.I.) of unionids at three Ohio River sites, 4, 1993–1997.

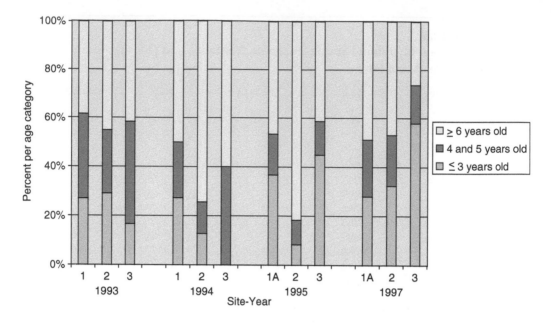

FIGURE 13.3 Percentage of young unionids at three Ohio River sites, 1993–1997.

immediately following the mine-water discharge, age classes were also very similar among the three sites. However, the percentage of *less than or equal to* three-year-olds and four- to five-year-olds at Site 2 declined during 1994 and 1995 before returning to a level comparable to the upstream site in 1997. At the far downstream mussel bed (Site 3), no unionids *less than or equal to* three years old were found in 1994, whereas four- to five-year-olds and greater than five-year-olds remained fairly constant. In contrast, the percentage of *less than or equal to* three-year-olds in 1995 and 1997 was greater than other sites.

Toxicity Testing

Acute toxicity of the mine water to four organisms resulted in fairly similar results for Cladocera and unionid mussel (L. Fragilis), but the three invertebrate species were much more sensitive to the mine water than the fathead minnow (Table 13.6). Results are provided in terms of total and ferrous iron concentrations at the statistically determined effect level, since iron was the primary toxic constituent of the mine water. An LC50 could not be calculated for the fathead minnow because sufficient mortality was not observed in either test of this species at up to 100% mine water. For *L. fragilis* and *D. magna*, repeated tests gave very similar results in terms of both total and ferrous iron. Total and ferrous iron LC50s for glochidial *L. fragilis* averaged 18.4 and 3.6 mg/L, respectively. Similarly, *D. pulex* LC50s were also 18.4 and 3.6 mg/L, respectively, while *D. magna* LC50s were slightly lower, averaging 11.7 and 2.2 mg/L, respectively.

In the initial 30-day chronic test of the Asian clam, both growth and cellulolytic enzyme activity were significantly impaired at 20% mine water, or 12.4 mg/L total iron and 2.54 mg/L ferrous iron. Therefore, the NOEC (no observable effect concentration) for this test was 10% mine water, where total and ferrous iron measurements averaged 5.79 and 0.83 mg/L, respectively (Table 13.6). In the second 30-day clam test, enzyme activity was impaired at only 10% mine water. Although the associated 0.62 mg Fe^{2+}/L was similar to the concentrations measured in Test 1, total iron was much higher (20.6 mg/L).

TABLE 13.6
Laboratory Toxicity Testing Endpoints: Mine-Water Iron Concentrations

	Chronic NOEC	
Species	**Total Iron (mg/L)**	**Ferrous Iron (mg/L)**
C. fluminea Test 1	5.79	0.83
Test 2	20.6	0.62
C. dubia Test 1	0.80	0.16
Test 2	5.79	0.83
Test 3	1.71	0.22
	Acute LC50	
L. fragilis Test 1	19.0[*]	3.73
Test 2	17.7[*]	3.49
D. magna Test 1	11.1[*]	2.12
Test 2	12.4[*]	2.37
D. pulex	18.4[*]	3.61
P. promelas Test 1	>61.1[*]	>12.3
Test 2	>45.8	>9.20

[a] Calculated concentration based on the percentage of stock mine water.

Ceriodaphnia dubia reproduction was significantly reduced in three separate chronic tests (NOEC concentrations of 4, 12, and 14%) with mine water. When combined with the 30-day clam test results, the mean NOECs were 0.53 ± 0.40 mg Fe^{2+}/L and 6.94 ± 9.90 mg total Fe/L.

DISCUSSION

The study objective was to determine if released mine water affected unionid mollusk communities in the Ohio River, and if so, the extent of the impact. The combined analysis of water quality, sediment quality, and unionid communities in the Ohio River upstream and downstream of Leading Creek (the source of impact) and toxicity testing of mine water indicate that mine-water constituents were not present at levels predicted to be lethal to adult unionids. However, the mine water may have impacted reproduction in adults and/or survival of young age classes as far as 2.25 km downstream of Leading Creek.

The permitted release of partially treated mine water to Leading Creek in 1993 was characterized by high concentrations of iron, along with elevated levels of manganese, nickel, and zinc. These results are similar to previous studies of acid mine drainage in Missouri (Parsons 1968) and Kentucky (Short et al. 1990), which reported water quality characterized by extremely high iron concentrations, low pH, and relatively high levels of copper, manganese, aluminum, lead, and zinc. The Leading Creek release resulted in a relatively consistent volume of mine water discharging to the Ohio River for a period of 21 days. Ohio River levels of pH and metals other than iron were not considered a threat to aquatic life; however, iron concentrations were significantly elevated above ambient levels for close to one mile (1.6 km) downstream of the Leading Creek confluence.

While both total and ferrous iron were significantly higher downstream than upstream of Leading Creek, only ferrous iron appeared to be higher than normal for this section of the Ohio River. The mean total iron concentration for the 21-day discharge was 25.6 mg/L within the mine-water plume that formed at the Ohio River bottom at a distance of 122 m downstream of Leading Creek, gradually decreasing to 1.34 mg/L at 4.7 km. Upstream ambient total iron levels averaged 0.27 mg/L during the study. Both the upstream and within-plume total iron values were similar to previous long-term monitoring for total iron 9.7 km downstream of the Leading Creek confluence that ranged from 0.22 to 13.8 mg/L and averaged 3.3 mg/L (Loeffelman et al. 1986). Therefore, the total iron concentrations associated with the mine-water discharge were high but not atypical for the Ohio River. Ferrous iron results, however, were atypical. Loeffelman et al. (1986) reported an ambient ferrous iron range of 0.010–0.246 mg/L, with a mean value of 0.070 mg/L, or approximately 2% of total iron concentrations. Ferrous iron concentrations in the Ohio River during the mine-water discharge were almost two orders of magnitude above ambient, ranging from 11.6 mg/L at the 122-m transect to 5.0 mg/L at the 1.3-km transect before decreasing to normal levels at 2.6 km downstream of Leading Creek.

Sediment sampling of the Ohio River in 1994 and 1995 found little evidence of residual effects of the mine-water discharge, as upstream and downstream metal concentrations were similar. A survey of sediment metal concentrations at 11 Ohio River sites from ORM 6.2 to ORM 531.5 reported mean concentrations of 2.98% iron, 40.1 mg Cu/kg, 1,570 mg Mn/kg, 51.5 mg Ni/kg, and 369 mg Zn/kg (Youger and Mitsch 1989). These values are reasonably close to those reported in the present study, given differences in analytical approach, with higher levels of iron and lower levels of manganese and zinc reported here. Available empirically-based sediment guidelines also support the conclusion that post-discharge sediment metal levels did not pose a residual threat to aquatic life, as these levels were below the consensus-based probable effect concentrations of 149 mg Cu/kg, 48.6 mg Ni/kg, and 459 mg Zn/kg (MacDonald et al. 2000).

Total iron toxicity endpoints measured on the mine water relative to available literature data support the contention that iron was the primary toxic constituent of the mine water. Loeffelman et al. (1986) reported total iron LC50 values of 14.4–24.2 mg/L for fathead minnows and 18.3 mg/L for rainbow trout (*Oncorhynchus mykiss*), while Khangarot and Ray (1989) obtained an LC50 of 7.2 mg/L for *D. magna*. These results compare favorably with the range of values obtained for the mine water of 11.1–61.1 mg/L total iron. A chronic value of 4.38 mg total iron/L for *D. magna* reported by Biesinger and Christensen (1972) falls within the range of values obtained here of 0.80–5.79 mg/L.

Few data are available of ferrous iron toxicity, even though ferrous iron represents the bioavailable fraction of total iron. Loeffelman et al. (1986) obtained acute LC values for ferrous iron of 3.7–9.2 mg/L for fathead minnows and 4.4 mg/L for rainbow trout, compared to a range of 2.12–12.3 mg/L obtained here for four different organisms. No studies of chronic ferrous iron toxicity were found in the available literature for comparison to the range of 0.16–0.83 mg/L obtained in this study (and previously reported in Milam and Farris (1998)) for *C. dubia* and the Asian clam.

Of particular interest for this study were the toxicity test results for the two bivalve species because of potential implications for mine-water effects on the Ohio River mussel beds. The *L. fragilis* glochidia were similarly sensitive to acute ferrous iron exposure as the typically metal-sensitive daphnid species. Glochidia are a sensitive stage of the unionid life cycle (Jacobson et al. 1997). Likewise, Asian clam sensitivity in chronic toxicity testing of ferrous iron was comparable to that measured for *C. dubia*. Asian clam ability to close up and thereby avoid toxic effects might be comparable to a young, thick-shelled Ambleminae. Loss of condition during this anaerobic phase might also be comparable to a young unionid.

A combination of water quality analysis and toxicity testing indicated that the levels of 5.6 mg/L total iron and 1.6 mg/L ferrous iron experienced at Site 2 could have affected unionids. Both values are below the acute LC found for glochidial *L. fragilis*, as well as *D. magna* and *D. pulex*. Thus, immediate mortality of even the most sensitive stage of unionids would not have been expected. However, these values are near or above the chronic values measured for *C. fluminea* and *C. dubia*, such that some long-term effects might be expected.

Consistent with these findings, no immediate mortality was noted in either behavioral observations or quantitative unionid samples at Site 2, only 2.3 km downstream of the Leading Creek confluence. If mine water had an acute effect, freshly dead shells and/or unusual behavior, such as animals lying prone on the substrate or gaping, may have been observed at this location. This was not the case. In fact, the three sampled beds were remarkably similar in these respects, particularly the upstream and near downstream beds. There was also no documented effect on Ohio River fish or invertebrates other than mussels. Unionids tend to close up tightly in response to environmental perturbation, and some species can survive in an anaerobic mode for long periods of time (Fuller 1974). However, no measurable change in density or mortality was observed at any of the sites sampled in 1994.

The reduction in percentage of young animals and increase in average age seen one and two years after the mine release seems to indicate that impact was limited to young unionids at Site 2, which is consistent with water quality analysis and chronic toxicity testing results for Asian clams. A marked decline in young mussels aged five years or less was documented at the near downstream mussel bed in 1994 and 1995, but not in 1993, within a month of the mine-water release. This finding indicates that the impact was either confined to young unionids that would have been less able to withstand a period of anaerobic metabolism associated with the mine release, or the impact took the form of sublethal inhibition of adult mussel reproductive potential that was not manifested until the following spawning season. The available evidence can support either explanation. Juvenile mussels were shown to be the most metal-sensitive life stage in a series of tests by Jacobson et al. (1997), using copper, and may have been selectively impacted by exposure to elevated iron concentrations. Alternatively, adult mussel reproduction may have been inhibited either through reduction of available energy reserves or direct inhibition of developing glochidia.

Research on the toxic mode of action of iron by Gerhardt (1992), using a mayfly, indicated that ferrous iron may damage cell membranes and prevent the absorption of nutrients. At the time of the mine-water release in late summer, many species of mussels have just completed spawning, and the females hold glochidia for release the following spring. These long-term brooders, or bradytictic mussels, include some of the most abundant species found at these sites, including *O. reflexa*, *P. alatus*, and *Lasmigona c. complanata* (white heelsplitter). Iron exposure during this period may have impacted either female energy reserves needed for brood development, or the glochidia themselves, thereby reducing the percentage of age one animals in 1994. The other common reproductive strategy employed by freshwater unionids is termed short-term brooders, or tachytictic, and includes the commonly collected *A. plicata* and *Q. quadrula*. These mussels spawn in the spring and early summer and release glochidia about a month later following a short intervening period when the parasitic glochidia are encysted on host fish. The late summer mine-water release could potentially have impacted very young juveniles of these species. Either of these mechanisms could have produced the observed reduction in the percentage of age one juveniles at Site 2 in 1994 compared to the other sites. However, the percentage of ages one through five decreased, suggesting that not just the 1993 class was affected by the mine-water release. Young unionids may have depleted available energy reserves before larger adults, resulting in a decline in the percentage of young animals and an increase in average age at Site 2 but not at other sites.

The near downstream mussel bed at Site 2 was situated between the 1.3 and 2.6-km water quality transects. Interpolation of total and ferrous iron concentrations during the 21-day period of significant mine-water presence in the Ohio River indicates that this mussel bed was exposed to estimated mean concentrations of 5.6 mg/L total iron and 1.6 mg/L ferrous iron. A comparison of these values to the toxicity data in Table 13.6 indicates that the measured concentration of ferrous iron was sufficient to cause the temporary reduction in young mussels at this site. Although mean ferrous iron concentrations within the first 1.3 km downstream of Leading Creek of up to 11.6 mg/L were high enough to have potentially caused an acute response in mussels, only scattered mussels and no mussel beds have been found in this portion of the river, thereby limiting potential exposure.

These findings, in addition to observations of no ferric iron coating on the mussels or substrates during this event, lead to the conclusion that temporarily elevated ferrous iron concentrations may have caused the observed reduction in young mussels in the near downstream bed. There are no published ferrous iron criteria for the protection of aquatic life, only an outdated total iron criterion of 1 mg/L that appeared in the USEPA (1976) Red Book. Loeffelman et al. (1986) presented extensive evidence that the 1 mg/L criterion is routinely exceeded in many healthy ecosystems and is in need of revision. These authors proposed a tentative ferrous iron criterion of 0.37 mg/L based on available data, which corresponds fairly well with the information presented here.

This study of the chemistry, toxicity, and bioassessment of a partially treated mine-water release on the Ohio River illustrates the value of using multiple lines of evidence to provide a reasoned assessment of impact, as well as the usefulness of laboratory toxicity data for interpreting, or potentially predicting, effects on field populations of aquatic organisms. With respect to unionid mussels, there are particular concerns with laboratory testing related both to their limited availability and the difficulties encountered with laboratory holding and testing. This study provides evidence that the Asian clam, *C. fluminea*, may be a useful surrogate that is widely available and amenable to testing. This approach was suggested by Foster and Bates (1978) and has been applied successfully in a number of studies (Doherty 1990). Another noteworthy aspect of this study is the observed delay in demonstrating an impact on mussels until the year following the exposure event. This finding argues for extended monitoring programs for unionids that encompass critical portions of the reproductive cycle or life history patterns to provide a full assessment of both short-term and long-term effects.

ACKNOWLEDGMENTS

The authors extend their thanks to C. D. Bishop and K. V. Wood for technical assistance in toxicity testing and sampling. American Electric Power's John E. Dolan Laboratory and Meigs Mine Coal Laboratory provided analytical services. EA Engineering, Science, and Technology performed water quality transect sampling.

REFERENCES

American Public Health Association, *Standard Methods for the Examination of Water and Wastewater*, 18th ed., Amer. Publ. Health Assoc., Amer. Water Works Assoc., Water Pollut. Control Fed., Washington, DC, 1992.

Biesinger, K. E. and Christensen, G. M., Effects of various metals on survival, growth, reproduction, and metabolism of *Daphnia magna*, *J. Fish. Res. Bd. Can.*, 29, 1691–1700, 1972.

Cherry, D. S., Van Hassel, J. H., Farris, J. L., Soucek, D. J., and Neves, R. J., Site-specific derivation of the acute copper criteria for the Clinch River, Virginia, *Human Ecol. Risk Assess.*, 8, 591–601, 2002.

Doherty, F. G., The asiatic clam, *Corbicula* spp., as a biological monitor in freshwater environments, *Environ. Monit. Assess.*, 15, 143–181, 1990.

Farris, J. L., Belanger, S. E., Cherry, D. S., and Cairns, J., Cellulolytic activity as a novel approach to assess long-term zinc stress to *Corbicula*, *Water Res.*, 23, 1275–1283, 1989.

Foster, R. B. and Bates, J. M., Use of freshwater mussels to monitor point source industrial discharges, *Environ. Sci. Technol.*, 12, 958–962, 1978.

Fuller, S. L. H., Clams and mussels (Mollusca: Bivalvia), In *Pollution Ecology of Freshwater Invertebrates*, Hart, C. W. and Fuller, S. L. H., Eds., Academic Press, New York, pp. 215–273, 1974.

Gardner, W. S., Miller, W. H., and Imlay, M. J., Free amino acids in mantle tissues of the bivalve *Amblema plicata*: Possible relation to environmental stress, *Bull. Environ. Contam. Toxicol.*, 26, 157–162, 1981.

Gerhardt, A., Effects of subacute doses of iron (Fe) on *Leptophlebia marginata* (Insecta: Ephemeroptera), *Freshw. Biol.*, 27, 79–84, 1992.

Green, R. H., Bailey, R. C., Hinch, S. G., Metcalfe, J. L., and Young, V. H., Use of freshwater mussels (Bivalvia: Unionidae) to monitor the nearshore environment of lakes, *J. Great Lakes Res.*, 15, 635–644, 1989.

Grippo, R. S. and Dunson, W. A., The body ion loss biomarker. 2. Field validation in coal mine-polluted streams, *Environ. Toxicol. Chem.*, 15, 1964–1972, 1996.

Haag, W. R., Berg, D. J., Garton, D. W., and Farris, J. L., Reduced survival and fitness in native bivalves in response to fouling by the introduced zebra mussel (*Dreissena polymorpha*) in western Lake Erie, *Can. J. Fish. Aquat. Sci.*, 50, 13–19, 1993.

Hamilton, M. A., Russo, R. C., and Thurston, R. V., Trimmed Spearman-Karber method for estimating median lethal concentrations, *Environ. Sci. Technol.*, 11, 714–719, 1977. Correction 12:417 (1978).

Jacobson, P. J., Neves, R. J., Cherry, D. S., and Farris, J. L., Sensitivity of glochidial stages of freshwater mussels (Bivalvia: Unionidae) to copper, *Environ. Toxicol. Chem.*, 16, 2384–2392, 1997.

Khangarot, B. S. and Ray, P. K., Investigation of correlation between physicochemical properties of metals and their toxicity to the water flea *Daphnia magna* Straus, *Ecotoxicol. Environ. Saf.*, 18, 109–120, 1989.

Loeffelman, P. H., Van Hassel, J. H., Arnold, T. E., and Hendricks, J. C., A new approach for regulating iron in water quality standards, In Aquatic Toxicology and Hazard Assessment, 8th Symposium, ASTM STP 981, Bahner, R. C. and Hansen, D. J., Eds., Amer. Soc. Test. Mat., Philapelphia, PA, pp. 137-152, 1986..

MacDonald, D. D., Ingersoll, C. G., and Berger, T. A., Development and evaluation of consensus-based sediment quality guidelines for freshwater ecosystems, *Arch. Environ. Contam. Toxicol.*, 39, 20–31, 2000.

Milam, C. D. and Farris, J. L., Risk identification associated with iron-dominated mine discharges and their effect upon freshwater bivalves, *Environ. Toxicol. Chem.*, 7, 1611–1619, 1998.

Neves, R. J. and Moyer, S. N., Evaluation of techniques for age determination of freshwater mussels (Unionidae), *Am. Malacol. Bull.*, 6, 179–188, 1988.

Parsons, J. D., The effects of acid strip-mine effluents on the ecology of a stream, *Arch. Hydrobiol.*, 65, 25–50, 1968.

SAS Institute, *SAS® User's Guide*, SAS Institute, Cary, NC, 1985.

Short, T. M., Black, J. A., and Birge, W. J., Effects of acid-mine drainage on the chemical and biological character of an alkaline headwater stream, *Arch. Environ. Contam. Toxicol.*, 19, 241–248, 1990.

Simmons, J. R. and Reed, G. M., Mussels as indicators of biological recovery zone, *J. Water Pollut. Control Fed.*, 45, 2480–2492, 1973.

[USEPA] U.S. Environmental Protection Agency, *Quality Criteria for Water*, U.S. Environmental Protection Agency, Washington DC, 1976.

[USEPA] U.S. Environmental Protection Agency, *Methods for Chemical Analysis of Water and Wastes*, U.S. Environmental Protection Agency, Cincinnati, OH, EPA 600/4-79-020, 1979.

[USEPA] U.S. Environmental Protection Agency, *Short-Term Methods for Estimating the Chronic Toxicity of Effluents and Receiving Waters to Freshwater Organisms* 2nd ed., U.S. Environmental Protection Agency, Washington, DC, 1989. EPA 600/4-89/001.

[USEPA] U.S. Environmental Protection Agency, *Methods for Measuring the Acute Toxicity of Effluents and Receiving Waters to Freshwater and Marine Organisms*, 4th ed., U.S. Environmental Protection Agency, Washington, DC, 1993. EPA 600/4-90/027F.

Younger, J. D. and Mitsch, W. J., Heavy metal concentrations in Ohio River sediments—longitudinal and temporal patterns, *Ohio J. Sci.*, 89, 172–175, 1989.

Zeto, M. A., Tolin, W. A., and Schmidt, J. E., The freshwater mussels (Unionidae) of the upper Ohio River, Greenup and Belleville Pools, West Virginia, *Nautilus*, 101, 182–185, 1987.

14 Case Study: Sensitivity of Mussel Glochidia and Regulatory Test Organisms to Mercury and a Reference Toxicant

Theodore W. Valenti, Donald S. Cherry, Richard J. Neves,
Brandon A. Locke, and John J. Schmerfeld

INTRODUCTION

Freshwater mussel populations have declined substantially in North America, and more than two-thirds of the identified species (*Unionidae*) are classified as extinct, endangered, threatened, or of special concern (Williams et al. 1993; Naimo 1995; Jacobson et al. 1997). Although exploitation from commercial over-harvest and the introduction of nonnative species have had substantial impacts (Williams et al. 1993; Yeager, Neves, and Cherry 1999), many declines are attributed to anthropogenic stresses that have eliminated or degraded the natural habitat of mussels (Keller and Zam 1991; Williams et al. 1993; Naimo 1995; Milam and Farris 1998; Henley and Neves 1999; Diamond, Bressler, and Serveiss 2002; Weinstein 2002). Scientists have addressed these potential risks by improving agricultural practices, waste management, and pollution monitoring in the United States, and consequently, water quality has substantially improved. Furthermore, the implementation of regulatory policies that are focused on preserving wildlife and the environment, such as the Endangered Species Act of 1973 and Clean Water Act of 1977, promotes the protection of not only native unionids, but also their habitat. However, despite clear progress, there is still concern about the future conservation of native mussels, as survey efforts have shown little recruitment (Neves and Widlak 1987; Breunderman and Neves 1993; Henley and Neves 1999).

Researchers have observed that, of the remaining diverse mussel assemblages, many are comprised primarily of older, adult mussels, and few have an abundance of young mussels present (Henley and Neves 1999; Weinstein 2001). These trends indicate that populations are unstable and declining. Conservationists are especially concerned because it may take years for young mussels currently residing in rivers to reach peak sexual maturity. The complex life history of unionids has made it difficult for researchers to determine the causes of reproductive failure. However, there is substantial evidence that pollution is a contributing factor, as several laboratory studies have documented that freshwater mussels, like most aquatic organisms, are more sensitive to contaminants during their early life stages than as adults (Naimo 1995; Jacobson et al. 1997; Keller and Ruessler 1997; Yeager, Neves, and Cherry 1999; Weinstein 2001).

Jacobson et al. (1997) conducted a comprehensive study that examined the effects of copper exposure on the various life stages of freshwater mussels. Their study compared the sensitivities of

Villosa iris glochidia that were brooded (still in the gills of a gravid adult), released (in the water column), and encysted (attached to a fish host). Released glochidia were impacted at lower copper concentrations (36–80 µg Cu/L) than encysted glochidia (greater than 400 µg Cu/L). No adverse effects were observed for any treatments in the brooded glochidia test; however, the highest concentration tested was only 19 µg Cu/L. Interestingly, released glochidia and juveniles had very similar tolerances, as 24-hour LC50 values for glochidia of *V. iris* and *Pyganodon grandis* were 36–80 and 46–347 µg Cu/L, respectively, while those for juveniles were 83 and 44 µg Cu/L. More important, the study provided clear evidence that early life stages of freshwater mussels have far lower acute contaminant exposure thresholds than adults, as the 96-hour LC50 for adults was greater than 1000 µg Cu/L.

Only a few other studies have examined the acute tolerances of glochidia and juvenile mussels of the same species, but most concur with Jacobson et al. (1997) and report that glochidia are as sensitive or more sensitive than juveniles in acute exposures. In a study examining the toxicity of ammonia, Augspurger et al. (2003) recorded higher tolerances for juveniles than glochidia, despite a longer exposure duration. The 96-LC50 values for juvenile pheasantshell mussels (*Actinonaias pectorosa*) and paper pondshell mussels (*Utterbackia imbecillis*) were 14.05 and 10.60 mg total ammonia as N/L, respectively, while the corresponding 48-hour value for glochidia were 3.76 and 5.85 mg total ammonia/L. Similarly, the mean 96-hour LC50 for the rainbow mussel (*V. iris*) was 6.75 mg total ammonia/L, and the 24-hour value for glochidia was 3.79 mg total ammonia/L. Keller and Ruessler (1997) examined the toxicity of malathion to early life stages of the pondshell (*U. imbecillis*), little spectaclecase (*Villosa lienosa*), and downy rainbow mussel (*Villosa villosa*), and also recorded substantially lower tolerances for glochidia than for juveniles.

Additional studies have also documented that glochidia are more acutely sensitive to contaminants than standard regulatory organisms used for Whole Effluent Toxicity (WET) testing, and US Environmental Protection Agency (USEPA) Water Quality Criteria (WQC). Cherry et al. (2002) compared the acute sensitivities of 17 species of freshwater organisms to copper. Four of the five most sensitive test organisms were freshwater mussel glochidia, while standard regulatory test organisms *Ceriodaphnia dubia* and *Pimephales promelas* ranked sixth (88µg Cu/L), and fourteenth (310 µg Cu/L), respectively. The Genus Mean Acute Values (GMAV) for glochidia of the four most sensitive mussels species ranged from 37 to 60 µg Cu/L. Studies that examined the toxicity of ammonia to early life stages of freshwater mussels also reported LC50 values that are within the ranges described for standard test organisms *C. dubia*, *P. promelas*, *Daphnia magna*, and *Oncorhynchus mykiss* (rainbow trout) (Goudreau, Neves, and Sheehan 1993; Mummert et al. 2003). Milam and Farris (1998) noted that glochidia of *Leptodea fragilis* were more sensitive than *P. promelas* to partially treated mine water but less sensitive than *D. magna* and *C. dubia*. However, their study contrasted the 24-hour acute glochidia LC50s with 48-hour acute LC50s for *D. magna* and 7-day fecundity EC_{50}s for *C. dubia*. Although the results of the aforementioned studies may influence freshwater regulatory policy, agencies are hesitant to accept test results because there is concern about the effectiveness of glochidia as test organisms in the laboratory.

Guidelines for conducting acute toxicity tests with early life stages of freshwater mussels were submitted to the USEPA in 1990 (USEPA 1990). The effort brought laboratory toxicity testing with freshwater mussels to the foreground of aquatic toxicology but failed to address several aspects essential for the development of a standard protocol. The primary criticism was the use of glochidia in toxicity tests that were obtained from gravid adults collected from rivers. There is concern that environmental variables, such as pollution or nutrient availability, may affect the ability of gravid females to produce fit offspring. The maturity of glochidia collected from different adults of the same species will likely vary, as not all individuals from a species have synchronized reproductive cycles. The time of season that mussels are obtained from the field may also influence maturity of glochidia, as unionids can be categorized into long- and short-term brooders (Jacobson et al. 1997). Unhealthy or immature glochidia are likely to be more susceptible to contaminant exposure

(Huebner and Pynnonen 1992; Goudreau, Neves, and Sheehan 1993; Jacobson et al. 1997), and their use in tests may lead to biased, false-positive results. Although verifying test organism health is a universal concern for all toxicological studies, it is especially problematic for research with glochidia because researchers are still unsure of appropriate methods. There have been substantial strides towards establishing acceptable test parameters and methodologies for glochidia tests (Chapter 5), but efforts will go unheeded unless better techniques for assessing the health of glochidia are developed.

Study Goals

The primary purpose of this study was to compare the sensitivities of glochidia from different species of freshwater mussels to mercury (Hg) by conducting laboratory tests with organic and inorganic mercury salts. Many freshwater systems are contaminated by mercury pollution, as anthropogenic sources, such as the incineration of medical wastes, disposal of mercury-laden material, industrial processing, pesticide use, and the burning of fossil fuels, have made it more available in ecosystems. Although most mercury is emitted in elemental or inorganic forms that are not highly toxic, several abiotic and biotic factors may facilitate the conversion of these forms into methylmercury (MeHg) in water (Barkay, Gillman, and Turner 1997; Wiener and Shields 2000; Mauro, Guimaraes, and Hintelmann 2002). This organic form of mercury is highly toxic to aquatic life and has been documented to bio-accumulate in food webs (Barkay, Gillman, and Turner 1997; French et al. 1999; Mason, Laporte, and Andres 2000; Wiener and Shields 2000; Mauro, Guimaraes, and Hintelmann 2002). The USEPA is currently reassessing the WQC for mercury, as researchers have become more aware of the threat it poses to humans and wildlife (Moore, Teed, and Richardson 2003). Fish Consumption Advisories (FCA) for mercury have been issued in nearly every US state (French et al. 1999; Mason, Laporte, and Andres 2000; Webber and Haines 2003). However, recent studies examining the sensitivities of freshwater organisms are sparse, and results from older studies may be flawed because technology for measuring a low concentration of mercury did not exist. Furthermore, there is little known about the sensitivity of freshwater mussels to mercury, despite documented declines in polluted water (Henley and Neves 1999; Beckvar et al. 2000). It is pertinent to address these voids because a more comprehensive species database will be needed to establish appropriate water standards.

Another objective of this study was to compare the mercury sensitivities of glochidia to those of standard regulatory organisms, *C. dubia*, *D. magna*, and *P. promelas*. Several studies have noted that glochidia are extremely sensitive compared to the larvae stages of other aquatic biota (Jacobson et al. 1997; Weinstein 2001; Weinstein and Polk 2001; Cherry et al. 2002). We wanted to determine if standard, freshwater, regulatory test organisms are adequate surrogate test organisms for assessing mercury exposure risks to glochidia. Environmental risk is often inferred by conducting toxicity tests with standard monitoring organisms that are sensitive to most toxicants. This approach should not be implemented for assessing risk to freshwater mussels until the relative tolerances of the respective genera are discerned.

The final goal of this study was to expose glochidia to sodium chloride (NaCl) to determine if it is an appropriate reference toxicant. Tests were conducted based on methods described in protocol for standard freshwater test organisms (USEPA 1993). Reference toxicity test measures are useful QA/QC assurances for standard test organisms because they enable researchers to evaluate the relative health of the test organisms, verify the acceptability of test conditions or procedures, and validate toxicity tests results. Reference toxicant tests are supposed to be conducted monthly at culturing facilities, and concurrently with acute and chronic WET testing with standard test organisms. Similar approaches have not been applied to glochidia, and the inadequacy of current methods for assessing the health of glochidia must be addressed for regulatory agencies to be willing to incorporate test results into environmental policy.

METHODS

Test Organisms

Gravid specimens of *Lampsilis fasciola* (Wavyrayed lampmussel), *V. iris* (Rainbow mussel), *Epioblasma capsaeformis* (Oyster mussel), and *Epioblasma brevidens* (Cumberland combshell) were obtained from the Virginia Polytechnic Institute & State University (VPI&SU) Aquaculture Center in Blacksburg, VA. Gravid adults of the various species were collected from the Clinch River, VA, and stored at the Buller Fish Hatchery in Marion, VA. Adult mussels were acclimated to laboratory conditions for at least 48 hours before the glochidia were harvested. Glochidia were extracted by gently prying open the valves of a gravid female, puncturing the gill tissue with a sterile, water-filled syringe, and then injecting water to flush individuals out. Glochidia were loaded into test chambers less than 2 hours after extraction.

Daphnids, *C. dubia* and *D. magna* (less than 24 hours old), were cultured at the VPI&SU Aquatic Toxicology Laboratory according to standard procedure (APHA, AWWA, and WEF 1998). Organisms were cultured in an 80:20 mixture of moderately hard, synthetic water (EPA[100]) (USEPA 1993) and filtered reference water at $25 \pm 1°C$ under a 16:8, light:dark photoperiod and were fed a diet of unicellular algae (*Selenastrum capricornutum*) and YCT (yeast/cereal leaves/trout chow). Fathead minnows were obtained from a commercial supplier (Aquatox, Inc., Hot Springs, AR).

Preparation of Mercury Test Solutions

Mercuric chloride (MC) and methylmercuric chloride (MMC) salts were used to create the inorganic and organic test solutions, respectively. Test concentrations were 8, 15, 30, 60, and 120 μg/L total Hg, plus a control, in all bioassays, except for some *C. dubia* and *D. magna* tests when the highest concentration, 120 μg/L, was replaced with the lower concentration of 4 μg/L total Hg.

Toxicity Tests

Glochidia. Because a protocol has yet to be established for glochidia bioassays, we attempted to adhere to the test design described in USEPA protocol (1993) for standard freshwater test organisms. The main modification was an increase in the number of test organisms per replicate. The small size of glochidia makes them difficult to monitor individually; therefore, researchers assessed viability for a sub-sample of individuals from each replicate. This approach provided a more accurate estimate of viability per replicate and also minimized problems from potential handling stress.

Glochidia were randomly distributed to 50-mL glass beakers filled with ∼35 mL of test solution. There were eight replicates of 50–100 glochidia for each treatment. Viability was assessed in four randomly selected replicates after 24 hours, and the remaining four replicates were assessed after 48 hours. Tests were conducted at $20 \pm 1°C$ under a 12:12, light:dark photoperiod.

Glochidia viability was assessed through a sodium chloride response test, similar to that described by Huebner and Pynnonen (1992), Goudreau, Neves, and Sheehan (1993), Jacobson et al. (1997), and Keller and Ruessler (1997). A sample of glochidia from a replicate was transferred with a fine-tip glass to a petri dish for observation using a dissecting scope. The total number of open and closed glochidia was recorded, and after which, a concentrated sodium chloride solution was added. Any glochidia closed prior to, or remaining open after, the addition of the salt solution were documented as functionally dead.

EPA test organisms. Acute 48-hour toxicity tests were conducted with *C. dubia*, *D. magna*, and *P. promelas* according to USEPA standard protocol (1993). Cladoceran bioassays were conducted in 50-mL glass beakers with approximately 35 mL of test solution. There were four replicates of five individuals for each treatment. *Pimephales* bioassays were conducted in 300-mL glass beakers filled with ~250 mL of test solution. There were two replicates of ten individuals for each concentration. Mortality was assessed after 24 and 48 hours. All tests were conducted at $20 \pm 1°C$ under a 12:12, light:dark photoperiod, and organisms were not fed during the tests.

REFERENCE TOXICANT TESTS

Reference toxicity tests were conducted with glochidia of *L. fasciola*, *E. capsaeformis*, and *E. brevidens*. Sodium chloride was used as the toxicant because it is the suggested contaminant for reference bioassays with standard freshwater regulatory test organisms (USEPA 1993). A 0.5 serial dilution was used to create treatments, which include a control, 0.5, 1.0, 2.0, 4.0, and 8.0 g NaCl/L diluent water; these are the same concentrations for *C. dubia* reference tests. Certified reference-grade sodium chloride was used as the toxicant, and EPA[100] was used as the diluent and control treatment. Viability of glochidia was assessed after 24 and 48 hours of exposure. Bioassays were conducted at $20 \pm 1°C$ under a 12:12, light:dark photoperiod.

Results of monthly acute sodium chloride reference toxicant tests at the VPI&SU Aquatic Toxicology Laboratory for NPDES permit tests with *C. dubia*, *D. magna*, and *P. promelas* were compiled for comparative purposes. Tests were conducted according to standard protocol (USEPA 1993) between January 2001 and August 2003.

WATER CHEMISTRY AND MERCURY ANALYSIS

Temperature was monitored twice daily. Dissolved oxygen, conductivity, and pH were measured for all in-water and out-water in the bioassays. Alkalinity and hardness were measured for the control and highest concentration for in-water. An Accumet® (Fisher Scientific, Pittsburgh, PA, USA) pH meter with an Accumet gel-filled combination electrode (accuracy less than ± 0.05 pH at 25°C) was used to measure pH. Dissolved oxygen and conductivity were measured with a 54A meter® and model 30 conductivity meter®, respectively, from Yellow Springs (Yellow Springs, OH, USA). Total hardness and alkalinity (as mg/L $CaCO_3$) were measured in accordance with APHA, AWWA, and WEF (1998) through colorimetric titrations.

Samples of in- and out-water from several replicates were combined for each treatment and prepared for Inductively Coupled Plasma (ICP) spectrometry according to USEPA (1991) standard methods. Trace metal-grade pure hydrochloric acid was used to reduce the sample pH to less than or equal to two. The prepared samples were refrigerated until analysis at the VA Tech Soil Laboratory (Blacksburg, VA).

DATA ANALYSIS

Toxicity test results were presented as LC50 values and were calculated by Spearman Karber analysis on computer software (Gulley 1993). All calculations based on nominal total mercury concentrations as treatments less than 15 µg Hg/L were below detection limits (BDL).

RESULTS

CONTROL SURVIVORSHIP

The combined mean glochidia viability in control treatments for all of the bioassays was greater than 89% for the species tested after 24 hours (Figure 14.1). Mean control survivorship remained greater than 80% after 48 hours for all species except *L. fasciola*, which declined to 78%. Overall, viability did substantially decrease with increased exposure time for all species except *V. iris*.

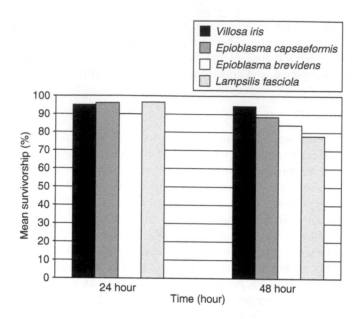

FIGURE 14.1 Glochidia control survivorship.

MERCURY SALT RESULTS

Mercuric chloride. Glochidia from the different species of freshwater mussels had similar tolerances to MC, as 24-hour and 48-hour LC values for *L. fasciola*, *E. capsaeformis*, and *E. brevidens*
ranged from 25–54 and 27–40 µg Hg/L, respectively (Table 14.1). Although not evident by the
LC50 values, viability decreased with increased exposure time in nearly every treatment. Survivorship remained high in the control (24 hours = greater than 89% and 48 hours = greater than 81%)
but was substantially reduced in treatments containing elevated concentrations of mercury. After
48 hours, 100% mortality was observed in treatments greater than or equal to 120 µg Hg/L.

Ceriodaphnia was far more sensitive to MC than *D. magna*, as the respective 48-hour LC50
values were 7 and 19 µg Hg/L (Table 14.1). Sensitivity increased with exposure time in both tests,
and the largest contrast in 24- and 48-hour LC50 values (90 and 15 µg Hg/L, respectively) was
observed with *D. magna*. Survivorship in the control remained 100% but was substantially reduced
in treatments with measurable concentrations of mercury for both species.

Methylmercuric Chloride

The LC50 values for glochidia of *E. capsaeformis* and *E. brevidens* exposed to MMC were substantially lower than those documented in MC tests. The LC50 values after 24 hours ranged from 21 to
26 µg Hg/L for the two species (Table 14.2). However, 48-hour LC50 values could not be calculated
because mortality was more than 50% in the lowest test treatment, 8 µg Hg/L. Therefore, these values
were reported conservatively as less than 8 µg Hg/L. *Villosa iris* glochidia were far more tolerant than
the two other species. A 24-hour LC50 could not be calculated because only 38% of the individuals
exposed to 120 µg Hg/L died; however, the value was reported as more than 120 µg Hg/L for
comparative purposes. After 48 hours, the LC50 for *V. iris* declined substantially to 43 µg Hg/L,
but was still five times higher compared to the values found for glochidia from the other species.

Ceriodaphnia was the most sensitive organism tested to MMC, as 100% mortality occurred in
treatments greater than or equal to 8 µg Hg/L, despite 100% survivorship in the control
(Table 14.2). The 48-hour LC50 could not be calculated in either *C. dubia* test because of high
mortality in low concentrations. Subsequently, these values were reported conservatively as less

TABLE 14.1
Comparative Acute Toxicity of Glochidia from Three Mussel Species and Two Daphnids to Mercuric Chloride

Organisms	Species	Concentration (µg Hg/L)	n	24-hour % Mortality	24-hour LC50 (95% CI)	n	48-hour % Mortality	48-hour LC50 (95% CI)
Glochidia	L. fasciola	Control	200	6	40 µg Hg/L (40–50)	200	19	40 µg Hg/L (30–40)
		5	200	4		200	17	
		10	200	4		200	15	
		15	200	6		200	16	
		30	200	7		200	10	
		60	200	9		200	30	
		120	200	85		200	100	
		250	200	100		200	100	
Glochidia	L. fasciola	Control	200	3	40 µg Hg/L (30–40)	n/a	n/a	n/a
		8	200	4				
		15	200	13				
		30	200	60				
		60	200	100				
		120	200	100				
Glochidia	E. capsaeformis	Control	50	4	25 µg Hg/L (22–25)	50	18	27 µg Hg/L (n/a)
		8	50	6		50	10	
		15	50	16		50	36	
		30	50	64		50	68	
		60	50	100		50	100	
		120	50	100		50	100	
Glochidia	E. capsaeformis	Control	100	3	54 µg Hg/L (49–60)	100	10	36 µg Hg/L (33–38)
		8	100	6		100	7	
		15	100	8		100	6	
		30	100	14		100	28	
		60	100	50		100	95	
		120	100	100		100	100	

(continued)

TABLE 14.1 *(Continued)*

Organisms	Species	Concentration (μg Hg/L)	n	24-hour % Mortality	24-hour LC50 (95% CI)	n	48-hour % Mortality	48-hour LC50 (95% CI)
Glochidia	E. brevidens	Control	100	11	47 μg Hg/L (42–53)	100	17	27 μg Hg/L (24–30)
		8	100	8		100	21	
		15	100	12		100	16	
		30	100	17		100	53	
		60	100	62		100	100	
		120	100	100		100	100	
Cladoceran	C. dubia	Control	20	0	11 μg Hg/L (10–12)	20	0	7 mg Hg/L (5–9)
		4	20	5		20	15	
		8	20	30		20	60	
		15	20	60		20	85	
		30	20	100		20	100	
		60	20	100		20	100	
Cladoceran	D. magna	Control	20	0	90 μg Hg/L (80–100)	20	0	19 μg Hg/L (17–22)
		8	20	0		20	5	
		15	20	5		20	40	
		30	20	5		20	80	
		60	20	15		20	100	
		120	20	80		20	100	

TABLE 14.2
Comparative Acute Toxicity of Glochidia from Three Mussels Species and Three Standard USEPA Test Organisms to Methylmercuric Chloride

Organisms	Species	Concentration (μg Hg/L)	N	24-hour % Mortality	24-hour LC50 (95% CI)	n	48-hour % Mortality	48-hour LC50 (95% CI)
Glochidia	*E. capsaeformis*	Control	50	4	21 μg Hg/L (17–24)	50	18	8 μg Hg/L (4–9)
		8	50	10		50	70	
		15	50	36		50	80	
		30	50	68		50	100	
		60	50	100		50	100	
		120	50	100		50	100	
Glochidia	*E. capsaeformis*	Control	100	3	26 μg Hg/L (23–28)	100	10	<8 μg Hg/L (n/a)
		8	100	4		100	49	
		15	100	13		100	100	
		30	100	60		100	100	
		60	100	100		100	100	
		120	100	100		100	100	
Glochidia	*E. brevidens*	Control	100	11	25 μg Hg/L (22–28)	100	17	<8 μg Hg/L (n/a)
		8	100	10		100	56	
		15	100	26		100	100	
		30	100	51		100	100	
		60	100	100		100	100	
		120	100	100		100	100	
Glochidia	*V. iris*	Control	326	6	>120 μg Hg/L	305	5	43 μg Hg/L (41–45)
		8	246	4		316	5	
		15	257	6		309	8	
		30	316	6		325	15	
		60	276	8		314	90	
		120	255	38		336	100	
Cladoceran	*C. dubia*	Control	20	0	30 μg Hg/L (20–30)	20	5.0	<8 μg Hg/L (n/a)
		8	20	10		20	100	

(continued)

TABLE 14.2 (Continued)

Organisms	Species	Concentration (µg Hg/L)	N	24-hour % Mortality	24-hour LC50 (95% CI)	n	48-hour % Mortality	48-hour LC50 (95% CI)
		15	20	15		20	100	
		30	20	60		20	100	
		60	20	90		20	100	
		120	20	100		20	100	
Cladoceran	C. dubia	Control	20	0	25 µg Hg/L (20–30)	20	0	<4 µg Hg/L (n/a)
		4	20	5		20	85	
		8	20	15		20	100	
		15	20	15		20	100	
		30	20	30		20	100	
		60	20	100		20	100	
Cladoceran	D. magna	Control	20	0	20 µg Hg/L (20–22)	20	0	18 µg Hg/L (15–21)
		8	20	0		20	5.0	
		15	20	0		20	15	
		30	20	95		20	100	
		60	20	100		20	100	
		120	20	100		20	100	
Cladoceran	D. magna	Control	20	0	> 60 µg Hg/L	20	0	15 µg Hg/L (11–19)
		4	20	0		20	0	
		8	20	0		20	5	
		15	20	5		20	45	
		30	20	15		20	100	
		60	20	35		20	100	
Fish	P. promelas	Control	20	0	120 µg Hg/L (n/a)	20	0	67 µg Hg/L (57–77)
		0.008	20	0		20	0	
		0.015	20	0		20	0	
		0.03	20	0		20	0	
		0.06	20	0		20	35	
		0.12	20	15		20	100	

than 4 and less than 8 μg Hg/L. *Daphnia* were also quite sensitive to MMC, as 48-hour LC50 values for the two trials were 18 and 15 μg Hg/L. *Pimephales promelas* was extremely tolerant to MMC exposure, as a 24-hour LC50 value could not be calculated due to only 15% mortality in the highest treatment; this value was expressed as more than 120 μg Hg/L. A 48-hour LC50 value was calculated for *P. promelas* that was considerably lower, 67 μg Hg/L, but remained markedly higher than values for the other species.

REFERENCE TOXICANT RESULTS

Glochidia. Glochidia of *L. fasciola*, *E. capsaeformis*, and *E. brevidens* had similar tolerances to sodium chloride, as the upper and lower 95% confidence limits for the 24- and 48-hour LC50 values nearly overlapped (Table 14.3). After 48 hours, control survivorship was extremely low in the *L. fasciola* bioassay (68%); therefore, the reported LC50 value of 2.25 g NaCl/L is considered unreliable. Control survivorship was more stable in the bioassays with the other species. As in the mercury tests, viability decreased with increased exposure times in most treatments. The average 48-hour LC50 for glochidia from all three species combined was 2.46 g NaCl/L (Figure 14.2).

Standard Regulatory Test Organisms

The three different standard regulatory organisms, *C. dubia*, *D. magna*, and *P. promelas*, had very distinct sodium chloride tolerances. The most sensitive species was *C. dubia*, as the average 48-hour LC50 was 2.33 g NaCl/L. Similar values for *D. magna* and *P. promelas* were 4.96 and 9.84 g NaCl/L (Figure 14.2).

WATER CHEMISTRY AND MERCURY CONCENTRATIONS

Water chemistry parameters and mercury concentration analysis results for the different test treatments are summarized in Table 14.4. Dissolved oxygen remained more than 5.0 mg/L in all bioassays. Other water parameters for in- and out-water did not differ substantially, except for total mercury concentration, which was substantially lower in out-water. Treatments less than or equal to 15 μg Hg/L were below detection limit. The in-water for treatments greater than or equal to 30 μg Hg/L were very close to nominal concentrations.

DISCUSSION

MERCURY TESTS

Glochidia sensitivities. Researchers have noted that glochidia may sporadically clasp or completely seal their valves when exposed to contaminants during laboratory toxicity tests. Effects induced by toxicants may also be less apparent if glochidia remain open. However, researchers can infer the viability of these individuals by exposing them to a noxious substance, such as sodium chloride, that is known to elicit this avoidance behavior (Huebner and Pynnonen 1992; Goudreau, Neves, and Sheehan 1993; Jacobson et al. 1997; Keller and Ruessler 1997). Several laboratory studies have reported that released glochidia have substantially lower viability in treatments containing elevated concentrations of contaminants (Huebner and Pynnonen 1992; Goudreau, Neves, and Sheehan 1993; Jacobson et al. 1997; Keller and Ruessler 1997; Cherry et al. 2002). These observations have incited speculation that contaminants may be attributing to the lack of recruitment in the water column by reducing the ability of glochidia to successfully attach to host fish. A decrease in this ability would inevitably lower the reproductive potential of impacted individuals. In our study, very low concentrations of mercury drastically affected the viability of glochidia. However, additional research is needed to more accurately determine species tolerances because impairment was observed in treatments with mercury concentrations BDL in this study. Regardless, this study

TABLE 14.3
Mussel Glochidia Acute Response to a Reference Toxicant, NaCl

Species	Concentration (g NaCl/L)	Temperature	N	24-hour % Mortality	24-LC50 (95% CI)	n	48-hour % Mortality	48-hour LC50 (95% CI)
L. fasciola	Control	22	200	3.5	3.08 g NaCl/L (2.91–3.26)	200	32	2.25 g NaCl/L (2.14–2.35)
	0.5	22	200	3.0		200	18	
	1.0	22	200	2.0		200	23	
	2.0	22	200	17.5		200	49.5	
	4.0	22	200	73.5		200	100	
	8.0	22	200	100		200	100	
E. capsaeformis	Control	20	50	6.0	2.71 g NaCl/L (2.54–2.88)	50	8.0	2.45 g NaCl/L (2.21–2.70)
	0.5	20	50	4.0		50	6.0	
	1.0	20	50	4.0		50	8.0	
	2.0	20	50	8.0		50	22	
	4.0	20	50	100		50	100	
	8.0	20	50	100		50	100	
E. brevidens	Control	20	50	8.0	2.68 g NaCl/L (2.50–2.88)	100	14	2.67 g NaCl/L (2.54–2.79)
	0.5	20	50	6.0		100	12	
	1.0	20	50	6.0		100	12	
	2.0	20	50	10		100	10	
	4.0	20	50	100		100	100	

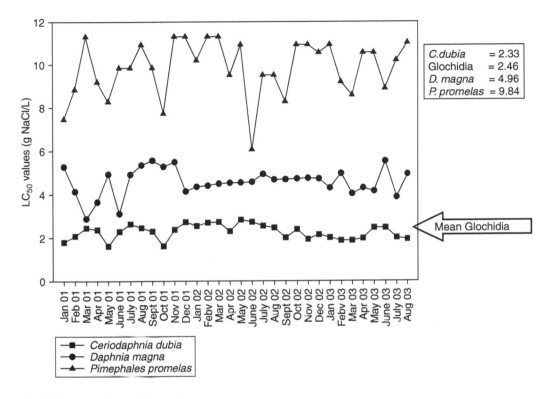

FIGURE 14.2 Reference data.

provided substantial evidence that released glochidia from some species of freshwater mussels are sensitive to mercury at concentrations that may be measured in the environment.

Interestingly, we also noted substantial interspecific variability in mercury tolerances among glochidia of different species, as individuals from *L. fasciola*, *E. capsaeformis*, and *E. brevidens* were highly sensitive to acute exposure but those from *V. iris* were not. Only a few other studies have conducted experiments with glochidia from numerous species of freshwater mussels; however, those that have often report substantial variability in species tolerances. Cherry et al. (2002) examined the effect of copper on glochidia from eight species of mussels, and reported mean LC50 values that ranged from 37 to 137 µg Cu/L. Keller and Ruessler (1997) conducted experiments on glochidia of six mussel species with the pesticide malathion and reported an even greater range, as the 48-hour LC50 value for the most sensitive species, *Lampsilis siliquoidea*, was 7 mg/L compared to 324 mg/L for the most tolerant species tested, *U. imbecillis*.

There was also a distinct difference in the toxicity of the different mercury salt forms, as glochidia from *L. fasciola*, *E. capsaeformis*, and *E. brevidens* were far more sensitive to MMC than to MC. Although the same total mercury concentrations were tested for both salt treatments, the 48-hour LC50 values were approximately three times lower for glochidia exposed to MMC. Several other studies have documented similar differences in the toxicity of mercury salts with test organisms other than glochidia. Baby and Menon (1987) observed that juvenile marine bivalves were more sensitive to mercury if exposed to the organic salt $(CH_3COO)_2$ Hg, than when exposed to the inorganic salt $HgCl_2$. Similarly, Wobeser (1975) reported that MMC is more toxic and accumulates faster than MC in the tissues of young age classes of rainbow trout. Biesinger, Anderson, and Eaton (1982) cited that *D. magna* in a chronic study excreted mercury slower when exposed to MMC than MC.

TABLE 14.4
Mean Water-Quality Data for the Acute Mercury and Reference Toxicant Tests—Samples from the Different Treatments Were Combined before Analysis

Test	Treatment	Conductivity (μmhos)	pH (su)	Alkalinity (mg/L as CaCO₃)	Hardness (mg/L as CaCO₃)	In Hg (μg/L)	Out Hg (μg/L)
All	Control	298 ± 9	7.78 ± 0.13	62.7 ± 4.4	83.5 ± 4.6	BDL	BDL
MC	4	297 ± 8	7.81 ± 0.11	n/a	n/a	BDL	BDL
	8	300 ± 3	7.80 ± 0.09	n/a	n/a	BDL	BDL
	15	294 ± 14	7.77 ± 0.14	n/a	n/a	BDL	BDL
	30	296 ± 8	7.83 ± 0.06	n/a	n/a	31.4 ± 8.0	26.6 ± 9.0
	60	301 ± 5	7.76 ± 0.18	n/a	n/a	63.2 ± 7.1	52.7 ± 18.0
	120	297 ± 11	7.81 ± 0.08	63.1 ± 5.7	86.2 ± 7.8	117.7 ± 21.6	98.4 ± 27.0
MMC	4	297 ± 12	7.78 ± 0.08	n/a	n/a	BDL	BDL
	8	302 ± 18	7.82 ± 0.14	n/a	n/a	BDL	BDL
	15	293 ± 7	7.81 ± 0.12	n/a	n/a	BDL	BDL
	30	298 ± 5	7.79 ± 0.09	n/a	n/a	32.9 ± 7.3	22.5 ± 14.0
	60	299 ± 11	7.81 ± 0.14	n/a	n/a	62.0 ± 18.4	46.7 ± 22.0
	120	294 ± 9	7.83 ± 0.12	61.9 ± 7.2	84.8 ± 5.6	133.6 ± 38.8	86.6 ± 41.0
NaCl	0.5	1218 ± 104	7.84 ± 0.06	n/a	n/a	n/a	n/a
	1	2154 ± 131	7.82 ± 0.04	n/a	n/a	n/a	n/a
	2	3884 ± 248	7.84 ± 0.11	n/a	n/a	n/a	n/a
	4	7160 ± 177	7.83 ± 0.12	n/a	n/a	n/a	n/a
	8	$14{,}570\pm342$	7.81 ± 0.13	62.6 ± 3.9	82.9 ± 5.8	n/a	n/a

Glochidia tolerance compared to standard test organisms. Both *C. dubia* and *D. magna* were more sensitive to MC than glochidia from any species of freshwater mussel tested, as 48-hour LC50 values for standard organisms were 7 and 19, respectively, compared to a range of 27–40 μg Hg/L for glochidia. The GMAV of 28 aquatic organisms exposed to MC are documented in the 1984 WQC for Hg (USEPA 1985), and of them, only four genera had mean LC50 values lower than the GMAV of 30 μg Hg/L for *Epioblasma* in our study. The documented GMAVs for the more sensitive genera are 2.6 for *Daphnia*, 10 for *Gammarus* (amphipod), 20 for *Chironomus* (midge), and 20 μg/L for *Faxonella* (crayfish). The GMAVs for other standard test organisms *Salmo gairdneri* (Rainbow trout) and *P. promelas* were 275 and 159 μg/L, respectively.

Both glochidia and standard regulatory organisms were more sensitive to MMC than MC. Methylmercury is more toxic than other forms because it is more available to biota and accumulates in aquatic food webs. Inorganic mercury is converted into methylmercury through natural microbial respiration in aquatic systems, and we assume that this conversion occurred during the bioassays. Though we did not measure methylmercury concentrations, MMC salt may have been more toxic than MC salt because a greater portion of the total measured mercury in solution already existed in the organic form. It will be important to have lower detection limits, and analyze both total mercury and methylmercury concentrations in future bioassays. Overall, this study suggests that aquatic organisms are highly sensitive to mercury, regardless of which salt is tested.

Sodium Chloride Glochidia Reference Test

A dose-dependent response was evident in all of the glochidia reference tests, as viability was substantially reduced in treatments with higher sodium chloride concentrations. Furthermore, glochidia were quite sensitive to sodium chloride, as LC50 values were near those recorded for *C. dubia*, which is currently regarded as one of the more sensitive standard test organisms. These observations support the use of sodium chloride as a reference toxicant for glochidia, but additional studies are needed to verify these results and improve to the precision of acceptable tolerance ranges for species.

Standard USEPA methods (USEPA 1991) require that reference toxicant tests be conducted concurrently with WET testing for both acute and chronic *C. dubia* and *P. promelas* bioassays. For results to be valid, endpoints of reference toxicant bioassays must be within an acceptable range for a given species. Researchers are able to infer the relative health of test organisms by comparing reference toxicity endpoints to established databases based on the premise that the healthy individuals of a species will have similar tolerances to a toxicant. Reference test endpoints that are below the acceptable species ranges suggest that organisms used in the test may have inferior health or that test conditions were not acceptable.

Results are only acceptable for standard test organisms if a certain control survivorship is maintained throughout tests; these thresholds are typically greater than or equal to 90% for acute and greater than or equal to 80% for chronic bioassays (USEPA 1994). Additional requirements have also been established for chronic *C. dubia* bioassays and include a minimum control survivorship of greater than or equal to 80%, greater than or equal to 60% of organisms in control treatments that have three broods within eight days, and surviving organisms in controls average greater than or equal to 15 neonates. Chronic *P. promelas* bioassay test results are only acceptable if control organisms survivorship is greater than or equal to 80% and average growth is greater than or equal to 2.5 mg/individual. These thresholds have been established by compiling the results of numerous trials and provide researchers with specific thresholds.

Several researchers have proposed similar validity endpoints for glochidia bioassays, such as greater than or equal to 90% after extraction or greater than or equal to 80% during the duration of the test (Keller 1993; Jacobson et al. 1997). Although these endpoints are useful and may potentially be incorporated into protocols for glochidia bioassays, additional QA/QC measures are essential for validating test results. We advocate conducting reference bioassays concurrently with other glochidia toxicant tests as a means for inferring the relative health of organisms used. Currently, additional research is needed to verify the acceptable sodium chloride tolerances of glochidia from different species of freshwater mussels before reference test endpoints can be used effectively as QA/QC measures. Lastly, it will be important to determine intra-specific variability within several mussel species and how this may vary seasonally.

ACKNOWLEDGMENTS

The authors thank John Schmerfeld, Cindy Kane, and Jess Jones for their active roles in the project. The research was supported by the U.S. Fish and Wildlife Service.

REFERENCES

American Public Health Association (APHA), American Water Works Association (AWWA), Water Environmental Federation (WEF), *Standard Methods for the Examination of Water and Wastewater*, 20th ed., American Public Health Association, Washington, DC, 1998.

Augspurger, T., Keller, A. E., Black, M. C., Cope, W. G., and Dwyer, F. J., Water quality guidance for protection of freshwater mussels (*Unionidae*) from ammonia exposure, *Environ. Toxicol. Chem.*, 22, 2569–2575, 2003.

Baby, K. V. and Menon, N. R., Salt forms of metals & their toxicity in the brown mussel, *Perna indica* (Kuriakose & Nair), *Indian J. Mar. Sci.*, 16, 107–109, 1987.

Barkay, T., Gillman, M., and Turner, R. R., Effects of dissolved organic carbon and salinity on bioavailability of mercury, *Appl. Environ. Microbiol.*, 4267–4271, 1997.

Beckvar, N., Salazar, S., Salazar, M., and Finkelstein, K., An in situ assessment of mercury contamination in the Sudbury River, Massachusetts, using transplanted freshwater mussels (*Elliptio complanata*), *Can. J. Fish. Aquat. Sci.*, 57, 1103–1112, 2000.

Biesinger, K. E., Anderson, L. E., and Eaton, J. G., Chronic effects of inorganic and organic mercury on *Daphnia magna*: Toxicity, accumulation, and loss, *Arch. Environ. Contam. Toxicol.*, 11, 769–774, 1982.

Breunderman, S. A. and Neves, R. J., Life history of the endangered fine-rayed pigtoe, *Fusconaia cuneolus* (Bivalvia: Unionidae), in the Clinch River, Virginia, *Am. Malacol. Bull.*, 10, 83–91, 1993.

Cherry, D. S., Van Hassel, J. H., Farris, J. L., Soucek, D. J., and Neves, R. J., Site-specific derivation of the acute copper criteria for the Clinch River, Virginia, *Hum. Ecol. Risk Assess.*, 8, 591–601, 2002.

Diamond, J. M., Bressler, D. W., and Serveiss, V. B., Assessing relationships between human land use and the decline of native mussels, fish, and macroinvertebrates in the Clinch and Powell River Watershed, USA, *Environ. Toxicol. Chem.*, 21, 1147–1155, 2002.

French, K. J., Scruton, D. A., Anderson, M. R., and Schneider, D. C., Influence of physical and chemical characteristics on mercury in aquatic sediment, *Water, Air, Soil Pollut.*, 110, 347–362, 1999.

Goudreau, S. E., Neves, R. J., and Sheehan, R. J., Effects of wastewater treatment plant effluents on freshwater mollusks in the upper Clinch River, Virginia, USA, *Hydrobiologia*, 252, 211–230, 1993.

Gulley, D. D., Toxstat Version 3.3, University of Wyoming, Department of Zoology and Physiology, Laramie, WY, 1993.

Henley, W. F. and Neves, R. J., Recovery status of freshwater mussels (Bivalvia: Unionidae) in the North Fork Holston River, Virginia, *Am. Malacol. Bull.*, 15, 65–73, 1999.

Huebner, J. D. and Pynnonen, K. S., Viability of glochidia of two species of *Anodonta* exposed to low pH and selected metals, *Can. J. Zool.*, 70, 2348–2355, 1992.

Jacobson, P. J., Neves, R. J., Cherry, D. S., and Farris, J. L., Sensitivity of glochidial stages of freshwater mussels (Bivalvia: Unionidae) to copper, *Environ. Toxicol. Chem.*, 16, 2384–2392, 1997.

Keller, A. E., Acute toxicity of several pesticides, organic compounds and wastewater effluent to the freshwater mussel, *Anodonta imbecillis, Ceriodaphnia dubia* and *Pimephales promelas, Bull. Environ. Contam. Toxicol.*, 51, 696–702, 1993.

Keller, A. E. and Ruessler, D. S., The toxicity of malathion to unionid mussels: Relationship to expected environmental concentrations, *Environ. Toxicol. Chem.*, 16, 1028–1033, 1997.

Keller, A. E. and Zam, S. G., The acute toxicity of selected metals to the freshwater mussel, *Anodonta imbecillis, Environ. Toxicol. Chem.*, 10, 539–546, 1991.

Mason, R. P., Laporte, J. M., and Andres, S., Factors controlling the bioaccumulation of mercury, methylmercury, arsenic, selenium, and cadmium by freshwater invertebrates and fish, *Arch. Environ. Contam. Toxicol.*, 38, 283–297, 2000.

Mauro, J. B. N., Guimaraes, J. R. D, and Hintelmann, H., Mercury methylation in macrophytes, periphyton, and water—comparative studies with stable and radio-mercury additions, *Anal. Bioanal. Chem.*, 374, 983–989, 2002.

Milam, C. D. and Farris, J. F., Risk identification associated with iron-dominated mine discharges and their effect upon freshwater bivalves, *Environ. Toxicol. Chem.*, 17, 1611–1619, 1998.

Moore, D. R. J., Teed, R. S., and Richardson, G. M., Derivation of an ambient water quality criterion for mercury: Taking account of site-specific conditions, *Environ. Toxicol. Chem.*, 22, 3069–3080, 2003.

Mummert, A. K., Neves, R. J., Newcomb, T. J., and Cherry, D. S., Sensitivity of juvenile freshwater mussels (*Lampsilis fasciola, Villosa iris*) to total and un-ionized ammonia, *Environ. Toxicol. Chem.*, 22, 2545–2553, 2003.

Naimo, T. J., A review of the effects of heavy metals on freshwater mussels, *Ecotoxicology*, 4, 341–362, 1995.

Neves, R. J. and Widlak, J. C., Habitat ecology of juvenile freshwater mussels (Bivalvia: Unionidae) in a headwater stream in Virginia, *Am. Malacol. Bull.*, 5, 1–7, 1987.

US Environmental Protection Agency, Ambient water quality criteria for mercury—1984, EPA 440/5-84-026, Criteria and Standard Division, USEPA, Washington, DC, 1985.

US Environmental Protection Agency, Proposed guide for conducting acute toxicity tests with the early life stages of freshwater mussels, Final Report, EPA Contract 68-024278, USEPA, Washington, DC, 1990.

US Environmental Protection Agency, Methods for determining metals in environmental samples, EPA/600/4-91/010, Environmental Monitoring and Support Laboratory, USEPA, Cinicinnati OH, 1991.

US Environmental Protection Agency, Methods for measuring the acute toxicity of effluents and receiving waters to freshwater and marine organisms, 4th ed., EPA/600/4-90/027F, Environmental Monitoring and Support Laboratory, USEPA, Cinicinnati OH, 1993.

US Environmental Protection Agency, Short-term methods for estimating the chronic toxicity of effluents and receiving waters to freshwater organisms, 3rd ed., EPA/600/4/91/002, Environmental Monitoring and Support Laboratory, USEPA, Cinicinnati OH, 1994.

Webber, H. M. and Haines, T. A., Mercury effects on predator avoidance behavior of a forage fish, golden shiner (*Notemigonus crysoleucas*), *Environ. Toxicol. Chem.*, 22, 1556–1561, 2003.

Weinstein, J. E., Characterization of the acute toxicity of photoactivated fluoranthene to glochidia of the freshwater mussel, *Utterbackia imbecillis*, *Environ. Toxicol. Chem.*, 20, 412–419, 2001.

Weinstein, J. E., Photoperiod effects of the UV-induced toxicity of fluoranthene to freshwater mussel glochidia: Absence of repair during dark periods, *Aquat. Toxicol.*, 59, 153–161, 2002.

Weinstein, J. E. and Polk, K. D., Phototoxicity of anthracene and pyrene to glochidia of the freshwater mussel *Utterbackia imbecillis*, *Environ. Toxicol. Chem.*, 20, 2021–2028, 2001.

Wiener, J. G. and Shields, P. J., Mercury in the Sudbury River (Massachusetts USA): Pollution history and synthesis of recent research, *Can. J. Fish. Aquat. Sci.*, 57, 1053–1061, 2000.

Williams, J. D., Warren, M. L., Cummings, K. S., Harris, J. L., and Neves, R. J., Conservation status of freshwater mussels of the United States and Canada, *Fisheries*, 18, 6–22, 1993.

Wobeser, G., Acute toxicity of methylmercuric chloride and mercuric chloride for rainbow trout (*Salmo gairdneri*) fry and fingerlings, *J. Fish. Res. Board Can.*, 32, 2005–2013, 1975.

Yeager, M. M., Neves, R. J., and Cherry, D. S., Competitive interactions between early life stages of *Villosa iris* (Bivalvia: Unionidae) and adult Asian clams (*Corbicula fluminea*), In *Proceeding of the First Freshwater Mollusk Conservation Society Symposium*, pp. 239–2594, 1999.

Index

Other Titles from the Society of Environmental Toxicology and Chemistry (SETAC):

Working Environment in Life-Cycle Assessment
Poulsen and Jensen, editors
2005

Life-Cycle Assessment of Metals
Dubreuil, editor
2005

Life-Cycle Management
Hunkeler, Saur, Rebitzer, Finkbeiner, Schmidt, Jensen, Stranddorf, Christiansen
2004

Scenarios in Life-Cycle Assessment
Rebitzer and Ekvall, editors
2004

Life-Cycle Assessment and SETAC: 1991–1999
(15 LCA publications on CD-ROM)
2003

Code of Life-Cycle Inventory Practice
de Beaufort-Langeveld, Bretz, van Hoof, Hischier, Jean, Tanner, Huijbregts, editors
2003

Life-Cycle Assessment in Building and Construction
Kotaji, Edwards, Shuurmans, editors
2003

Community-Level Aquatic System Studies—Interpretation Criteria (CLASSIC)
Giddings, Brock, Heger, Heimbach, Maund, Norman, Ratte, Schäfers, Streloke, editors
2002

Interconnections between Human Health and Ecological Variability
Di Giulio and Benson, editors
2002

Life-Cycle Impact Assessment: Striving towards Best Practice
Udo de Haes, Finnveden, Goedkoop, Hauschild, Hertwich, Hofstetter, Jolliet,
Klöpffer, Krewitt, Lindeijer, Müller-Wenk, Olsen, Pennington, Potting, Steen, editors
2002

Silver in the Environment: Transport, Fate, and Effects
Andren and Bober, editors
2002

Test Methods to Determine Hazards for Sparingly Soluble Metal Compounds in Soils
Fairbrother, Glazebrook, van Straalen, Tararzona, editors
2002

Avian Effects Assessment: A Framework for Contaminants Studies
Hart, Balluff, Barfknecht, Chapman, Hawkes, Joermann, Leopold, Luttik, editors
2001

SETAC

**A Professional Society for Environmental Scientists and Engineers and Related Disciplines
Concerned with Environmental Quality**

ociety of Environmental Toxicology and Chemistry (SETAC), with offices currently in North America and Eu-
is a nonprofit, professional society established to provide a forum for individuals and institutions engaged in
udy of environmental problems, management and regulation of natural resources, education, research and de-
ment, and manufacturing and distribution.

fic goals of the society are:

Promote research, education, and training in the environmental sciences.

Promote the systematic application of all relevant scientific disciplines to the evaluation of chemical hazards.

Participate in the scientific interpretation of issues concerned with hazard assessment and risk analysis.

Support the development of ecologically acceptable practices and principles.

Provide a forum (meetings and publications) for communication among professionals in government, business, academia, and other segments of society involved in the use, protection, and management of our environment.

e goals are pursued through the conduct of numerous activities, which include:

Hold annual meetings with study and workshop sessions, platform and poster papers, and achievement and
merit awards.

Sponsor monthly and quarterly scientific journals, a newsletter, and special technical publications.

Provide funds for education and training through the SETAC Scholarship/Fellowship Program.

Organize and sponsor chapters to provide a forum for the presentation of scientific data and for the inter-
change and study of information about local concerns.

Provide advice and counsel to technical and nontechnical persons through a number of standing and ad hoc
committees.

AC membership currently is composed of more than 5,000 individuals from government, academia, business,
public-interest groups with technical backgrounds in chemistry, toxicology, biology, ecology, atmospheric sci-
s, health sciences, earth sciences, and engineering. If you have training in these or related disciplines and are
ged in the study, use, or management of environmental resources, SETAC can fulfill your professional affiliation
s.

members receive a newsletter highlighting environmental topics and SETAC activities, and reduced fees for the
ual Meeting and SETAC special publications. All members except Students and Senior Active Members receive
thly issues of *Environmental Toxicology and Chemistry* (*ET&C*) and *Integrated Environmental Assessment and Man-
ent* (*IEAM*), peer-reviewed journals of the Society. Student and Senior Active Members may subscribe to the
nal. Members may hold office and, with the Emeritus Members, constitute the voting membership.

u desire further information, contact the appropriate SETAC Office.

1010 North 12th Avenue
Pensacola, Florida 32501-3367 USA
T 850 469 1500 F 850 469 9778
E setac@setac.org

Avenue de la Toison d'Or 67
B-1060 Brussels, Belgium
T 32 2 772 72 81 F 32 2 770 53 83
E setac@setaceu.org

www.setac.org

Environmental Quality Through Science®

SETAC

Environmental Quality Through Science

Milton Keynes UK
Ingram Content Group UK Ltd.
UKHW052021071024
449327UK00027B/2373